HUMAN GROWTH
AND DEVELOPMENT

In memory of James Mourilyan Tanner (1920–2010):
teacher, mentor, colleague and friend.

HUMAN GROWTH AND DEVELOPMENT

SECOND EDITION

Edited by

NOËL CAMERON

Centre for Global Health and Human Development,
School of Sport, Exercise and Health Sciences,
Loughborough University, Leicestershire

and

BARRY BOGIN

Centre for Global Health and Human Development,
School of Sport, Exercise and Health Sciences,
Loughborough University, Leicestershire

ELSEVIER

Amsterdam • Boston • Heidelberg • London • New York • Oxford
Paris • San Diego • San Francisco • Singapore • Sydney • Tokyo

Academic Press is an imprint of Elsevier

Academic Press is an imprint of Elsevier
32 Jamestown Road, London NW1 7BY, UK
225 Wyman Street, Waltham, MA 02451, USA
525 B Street, Suite 1800, San Diego, CA 92101-4495, USA

First edition 2002
Second edition 2012

British Library Cataloguing-in-Publication Data
A catalogue record for this book is available from the British Library

Library of Congress Cataloging-in-Publication Data
A catalog record for this book is available from the Library of Congress

ISBN: 978-0-12383882-7

For information on all Academic Press publications
visit our website at elsevierdirect.com

Typeset by TNQ Books and Journals Pvt Ltd.
www.tnq.co.in

Printed and bound by CPI Group (UK) Ltd, Croydon, CR0 4YY
Transferred to digital print 2012

Working together to grow
libraries in developing countries

www.elsevier.com | www.bookaid.org | www.sabre.org

ELSEVIER   BOOK AID International   Sabre Foundation

CONTENTS

16. Saltation and Stasis   **415**

Michelle Lampl

17. Lectures on Human Growth   **435**

Peter C. Hindmarsh

18. Body Composition During Growth and Development   **461**

Babette S. Zemel

CONTRIBUTORS

Stephen Bailey
Department of Anthropology, Tufts University, Medford, MA 02155, USA

Wim Vanden Berghe
Department of Biomedical Sciences, PPES Lab Protein Science, Proteomics and Epigenetic Signaling, Antwerp University, B-2610 Wilrijk, Belgium

Barry Bogin
Centre for Global Health and Human Development, School of Sport, Exercise and Health Sciences, Loughborough University, Leicestershire LE11 3TU, UK

Noël Cameron
Centre for Global Health and Human Development, School of Sport, Exercise and Health Sciences, Loughborough University, Leicestershire LE11 3TU, UK

T.J. Cole
UCL Institute of Child Health, University College London, London WC1N 1EH, UK

Stefan A. Czerwinski
Division of Epidemiology, Lifespan Health Research Center, Department of Community Health, Boonshoft School of Medicine, Wright State University, Dayton, OH 45420, USA

Ellen W. Demerath
Division of Epidemiology and Community Health, School of Public Health, University of Minnesota, Minneapolis, MN 55454, USA

Peter T. Ellison
Department of Human Evolutionary Biology, Harvard University, Cambridge, MA 02138, USA

Roland Hauspie
Laboratory of Anthropogenetics, Vrije Universiteit Brussel, B-1050 Brussels, Belgium

Peter C. Hindmarsh
Developmental Endocrinology Research Group, Institute of Child Health, University College London, London WC1N 1EH, UK

Kristen L. Knutson
Department of Medicine, University of Chicago, Chicago, IL 60637, USA

Christopher W. Kuzawa
Department of Anthropology and Cells 2 Society, The Center on Social Disparities and Health, Northwestern University, Evanston, IL 60208, USA

Michelle Lampl
Predictive Health Institute, Atlanta, GA 30308, USA and Department of Anthropology, Emory University, Atlanta, GA 30322, USA

Horacio Lejarraga
University of Buenos Aires at Hospital Garrahan, 1245 Buenos Aires, Argentina

William R. Leonard
Department of Anthropology, Northwestern University, Evanston, IL 60208, USA

Robert M. Malina
Department of Kinesiology and Health Education, University of Texas at Austin, TX 78712, USA and Department of Kinesiology, Tarleton State University, Stephenville, TX 76402, USA

Geert R. Mortier
Department of Medical Genetics, Antwerp University Hospital and University of Antwerp, B-2650 Edegem (Antwerp), Belgium

Nicholas G. Norgan
Department of Human Sciences and Loughborough University, Leicestershire LE11 3TU, UK

Meredith W. Reiches
Department of Human Evolutionary Biology, Harvard University, Cambridge, MA 02138, USA

Marcia L. Robertson
Department of Anthropology, Northwestern University, Evanston, IL 60208, USA

Mathieu Roelants
Laboratory of Anthropogenetics, Vrije Universiteit Brussel, B-1050 Brussels, Belgium

Ron G. Rosenfeld
Oregon Health and Science University, Portland, OR 97239, USA

Lawrence M. Schell
Department of Anthropology and Department of Epidemiology and Biostatistics, University at Albany, State University of New York, Albany, NY 12222, USA

J. Josh Snodgrass
Department of Anthropology, University of Oregon, Eugene, OR 97403, USA

Richard H. Steckel
Economics, Anthropology and History Departments, Ohio State University, Columbus, OH 43210, USA

Bradford Towne
Division of Epidemiology, Lifespan Health Research Center, Department of Community Health, Boonshoft School of Medicine, Wright State University, Dayton, OH 45420, USA

Babette S. Zemel
Perelman School of Medicine, University of Pennsylvania, Division of Gastroenterology, Hepatology and Nutrition, The Children's Hospital of Philadelphia, Philadelphia, PA 19104, USA

The first edition of *Human Growth and Development* is now 10 years old. Given the pace of discovery in the biological, medical and social sciences relating to human growth, a new edition is needed. To illustrate the pace of discovery, this second edition includes several new chapters. The first complete sequence of the human genome was published in 2003, and it soon became clear that knowing this sequence was not sufficient to understand the action of the DNA sequence. Actions above the level of the gene, called epigenetics, proved to be at least as important as the genes themselves. This second edition includes a new chapter entitled "Genomics, Epigenetics and Growth", which reflects these two interrelated areas of "new" research. Research into the developmental origins of health and disease (DOHaD) has been carried out since the early twentieth century, but has intensified since the 1990s. Thus, this edition includes a new chapter entitled "Early Environments, Developmental Plasticity and Chronic Degenerative Disease" to review the important developments in this field. There are other new chapters entitled "Leg Length, Body Proportion, Health and Beauty" and "Physical Activity as a Factor in Growth and Maturation" to review the latest research in those areas.

Existing chapters in the first edition have all been updated and some have new authors, for example the chapter "Endocrine Control of Growth" is, in essence, a new chapter with a new author, and the author of the first edition endocrinology chapter is now the new author of "Endocrine Disorders of Growth". The second edition also has a new editor, who helped to guide the inclusion of the new chapters.

USING THIS BOOK

The philosophy of how to use this book has not changed between editions. The chapters or lectures within this volume have been designed so that a "core" course can be extracted that provides information on the most important issues. For example, assuming that the first introductory chapter is always included, a class of human biologists or anthropologists would also need the chapters on infancy and childhood, adolescence, puberty and evolution to understand the basic biology. To these could be added the chapters on social, economic and environmental issues and the methodological chapters to equip them with the skills for fieldwork. Preclinical and/or clinical students would need to understand the basic biology but also include chapters on genomics, epigenetics, genetic epidemiology, endocrinology, growth disorders and assessment procedures. According to need, other chapters on nutrition, saltation and stasis, developmental origins, physical activity, brain development and body composition may be included. In this way a series of lectures can be created to cater for the needs of a variety of audiences, e.g. medical, allied medical disciplines (physiotherapy, occupational therapy, nursing, etc.), dentistry, anthropology, human biology, education, sports science, sociology, psychology and any other course dealing with children that will necessarily include information on human growth and development. Almost all chapters carry their own reference list or bibliography divided into three sections: references of the literature cited in the chapter; suggested reading of the most useful items for students; and Internet resources, often with high-quality illustrations, videos and animations to enhance learning.

Final-year students, graduates and those who have wandered in the vale of academe for many years will appreciate the old adage that "organizing academics is like herding cats". Their very independence of thought and action is what makes them the free thinkers that they are. Thus, to get them all to conform to a specific style is not even a remote possibility. This results in a series of lectures that vary in format. Some lecturers have chosen to lecture as they would present a scholarly textbook chapter while others have been more expansive and less formal. In any case we consider this variability to be a strength. The student will not be faced by a stereotyped series of lectures, just as in the university lecture theater no two lecturers are the same.

Even so, the contributors were requested to design their chapter as a lecture that can be given in approximately 60 minutes. The limited time for each lecture is based on the duration of a normal university lecture of approximately 60 minutes and forces the lecturer to focus on the essential information. All chapters open with an Introduction and most end with a Summary and/or Conclusion. These may serve as memory aids of the most important information, once the entire chapter has been read. Through the reference list the lecturer may guide the students towards extending their knowledge.

THE LECTURES AND LECTURERS

The first four lectures provide the core of a course on human growth in which the biological process of growth from birth to adulthood is described. The first chapter, by Noël Cameron, forms an introduction to the pattern and biology of human growth and development; the major areas that will be covered by the following 20 chapters. This broad overview is very much reflected by his breadth of experience and research in human growth and development. Cameron completed his PhD supervised by James Tanner at the Institute of Child Health at London University. Concurrently he acted as the "clinical auxologist" for Tanner's growth disorder clinics at the Hospital for Sick Children, Great Ormond Street, London, in which he assessed the growth and skeletal maturation of each child attending the clinics. With this dual role he was receiving probably the best available education and experience in the research techniques applied to both normal and abnormal growth. A personal statement by Cameron later in this Introduction provides more background to his education and interests.

Professor Horacio Lejarraga from Buenos Aires, who provides the second lecture on growth in infancy and childhood, is a pediatrician with an interest in child growth that extends over the past 40 years. Having qualified in medicine, he took a PhD under James Tanner's supervision in London before returning to Argentina to develop an awareness of the importance of human growth amongst the pediatric community. As former President of Argentina's 14,000-strong Society of Pediatrics he was responsible for Argentina's national growth studies and growth reference charts. Thus, while covering the basic pattern of growth in infancy and childhood he also brings a clinical perspective to preadolescent growth.

Chapter 3 is provided by Professor Roland Hauspie and Mathieu Roelants from the Free University of Brussels. Professor Hauspie is recognized as an international expert on the mathematical modeling of the human growth curve and plays a prominent role in European auxology. Dr Roelants specializes in human growth and youth health and is well known for his expertise in growth chart construction and use. Their knowledge of the pattern, magnitude, duration and variability of adolescent growth is considerably enhanced by modeling techniques and these are expertly described in their lecture.

Peter Ellison is Professor at the Department of Human Evolutionary Biology, Harvard University. He is an anthropologist with an international reputation for his research on reproductive biology. His many articles and books (e.g. Ellison, 2001[1]) have established him as the leading reproductive physiologist of his generation. Meredith Reiches is a PhD student in the Ellison laboratory and lead author on recent articles about the evolution of human growth, energy adaptation and kinship. Chapter 4, on puberty, reflects these research interests in addition to demonstrating their strong reputations as communicators and teachers.

Chapter 5 is the first to address the control of the process of growth through the endocrine system. Ron G. Rosenfeld is Professor of Pediatrics, Oregon Health and Science University. His chapter reviews the state of the art of some of the principal hormones regulating growth, from the hypothalamic−pituitary pathways releasing growth hormone to actions of insulin-like growth factor (IGF) and signal transducer and activator of transcription (STAT).

Growth is an energy-demanding process and the topic of "Nutrition and Growth" is treated in Chapter 6 by Nicholas Norgan and the editors of this book. Dr Norgan was a reader in Human Biology and a colleague of Noël Cameron at Loughborough University when the first edition was prepared and published. As a new lecturer in the early 1970s, Dr Norgan was also one of Cameron's first university teachers. Dr Norgan died in 2006. He was a human biologist specializing in human energetics and body composition. His international reputation was founded on population studies and surveys of diet, nutritional status, anthropometry, physical activity and energy expenditure conducted in the UK, Europe, India, Australia and Papua New Guinea. The preamble to Chapter 6 describes further his contributions to this book and to science.

Chapter 7 on "Genomics, Epigenetics and Growth" is authored by Geert Mortier, and Wim Vanden. Geert Mortier is Professor in Medical Genetics at the University of Antwerp and the Ghent University. He is recognized as an international expert in skeletal dysplasias which are genetic disorders of the skeleton often resulting in disproportionate short stature. Wim specializes in medical genetics, proteomics and epigenetic signaling. As mentioned above, epigenetic control of growth, meaning above the level of the gene, is proving to be at least as important as genetic control. The study of epigenetics and growth enters a realm of biological interactions among nutrients, behavior, emotional states and genes.

When they authored their first edition chapter on "Genetic Epidemiology of Growth and Development" Drs Brad Towne, Ellen Demerath and Stefan Czerwinski worked within one of America's leading centers for human growth research at Wright State University and, in many respects, formed the "rising stars" of our dream team of lecturers. Since then, they have all proved their potential. Demerath is currently at the University of Minnesota. Towne and Czerwinski remain at Wright Sate. Ellen Demerath and Stefan Czerwinski are thoroughbred anthropologists coming from excellent stables: the former from Harvard (under the influence of Peter Ellison, Chapter 4) and the University of Pennsylvania and the latter from the University of New York and the laboratory of Lawrence M. Schell (Chapter 10). Professor Brad Towne is a physical anthropologist by undergraduate training but an epidemiological and statistical geneticist by postgraduate experience and international reputation. Following a postdoctoral position in the early 1990s with Dr John Blangero's team at the highly respected South-West Foundation in Texas, he continues to be closely associated with their work. They provide an excellent and detailed chapter that thoroughly introduces the theory and methods of auxological genetics.

Chapters 9 and 10 provide two lectures on factors that affect human growth through the physical environment and the socioeconomic environment. Research at the intersection of economics and human biology is now a thriving field and we are fortunate to have one of the founders of this field as the author of Chapter 9, "Social and Economic Effects on Growth". Richard Steckel is Distinguished Professor of Economics, Anthropology and History at Ohio State University. He is the author of many books and articles that have established the methods of research in economic history, especially those making use of physical growth and maturation data to illuminate social and political conditions of populations that lived before the advent of modern economic statistics. Lawrence Schell and Kristen L. Knutson return as authors of the chapter "Environmental Effects on Growth", along with a new coauthor Stephen Bailey. Professor Schell has a distinguished academic career in anthropology and public health. His interest in human growth and particularly in the effects of environmental stressors began almost 40 years ago while studying for his PhD under the supervision of Francis E. Johnston at the University of Pennsylvania. At that time, the effect of noise-induced stress on the growth, health and well-being of infants living near airports was his primary concern. Now he is recognized as the leading authority on the effect of environmental pollutants on human growth. Dr Knutson was a PhD student of Schell, and she is now a biomedical anthropologist at University of Chicago specializing in sleep and cardiovascular health. Dr Bailey of Tufts University studied at the University of Michigan under Professor Stanley M. Garn (1922–2007), who was one of the leading growth researchers of the twentieth century. Bailey went on to a productive career in the fields of biological and nutritional anthropology, growth and body composition, and field methodology in such places as Latin America, China and the Southwestern USA.

Chapter 11 is by coeditor Barry Bogin. This chapter updates the research bearing on the evolution of the pattern of human growth, and Bogin's interpretation of that research. Bogin's evolutionary and biocultural approach to human growth and development was first published in *Patterns of Human Growth* in 1988 and updated in the second edition of that book in 1999. Since then, the field of human life history evolution has come into its own, with impacts on the direction of basic and applied research in anthropology, psychology, economics, evolutionary medicine and other areas. In the current volume Bogin takes time to explain the pattern of human growth in relation to mammalian life history in general, and specifically how human life history is related to evolutionary developments in human technology and social organization.

Chapters 12 and 13 are two new lectures for this second edition. "Early Environments, Developmental Plasticity and Chronic Degenerative Disease" is provided by Christopher W. Kuzawa, and "Leg Length, Body Proportion, Health and Beauty" is by Bogin. Dr Kuzawa is a biological anthropologist at Northwestern University. His lecture highlights the DOHaD approach, which is strongly influenced by life history biology. One message of DOHaD research is summarized in the old comic adage, "the past determines the future … so be very careful what you do in your past!" The findings of DOHaD research show that determinants of health cross several generations, so we must also be careful of what our parents and grandparents have done. Body proportions have long been studied, with references going back to ancient writings of Egyptians, Greeks and others of the Classical world. Much of the pre-twentieth century concern is related to art and "beauty", but today there are biomedical impacts of variation in the length of legs relative to total stature.

Changes in body shape during growth and physical performance have well-known interactive effects, which are the theme of Chapter 14, "Physical Activity as a Factor in Growth and Maturation", by Robert M. Malina. Professor Malina holds two PhDs, in Anthropology and Physical Education, and is recognized as the world authority in the relationship between exercise and growth.

The last of the new chapters is number 15, "Comparative and Evolutionary Perspectives on Human Brain Growth" by William R. Leonard, J. Josh Snodgrass and Marcia L. Robertson. All are anthropologists; Professor Leonard and Dr Robertson at Northwestern University and Dr Snodgrass at the University of Oregon. This lecture continues the life history evolution themes of earlier lectures and adds to the previous and following lectures on nutrition, physical activity and body composition. Leonard and Robertson helped to pioneer research on human brain development and metabolism, and their implications for human physical and social development.

Chapters 16 to 18 provide insight into specific topics within auxology that the editors believe should not be excluded from a thorough consideration of the science. The achievement of growth through the process of saltation and stasis was first demonstrated by Dr Michelle Lampl and colleagues in 1992.[2] This discovery was one of the most profound contributions to auxology in the latter part of the last century. It has given focus to the way in which we think about the control of human growth and answered many questions about the relationship between growth at a molecular, cellular, tissue, organ and whole-body level. The author of Chapter 17, Professor Peter Hindmarsh, is coauthor of the core text on pediatric endocrinology,[3] and has for some time been the leading pediatric endocrinologist at London University's Institute of Child Health. He thus brings both a biological and clinical approach to the endocrinology of growth and growth disorders. Dr Babette S. Zemel from the University of Pennsylvania, who writes about "Body Composition During Growth and Development" (Chapter 18), brings the expertise of both the anthropologist and clinical scientist to bear on this important area. Her early training was in human growth under the supervision of Francis E. Johnston at the University of Pennsylvania. Following her PhD studies in Papua New Guinea she became increasingly involved in the assessment of the growth and development of children with clinical disorders. Her contributions to our knowledge of the changes in the body composition of children compromised by the process of disease and disorder are internationally recognized.

Finally, Chapters 19 to 21 describe the methodological basis to research in human growth and development; how we assess growth and maturation and how we convert these data to usable growth references and standards. In two chapters, Noël Cameron lectures on the methods and equipment required for "The Measurement of Human Growth" and "Assessment of Maturation". In the final chapter, Tim J. Cole explains the ways in which "Growth References and Standards" are constructed and used. Professor Cole is Britain's leading specialist in the statistical analysis of growth data. His LMS method for creating the centiles required for growth reference charts has been accepted throughout the world and was used to produce the World Health Organization's 2006 growth charts recommended for global use.

EDITORS' PERSONAL STATEMENTS

Noël Cameron

This book has its origins in my lectures on human growth and development given to undergraduates attending British and South African universities over the past 35 years.

In 1976, while studying for my doctoral degree under the supervision of the late Professor James Tanner (1920–2010) at London University's Institute of Child Health, I was asked to give an annual series of lectures to Biological Anthropology students at Cambridge University. The late Professor William Marshall (1929–1984), who had been a "reader" in Tanner's department before taking the Professorship and Chair in Human Biology at Loughborough University, had previously given these lectures. At about the same time the geneticist Professor Alan Bittles, then a Senior Lecturer in Human Biology at Chelsea College, London University, asked me to provide a similar series of lectures to his students.

Thus faced with the prospect of two series of lectures I searched the available literature to determine what I could use as sources to write the lectures and what I could recommend to students. In the 1970s there was a number of texts, almost exclusively from America, describing the growth and development of children. Ernest Watson and George Lowrey, pediatricians at the University of Michigan Medical School, had first published *Growth and Development of Children* in 1951.[4] The physical anthropologist and human biologist Stanley Garn, then Chairman of the Physical Growth Department at the Fels Research Institute in Ohio, collaborated with the Israeli pediatrician Zvi Shamir, from Jerusalem, to publish *Methods for Research in Human Growth* in 1958.[5] Donald Cheek of Johns Hopkins University published *Human Growth: Body Composition, Cell Growth, Energy, and Intelligence* in 1968,[6] and Wilton Krogman, recently retired from the University of Pennsylvania, published his very useful *Child Growth* in 1972.[7] In 1966, a landmark work edited by Frank Falkner was destined to be the forerunner to a number of more recent texts in similar style. It was simply called *Human Development* and was, I think, the first volume to use different "authorities" (29 in this case) to provide the breadth and depth required to understand this most diverse of subjects.[8] However, with some notable exceptions, almost all of these volumes had been written by pediatricians interested in the clinical aspects of the subject rather than the biology of growth. In addition to the usual descriptions of the pattern of human growth, they were replete with diagnostic criteria and assessment procedures. They had little in the way of discussion about broader topics and the biological and conceptual basis of growth and development. In the UK, Tanner's *Growth at Adolescence*, first published in 1959 but in a second edition in 1962,[9] was, as it is now, *the* classic core text to be supplemented by a variety of scholarly scientific reviews and research papers to cover preadolescent growth and some other areas in greater depth. Later, his *Foetus into Man* (1978)[10] partially made up for this deficit but it was to some extent an introductory text and there was still the need for greater depth to be added through specific references.

In the same year (1978), Frank Falkner collaborated with his friend and previous colleague Jim Tanner to edit the three-volume series *Human Growth: A Comprehensive Treatise*, which was an excellent library resource but was far too expensive for the undergraduate or graduate student.[11] Clearly, by the 1970s many of the earlier texts were becoming dated and my solution was to cite a variety of individual chapters from these various authorities and supplement them with more up-to-date research papers.

During my sojourn in South Africa, between 1984 and 1997, I lectured to large classes of 400 or more students studying medicine and the allied medical disciplines (dentistry, physiotherapy, occupational therapy and nursing) in addition to smaller classes of medical science students. Within the large formal lecture classes there was a relatively restricted opportunity for discussion but the need to portray the biology of human growth in an immediate and vivid way in a short series of five or six lectures. The smaller medical science classes allowed me the freedom to "discuss" rather than "teach" human growth and development and to do so in an expansive series of 15 lectures covering half the academic year. By this time I was invariably recommending Tanner's *Growth at Adolescence* and *Foetus into Man* in addition to specific contributions from Falkner and Tanner's *Comprehensive Treatise*. Barry Bogin's *Patterns of Human Growth* became an accepted alternative text for this audience on its publication in 1988.[12] However, like *Foetus into Man*, it suffered from being written from the perspective and knowledge of a single author and thus lacked the breadth, and at times the depth, to be universally recommended.

Out of these experiences came the awareness that a coursework reference text was needed for undergraduate and graduate students but that no single scientist could hope to properly cover the different aspects of human growth and development with the breadth and depth required. Rather, what was needed was a team of lecturers and, if the text was to be the best possible text, this team would have to be recognized international experts in their fields of interest. They would indeed be a "dream team" that would, in effect, be invited into the lecture theater to provide a one-hour discourse on their subject. The target audience was the senior undergraduate student and/or immediate postgraduate student, i.e. the American "graduate student". Thus, a basic understanding of human biology was expected: human evolution, Mendelian genetics, anatomy, physiology and descriptive statistics. The text would not only cover the important issues in human growth and development but also allow the students the freedom to investigate the subject further through a good annotated reference list and a variety of recommended websites.

Thus this particular volume was conceived.

As editor I have had the mostly pleasurable experience of seeking some degree of rationality; of attempting to create an ordered series of lectures that will be of major benefit to students and lecturers alike. I thank all of the contributors for their willingness to cooperate in this venture and appreciate that most have been under considerable pressure but have nevertheless been timeous and gracious in their dealings with me. I thank my friends and colleagues within the science of auxology who have encouraged me to complete this task and hope that their confidence in my ability to produce a worthwhile volume has not been misplaced. Finally, I thank my partner, Anette, and my children, Jamie, Beth and Fin, for their forbearance of my ready willingness to leave them in search of new audiences for my research. This book is dedicated to them, for it would not have been possible without them.

Barry Bogin

I came to the study of human growth and development as a student of Professor Francis (Frank) Johnston, then at Temple University, Philadelphia. While at Temple I participated in seminars led by Frank, Professor Kuno Beller, a developmental psychologist, and Professor Ralph Hillman, a developmental geneticist. As an undergraduate at Temple University I had worked in the laboratory of developmental biologist Dr Richard Miller. Miller studied fertilization in the invertebrate group called Ctenophora. Watching these small, sessile, hydra-like organisms reproduce and grow may have been the initial spark for a career in growth and development.

My PhD thesis research started me on a journey as a biocultural anthropologist. In 1974 I set off for fieldwork in Guatemala to study seasonal variation in rates of growth of height and weight. I measured 245 boys and girls of various ages once a month for one year. All of them attended an expensive private school, so they were well nourished and healthy, and most of them displayed a faster rate of growth in height in the sunny, dry season and a slower rate in the cloudy, rainy season. I chose this privileged group to measure because 98% of Guatemalan children suffer from poverty and its attendant undernutrition, poor health and poor growth. Just before I left Guatemala in 1976 to complete my PhD, I began to understand some of the social, economic and political reasons why the 98% suffer so much. After completing my thesis in 1977 my attention turned to the interactions of socioeconomics, politics, ethnicity and biology in relation to human physical growth. That interest turned into a passion for the emerging field of biocultural anthropology.

Civil war in Guatemala between 1976 and 1996 forced nearly one million Guatemalans to flee their country. By a series of academic accidents I found groups of Guatemala Maya families living in California and Florida. This started an ongoing series of biocultural studies of the growth and development of these Maya children in the USA. The living Maya are the biocultural descendants of the pre-Columbian population of Central America. Today, the 6–7 million living Maya constitute the largest group of Native Americans. Since the Conquest (by Europeans in the year AD 1500), the Maya have suffered physically, socially and economically. More than 60% of Maya children and youth living in Guatemala and southern Mexico are stunted; that is, they have very short stature for their age. This makes the Maya the most stunted population in world. Our studies show that their short stature is not genetic, because the Maya growing up in the USA are, on average, 11 cm taller than their age-mates in Guatemala. That difference appears in less than one generation. But, the Maya in the USA are also much heavier, by an average of 12 kg.

My most recent research focuses on the theoretical and applied aspects of the worldwide overweight/obesity epidemic. In this work our research team takes evolutionary, life history and biocultural approaches to test hypotheses about the causes of what is termed the nutritional dual burden. This is defined as the coexistence of stunting and overweight in the same person, members of the same family (such as an overweight mother and her stunted child), or a community with many of its members stunted while others are overweight or obese. Our current fieldwork site is the Maya neighborhoods of Merida, Mexico. Merida is the largest city (about 1,000,000 inhabitants) in the State of Yucatan. We are partners with Mexican anthropologists and human ecologists as well as the Maya community. Together we learn from each other and hope to find ways to combat the causes and serious health consequences of the nutritional dual burden.

When Noël Cameron invited me to help coedit this second edition I eagerly accepted. One motivation was to be associated with our "dream team" of lecturers. Forced to read and reread the chapters as they progressed through various drafts, I knew that I would learn a great deal. I did not expect this learning to be as agreeable and satisfying as it has been. Along with Noël, I thank all of our contributors and also express thanks to Mary Preap and her editorial team at Elsevier. To my colleague, partner and muse Inês Varela-Silva and our daughter Isabel (a study of growth in motion), I offer much love.

CODA

The editors hope that students will find within these pages a biological story that will excite and fascinate them as it has done us for the last four decades. The process of growth and maturation is one that every living thing in the history of our planet has experienced. We do not think that the complexity of that process has reached or will reach an end point with *Homo sapiens*, because the process of human growth is constantly dynamic and constantly changing in response to the changes in the environment, both global and local, in which we live. Indeed, it is this plasticity, resulting in the wonderfully varied species which we see around us, that makes the process of human growth so fascinating.

Noël Cameron

Barry Bogin

Loughborough

REFERENCES

1. Ellison PT. *On fertile ground*. Cambridge, MA: Harvard University Press; 2001.
2. Lampl M, Veldhuis JD, Johnson ML. Saltation and stasis: a model of human growth. *Science* 1992;**158**:801—3.
3. Brook CGD, Hindmarsh P. *Clinical paediatric endocrinology*. 4th ed. Oxford: Blackwell Science; 2001.
4. Watson EH, Lowrey GH. *Growth and development of children*. Chicago, IL: Year Book; 1951.

5. Garn SM, Shamir Z. *Methods for research in human growth*. Springfield, IL: CC Thomas; 1958.
6. Cheek DB. *Human growth: body composition, cell growth, energy, and intelligence*. Philadelphia, PA: Lea & Febiger; 1968.
7. Krogman WM. *Child growth*. Ann Arbor, MI: University of Michigan Press; 1972.
8. Falkner F. *Human development*. Philadelphia, PA: WB Saunders Co; 1966.
9. Tanner JM. *Growth at adolescence*. 2nd ed. Oxford: Blackwell Scientific Publications; 1962.
10. Tanner JM. *Foetus into man*. London: Open Books; 1978.
11. Falkner F, Tanner JM. *Human growth: a comprehensive treatise*. New York: Plenum; 1978.
12. Bogin B. *Patterns of human growth*. Cambridge: Cambridge University Press; 1988.

CHAPTER 1

The Human Growth Curve, Canalization and Catch-Up Growth

Noël Cameron

Centre for Global Health and Human Development, School of Sport, Exercise and Health Sciences, Loughborough University, Leicestershire LE11 3TU, UK

Contents

1.1 INTRODUCTION

Human growth and development are characterized and defined by the way in which we change in size, shape and maturity relative to the passage of time. In order to understand this biological process it is fundamentally important to understand the terminology used to describe the process and the way in which it is measured and assessed. It is also important to appreciate the historical context within which the study of human growth and development has its roots.

1.2 HISTORICAL BACKGROUND

This introduction to the curve of human growth and development begins in the "age of Enlightenment" in eighteenth century France. Between the death of Louis XIV in 1715

N. Cameron & B. Bogin (eds): Human Growth and Development, Second edition.
ISBN 978-0-12-383882-7, Doi: 10.1016/B978-0-12-383882-7.00001-5

and the coup d'état of 9 November 1799 that brought Napoleon Bonaparte to power, philosophy, science and art were dominated by a movement away from monarchial authority and dogma and towards a more liberal and empirical attitude.[1] Its philosophers and scientists believed that people's habits of thinking were based on irrationality, polluted by religious dogma, superstition, and overadherence to historical precedent and irrelevant tradition. The way to escape from this, to move forward, was to seek for true knowledge in every sphere of life, to establish the truth and build on it. People's minds were, literally, to be "enlightened".[2] Its prime impulse was in pre-Revolutionary France within a group of mostly aristocratic and bourgeois natural scientists and philosophers that included Rousseau, Voltaire, Diderot and Georges Louis LeClerc, the Compte de Buffon (Figure 1.1). Their contributions to Diderot's Encyclopedia — the first literary monument to the Enlightenment — earned them the collective title of "the Encyclopedists".

Buffon was born on 7 September 1707 at Montbard in Bourgogne, in central France. His father, Benjamin-Francois Leclerc, described by the biographer Franck Bourdier as

Figure 1.1 Georges-Louis Leclerc, Compte de Buffon (1707—1788).

"un homme sans grand charactère", was a minor parliamentary official in Burgundy and was married to an older woman, Anne-Christine Marlin.[3] In 1717 Anne-Christine inherited a considerable fortune from an extremely wealthy uncle, Georges Blaisot, which allowed Monsieur Leclerc to buy the land of Buffon and the "châtellenie" of Montbard. Georges Louis was educated by the Jesuits at the Colleges de Godran, where he demonstrated an aptitude for mathematics. In 1728 he moved to the University of Angers and thence suddenly to England following a duel with an officer of the Royal-Croates over "une intrigue d'amour". He traveled in Switzerland, France and Italy during the next four years, returning to Dijon on the death of his mother in 1732. Much against his father's wishes he inherited his mother's estate at Montbard and from then on divided his time between Paris and the country pursuing his interests in mathematics, natural science and silviculture. By the age of 32 he was recognized as the premier horticulturist and arborist in France and was appointed by King Louis XV as the director of the Jardin du Roi in 1739. This position was the equivalent of being the chief curator of the Smithsonian, or the British Museum of Natural History — it was the most prestigious governmental scientific position in the "natural sciences" that Buffon could have obtained. During the next few years Buffon started to work on an immense project that was to include all that was known of natural history. *Histoire Naturelle, Générale et Particulière* would be a vast undertaking but one that Buffon, who from all accounts was a man of no small ego, appeared to relish and which by his death in 1788 was composed of 36 volumes. There were 15 volumes on quadrupeds (1749—1767), nine on birds (1770—1783), five on minerals (1783—1788) and seven supplementary volumes. Eight further volumes prepared by E. de Lacepede were added posthumously between 1788 and 1804 and included two volumes on reptiles (1788—1789), five on fish (1798—1803) and one on *Cetacea* (1804). However, it is the supplement to Volume 14, published in 1778, that is of particular interest.

Within this supplement on page 77 there is the record of the growth of a boy known simply as "De Montbeillard's son". The friendship between Philibert Geuneau De Montbeillard (1720—1785) and Buffon had been secured by a common interest in the natural sciences. Buffon had been working closely for many years with his younger neighbor from Montard, Louis-Jean-Marie Daubenton (1716—1799), whose statue now adorns the Parc Buffon in Montard (while Buffon's statue is to be found in the Jardin des Plantes in Paris). Daubenton had graduated in Medicine at Reims in 1741 and returned to Montard to set up practice as a physician. This coincided with Buffon's initial preparations for the first volumes of *Histoire Naturelle* and in 1742 he invited Daubenton to provide a series of anatomical descriptions of animals. Daubenton's subsequent descriptions of 182 species of quadruped that appeared in the early volumes of *Histoire Naturelle* established him as the foremost comparative anatomist of his day. However, De Montbeillard was to replace Daubenton in Buffon's affections and between 1770 and 1783 De Montbeillard coauthored the nine volumes of *Histoire Naturelle* devoted to birds. He was also a correspondent of Diderot and clearly recognized as one of the Encyclopedists. Given the desire of these central scientific figures of the Enlightenment to measure and describe the natural world it is not too surprising that De Montbeillard would take an empirical interest in the growth of his own son. Nor is it inconceivable that his friend and colleague Buffon would wish to include this primary evidence of the course of human growth within his *opus magnum*.

De Montbeillard had been measuring the height of his son about every 6 months from his birth in 1759 until he was 18 years of age in 1777. The boy's measurements of height were reported in the French units of the time — pieds, pouces and lignes — which correspond roughly to present-day units as a foot, an inch and the 12th part of an inch. (Tanner,[4] p. 470, notes that, "The Parisian *pied*, or foot, divided into 12 pouces, or inches, each divided into 12 lignes, was longer than the English foot. Isaac Newton … found 1 *pied* equal to 12.785 inches, but the later official conversion, on the introduction of the metre, gave it as 12.7789 inches. The *pouce*, then, equals 2.71 cm whereas the English inch equals 2.54 cm".)

Richard E. Scammon (1883–1952), of the Department of Anatomy and the Institute of Child Welfare at the University of Minnesota, translated these measurements into centimeters and published his results in 1930 in the *American Journal of Physical Anthropology* under the title of "The first seriatim study of human growth" and thus for the first time we were able to look upon the growth of De Montbeillard's son in the form of a chart.[5]

1.3 THE DISTANCE CURVE OF GROWTH

By joining together the data points at each age, Scammon produced a curve that described the height achieved at any age that became known as a "height distance" or "height-for-age" curve (Figure 1.2). The term "distance" is used to describe height achieved because it is easy to visualize and understand the fact that a child's height at any particular age is a reflection of how far that child has progressed towards adulthood. It embodies the sense of an ongoing journey that we are, as it were, interrupting to take a "snapshot" at a particular moment in time. The resulting curve is interesting for a number of reasons. First, when growth is measured at intervals of 6 months or a year, the resultant curve is a relatively smooth and continuous process; it is not characterized by periods of no growth and then by dramatic increases in stature. Second, growth is not a linear process; we do not gain the same amount of height during each calendar year. Third, the curve of growth has four distinct phases (or perhaps five if the mid-growth spurt is included; see below) corresponding to relatively rapid growth in infancy, steady growth in childhood, rapid growth during adolescence and very slow growth as the individual approaches adulthood. Fourth, growth represents a most dramatic increase in size; De Montbeillard's son, for instance, grew from about 60 cm at birth to over 180 cm

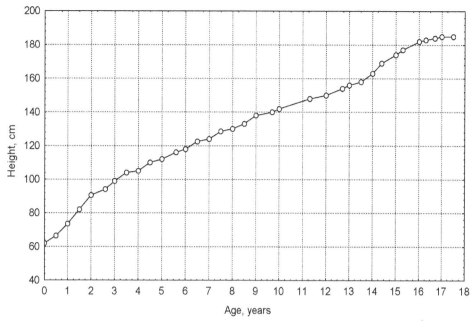

Figure 1.2 The growth of De Montbeillard's son 1759–1777: distance. *(Source: Tanner.[6])*

at adulthood. The majority of that growth (more than 80%) occurs during infancy and childhood, but perhaps the most important physical changes occur during adolescence. Fifth, humans cease growing, or reach adult heights, during the late teenage years at 18 or 19 years of age.

The pattern of growth that can be seen from this curve is a function of the frequency of data acquisition. For instance, if we were to measure a child only at birth and at 18 years we might believe, by joining up these two data points, that growth was a linear process. Clearly, the more frequently we collect data the more we can understand about the actual pattern of growth on a yearly, monthly, weekly or even daily basis. Naturally, such high-frequency studies are logistically very difficult and thus there are only a very few in existence. Perhaps the most important are those of Dr Michele Lampl, who was able to assess growth in length, weight and head circumference on a sample of 31 children on daily, twice-weekly and weekly measurement frequencies.[7] The resulting data demonstrated that growth in height may not be a continuous phenomenon but may actually occur in short bursts of activity (saltation) that punctuate periods of no growth (stasis) (see Chapter 16). However, the data for De Montbeillard's son were collected approximately 6-monthly and thus at best they can only provide information about the pattern of growth based on a half-yearly or yearly measurement frequency.

It is clear that the pattern of growth that results from these 6-monthly measurements is in fact composed of several different curves. During "infancy", between birth and about 5 years of age, there is a smooth curve that can be described as a "decaying polynomial" because it gradually departs negatively from a straight line as time increases. During childhood, between 5 and about 10 years of age, the pattern does not depart dramatically from a straight line. This pattern changes during adolescence, between about 10 and 18 years of age, into an S-shaped or sigmoid curve reaching an asymptote at about 19 years of age.

The fact that the total distance curve may be represented by several mathematical functions allows mathematical "models" to be applied to the pattern of growth. These models are, in fact, parametric functions that contain constants or "parameters". Once an appropriate function that fits the raw data has been found the parameters can be analyzed, revealing a good deal about the process of human growth (see Chapter 3). For instance, in the simplest case of two variables such as age (x) and height (y) being linearly related between, say, 5 and 10 years of age (i.e. a constant unit increase in age is related to a constant unit increase in height), the mathematical function $y = a + bx$ describes their relationship. The parameter a represents the point at which the straight line passes through the y-axis and is called the intercept, and b represents the amount that x increases for each unit increase in y and is called the regression coefficient. The fitting of this function to data from different children and subsequent analysis of the parameters can provide information about the magnitude of the differences between the children and lead to further investigations of the causes of the differences. Such "time series analysis" is extremely useful within research on human growth because it allows the reduction of large amounts of data to only a few parameters. In the case of De Montbeillard's son there are 37 height measurements at 37 different ages. Therefore, there are 74 data items for analysis. The fitting of an appropriate parametric function, such as the Preece—Baines function,[8] which will be discussed later (see Chapter 3), reduces these 74 items to just 5. Because of their ability to reduce data from many to only a few data items, such parametric solutions are said to be parsimonious and are widely used in research into human growth.

1.4 THE VELOCITY GROWTH CURVE AND GROWTH SPURTS

The pattern created by changing rates of growth is more clearly seen by actually visualizing the rate of change of size with time, i.e. "growth velocity" or, in this particular case, "height velocity". The term "height velocity" was coined by Tanner[9] and was based on the writings of Sir D'Arcy Wentworth Thompson (1860—1948). D'Arcy Thompson was a famous British natural scientist and mathematical biologist who published a landmark biology text entitled "On growth and form" in 1917, with a second edition in 1942.[10,11] Thompson's core thesis was that structuralism underpinned by the laws of physics and mechanics was primarily responsible for variation in size and shape within phylogeny (i.e. within the evolutionary development of our species; see Chapter 11). Considering allometry, the impact on the whole organism of varying growth rates of different body parts, D'Arcy Thompson wrote the oft-quoted passage, "An organism is so complex a thing, and growth so complex a phenomenon, that for growth to be so uniform and constant in all the parts as to keep the whole shape unchanged would indeed be an unlikely and an unusual circumstance. Rates vary, proportions change, and the whole configuration alters accordingly". Within the second edition (p. 95) Thompson wrote that while the distance curve, "showed a continuous succession of varying *magnitudes*", the curve of the rate of change of height with time, "shows a succession of varying *velocities*. The mathematicians call it a *curve of first differences*; we may call it a curve of the rate (or rates) of growth, or more simply a *velocity curve*". The velocity of growth experienced by De Montbeillard's son is displayed in Figure 1.3. The *y*-axis records height gain in cm/year, and the *x*-axis chronological age in years. It can be seen that following birth two relatively distinct increases in growth rate occur at 6—8 years and again at 11—18 years. The first of these "growth spurts" is called the juvenile or mid-growth spurt (see Chapter 2) and the second is called the adolescent growth spurt (see Chapter 3).

There is, in fact, another growth spurt that cannot be seen because it occurs before birth. Between 20 and 30 weeks of gestation the rate at which the length of the fetus increases reaches a peak at approximately 120 cm/year, but all that can be observed

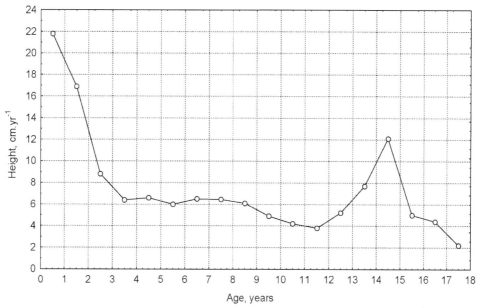

Figure 1.3 The growth of De Montbeillard's son 1759—1777: velocity. *(Source: Tanner.[6])*

postnatally is the slope of decreasing velocity lasting until about 4 years of age. Similarly, increase in weight also experiences a prenatal spurt but a little later, at 30–40 weeks of gestation. Of course, information on the growth of the fetus is difficult to obtain and relies largely on two sources of information: extrauterine anthropometric measurements of preterm infants and intrauterine ultrasound measurements of fetuses. Ultrasound assessments of crown–rump length indicate that growth is smooth and rapid during the first half of pregnancy. Indeed, it is so smooth between 11 and 14 postmenstrual weeks, when the growth velocity is 10–12 mm/week, that gestational age can be calculated from a single measurement to within ±4.7 days. The 95% error band when three consecutive measurements are taken is ±2.7 days. Intrauterine growth charts for weight demonstrate that growth over the last trimester of pregnancy follows a sigmoid pattern and thus, like the sigmoid pattern reflected in height distance during adolescence, will also demonstrate a growth spurt when velocity is derived. The spurt should reach a peak at about 34–36 weeks. Why should the fetus be growing so quickly in terms of weight at this time? Results from an analysis of 36 fetuses in the mid-1970s demonstrated that between 30 and 40 postmenstrual weeks fat increases from an average of 30 g to 430 g. This dramatic accumulation of fat is directly related to the fact that fat is a better source of energy per unit volume, releasing twice as much energy per gram as either protein or carbohydrate. Thus, a significant store of energy is available to the fetus for the immediate postnatal period.

While the prenatal spurt and juvenile growth spurt may vary in magnitude, they seem to occur at roughly the same age both within and between the sexes. The adolescent growth spurt, however, varies in both magnitude and timing within and between the sexes; males enter their adolescent growth spurt almost 2 years later than females and have a slightly greater magnitude of height gain. The result is increased adult height for males, mainly resulting from their 2 years of extra growth prior to adolescence. At the same time, other skeletal changes are occurring that result in wider shoulders in males and, in relative terms, wider hips in females. Males demonstrate rapid increases in muscle mass and females accumulate greater amounts of fat. Their fat is distributed in a "gynoid" pattern, mainly in the gluteofemoral region, rather than in the "android" pattern with a more centralized distribution characteristic of males (see Chapter 18). Physiologically, males develop greater strength and lung capacity. Thus, by the end of adolescence a degree of morphological difference exists between the sexes; males are larger and stronger and more capable of hard physical work. Such "sexual dimorphism" is found to a greater or lesser extent in all primates and serves as a reminder that these physical devices had, and perhaps still have, important sexual signaling roles (see Chapter 11).

In addition to dramatic growth during adolescence, increased adult size in males is achieved because of the extended period of childhood growth. This period of childhood is peculiar to the *human* child and its existence raises important questions about the evolution of the *pattern* of human growth. Professor Barry Bogin argues (see Chapter 11) that humans have a childhood because it creates a reproductive advantage over other species through the mechanism of reduced birth spacing and greater lifetime fertility. In addition, slow growth during childhood allows for "developmental plasticity" in sympathy with the environment, with the result that a greater percentage of human young survive than the young of any other mammalian species.

1.5 OTHER PATTERNS OF GROWTH

The pattern of growth in height, as demonstrated by De Montbeillard's son, is only one of several patterns of growth that are found within the body. Figure 1.4 illustrates the major differences in pattern as exemplified by neural tissue (brain and head), lymphoid tissue (thymus, lymph nodes, intestinal lymph masses) and reproductive tissue (testes, ovaries, epididymis, prostate, seminal vesicles, fallopian tubes) in addition to the general growth curve of height or weight and some major organ systems (respiratory, digestive, urinary). The data on which this figure is based are old, having originally been published by R.E. Scammon in 1930,[14] but they are sufficient to demonstrate that lymphoid, neural and reproductive tissue have very different patterns of growth from the general growth curve that was initially observed. Neural tissue exhibits strong early growth and is almost complete by 8 years of age, whereas reproductive tissue does not really start to increase in size until 13 or 14 years of age. The lymphatic system, which acts as a circulatory system for tissue fluid and includes the thymus, tonsils and spleen in addition to the lymph nodes, demonstrates a remarkable increase in size until the early adolescent years and then declines, perhaps as a result of the activities of sex hormones during puberty (see Chapters 4 and 5). The majority of the author's interest in this and other chapters on growth concerns the pattern of growth as exhibited by height and weight, i.e. the "general" pattern in Figure 1.4. It is clear, however, that research on the growth of neural tissue must be targeted at fetal and infant ages and research on the growth of reproductive tissue on adolescent or teenage years when growth is at a maximum.

1.6 GROWTH VERSUS MATURITY

Although this discussion has concentrated on the growth of one boy in eighteenth century France, De Montbeillard's son, it is now evident that his curves of growth (i.e. distance and velocity) reflect patterns that are found in all children who live in normal environmental circumstances. We may differ in the magnitude of growth that occurs, as is evident from our varying adult statures, but in order to reach our final heights we have all experienced a similar pattern of human growth to a greater or lesser degree. It is evident that growth in height is not the only form of somatic growth that occurs in the human body. This chapter has already discussed the fact that as we experience the process of growth in linear dimensions, i.e. as we get taller, we also experience other forms of growth. We get heavier, fatter and more muscular, and we experience changes in our

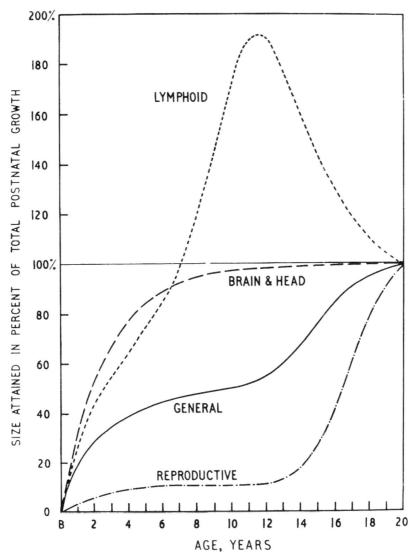

Figure 1.4 Growth curves of different parts and tissues of the body, showing the four main types: *lymphoid* (thymus, lymph nodes, intestinal lymph masses); *brain, neural tissue and head* (brain and its parts, dura, spinal cord, optic system, cranial dimensions); *general tissue* (whole-body linear dimensions, respiratory and digestive organs, kidneys, aortic and pulmonary trunks, musculature, blood volume); *reproductive tissue* (testes, ovary, epididymis, prostate, seminal vesicles, fallopian tubes. (Source: *Tanner.*[12])

body proportions. In addition, we become more "mature" in that we experience an increase in our functional capacity with advancing age that may be evidenced in our increasing ability to undertake physical exercise in terms of both magnitude and duration (see Chapter 14). Although "growth and development" tend to be thought of as a single biological phenomenon, both aspects have distinct and important differences. "Growth" is defined as an increase in size, while "maturity" or "development" is an increase in functional ability. The endpoint of growth is the size attained by adulthood, roughly corresponding to growth rates of less than 1 cm/year, and the endpoint of maturity is when a human is functionally able to procreate successfully, but not simply to be able to produce viable sperm in the case of males and viable ova in the case of females. Successful procreation in a biological sense requires that the offspring survive so that they themselves may also procreate. Thus, successful maturation requires not just biological maturity but also behavioral and perhaps social maturity.

The relationship between somatic growth and maturity is perhaps best illustrated by Figure 1.5. The figure shows three boys and three girls who are of the same ages within gender; the boys are exactly 14.75 years of age and the girls 12.75 years of age. The most striking feature of this illustration is that even though they are the same age they demonstrate vastly different degrees of maturity. The boy and girl on the left are relatively immature compared to those on the right. In order to be able to make these distinctions in levels of maturity we must be using some assessments of maturation, or "maturity indicators" (see Chapter 20). These may well include the obvious development of secondary sexual characteristics (breasts and pubic hair in girls and genitalia and pubic hair in boys), in addition to dramatic changes in body shape, increases in muscularity in males and increases in body fat in females. If we look carefully we will also see that distinct changes in the shape of the face also occur, particularly in boys, which result in "stronger" or more robust features compared to the rather soft outline of the preadolescent face. However, the maturity indicators used to assess maturation for clinical and research purposes are constrained by the fact that they must demonstrate "universality" — they must appear in the same sequence within both sexes — and similarity in both beginning and end stages. Because our size is governed by factors other than the process of maturation it is not possible to use an absolute size to determine maturation. Although it is true in very general terms that someone who is large is likely to be older and more mature than someone who is small, it is apparent from Figure 1.5 that as the two individuals approach each other in terms of age that distinction becomes blurred. Therefore, the appearance and *relative* size of structures, rather than their *absolute* size, are used to reflect maturity. The most common maturity indicators are secondary sexual development, skeletal maturity and dental maturity (see Chapter 20).

1.7 THE CONTROL OF GROWTH

It is clear that the process of human growth and development, which takes almost 20 years to complete, is a complex phenomenon. It is under the control of both genetic and environmental influences that operate in such a way that at specific times during the period of growth one or the other may be the dominant influence. At conception we

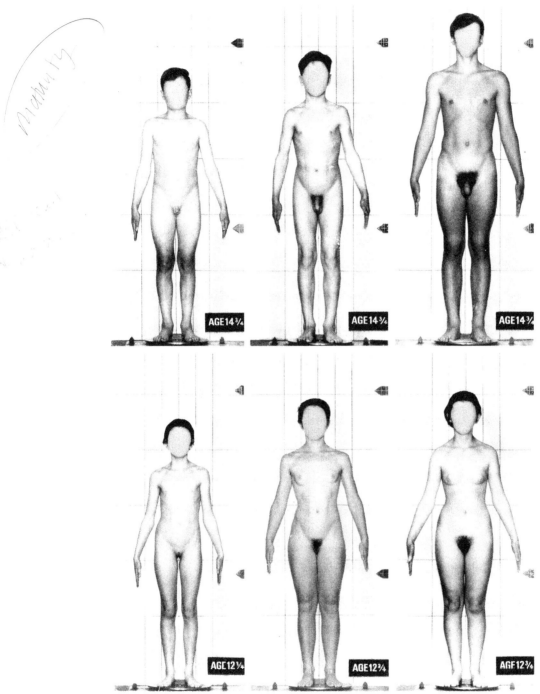

Figure 1.5 Three boys and three girls photographed at the same chronological ages within sex; 12.75 years for girls and 14.75 years for boys. *(Source: Tanner.[13])*

obtain a genetic blueprint that includes our potential for achieving a particular adult size and shape. The environment will alter this potential. Clearly, when the environment is neutral, when it is not exerting a negative influence on the process of growth, then the genetic potential can be fully realized. However, the ability of environmental influences to alter genetic potential depends on a number of factors, including the time at which they occur, the strength, duration and frequency of their occurrence, and the age and gender of the child (see Chapters 8 and 9).

The control mechanism that environmental insult will affect is primarily the endocrine system. The hypothalamus or "floor" of the diencephalon situated at the superior end of the brainstem coordinates the activities of the neural and endocrine systems. In terms of human growth and development its most important association is with the pituitary gland, which is situated beneath and slightly anterior to the hypothalamus. The rich blood supply in the infundibulum, which connects the two glands, carries regulatory hormones from the hypothalamus to the pituitary gland. The pituitary gland has both anterior and posterior lobes and it is the anterior lobe or "adenohypohysis" that releases the major hormones controlling human growth and development: growth hormone, thyroid-stimulating hormone, prolactin, the gonadotrophins (luteinizing and follicle-stimulating hormone) and adrenocorticotrophic hormone (see Chapters 4, 5 and 17). Normal growth is not simply dependent on an adequate supply of growth hormone but is the result of a complex and at times exquisite relationship between the nervous and endocrine systems. Hormones rarely act alone but require the collaboration and/or intervention of other hormones in order to achieve their full effect. Thus, growth hormone causes the release of insulin-like growth factor-1 (IGF-1) from the liver. IGF-1 directly affects skeletal muscle fibers and cartilage cells in the long bones to increase the rate of uptake of amino acids and incorporate them into new proteins, and thus contributes to growth in length during infancy and childhood. At adolescence, however, the adolescent growth spurt will not occur without the collaboration of the gonadal hormones: testosterone in boys and estrogen in girls.

There is ample evidence from research on children with abnormally short stature that a variety of environmental insults will disturb the endocrine system, causing a reduction in the release of growth hormone. However, other hormones are also affected by insult and thus the diagnosis of growth disorders becomes a complex and engrossing series of investigations that increasingly requires an appreciation of both genetic and endocrine mechanisms (see Chapters 7 and 17).

1.8 GROWTH REFERENCE CHARTS

The growth of De Montbeillard's son was interesting not only because he depicts a normal pattern of growth but also because he achieved an adult height that was over 180 cm or about 6 feet. He was actually quite tall for a Frenchman in the eighteenth century. How do we know that someone is "tall" or "short"? What criteria do we use to allow us to make such a judgment? Those not involved in the study of human growth will make such a judgment based on their exposure to other people. If, for instance, they have only lived among the pygmies of Zaire then anyone over 165 cm or 5 feet 5 inches would be very tall. If, on the other hand, they had only lived among the tall Nilotic tribesmen of north Africa then anyone less than 175 cm or 5 feet 9 inches would be unusually small. Most of us live in regions of the world in which the majority of people have adult heights that are between these extremes and view average adult heights at about 178 cm (5 feet 10 inches) for males and 164 cm (5 feet 5 inches) for females as being "normal". There is, of course, a range of adult heights about these average values and that range provides an estimate of the normal variation in adult stature. Beyond certain points in that range we begin to think of an individual's height as being either "too tall" or "too short". This is also true of the heights of children during the process of growth. At any age from birth to adulthood there is a range of heights that reflects the sizes of normal children, i.e. children who have no known disease or disorder that adversely affects height (e.g. bone dysplasias, Turner syndrome). In order to assess the normality or otherwise of the growth of children, growth reference charts are used. These charts depict both the average height to be expected throughout the growing years (typically from birth to 18 years), and the range of normal heights, in the form of percentile or "centile" distributions.

Figure 1.6 is an example of such a reference chart. It depicts the normal range of heights for British boys from 4 to 18 years. The normal range is bound by outer centile limits of the 0.4th and 99.6th centiles. Thus, "normal" heights are thought of as heights that fall between these limits although, of course, 0.8% of normal children will have heights below the 0.4th centile or above the 99.6th centile (see Chapter 21). The illustrated centiles have been chosen because they each equate roughly to 0.67 Z-scores or standard deviation scores (SDS) from the 50th centile or average values. Thus, the 75th and 25th centiles are ± 0.675 Z-scores above or below the mean, the 90th and 10th centiles are ± 1.228 Z-scores, and the 98th and 2nd centiles are ± 1.97 Z-scores. Their importance is that they provide not only a reasonable point at which to investigate possible abnormalities of growth but also reasonable guidelines for how growth is expected to proceed within the normal range. It has been generally accepted that a child whose growth exhibits a movement of $+0.67$ Z-scores is exhibiting a clinically significant response to the alleviation of some constraining factor (see Catch-up growth, Section 10).[15] Thus, the movement of a child's height or weight upwards through the centiles from the 10th to the 25th or downwards from the 98th to the 75th can be viewed by clinicians as being more than simply a chance occurrence.

Children who do not have constraints upon their growth exhibit patterns of growth parallel to the centile lines prior to adolescence. However, as the adolescent growth spurt takes place they will depart from this parallel pattern and all adolescents will demonstrate

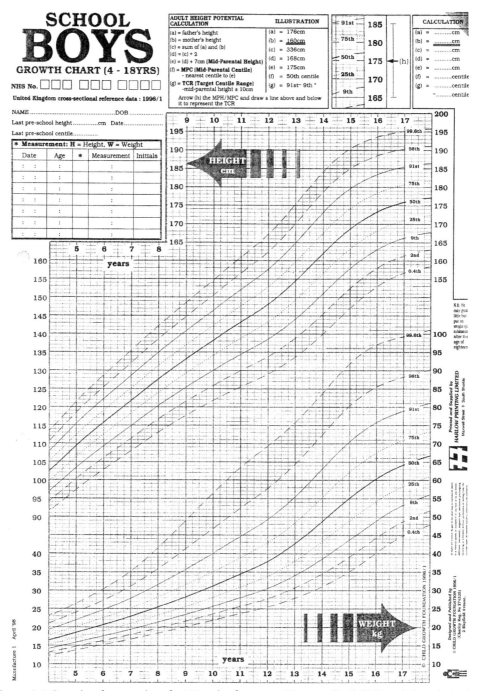

Figure 1.6 Growth reference chart for UK males from 4 to 18 years. (© *Child Growth Foundation.*)

"centile crossing". If they are "early developers" their height-for-age curve will rise through the centiles before their peers and level off early as they achieve their adult stature. "Late developers", on the other hand, will initially appear to fall away from their peers, as the latter enter their growth spurts, and then accelerate into adolescence rising through the centile lines when their peers have ceased or nearly ceased growing. Even the child who enters their growth spurt at the average age for the population will cross centile lines. This is because the source data for these reference charts were collected in cross-sectional studies; studies in which children of different ages were measured on a single occasion. They thus reflect the average heights, weights, etc., of the population rather than the growth of an individual child. If one were able to undertake a growth study of the same children over many years (a longitudinal study) one could theoretically adjust the data so that it illustrated the adolescent growth spurt of the average child, i.e. the child experiencing the adolescent growth spurt at the average age. In such a hypothetical situation the growth curve of the average child would fall exactly on the 50th centile line. But that is not the case with growth reference charts based on cross-sectional data; the average child will initially fall away from the 50th centile line as he or she enters the growth spurt and then cross it at the time of maximum velocity (peak velocity) before settling back onto the 50th centile as he or she reaches adult height.

Figure 1.7 illustrates the typical growth patterns exhibited by early, average and late developers. The early developing girl (E) accelerates into adolescence at about 8 years of age, some 2 years before the average, and rapidly crosses centile lines to move from just above the 50th to the 90th centile. However, her growth slows at about 13 years and her height centile status falls back to the 50th centile. Conversely, the late developer (L) is almost 13 years old before she starts to accelerate and that delay causes her height centile status to fall from the 50th to below the 10th centile before rising to the 50th centile as she approaches adulthood. Finally, the average girl initially falls away from the 50th centile but then accelerates through it at the average age of peak velocity before following the 50th centile as adulthood is reached.

1.9 CANALIZATION

Figure 1.7 demonstrates more than simply the crossing of centiles by early and late developers. It also tells us something about the control of human growth. These are not hypothetical curves. They are the growth curves of real children who were measured on a 6- or 3-monthly basis throughout childhood and adolescence.[16] Note that during childhood they are growing on or near to the 50th centile and after the deviations brought about by their adolescent growth spurts they return to that same centile position in adulthood. Such adherence to particular centile positions is found time and again when one studies the growth of children. Indeed, it is true to say that all children, when in an environment that does not constrain their growth, will exhibit a pattern of growth

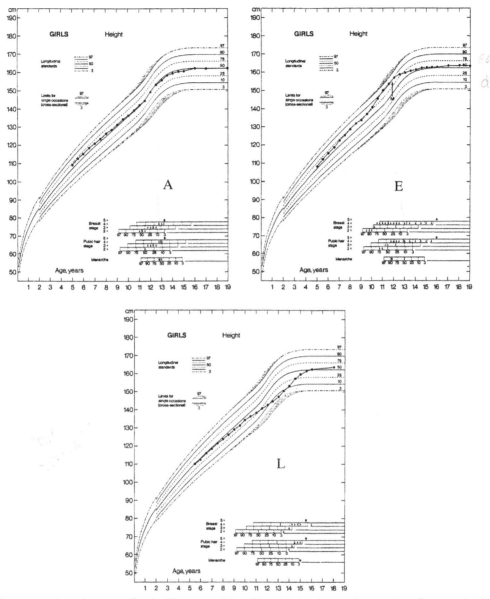

Figure 1.7 Growth curves of early (E), average (A) and late (L) developers. (Source: *Data from numbers 35, 38 and 45 in Tanner and Whitehouse.[16]*)

that is more or less parallel to a particular centile or within some imaginary "canal". This phenomenon was described by the British geneticist C.H. Waddington (1905–1975) in 1957[17] and has been termed "canalization" or "homeorrhesis". It is most likely that this pattern is genetically determined and that growth is target seeking in that we have a genetic potential for adult stature and the process of growth, in an unconstrained environment, takes us inexorably towards that target.

1.10 CATCH-UP GROWTH

However, it is a truism to say that none of us has lived or been brought up in a completely unconstrained environment. Towards the end of our intrauterine life our growth was constrained by the size of the uterus. During infancy and childhood we succumbed to a variety of childhood diseases that caused us to lose our appetite and at those times our growth would have reflected the insult by appearing to slow down or, in the more severe case, to cease.

Waddington[17] likens growth (or, as we now think of it, cellular decision making during growth) to the movement of a ball rolling down a valley floor. The sides of the valley keep the ball rolling steadily down the central course (point A in Figure 1.8). If an insult occurs it tends to push the ball out of its groove or canal and force it up the side of

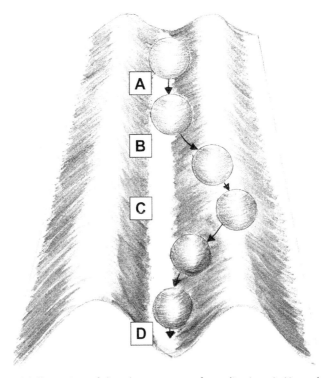

Figure 1.8 A pictorial illustration of the phenomenon of canalization. A: Normal canalized growth; B: point at which an impact causes the ball to deviate up the side of the valley; C: alleviation of the impact and a return to the valley floor; D: resumption of normal canalized growth.

the valley (point B). The amount of deviation from the predetermined pathway will depend on the severity and duration of the insult. However, any insult will cause a loss of position and a reduction in growth velocity as the ball is confronted by the more severe slope of the valley wall. The magnitude of the loss of velocity will also depend on the severity and duration of the insult. Thus, a small insult of short duration will cause a slight shift onto the valley sides which will entail a minor change in velocity. The alleviation of the insult will result in a rapid return to the valley floor at an increased velocity (point C). Having reached the floor, normal growth velocity is resumed (point D).

This analogy may be seen to apply appropriately to the process of human growth. Figure 1.9 shows the growth chart of a girl who has suffered from celiac syndrome.[16] In this condition there is an abnormality of the lining of the gut and food cannot be absorbed,

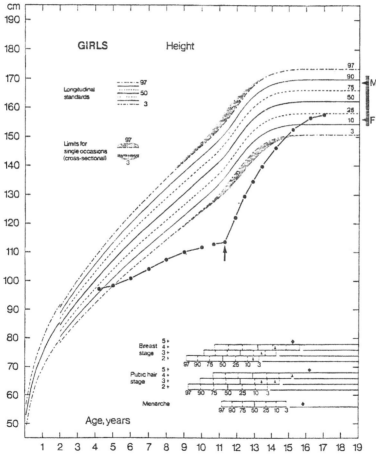

Figure 1.9 Catch-up growth exhibited by a child with celiac syndrome. (Source: *Data from number 102 in Tanner and Whitehouse.[16]*)

resulting in the child being starved. The result in terms of growth is that the height velocity is gradually reduced as the malnutrition becomes more and more severe. The reduction in height velocity means that the height distance curve leaves the normal range of centiles and the child becomes abnormally short for her age. So, at the age of almost 12 years she is the average height of a 5-year-old. On diagnosis the child is switched to a gluten-free diet which alleviates the malabsorption. Recovery of height velocity is rapid and jumps from 1 cm/year to 14 cm/year, returning the child to the normal range of centiles within 3 years. Indeed, this girl ends up within the range of heights one would expect given the heights of her parents. So she demonstrates complete catch-up growth in that she returns to the centile position from which she most probably started.

Catch-up growth is, however, not always complete and appears to depend on the timing, severity and duration of the insult. This appears to be particularly true in the treatment of hormone deficiencies. Initial diagnosis is often delayed until the child is seen in relation to other children and the deficiency in stature becomes obvious. Usually a hormone deficiency, e.g. growth hormone deficiency, is also accompanied by a delay in maturation. Response to treatment appears to depend on some pretreatment factors such as chronological age, height, weight and skeletal maturity, i.e. on how long the child has been deficient, how severe the deficiency in height and weight are, and by how much the maturity has been affected.

1.11 SUMMARY AND CONCLUSIONS

This chapter forms an introduction to the study of human growth and development. The curve of human growth has been a changing characteristic of our genus as we have evolved over the past two million years. It has changed in duration and magnitude as we have been freed of the environmental constraints that have affected us throughout our evolution, but its major characteristics have remained unaltered. That curve reflects two major stages during which adjustment to final size and shape are the direct consequences. Infancy and adolescence are times of adjustment and assortment. More than 50% of infants exhibit either catch-up or catch-down growth during the first 2 years of life.[15] These adjustments have long-term consequences in terms of final size, shape, morbidity and perhaps mortality. The timing of the adolescent growth spurt, its magnitude and its duration are also fundamentally important in terms of healthy and successful survival. The biological phenomena of canalization and catch-up growth dictate the magnitude, duration and ultimate success of these alterations and adjustments.

No one would argue that environmental constraints on growth through the processes of famine and disease have been constant influences, not only on our survival, but also on our size and shape at any age. The past millennia of environmental insults have resulted in the survival of representatives of the species *Homo sapiens* who are adapted and adaptable to their environment. We are the survivors and as such we use survival strategies to ensure that we continue the species. One of the most powerful of these strategies is the plasticity of our growth and development. Throughout the following chapters you will learn how that plasticity is inherited, controlled and expressed — it is a fascinating story about the most fundamental biological phenomena of our species.

REFERENCES

1. http://www.ucmp.berkeley.edu/history/evolution
2. McLeish K, editor. *Bloomsbury guide to human thought*. London: Bloomsbury Press; 1993.
3. Bourdier F. Principaux aspects de la vie et de l'oevre de Buffon. In: Heim R, editor. *Buffon*. Paris: Publications Françaises; 1952. p. 15–86.
4. Tanner JM. *History of the study of human growth*. New York: Academic Press; 1988.
5. Scammon RE. The first seriatim study of human growth. *Am J Phys Anthropol* 1927;**10**:329–36.
6. Tanner JM. *Growth at adolescence*. 2nd ed. Oxford: Blackwell Scientific Publications; 1962.
7. Lampl M, Veldhuis JD, Johnson ML. Saltation and stasis: a model of human growth. *Science* 1992;**58**:801–3.
8. Preece MA, Baines MK. A new family of mathematical models describing the human growth curve. *Ann Hum Biol* 1978;**5**:1–24.
9. Tanner JM. Some notes on the reporting of growth data. *Hum Biol* 1951;**23**:93–159.
10. Thompson D'Arcy W. *On growth and form*. Cambridge: Cambridge University Press; 1917.
11. Thompson D'Arcy W. *On growth and form*. Revised ed. Cambridge: Cambridge University Press; 1942.
12. Tanner JM. *Growth at adolescence*. Oxford: Blackwell Scientific Publications; 1955.
13. Tanner JM. Growth and endocrinology of the adolescent. In: Gardner L, editor. Endocrine and genetic diseases of childhood. 2nd ed. Philadelphia, PA: WB Saunders.
14. Scammon RE. The measurement of the body in childhood. In: Harris JA, Jackson CM, Patterson DG, Scammon RE, editors. *The measurement of man*. Minneapolis, MN: University of Minnesota Press; 1930. p. 171–215.
15. Ong KL, Ahmed ML, Emmett PM, Preece MA, Dunger DB. Avon Longitudinal Study of Pregnancy and Childhood Study Team. Association between postnatal catch-up growth and obesity in childhood: prospective cohort study. *BMJ* 2000;**320**:967–71.
16. Tanner JM, Whitehouse RH. *Human growth and development*. London: Academic Press; 1980.
17. Waddington CH. *The strategy of the genes*. London: Allen and Unwin; 1957.

SUGGESTED READING

Alder K. *The measurement of all things*. New York: Free Press; 2002.
> *A fascinating account of the development of a uniform system of measurement against the background of revolutionary France and the acceptance of arbitrary systems of measurement resulting from papal and monarchial dictates.*

Bogin BA. *Patterns of human growth*. 2nd ed. Cambridge: Cambridge University Press; 1999.
> *Barry Bogin's landmark volume on the patterns of human growth is a personal journey through the bicultural approach to human growth and development. It is well written and accessible, as befits a coeditor to the current volume, and is an extremely useful additional reference text.*

Thompson D'Arcy W. *On growth and form*. Abridged ed. Bonner JT, editor. Cambridge: Cambridge University Press; 1961.
> *D'Arcy Thompson's classic volume in an abridged and thus accessible form. None of the poetry of Thompson's writing has been lost in this edition and for the serious scholar of human biology this is an important volume.*

Tanner JM. *Fetus into man*. Cambridge, MA: Harvard University Press; 1990.
> *Fetus into Man was the first of Jim Tanner's volumes aimed at a general audience rather than at human biologists. It explains physical growth and development from birth to adulthood precisely and elegantly and as a general book for background reading it is invaluable.*

Tanner JM. *A history of the study of human growth.* Cambridge: Cambridge University Press; 1981.

> *A historical account of the study of human growth which will be of interest to postgraduate students wishing to understand the context within which the growth of children has become a core aspect of human biology, education, psychology, sociology, pediatrics and economic development.*

INTERNET RESOURCES

A French website providing access to Histoire Naturelle: http://www.buffon.cnrs.fr/

> *There are a variety of websites that provide a wealth of information on human growth and development. However, it is very important to distinguish between the types of information required before accepting the text of any particular site. Many are concerned primarily with psychological, emotional or social development and include little or no physical development. Others relate to growth disorders or clinical aspects of growth and yet others are aimed at parents. It is always important to rigorously check the provenance of the website and ensure that it is from a reputable scientific institution.*

CHAPTER 2

Growth in Infancy and Childhood: A Pediatric Approach

Horacio Lejarraga
University of Buenos Aires at Hospital Garrahan, 1245 Buenos Aires, Argentina

Contents

N. Cameron & B. Bogin (eds): Human Growth and Development, Second edition.
ISBN 978-0-12-383882-7, Doi: 10.1016/B978-0-12-383882-7.00002-7

2.1 INTRODUCTION

At birth the infant leaves the uterus, where he or she lived in a protected locus under quite stable conditions, and is delivered into a postnatal environment characterized by extremely varied and changing situations. The infant will be subject to intense and continuous physiological demands, which will require consequent adaptive responses during the first years of life. At the same time, besides the dramatic changes in size, during the first postnatal years the child experiences profound developmental changes in the central nervous system, providing the physiological basis for psycho-motor development, the most intense intellectual adventure of the human being: the course of changes in sensorimotor response, intelligence, and, in particular, language. This complex process is possible because the elapsed time between weaning and the onset of puberty is exceptionally long in humans compared to other primates. This period allows brain development and socialization. During the school years and before puberty the child acquires a large proportion of his or her cultural "inheritance".

Infancy is a high-velocity, rapidly decelerating, nutrition-dependent phase of growth, followed by a more stable, slowly decelerating growth hormone-dependent phase during preschool and school years. Both phases can be expressed by different mathematical functions. At birth, physical size is strongly related to prenatal growth, which in turn expresses the fetus's intrauterine living conditions and, consequently, the size of the newborn does not express the size genetically determined by the parents. During the first 2 years, genes expressing parental size become activated and a considerable proportion of healthy children shift linear growth, changing percentiles on distance charts, until they achieve, at around the second year of life, their genetically determined location on the percentiles. This phenomenon is an example of *canalization*, which was explained in Chapter 1.

Growth evolving under these circumstances becomes:

- a central subject in pediatric practice, because some of the central tasks of the pediatrician are to promote the positive growth of children, to detect in an opportune way any abnormal growth deviation, and to restore growth as much as possible. Infancy and childhood are sensitive periods in human life; interference with the growth processes in early years may have long-term consequences for later maturation and health, and opportune pediatric interventions can have a strong preventive impact. Pediatric surveillance and the promotion of normal growth in infancy and childhood come along with the knowledge of its physiological basis, and the skills for the performance of adequate anthropometric measurements
- an important objective in child health programs, because the ultimate goals of such programs are not simply to reduce infant mortality, but to provide health care and environmental conditions to allow children to grow and develop normally

- a relevant health indicator because it can be easily monitored and included in surveillance programs, an excellent indicator of general health and nutritional status, both in population groups and in individuals. In clinical practice, some conditions can be recognized because growth delay is the first clinical sign, as is the case with malnutrition, acquired hypothyroidism and celiac disease. It is also an excellent general indicator for the long-term health care of children with chronic diseases
- a criterion of success in therapeutic interventions, and may be a valuable aid to take therapeutic decisions such as surgical interventions in some chronic diseases. Drug treatments such as steroids or immune suppressing agents can seriously affect growth, and their therapeutic action should be constantly balanced against their impact on growth
- a highly motivating issue for health education at the clinical level and in the community, and it strengthens the bonds among patients and the members of the health team. All centers where children receive medical care should be provided with the appropriate personnel, anthropometric instruments and growth charts for the proper assessment of child growth and community education.

2.2 THE GROWTH CURVE

If the height of a child measured yearly, from birth to maturity, is plotted on a graph, a curve like that shown in Figure 2.1(a) will be obtained. Any other normal child measured in the same way would show a curve with the same shape. This shape is a primate characteristic (see Chapter 19). Humans may differ in height, but the shape of their curve would be the same. Some characteristics of child growth are best shown if instead of plotting height attained at a given age, the yearly increments, in terms of centimeters per year, are plotted as shown in Figure 2.1(b).

Infants grow very quickly during the first year of life, at approximately 25 cm/year, and during the first half of the first year velocity is even faster, around 30 cm/year. A steep and continuous deceleration can be observed from birth up to the third year. Thereafter, there is a much milder decay in velocity during childhood and before the adolescent growth spurt. During this spurt, height velocity is approximately 9.5 cm/year in boys and 8.5 cm/year in girls.

Growth curves expressing height (or weight or any other measurement) attained at a given age are called *distance curves*, whereas those expressing increments in a given period are called *velocity curves*.

At birth, sex differences are present, but are small, of about 1 cm in length in favor of boys by about 5 years of age. The mean adult sex difference in height of about 12.5–13.0 cm mainly develops before the adolescent growth spurt, as boys start the spurt

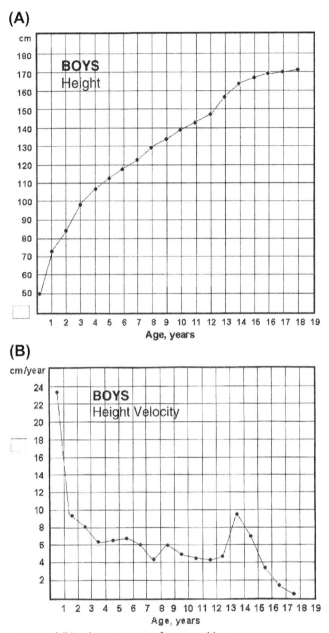

Figure 2.1 (a) Distance and (b) velocity curves of a normal boy.

2 years later than girls. These 2 years of extra childhood growth, at about 5 cm/year, account for most of the adult difference in height. In addition, boys grow slightly faster during the spurt and acquire about another 2−3 cm of adult stature relative to girls (see Chapter 3).

Velocity curves show certain information that is not apparent in the distance curve. Three major phases of growth curve can be identified: a rapidly decelerating phase, present from birth to approximately 2–3 years, a slowly decelerating phase from 3 years up to the onset of the adolescent growth spurt, and the adolescent spurt itself.

The three phases can be expressed in mathematical terms. This expression constitutes an important field of research in itself, serving several purposes: (1) for estimating the continuous growth process from discontinuous measurements; (2) for eliminating measurement errors; (3) for summarizing growth data with a limited number of parameters; (4) for estimating functions and parameters with biological meaning and so interpolating values and identifying precise milestones of the growth process; and (5) for estimating the average growth of population groups.

There have been many attempts to express the growth curve in mathematical terms, many of which have succeeded in finding an expression that summarizes growth of the whole postnatal period,[1] and recently new methods have been developed for the analysis of growth data.[2] However, for pediatric purposes Johan Karlberg's proposal will be described here. He designed a mathematical model called the infancy–childhood–puberty (ICP) model.[3] Figure 2.2 shows the decomposition of the growth curve from birth to maturity into three parts.

The infancy component is expressed with an exponential function that describes a rapidly decelerating phase:

$$Y = a_i + b_i(1 - \exp c_c t) \qquad [2.1]$$

This period is called the nutrition-dependent phase of growth, because nutrition is the main factor influencing it. For example, breast feeding makes a difference in the growth curve compared to formula feeding.

Figure 2.3 shows different studies on average growth curves of children receiving human milk during the first months of life, compared to the growth of children fed with formula.[4] During the first 4–6 months, breast-fed babies are longer, but thereafter there is a shift in the mean growth curves in favor of formula-fed infants, and this difference lasts until about 2 years of age. This is an important fact to remember at the time of building growth references, but there are studies showing that the difference in size due to breast feeding fades away after the second or third year of life, there being no difference between the two groups of children thereafter. The same is true with regard to growth in weight and in weight for height.

Another common example of nutritional influences on growth is that concerning the weaning process, and its contemporary introduction of other semi-solid and solid foods in the infant diet, a process called complementary feeding. A widespread phenomenon seen in many children with inadequate and insufficient complementary feeding is growth delay starting between the sixth and the 12th months of life, when breast feeding is no longer enough to satisfy the growing needs of the infant. When this happens, the

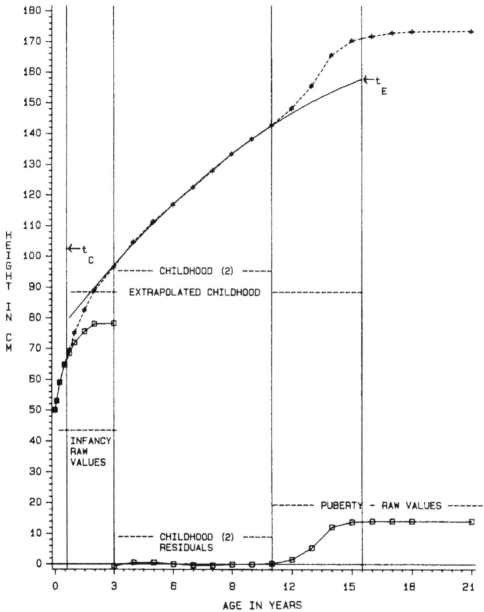

Figure 2.2 The three phases of postnatal growth according to the infancy–childhood–puberty (ICP) model. (Source: *Karlberg.*[7])

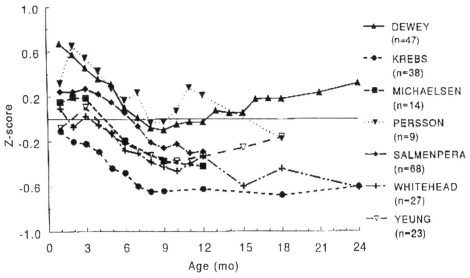

Figure 2.3 Growth curves (supine length *Z*-scores) of breast-fed children from different studies compared to NCHS standards. *(Source: WHO.[4])*

child shows growth curves in weight and in height similar to those shown in Figures 2.4 and 2.5. Child "a" grew normally in weight until the age of 7 months, when her weight curve started to shift, showing a clear growth delay.[5]

Her growth in height (Figure 2.5a), however, showed a normal growth curve. The anamnesis of her mother revealed that her complementary feeding and caloric intake were insufficient for her age.

In the case of curve of child "b", the delay in growth was evident in weight (Figure 2.4, curve b) as well as in height (Figure 2.5, curve b). Besides the insufficient caloric intake, there were many other environmental factors affecting growth. Pediatric intervention was not followed by a catch-up phenomenon in either case, but at least it halted the progressive deterioration of the growth curve.[6]

The childhood component is a second degree polynomial function that describes height velocity as following a gradually decelerating course that continues until the end of growth, and can be summarized with the formula:

$$Y = a_e + b_e t + c_e t^2 \qquad [2.2]$$

which expresses a gradually decelerating growth process.

The puberty component is modeled with a logistic expression describing additional growth induced by sex hormones that produces acceleration up to a peak height velocity and then a final period of deceleration until the end of growth:

$$Y = a_p / (1 + A[-b_p(t - tv)]) \qquad [2.3]$$

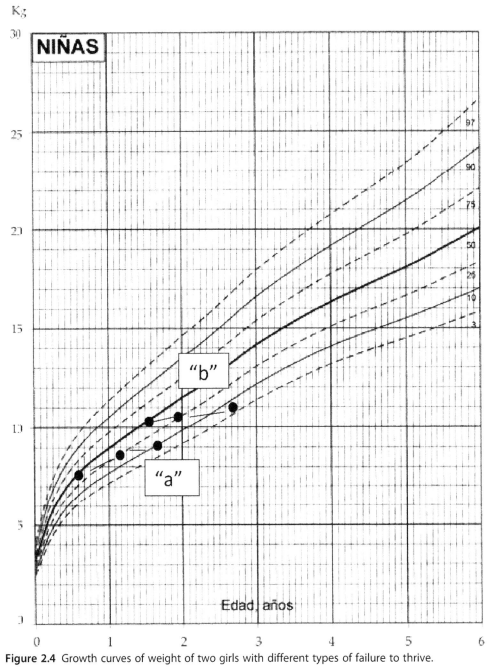

Figure 2.4 Growth curves of weight of two girls with different types of failure to thrive.

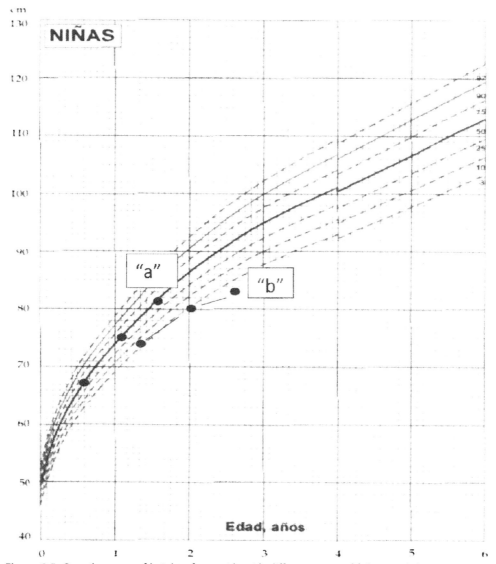

Figure 2.5 Growth curves of height of two girls with different types of failure to thrive.

(See the discussion by Cole and colleagues[2] of various methods of modeling the human growth curve.)

2.3 CHANGES IN BODY FAT AND OTHER TISSUES

Not all the body components grow with the same pattern as height. The brain, for example, shows a different growth curve, with a steep rise during the first years of life, achieving almost 85% of its adult size at the ages of 2—3 years. During the first postnatal years, an extraordinary process of brain development takes place, with the completion of cell replication of glial cells (neurons stop replicating before birth), organization of different cell centers and synaptogenesis.[6]

Growth of fat follows a completely different pattern, as shown in Figure 2.6, where the growth of subcutaneous skinfolds of boys and girls is plotted against age.[7]

The slopes of the curves at different ages are similar to those for the growth of total body fat (see Chapter 18). Fat tissue exhibits an important increase in the first year of life; the baby is born with a relatively small amount of fat, and this increase changes substantially the baby's body composition. The infant is thus equipped with an energy reserve which helps it to cope with the most dangerous period of nutritional scarcity in human evolution: weaning. Breast milk is no longer enough for normal growth after the sixth month, and the child needs complementary nutritional intake. After the first year of life, the skinfold curve shows a slow decrease taking place during the following 4—5 years, coincidently with the period in which the child also shows important development in motor control and physical activity.

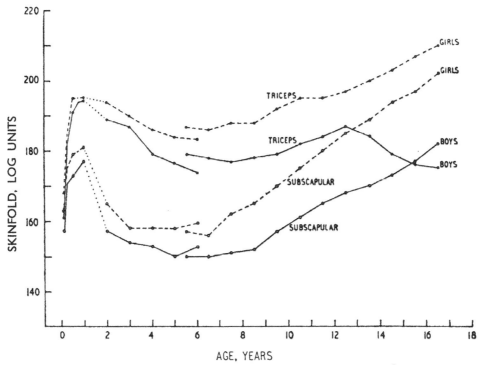

Figure 2.6 Distance curves of triceps and subscapular skinfolds. (Source: *Tanner.*[7])

Weaning should not be considered as a sudden punctuated event. Rather, it is a process of gradual incorporation of complementary solid foods into the diet. Initially, it complements breast milk, but after several months, between 1 and 2 years of age, it replaces it. The process involves the activation of a multiplicity of *family functions*. These functions include: (1) the selection of adequate food in terms of quality and quantity; (2) the development of *maternal technologies*, i.e. maternal skills for feeding the child, for looking after him or her, for detecting risk situations, for developing alarm signs, for promoting and stimulating the development of sensorimotor skills, including the eating of solid foods; and (3) the existence of a family structure to protect and allow the mother to perform all those functions.

The second year of life coincides with a decline in subcutaneous fat or skinfold thickness, and the baby changes his or her initial plump, baby-like appearance towards a leaner, slender body build. The child is now able to walk and run, performing intense motor activity and expending energy. The combination of increased activity and decreased fat is a source of parental concern in cases where the child does not eat the amount of food the parents believe they should eat ("poor appetite").

In school years, skinfolds and percentage of total body fat remain quite stable, as does height velocity. Body fat continues to be greater in girls than in boys, a genetic difference that holds true in skinfold thickness and percentage of body fat. At adolescence the process of fat centralization takes place much more predominantly in males, characteristic of adult fat distribution, a variable associated with risk factors for cardiovascular disease.[8]

2.4 MEASUREMENTS OF CLINICAL IMPORTANCE

The three most important anthropometric measurements to be taken in pediatric practice are height or supine length, weight and head circumference. These measurements describe different body components and their changes have different biological significance.

2.4.1 Height

Height is a linear measurement. Although it mainly indicates the length of long bones of the lower limb and the irregular bones of the spine and the flat bones of the skull, it is also an indirect indicator of the growth of total lean body mass. In clinical pediatrics, changes in height growth need a rather long period of observation to be detected (3 months or more), depending on the age of the child.

2.4.2 Weight

Weight is a three-dimensional measurement that includes both lean body mass and body fat. The relative proportions and distributions of lean and fat components depend on age, sex, and other environmental and genetic factors. Weight is a very sensitive measurement, in the sense that it can change from one day to another owing to very minor alterations in body composition; for example, the weight loss in infants during a common cold is an expected observation. A change in weight in the short term does not tell us which particular tissue is being affected, since changes in body weight can be secondary to changes in body water (dehydration or overhydration), muscle mass (muscle hypertrophy by training, muscle atrophy), total lean body mass (wasting), fat (obesity, malnutrition), and so on. Weight changes can also be secondary to changes in stature, as is the case in growth retardation or stunting.

In persistent food scarcity the first affected measurement is weight, but if the conditions persist, growth in height is also affected.

2.4.3 Weight/Height Ratios

Some children have a low weight, not because they are lean but because they are short. Others are heavy, not because they are fat but because they are very tall. This is a limitation to be considered when wanting to assess the amount of body fat as an indicator of nutritional status. One way to evaluate body fat directly is by measuring subcutaneous fat using skinfolds, and this method is encouraged, together with appropriate reference charts or growth standards. One way to evaluate it indirectly is by assessing weight in relation to height. In order to do this, the usefulness of a number of weight/height ratios or indices has been investigated. It is important that weight per unit height is assessed independently of height, i.e. weight *regardless* of height, and thus the most efficient ratio would be that one which has the least correlation with height. Table 2.1 shows correlation coefficients between standing height and several weight/height ratios.[9] The lowest figure is that corresponding to weight (kilograms) divided by the square of height (meters): $BMI = W/H^2$. This is called the Quetelet's index, or body mass index (BMI).[9] BMI is now widely used as an indicator of nutritional status. Initially, it was mainly used for assessing overweight (BMI above 25 kg/m^2) and obesity (BMI above 30 kg/m^2) (see Chapter 14), but now it is also possible to use it for assessing undernutrition (wasting or thinness).[10]

Table 2.1 Pearson's correlation coefficients between weight or height and several weight/height indices[9]

Index	With weight	With height
Weight/height	0.729	−0.213
Weight/height^2	0.861	−0.002
Weight/height − 100	0.771	−0.135
Weight$^{1/3}$/height	0.672	−0.295
Weight/height^3	0.679	−0.283

2.4.4 Wasting and Stunting

In order to assess population health and well-being it is important to identify appropriate indicators of malnutrition for epidemiological rather than for clinical use. The most widespread and important of these were developed by Professor John Waterlow and his colleagues, who coined two terms that were very useful in assessing the prevalence of malnutrition in population groups: wasting and stunting.[11] Wasting is characterized by low weight for height. The term described children with "acute malnutrition", who were extremely lean, with a high risk of becoming ill or dying because of deficit in the immune system. Population groups with high prevalences of wasting are in critical need of medical and nutritional intervention.

Stunting refers to children with short height for age and a normal weight for height. They have "chronic malnutrition", but this is perhaps a misnomer because the cause of their stunting is multifactorial, and nutritional rehabilitation does not produce a catch-up in height in the majority of the cases. In many circumstances, these children seem to have suffered growth delay in their early years, and their short stature is a consequence of this early insult.

In clinical practice the diagnosis of malnutrition, and its precise characteristics and etiology in each child, must be built on a clinical basis as well as on anthropometric measurements.

2.4.5 Head Circumference

Head circumference is a good expression of brain growth. Its main applications in pediatrics are the early detection of hydrocephalus in infants (due to congenital blockage of spinal fluid circulation) and, in older children, the diagnosis of macrocephaly, and in cases of acquired hydrocephalus or disproportionate short stature (see below). It should be measured systematically in all pediatric appointments, especially in the first years of life.

2.5 SOME PARTICULAR FEATURES OF GROWTH DURING INFANCY AND CHILDHOOD

The following paragraphs show some features of growth during infancy and childhood.

2.5.1 Parental Size and Size at Birth

Tall parents tend to have tall children and short parents tend to have short children. This is so because a considerable part of the individual variation in stature is due to genetic factors. This rather close relationship between the height of children and that of their parents becomes a useful tool for growth assessment. However, this relationship is not present at birth, but after the age of approximately 2 years. Figure 2.7 shows the

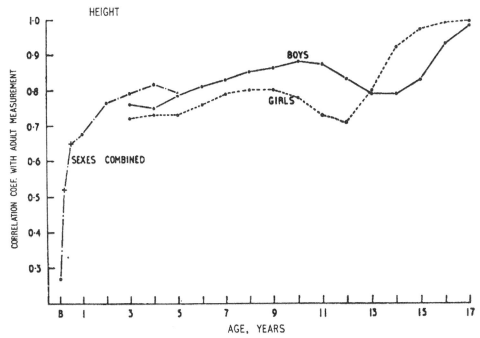

Figure 2.7 Correlation coefficients between adult height and height of the same individuals as children. (Source: *Tanner.[7]*)

correlation coefficients between the adult height of children and height of the same children at each age from birth onwards.[7] Size at birth is poorly correlated with adult size; the values rise steeply during the first 2 years to a level of about 0.80, then become quite stable with very small variation until adulthood. The single major variation observed in the figure at adolescence is due to different maturation rates (see Chapter 20).

The influence of maternal size on size at birth is best understood with an experiment carried out by Walton and Hammond.[12] Reciprocal crosses were made between a Shetland pony and a shire horse.[12] Foals sired by Shetland stallions on shire mares had the same birth weight as foals of pure maternal breed. Foals sired by shire stallions on Shetland mares had almost the same birth weight as foals of pure maternal breed. But the size at birth of foals from Shetland dams was considerably smaller than that from shire dams. Thus, maternal regulation in the horse is predominant in determining the rate of intrauterine growth.

The same applies to human fetal growth. This is why at birth, babies' size is much more related to maternal size and intrauterine living conditions. As the child grows older, at the age of about 2–3 years, the genes regulating adult size achieve their full expression. Thus, not all genes regulating body size express themselves at birth and, consequently, when a pediatrician has to deal with an infant of small size, rather than measuring the parents, he or she should make enquiries about pregnancy, maternal health and size at birth. The relationship between the child's height and parental height should be looked at after the second year of life.

2.5.2 Shifting Linear Growth During Infancy: an Example of Catch-Up and Catch-Down

If size at birth and the first year of life depends on prenatal and maternal conditions, and size at later ages depends on the genetic background of the child, then in many cases, the growth curve must change percentiles and indeed this is what actually happens in normal children. Smith, in 1976, found that about two-thirds of healthy children shifted percentiles upward or downward during the first 2 years of life.[13] Babies from genetically tall parents, but whose fetal growth was constrained, once in the extrauterine environment, free from this constraint, express their genetic potential by crossing percentiles upwards (catch-up growth). Conversely, those who grew under very favorable prenatal conditions from genetically short parents crossed percentiles downwards after birth (catch-down growth). These are both examples of *canalization* (see Chapter 1). From the pediatric point of view, this means that not all deviations from the original percentiles in the first 2 years of life are pathological. The differential diagnosis may be difficult, but obstetric history, as well as clinical examination and assessment of growth velocity, can be of great help.

2.5.3 Growth as Measured in Very Short Periods

Up to the 1990s it used to be assumed that growth was a smooth, continuous process, which proceeded without sudden arrests or accelerations. However, since the pioneering work of Michelle Lampl,[14] it is known that when measured in short periods, growth is not a smooth process, but proceeds in terms of jumps (saltation) and periods or arrested growth (stasis). Figure 2.8 shows daily measurements of supine length and head circumference taken on an infant 5 days a week during 6 months. Three phases can be identified: saltation, stasis and continuous growth.[15] It can also be observed that there seems to be a certain degree of coordination in the occurrence of those phases between both measurements over time.

2.5.4 Mean Childhood or Juvenile Growth Spurt

The plateau in growth velocity occurring between 5 and 10 years of age is interrupted at the age of 6—8 years in both sexes by a small growth spurt called the mid-growth spurt or juvenile growth spurt. Some authors claim that this spurt is present only in boys, but others have found it in both sexes.

Figure 2.9, from the work of James Tanner and Noël Cameron in London children, illustrates a clear mid-growth spurt in calf circumference velocity in both sexes, followed

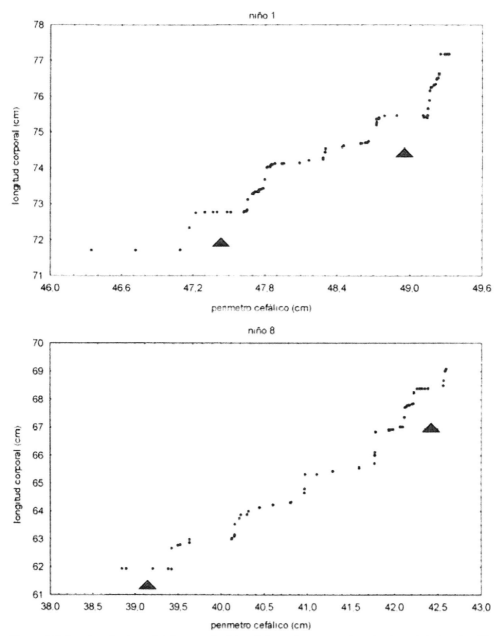

Figure 2.8 Synchronization of daily growth of head circumference and supine length in two healthy infants. Triangles indicate periods of synchronized stasis and saltation. (Source: *Personal data.)*

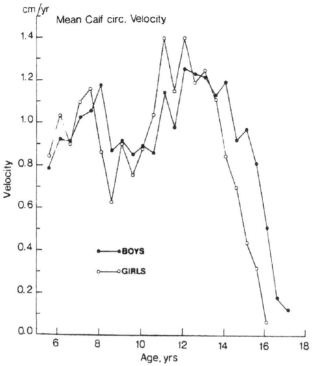

Figure 2.9 Mid-growth spurt in calf circumference velocity. *(Source: Tanner and Cameron.[16])*

by the adolescent growth spurt, of greater magnitude. The spurt shows no sex differences, and is uncorrelated with the timing and magnitude of the adolescent spurt. It has been attributed to adrenarche, characterized by an increase in the secretion of androgenic hormones of adrenal origin (see Chapter 5).[16]

2.5.5 Early Growth and Nutritional Experiences, and Later Outcome in Health and Maturation

In recent decades, considerable advances in knowledge have taken place regarding the multiple relationships between early nutritional experiences and long-term growth, maturation and health.

In the 1980s, David Barker of Southampton University described what would later be called the Barker hypothesis, when he found a strong relationship between low birth weight, and higher mortality risk from cardiovascular disease. Later, this association extended to the risk for metabolic syndrome and high cholesterol levels, besides the risk of heart disease.[17,18]

Not only size at early ages, but also variations in the rate of weight gain influence later maturational and health events. A rapid weight gain in infancy is associated with a higher risk of obesity,[19] insulin resistance and β-cell dysfunction in later years, which is in turn associated with a higher risk of cardiovascular disease in adult life.[20]

Rapid weight gain in infancy may also have consequences on later physical maturation. Girls showing increased weight gain in the first 2 years of life have earlier menarcheal age.[21] Insulin resistance, exaggerated adrenarche and reduced levels of sex hormone binding globulin may explain this relationship: the elevated levels of insulin-like growth factor-1 (IGF-1), adrenal androgens, aromatase activity and "free" sex steroids could promote the activity of the gonadotrophin-releasing hormone (GnRH) pulse generator and thus early puberty.[22] These metabolic and hormonal conditions (specially elevated IGF-1) associated with early rapid growth may also lead to tall stature and advanced skeletal maturity.[23-25] On this basis, the auxological status of children at adolescence may be the result of early nutritional, metabolic and growth experiences. Infancy, childhood and adolescence therefore behave as a critical period for health status in later life.[26]

2.6 GROWTH PROBLEMS IN INFANCY AND CHILDHOOD

One of the most common causes of pitfalls in the pediatric study of growth problems is the ambiguity in initially defining the growth problem itself. Failure in identifying whether the child is too short, too light or not growing at a normal velocity may be misleading with regard to further actions (laboratory studies, tests, etc.). Parents are usually unspecific in stating the problem, and they may not discriminate among phrases such as: "the child is not growing properly", "the child is too short" or "the child is too light". These three diagnoses have essentially different clinical meanings and deserve different clinical approaches. Consequently, it becomes necessary when dealing with children with alleged growth problems to arrive at a correct auxological diagnosis before proceeding with other actions. The clinician can then search for possible causes.

In practice, when a pediatrician sees a child with an apparent growth problem, he or she should ask two main questions:
- Is the child's size normal for his or her age?
- Is the child growing at a normal velocity for his or her age?

2.6.1 Is the Child's Size Normal for his or her Age?

In order to answer the first question, the child should be appropriately measured and the height plotted against distance charts, such as the one shown in Figure 2.10, in this case, a chart for Argentinian children.[27]

In this recent version of these charts, ±2 standard deviations (SDs) of the age at onset of puberty events (Tanner stages B2, G2, PH2 and menarche) are included. The relative merits of national charts as opposed to international charts and references as opposed to standards is a controversial issue, and more information is given in Chapter 21. Yet there

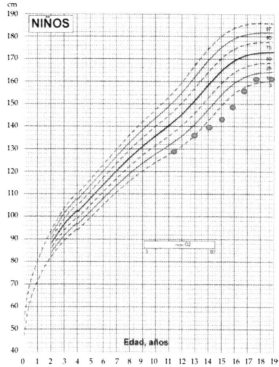

Figure 2.10 Growth curve of a growth with delayed maturation. See the adolescent growth spurt taking place at the age of 16 years.

are countries in which the general consensus of opinion is that national charts, based on selected samples of healthy children who have not been exposed to growth constraints, are appropriate as references to reflect growth as it should be in the best possible circumstances within the given national environment. They are thus appropriate for the clinical monitoring of individual children. In the case of Argentina, it was decided to use World Health Organization (WHO) data of breast-fed children from 0 to 2 years and to keep using national data thereafter.

When height or weight (or any other measurement) falls within the percentile lines, the child is classified as having a normal height or weight (or other dimension). Most of the countries set the lower normal limit for clinical use as the 3rd centile, but this is a convention. By measuring normal children and those who are known to have a growth problem, two distributions will be obtained, relating to each other in the way shown in Figure 2.11.

There is an overlapping of the opposite tails of the two distributions. The tallest of those children with short stature (it is not recommended that the term "dwarf" is used) have the same stature as the shortest normal children. So, if one chooses a higher centile

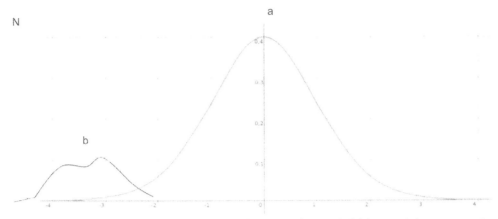

Figure 2.11 Frequency distribution (Z-scores) of heights of normal children and those presenting with a growth disorder. a: Normal children; b: growth disorders. This figure is reproduced in the color plate section.

(in order to be more sensitive and so select the majority of abnormal children), such as the 10th centile, 10% of normal children will be classified as abnormally short. Conversely, if one chooses a lower centile as the lower normal limit in order to be highly specific, and so select the highest proportion of normal children, such as the 1st centile, some abnormally short children will be classified as normal.

Hence, the lower limit to be chosen depends on the consequences that the diagnosis is going to bring to the child.[28] In the case of some nutritional support within social programs, for example, to be given to low-weight children, the aim is to be highly sensitive, and to cover low-weight children who might suffer for any nutritional deficit. Therefore, a high percentile such as the 10th centile would be chosen. In clinical pediatrics, where patients will be referred or complicated laboratory hormonal studies performed in short children, the aim is to be highly specific, to save resources and unnecessary studies. Thus, the choice would be a lower percentile centile, such as the 3rd centile or even lower, as the British references have recently proposed (the 0.4th percentile),[29] with the purpose of being more specific (fewer studies, fewer referrals, etc.). In Argentina, the lowest limit has been set as the 3rd centile, knowing that with this decision, 3% of the normal population will be classified as abnormally short.

The position of the measurement on the chart can also be expressed as a standardized score [Z-score or Standard Deviation Score (SDS)]. However, for clinical use centiles are preferred because they express the *probability* that a child has to belong to a normal or to an abnormal population. SDS are less meaningful when they represent measurements that are not normally distributed. Of course, when the measurement is beyond the extreme percentile limit, it becomes necessary to use SDS. A method for transforming measurements into SDS is available and capable of estimating scores from British, US,

WHO and Argentinian references (see below: Internet sites, LMS Growth Program, in English and Spanish).

Having answered the first question, the second one is addressed.

2.6.2 Is the Child Growing at a Normal Velocity for his or her Age?

To answer this question, at least two measurements are needed, separated by an appropriate period of time. In Chapter 21, the way growth velocity is calculated and assessed on velocity charts is explained in detail.

Velocity charts indicate whether the child is growing too quickly, too slowly or at a normal speed. In clinical work, pediatricians can estimate growth velocity indirectly on distance charts, by the slope of the distance curve compared to that of the percentiles lines. As shown in Figure 2.12, when the slope of the distance curve parallels those of the centile lines, growth velocity is said to be normal, as in curve a. When the curve falls away from the centile lines, velocity is too slow, as in curve b, and when it climbs up the centile lines, velocity is too fast, as shown in curve c.

Other examples in the figure can be illustrative. Child d's height is below the 3rd centile; therefore, his stature is abnormally short. However, the following measurements conform to a curve whose slope is parallel to those of the centiles lines, so it can be said that this child has short stature, but is growing at a normal growth velocity. Conversely, the child with curve b shows at all times normal stature; however, the slope of his distance curve is falling away and thus it can be assumed that the child is growing at a slow rate; this is the situation that deserves the diagnosis of *growth retardation*.

From these examples, six situations can clearly be distinguished: normal, tall or short stature combined with normal, slow or fast growth velocity.

Size attained and growth velocity are both rather independent concepts. Size attained can be considered a cross-section of the continuous growth process that has taken place from birth (or better, from conception) up to the moment in which the child is measured. It is like a photograph; it is the result of the algebraic sum of all increments, eventual decrements, arrests and compensatory accelerations taking place during the child's life. If we see a 9-year-old with a normal height, we can say something like "we do not know what happened to the child in the past, but we can say that all that could ever have happened to him did not significantly affect his growth in height, or if there was an injury to his height, it was subsequently 'compensated' ". Height attained tells us nothing about what is going on with the child's health at the time the child is being measured.

On the contrary, growth velocity is a dynamic concept. It is a measurement that is strongly independent from height, and provides information on what is happening during the period in which the measurements were taken. For example, in Figure 2.12, curve d shows that although the child has short stature, his growth is absolutely normal and it proceeds at a normal speed; consequently, it can be stated that growth was not affected during the period. Curve b, in contrast, it shows a normal height at all times, but

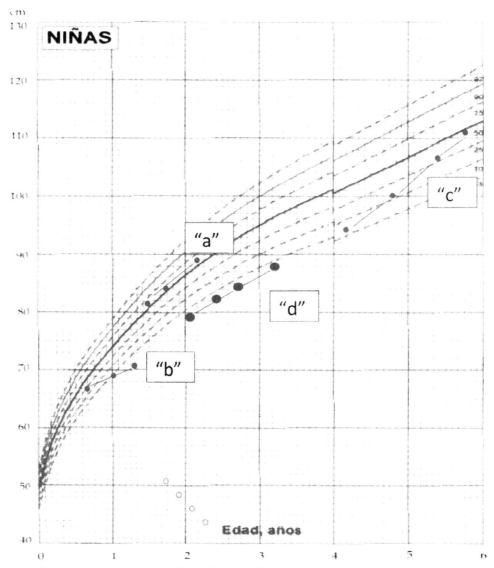

Figure 2.12 Schematic examples of growth curves for height.

an abnormally slow velocity. Consequently, it must be assumed that there is something wrong with the growth of this child and he should be studied.

Short stature can be an expression of a current disease, or a late consequence of a former disease. The child with short stature may or may not be healthy at the time he is measured. Conversely, a child with slow growth *always* has a health problem affecting his growth.

To assess growth velocity in schoolchildren, measurements must be taken separated by at least 6-month intervals. Shorter periods may be misleading, taking into account the measurement error of about 1 or 2 cm (in pediatric clinics) compared to the relatively small increments taking place at that age (around 5—6 cm per year).

In infants, clinicians cannot wait 6 months to decide whether there is a something wrong with the growth of their patients. Instead, they must pay attention to short changes in weight. A few years ago, the WHO produced standards for monthly and bimonthly weight increments, which are available on the Internet (see Internet sites: Weight increments in infancy).

Children do not always come to the office at exact monthly intervals, and this is a problem for assessing weight increments in periods of a few days. It must be taken into account that normal individual variations of weight increments are much greater when measured in a period of a few days than when measured at monthly intervals.

Both growth delay and short stature are quite common causes for consultation in pediatric practice. However, they are unspecific clinical signs, because growth impairment may have a great variety of underlying causes. The majority of the causes of growth problems at primary care level can be identified without sophisticated studies.

In general, the clinical approach to growth problems in the first 1 or 2 years of life is different from that regarding problems appearing at later ages. In the first years of life, small size or short stature is related to problems in the prenatal or perinatal period. Infants with growth delay can be included within a diagnosis of "failure to thrive", which is generally of multifactorial origin, and the underlying causes should be identified as soon as possible.

In contrast, in the case of short stature or growth delay in the school years, the underlying cause is generally unifactorial, and if the child is asymptomatic, the necessary time, including the assessment of growth velocity, can be taken to find out the cause.

Long-term consequences of growth delay in infancy may be serious. The obvious result of growth retardation during this period of life is short stature. The risk of final height being adversely affected as a consequence of growth delay is higher if the delay takes place during infancy, because growth velocity is very high at this age.

There are also other consequences of growth delay. Martorell and colleagues have described cognitive and functional impairments in adulthood as a result of stunting between 3 and 6 years of age in a sample of impoverished Guatemalan children.[30]

Growth retardation during the prenatal period and in the first year of life is associated with a variety of adult conditions including cardiovascular disease, stroke, hypertension, non-insulin-dependent diabetes mellitus and low bone mineral density. The mechanisms underlying this association may be related to early metabolic changes in hormone interactions and modalities of tissue response to these hormones, with persisting and long-term consequences in adult life. This point has been described earlier in this chapter.

In childhood, growth problems may have an important impact on psychosocial adjustment. The problem itself, and the accompanying clinical and laboratory studies directed to the identification of the problem, may interfere with school performance, sports and social integration. The psychological impact of short stature is not linearly related to the severity of the growth deficit but depends on individual characteristics. The author has seen children with familiar short stature with mild height deficit (for example, -2.2 SD) deeply concerned about the problem, and others with -3.00 with very good social integration.

As in other phases of growth, the chance of experiencing complete catch-up depends on the duration of action of the factors causing growth delay, their intensity, the age at which they occur and the individual impetus to grow.

The description of the causes of the growth problem varies according to the specific problem: different approaches are taken to study causes of *short stature* and the causes of *growth delay*. Taking into account that in a clinical setting short stature is the most frequent problem, the most important causes of this problem are described here.

2.7 CAUSES OF SHORT STATURE

Two large groups of children may be differentiated: those with and those without evident signs of malformation, deformation and/or alteration of body proportions.

2.7.1 Malformations, Deformations or Alterations of Body Proportions

In these cases, short stature is only a part of the clinical picture. This group includes children with congenital malformations or skeletal dysplasias, some of neonatal onset, for example, achondroplasia or osteogenesis imperfecta. Others are not so severe and become evident in the second or third year of life, such as dyschondrosteosis or hypochondroplasia. There are guidelines for helping pediatricians to achieve clinical orientation towards skeletal dysplasias.[31] In these cases, not only general growth charts, but also conditioned charts are useful, for example head circumference in relation to stature, for detecting mild forms of macrocephaly (relative microcephaly).[32] Congenital malformations and dysmorphic syndromes are also included within this group, such as Noonan syndrome chromosomal alterations, congenital rubella syndrome, Turner's syndrome and fetal alcohol syndrome. Some syndromes evolve with a normal growth velocity in childhood, such as Silver–Russell syndrome (Figure 2.10, curve b),[33] but others grow at a low speed, as for example in Turner's syndrome.

Children with skeletal dysplasias, dysmorphic syndromes and chromosomal disorders are well described in books dedicated to their clinical identification[34–36] and in databanks (see Internet databanks). Searching devices based on information databases may be helpful in the diagnosis (see below: Internet sites, Database on genetic syndromes). Some syndromes are quite frequent, for example Turner's syndrome has an incidence of one in every 2500 female births, whereas others are rarer, for example achondroplasia (one case in every 26,000 births).

In the absence of dysmorphism or alterations in body proportions, another group of children may be defined as those having extreme normal variations.

2.7.2 Extreme Normal Variations

Two conditions are included in this group: familial short stature (FSS) and delayed maturation.

The diagnosis of FSS is made in normal children older than 2 years, with no underlying disease, a normal physical examination, mild short stature (between −2.00 and −3.00 SDS; when the child's height is below −3.00 SDS, the diagnosis of FSS should be considered with extreme caution and put in doubt), and where the height of the child adjusted to that of the parents is within normal limits.

The child's height can be adjusted to that of the parents by subtracting the mid-parental height (expressed in SDS) from the child's height (expressed in SDS). For example, in a child whose height SDS is −2.5 and the father's and mother's height are −0.98 SDS and −1.60 SDS, respectively, the child's adjusted height (AH) is:

$$AH = -2.50 - \frac{[-0.98 + (-1.60)]}{2} = 1.21 \text{ SDS} \qquad [2.4]$$

This figure is within normal limits (±2 SD either side of the distribution).

The indicator, called the "genetic target", is estimated by calculating the *corrected mid-parental height* (MPH). This indicator is corrected for the sex of the child under consideration by adding 13 cm to the height of the mother if the child is a boy or subtracting 13 cm from the height of the father if the child is a girl. The figure of 13 cm is the mean sex difference in adult height in the normal population. For example, if the father's and mother's height are 167.0 cm and 158.0 cm, respectively, the corrected mid-parental height (CMPH), if the patient is a girl, would be:

$$CMPH = \frac{167 - 0 + (158.0 + 13)}{2} = 169.0 \text{ cm} \qquad [2.5]$$

This is the genetic target. Adding and subtracing (±) 7.5 cm gives the genetic range: 178.5/152.5 cm. This is the adult height range of 95% of possible daughters of the couple. The predicted height of the patient, by any of the available methods used in pediatrics,[37−39] should be within this range (if she has normal growth).

The height of parents should be measured in the clinic, since parental height reported by hearsay has proven to be quite unreliable and consequently the estimated genetic range is expected to be wider.[40]

Much care should be taken, however, when the parent's height is below normal limits from population standards. The clinician may be dealing with some pathological condition in the parent inherited by the child. This may happen in the case of dys-chondrosteosis or hypochondroplasia, both with mendelian, autosomic dominant inheritance, in the case of sex-linked inheritance such as familial hypophosphatemic rickets, or in conditions inherited as polygenic inheritance, such as celiac disease.

The other clinical condition presenting with extreme variation of normal is delayed maturation. This diagnosis is made in normal children with delayed bone age, delayed puberty or both, with no other abnormal features and a normal height velocity (see curve "a" in Figure 2.10). The syndrome is more frequent in boys than in girls. There is usually a family history of delayed puberty that is evident by a mother's late menarcheal age. Age of onset of puberty in the father is usually more difficult to determine. Predicted height in these cases is within normal limits and within the genetic range. Physical examination is normal and growth velocity is normal during childhood. At pubertal ages, if the child has no signs of puberty, growth velocity is normal for prepubertal values. Prognosis is good, since these children achieve a normal height. However, since the adolescent growth spurt is experienced at later ages, there is a period in which children are shorter than their peers and they may also have delayed sexual development. These features are often the cause of social maladjustment. Emotional support from both the family and the pediatrician is necessary to help children through this period. If the psychological impact of the condition is important and puberty has not started, induction of puberty with small doses of androgens in boys or small doses of estrogens in girls is possible and may be very effective in stimulating the hypothalamus to start puberty.[41]

Diagnosis of delayed maturation in young children (preschool years) should be made with extreme caution, as there are pathological conditions with delayed bone age and no apparent clinical signs. These cases usually evolve with growth retardation (slow growth velocity).

2.7.3 Perinatal Problems

Growth during the perinatal (prenatal and neonatal) period is very fast, and because of this, growth retardation at this stage may have serious consequences on size at later ages. Two main types of growth problem may be identified during this period: intrauterine growth retardation (IUGR) and growth retardation of children who are born prematurely and have postnatal growth delay at a postconceptional age in which the majority of children grow in utero.

IUGR is diagnosed when weight or length at birth is low for the infant's gestational age. To assess growth during this period, it is necessary to use appropriate growth charts that describe size in weight or length at each gestational age, such as that shown in Figure 2.13.[42] On the x-axis, postconceptional age in weeks is shown. Note that normal pregnancy lasts for 40 postconceptional weeks (actually, between 37 and 42 weeks). On the y-axis, the figure shows weight, height and head circumference at different

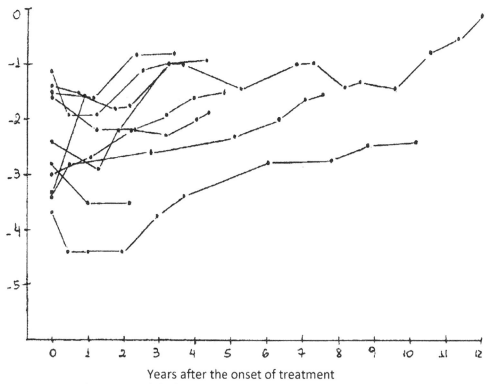

Figure 2.13 Height *Z*-scores of children with nephrogenic diabetes insipidus (NDI) from the onset of treatment onwards.

gestational ages, from the 24th up to the 90th centiles, which is equivalent to 1 year of post-term, postnatal age.

Many children with IUGR show catch-up growth after birth. In the majority of children, this catch-up growth is present immediately after birth during the first and second years, as shown in Figure 2.13. A few children show catch-up growth at later ages (5 or 6 years of age), and others (around 20%) show a partial catch-up growth or do not catch up in growth at all.

Factors associated with the possibility of postnatal catch-up growth are not well determined, but are very likely to relate to several factors: (1) the nature of the damaging agents; (2) its timing of occurrence; and (3) its duration. For example, in the case of an injury acting in the first trimester of pregnancy, in the embryonic period, as happens with congenital rubella, the impact on growth is very serious: the child is born with reduced birth weight, reduced birth length and reduced head circumference (sometimes described as harmonic IUGR). These children do not catch up, grow below the normal distance limits at a normal velocity, and usually show at school age the same height deficit as shown at birth.[43]

In the case of mild growth impairment occurring in the last weeks of pregnancy, as happens with twin pregnancy or minor acute diseases affecting the mother during the last weeks of pregnancy, the baby is born with normal birth length, normal head circumference and low birth weight (sometimes called disharmonic IUGR). In these cases, catch-up is feasible and usually occurs in the first months of postnatal life. In between these two extreme examples, there are many intermediate situations such as the case of children born to mothers with toxemia in the last trimester of pregnancy. These children have normal head circumference, low birth weight and short birth length. They usually do not show complete catch-up, and they may have short stature with normal growth velocity at school ages.

The second type of perinatal problem associated with short stature during school years or adolescence is that suffered by some children born preterm, before the 37th week of gestational age. Gestational age is calculated as completed weeks elapsed from the first day of the last menstrual period up to the day of birth. Menstrual period is an indirect indicator of the time of conception, which is why this age is called *postconceptional age*. This age is marked in the chart until 92 weeks, exactly 1 year after term, i.e. 40 weeks, or full-term gestation plus 52 postnatal weeks.

The mechanism by which these children may show short stature is different from those with IUGR. Preterm babies may be born with a normal birth weight and birth length for gestational age, that is, having had normal prenatal growth. However, postnatally, when birth takes place very early, birth weight is very low, babies are looked after under intensive care conditions and it is very difficult to keep them alive, to feed them and to maintain homeostasis. Usually these babies show an initial weight loss in the incubator at the neonatal intensive care units, some are kept alive with mechanical respiration, weight is kept constant, and thereafter, at some point, the baby starts gaining weight. Curve "a" in Figure 2.13 shows the postnatal growth in weight of a child born at 32 weeks' gestational age with a birth weight of 1400 g. This weight is located on the 25th centile, so the child has an adequate weight for gestational age. The baby spent 3 months in an incubator, during which time he had many complications (feeding, respiratory and metabolic problems). His growth curve could not be kept within the centile lines and his growth rate was slower than normal for the period 28–41 postconceptional weeks. Thereafter, when he became easier to feed, he started recovering weight and, finally, experienced a complete catch-up growth in weight. In this boy (as in many others) three phases of growth can be identified: the first period of weight loss and growth retardation, followed by a second period of catch-up, followed by a third period of normal growth rate. Not all preterm babies grow this way. Some, like the one represented by curve "b" (growth in supine length), may not catch up and remain with a height deficit during the rest of their infancy and childhood. These children may complain at later ages of short stature. Their clinical examination and growth velocity at the time of consultation (school years or adolescence) are normal, and unless a careful clinical perinatal history is taken, the cause of the short stature may remain obscure.

In the area of public health, standard and comparable ways of assessing postnatal growth should be developed as possible indicators of quality of care as an important complement to data on neonatal mortality. Immediate postnatal growth of newborn groups in institutions is a good indicator of the quality of perinatal care and, in the author's opinion, these indicators should be widely used in maternity units.

The incidence of short stature because of perinatal problems is increasing throughout the world. Progress in the quality of neonatal care and progress in technology make it possible for many children with very low birth weight to survive. Thirty years ago, babies weighing 700 g at birth were not viable. Nowadays, they do survive, and some of them contribute to the high prevalence of short stature at school age because of perinatal problems.

2.7.4 Malnutrition

Primary malnutrition is a consequence of reduced nutritional intake (predominantly calories, protein or both), strongly associated with unfavorable socioeconomic conditions. Kwashiorkor (predominantly protein deficiency) and marasmus (predominantly calorie deficiency) are perhaps the most frequent causes of growth impairment in the world. They are frequently observed in developing countries and in excluded and marginal groups in inner cities in developed countries. The severity of the growth deficiency is proportional to the severity of the nutritional deficit.

Depending on the duration and quality of protein and energy deficiency and its relative impact on weight and height, two main types of malnutrition can be recognized: stunting and wasting.[11] Wasting, also called emaciation, assessed by weight for height, and nowadays by BMI, is the expression of the present nutritional state and recent food intake. It is associated with a high risk of disease and death. Stunting, assessed by height for age, is an indicator of past nutrition, and is not necessarily associated with a greater risk of disease and death. These two categories have been identified mainly as public health indicators, and are not convenient for use in clinical work. Stunting is not due only to nutritional chronic deficiencies, and does not necessarily reflect an actual, current lack of nutrients. If we overfeed these children, we may produce overweight with no changes in height. In realistic terms, stunting in population groups is not only a consequence of nutritional factors, but also associated with poor health, low birth weight and other environmental conditions under which they have lived for generations.

In the first paragraphs of this chapter, the later consequences of poor nutritional experiences during the first year of life were mentioned. Not only the quantity, but also the quality of nutrient intake at early ages can influence later body composition. A high protein/low energy intake in infancy is associated with a higher degree of fatness in late childhood. This association may, in turn, be related to the influence of early diet on hormone secretion.[44]

2.7.5 Psychosocial Deprivation

Type I psychosocial deprivation applies to infants with non-organic failure to thrive. Maternal deprivation, lack of adequate nutrition and other factors may intervene in this entity in combination with a deficit in swallowing function, deficiency of micro-nutrients, and poor "maternal technologies" or child-rearing practices. There may be reduced growth velocity in weight, height or both. Type II psychosocial deprivation applies to children older than 4–5 years, in whom nutritional deficiencies are not apparent and the underlying mechanism is thought to be severe emotional or psychiatric disorders. Secondary growth hormone deficiency has been detected in some cases.[45]

2.7.6 Chronic Diseases

Practically any chronic disease may have an impact on physical growth.[46] The most frequent entities in practice are: severe asthma, malabsorption (celiac disease, chronic inflammatory bowel, cystic fibrosis), congenital heart disease (with cardiac failure or with right to left shunt), chronic anemia, metabolic acidosis, chronic pulmonary disease, chronic infections (tuberculosis, AIDS), and gross central nervous system disorders. The mechanisms underlying growth delay in these entities are varied, and most of them converge into a few basic causes: lack of adequate nutritional intake, metabolic acidosis, metabolic disorders, hypoxia, protein loss and, not infrequently, the treatment itself (see Drugs, below). In the majority of them, growth velocity is low, but during some periods it can be normal.[47]

With the progress of therapeutic resources in medicine and the consequent reduction in mortality, the prevalence of children with chronic disease is continuously increasing. In this context, physical growth becomes an excellent clinical indicator. If growth is normal, the pediatrician knows that the condition is not interfering with the process and, eventually, that treatment is being effective in allowing normal growth. For example, in the case of acute leukemia, the survival rate is now very high (above 90%),[48] and it is growth that matters to patients and to pediatricians.

With this and many other important therapeutic achievements in other chronic diseases, one can say that: "… [many of these] children do not die any more, but … how do they grow?" This presents a new challenge for the clinical management of children surviving diseases that were lethal in the near past.

2.7.7 Drugs

Some drugs are well known for their negative impact on growth. The main ones are adrenal steroids, used in asthma, nephritic syndrome, leukemia, lupus and many other conditions. A dose greater than the physiological secretion delays growth and the magnitude of the delay is proportional to the dose. Doctors know that there is a high price to pay in treating asthma with steroids. In the case of budesonide, for example, a dose greater than 400 µg/day may inhibit growth,[49] but studies show that growth deficit can be recovered after the treatment is discontinued.[50] In these cases, a balance between the need to keep a patient asymptomatic and the need to allow normal growth has to be continuously considered.

Other drugs, such as cytostatics, can also affect growth. Cranial irradiation used in the treatment of leukemia and tumors of the central nervous system can damage hypothalamic function. Such damage may impair the release of both hypothalamic and pituitary hormones, including growth hormone.

2.7.8 Endocrine Disorders

Endocrine disorders of growth are fully described in Chapter 5. All of these disorders run with a slow growth rate, and the pediatrician must suspect the existence of an endocrine disorder whenever growth velocity is subnormal.

2.7.9 Idiopathic Short Stature

This is an ill-defined condition, diagnosed in the cases of short stature (height −2.0 SD below the corresponding mean) "… for a given age, sex and population group, without evidence of systemic, endocrine, nutritional or chromosomal abnormalities".[51]

There has been a consensus for reaching this diagnosis, consisting of the following conditions: (1) height below the third percentile; (2) short stature of parents; and (3) no identified cause after routine investigations: endocrine [growth hormone deficiency, hypothyroidism, adrenal steroids (exogenous or endogenous), malabsorption, metabolic or nutritional disorders, etc.].

2.7.10 Miscellanea

As stated at the beginning of this section, any chronic disease can impact growth and produce short stature if its action lasts long enough. Rare growth problems that the author has seen during the past 20 years in the growth clinic of the Department of Growth and Development at Hospital Garrahan in Buenos Aires include familial hypophosphatemic rickets, nephrogenic diabetes insipidus,[52] congenital heart disease, failure to thrive, congenital benign myopathy, metabolic acidosis and severe lesions of the central nervous system.

2.8 SUMMARY

Growth is one of the central objectives of pediatric practice: the promotion of a positive growth and prevention of eventual deviations. It is also an excellent indicator of child health both at an individual clinical level and in population groups. Growth in infancy is a fast, rapidly decelerating process, a true "critical period", since certain auxological and nutritional experiences, such as low weight at birth and rapid weight gain during the first year of life, can have long-term consequences on the rate of maturation, growth during adolescence, and health and disease in adulthood.

Diseases affecting growth can be depicted at a clinical level by size attained or growth velocity. The most common causes of short stature are extreme variations of normality (familial short stature and delayed maturation). Progress made in the treatment of many conditions that were formerly lethal contributes to the increased prevalence of chronic disease in children and the survival of very low birth weight infants. An increased prevalence of short stature or growth delay in childhood due to these two conditions may be observed in everyday clinical practice.

REFERENCES

1. Hauspíe RC, Cameron N, Molinari L, editors. *Methods in human growth research*. London: Cambridge University Press; 2004.
2. Cole TJ, Donaldson MD, Ben-Shiono Y. SITAR — a useful instrument for growth curve analysis. *Int J Epidemiol* 2010;**39**:1558—66.
3. Karlberg JA. Biologically-oriented mathematical model (ICP) for human growth. *Acta Paediatr Scand* 1989;**350**:70—94.
4. World Health Organization Working Group on Infant Growth. *An evaluation of infant growth*. Geneva: World Health Organization; 1994.
5. Breitman F, del Pino M, Fano V, Lejarraga H. Crecimiento de lactantes con retardo del crecimiento no orgánico. *Arch Argent Pediatr* 2005;**103**:110—7.
6. Synaptogenesis Jin Y. The online review of C. elegans. *Biology*, www.wormbook.org/chapter/www_synaptogenesis/synaptogenesis.html; 2006. Wormbook.
7. Tanner JM. *Growth at adolescence*. 2nd ed. London: Blackwell; 1962.
8. Cnoi D, Wareham N, Luben R, Welch A, Bingham S, Day MN, et al. Serum lipid concentrations in relation to anthropometric indices in relation to central and peripheral fat distribution in 2021 British men and women. Results of the EPIC—Norfolk population-based cohort study. *Atherosclerosis* 2006;**189**:420—37.
9. Lejarraga H, Abeyá E, Andrade J. Body mass index in a national representative sample of 88,861 18 year old boys. Argentina 1987 (abstract). *Pediatr Res* 1993;**33**:663.
10. Cole TJ, Flegal KM, Nicholls D, Jackson AA. Body mass index cut offs to define thinness in children and adolescents: international survey. *BMJ* 2007;**335**:194.
11. Waterlow JC, Buzina RJ, Keller W, Nichaman MZ, Tanner JM. The presentation and use of height and weight data for comparing the nutritional status of groups of children under the age of 10 years. *Bull World Health Organ* 1977;**55**:489—98.
12. Walton A, Hammond J. Maternal effects on growth and conformation in shire horse—Shetland pony crosses. *Proc R Soc* 1938;**15B**:311.
13. Smith DW, Truog W, Rogers JE, Reizer LJ, Skinner AL, McCann JJ, et al. Shifting linear growth during infancy. Illustration of genetic factors in growth form fetal life through infancy. *J Pediatr* 1976;**89**:223—30.
14. Lampl M, Veldhuis JD, Johnson ML. Saltation and stasis. A model of human growth. *Science* 1992;**158**:801—3.
15. Caíno S, Kelmansky D, Adamo P, Lejarraga H. Short term growth in head circumference and its relationship with supine length in healthy infants. *Ann Hum Biol* 2010;**37**:108—16.

16. Tanner JM, Cameron N. Investigation of the mid-growth spurt in height, weight and limb circumference in single-year velocity data from the London, 1966–67 growth survey. *Ann Hum Biol* 1980;**7**:565–77.

17. Barker D, Winter P, Fall C, Simmonds S. Weight in infancy and death from ischaemic heart disease. *Lancet* 1989;**ii**:577–80.

18. Barker D, Osmond C, Golding J, Kuh D, Wadsworth M. Growth in utero, blood pressure in childhood and adult life, and mortality from cardiovascular disease. *BMJ* 1989;**298**:564–7.

19. Ong KK, Loos RJF. Rapid infancy weight gain and subsequent obesity: systematic reviews and hopeful suggestions. *Acta Paediatr* 2006;**95**:904–8.

20. Ong KK, Dunger DB. Birth weight, infant growth and insulin resistance. *Eur J Endocrinol* 2004;**151**:U131.

21. Ong K, Sloboda D, Hart R, Doherty D, Pennell C, Hickey M. Age at menarche: influences of prenatal and postnatal growth. *J Clin Endocrinol Metab* 2007;**92**:46–50.

22. Dunger DB, Ahmed ML, Ong KK. Early and late weight gain and the timing of puberty. *Mol Cell Endocrinol* 2006;**245–55**:140–5.

23. Demerath EW, Jones LL, Hawley NL, Norris SA, Pettifor JM, Duren D, et al. Rapid infant weight gain and advanced skeletal maturation in childhood. *J Pediatr* 2009;**155**:355–61.

24. Cameron N, Demerath EW. Growth, maturation and the development of obesity. In: Johnston FE, Foster GD, editors. *Obesity, growth and development*. London: Smith-Gordon; 2001.

25. Cameron N, Pettifor J, De Wet T, Norris S. The relationship of rapid weight gain in infancy to obesity and skeletal maturity in childhood. *Obes Res* 2003;**11**:457–60.

26. Cameron N, Demerath EW. Critical periods in human growth: relationship to chronic diseases. *Yearb Phys Anthropol* 2002;**45**:159–84.

27. Lejarraga H, del Pino M, Fano V, Caino S, Cole TJ. Referencias de peso y estatura desde el nacimiento hasta la madurez para niñas y niños argentinos. Incorporación de datos de la OMS de 0 a 2 años, recálculo de percentilos para obtención de valores LMS. *Arch Argent Pediatr* 2009;**107**:126–33.

28. Tanner J. The evaluation of physical growth and development. In: Holzel A, Tizard JPM, editors. *Modern trends in pediatrics*. 2nd series. London: Butterworth; 1958. p. 325–44.

29. Wright CM, Williams AF, Elliman D, Bedford H, Birks E, Breths G, et al. Using the new UK-WHO charts. *BMJ* 2010;**340**:1140.

30. Martorell R, Rivera J, Kaplowitz H, Pollit E. Long term consequences of growth retardation during early childhood. In: Hernandez M, Argente J, editors. *Human growth: basic and clinical aspects. International Congress Series 973*. London: Excerpta Medica; 1992. p. 143–50.

31. Lejarraga H, Fano V. Skeletal dysplasias. A pediatric approach. In: Nicoletti I, Benso L, Gilli G, editors. *Physical and pathological auxology*. Firenze: Centro Study Auxologici; 2004. p. 352–96.

32. Saunders C, Lejarraga H, del Pino M. Assessment of head size adjusted for height. An anthropometric tool for use in clinical auxology based on Argentinian data. *Ann Hum Biol* 2006;**33**:415–23.

33. Tanner JM, Lejarraga H, Cameron N. The natural history of the Silver Russell syndrome. A longitudinal study of thirty-nine cases. *Pediatr Res* 1975;**9**:611–23.

34. Jones KL. *Smith's recognizable patterns of congenital malformations*. 6th ed. Philadelphia, PA: Saunders; 2006.

35. Spranger JW, Brill PW, Poznaski A. *Bone dysplasias. An atlas of genetic disorders of skeletal development*. 2nd ed. London: Oxford University Press; 2002.

36. Cassidy SAB, Allanson JE. *Management of genetic syndromes*. 2nd ed. New York: John Wiley & Sons; 2005.

37. Bayley N, Pinneau SRT. Tables for predicting adult height from skeletal age: revised for use with the Greulich and Pyle standards. *J Pediatr* 1952;**40**:423–41.

38. Roche AF, Wainer HW, Thissen D. The RWT method for prediction of adult stature. *Pediatrics* 1975;**56**:1026–53.

39. Tanner JM, Healy M, Goldstein H. Assessment of skeletal maturity and prediction of adult height *(TWIII method)*. 3rd ed. London: Saunders; 2001.

40. Lejarraga H, Laspiur M, Adamo P. Validity of reported parental height in growth clinics in Buenos Aires. *Ann Hum Biol* 1995;**22**:163–6.

41. Bridges N. Disorders of puberty. In: Brook CGD, Hindmarsh PR, editors. *Clinical pediatric endocrinology*. 4th ed. London: Blackwell Sciences; 2001. p. 165–79.

42. Lejarraga H, Fustiñana C. Estándares de peso, longitud corporal y perímetro cefálico desde las 26 hasta las 92 semanas de edad postmenstrual. *Arch Argent Pediatr* 1986;**84**:210−4.

43. Lejarraga H, Peckham C. Birth weight and subsequent growth of children exposed to rubella infection in utero. *Arch Dis Childhood* 1974;**49**:50−8.

44. Roland Cachera MF, Deheeger M, Akrout M, Bellisle F. Influence of macronutrients on adiposity development: a follow up study on nutrition and growth from 10 months to 8 years of age. *Int J Obes* 1995;**19**:1−6.

45. Albanese A, Hamill G, Jones J, Skuse D, Matthews DR, Stanhope R. Reversibility of physiological growth hormone secretion in children with psychosocial dwarfism. *Clin Endocrinol* 1994;**40**:687−92.

46. Lejarraga H. Growth in chronic diseases. In: Gilli G, Schell LM, Benso L, editors. *Human growth from conception to maturity. International association for human auxology. Advances in the Study of Human Growth and Development, No. 4*. London: Smith-Gordon & Nishimura; 2002. p. 189−206.

47. Lejarraga H, Caíno S, Salvador A, De Rosa S. Normal growth velocity before diagnosis of celiac disease. *J Pediatr Gastroenterol Nutr* 2000;**30**:552−6.

48. Baruchel A, Leblanc T, Auclerc MF, Schaison G, Leserger G. Towards cure for all children with acute lymphoblastic leukemia? *Bull Acad Natl Med* 2009;**193**:1509−17.

49. Sharek PJ, Bergman DA. The effect of inhaled steroids on the linear growth of children with asthma: a meta analysis. *Pediatrics* 2000;**106**:8−12.

50. Anthracopoulos MB, Papadimitriou A, Panagiotakos DB, Syridou G, Giannakopoulou E, Fretzayas A, et al. Growth deceleration of children on inhaled corticosteroids is compensated for after the first 12 months of treatment. *Pediatr Pulmonol* 2007;**42**:465−70.

51. Cohen P, Rogol AD, Deal CL, Saenger P, Reiter EO, Ross JL, et al. on behalf of the 2007 ISS Consensus Workshop participants. Consensus statement on the diagnosis and treatment of children with idiopathic short stature. *J Endocrinol Metabol* 2008;**93**:4210−7.

52. Lejarraga H, Caletti MG, Caíno S, Jimènez A. Long term growth of children with nephrogenic diabetes insipidus. *Pediatr Nephrol* 2008;**23**:2007−12.

INTERNET RESOURCES

Database of World Health Organization on weight increments in infancy (breast fed until at least the 4th month of life): http://www.who.int/childgrowth/standards/velocity/tr3chap_3.pdf

Database on genetic syndromes (McCusick) − online mendelian inheritance in man: http://www.ncbi.nlm.nih.gov/omim

Database on the European Skeletal Dysplasia Network: http:/www.esdn.org

A program for transforming growth measurements into SD scores with US, UK and WHO references: www.garrahan.gov.ar/tdecrecimiento

LMS Growth Program − Argentine references with instructions in Spanish: www.LMSgrowth

CHAPTER 3

Adolescent Growth

Roland Hauspie, Mathieu Roelants
Laboratory of Anthropogenetics, Vrije Universiteit Brussel, B-1050 Brussels, Belgium

Contents

3.1 INTRODUCTION

Adolescence is the period of transition from childhood to adulthood. There is no single event that marks the boundaries of this period of major psychological and biological changes, but it is generally assumed that adolescence starts with the onset of puberty and spans the "teenage" years. Puberty is perhaps the most visible sign, with the appearance of secondary sexual characteristics and a notable growth spurt. Puberty refers to the developmental phase where the child will reach reproductive capacity, and covers approximately the first half of adolescence. At the end of this period, most children will have reached their final or mature size for most skeletal dimensions, and may have surpassed their parents in height when a positive secular trend exists. This chapter will cover the physical growth (change in size) during adolescence. Children increase considerably in size during this period, and the timing (tempo) of this increase is variable. As a consequence, large differences in size may occur between children of the same chronological age. Another consequence is that the growth pattern of individual adolescents will generally differ from the average growth curve, a phenomenon that is closely related to the features of longitudinal and cross-sectional growth charts. The chapter will conclude with an analysis of the different growth patterns observed in boys and girls during adolescence, a period during which substantial size differences develop between the sexes. It will also demonstrate how mathematical models can help to

N. Cameron & B. Bogin (eds): Human Growth and Development, Second edition.
ISBN 978-0-12-383882-7, Doi: 10.1016/B978-0-12-383882-7.00003-9

describe individual growth patterns and how they can contribute to the analysis of the dynamics of the growth process.

3.2 THE ADOLESCENT GROWTH CYCLE

Growth at adolescence is characterized by the presence of a pubertal growth spurt for many skeletal dimensions and for weight. This spurt is the result of hormonal changes that occur during sexual maturation. The timing, duration and magnitude of the pubertal growth spurt vary considerably between populations and between individuals within a population. The effect of these variations in tempo will be discussed later. This section will focus on three major patterns of growth during adolescence, namely growth in height, weight and body mass index (BMI), and head circumference.

A typical example of the growth in height between 1 month and 18 years of age is shown in Figure 3.1 for a girl from the Belgian Growth Study of the Normal Child.[1−3] The upper curve is a plot of height for age (distance curve), while the lower curve shows the corresponding increments in height, scaled to a whole year (annual growth velocity curve). Annual growth velocities or yearly increments are the average velocity over the considered interval. In physics, the term "velocity" would refer to the first derivative of a smooth distance curve, i.e. instantaneous velocity. In growth research, this concept is difficult to work with, because studies of the size of body segments at daily or weekly intervals, with high-precision techniques (e.g. knemometry of the knee−heel distance with a measurement error of about 0.1 mm), show that the underlying growth process is, at a microlevel, not as smooth as is usually assumed (see Chapter 16).[4,5] For most body dimensions, instantaneous velocity is also impractical, because the measurement error would exceed the observed change in growth. Despite this subtle difference, the terms "increment" and "velocity" are often used intermixed, and generally refer to annual velocity or increments. The horizontal bars in the velocity chart indicate the length of the intervals over which the increments were calculated. It is common practice to calculate increments only from measurements not less than 0.85 years and not more than 1.15 years apart, and to scale them to *whole-year* increments (cm/year) by dividing the difference between the two measurements by the duration of the interval. Increments calculated over a shorter interval can reflect seasonal variation and are relatively more affected by measurement error.[6] Increments over intervals larger then 1 year would smooth out age trends.

The growth pattern in height is characterized by a gradually decreasing (sometimes more or less constant) velocity during infancy and childhood, until there is a substantial acceleration that marks the beginning of the pubertal growth spurt in the early teens. Before puberty, considerably smaller prepubertal or mid-childhood spurts may be observed, but not in every child (see Chapter 2).[7,8] The pubertal growth spurt is, however, a persistent feature of the normal growth curve, which will always be observed when measurements are taken at reasonable intervals.

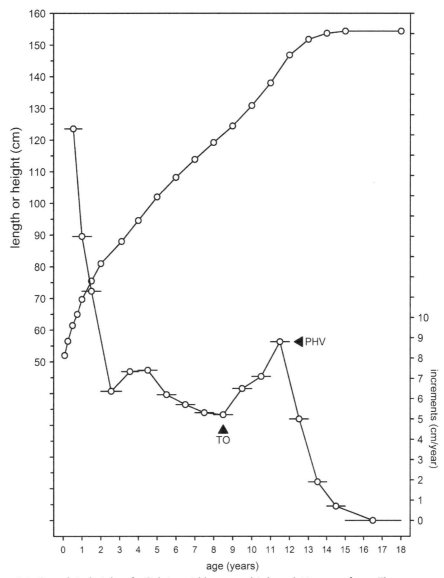

Figure 3.1 Growth in height of a Belgian girl between birth and 18 years of age. The upper curve is a plot of height for age (distance curve), and the lower curve shows the corresponding annual height increments (velocity curve). Horizontal bars indicate the length of the interval over which the increment was calculated. TO: take-off; PHV: peak height velocity. (Girl no. 29 from the Belgian Growth Study of the Normal Child.[1-3])

The age at minimal velocity before puberty is considered as the onset of the pubertal growth spurt, i.e. age at take-off. The age at take-off varies considerably between populations, between individuals within a population (the standard deviation is about 1 year) and between sexes (the pubertal spurt starts on average 2 years later in boys than in girls). Maximum velocity in height [or peak height velocity (PHV)] is generally reached 3–3.5 years after the onset of the growth spurt. At this age, boys can grow more than 11 cm per year, and girls more than 9 cm per year. Age at take-off and age at peak velocity are useful milestones for the study of the timing and duration of the adolescent growth spurt. After having reached a peak, the growth velocity decreases rapidly, inducing the end of the growth cycle at full maturity, which occurs around 16–17 years in girls and near 18–19 years in boys in western populations. After this age, growth may continue until the late twenties, but the additional gain in height will be small. There is a wide variation between populations, between individuals and between the two sexes as to the attained size at each age, the timing of events such as adolescent growth spurt and the age at which mature size is reached. The growth curve of height shown in Figure 3.2 is typical for all postcranial skeletal dimensions of the body.

The pattern of growth in weight is different from that of height because the start of the adolescent growth spurt in weight does not correspond to the age at minimal velocity in weight before puberty. Most children show the lowest annual increase in weight in late infancy or early childhood, i.e. around 2–3 years of age.[9,10] Thereafter, growth in weight slowly, but steadily, accelerates until a sudden rapid increase in weight velocity marks the onset of the pubertal spurt in weight. The typical features of growth in weight and weight velocity are illustrated in Figure 3.2. The sudden increase in the growth velocity in weight between childhood and puberty, which marks the onset of the growth spurt, occurs in this example at about 11.5 years of age. The precise location of this point is, however, more problematic and subjective than estimating take-off for height, because there is no real point of inflection.

Closely related to weight is the BMI, which is an anthropometric index of weight (in kilograms) divided by the height (in meters) squared (kg/m^2). Before adolescence, the BMI curve fluctuates, with a sharp increase in early infancy, and subsequently a decline in early childhood, until a local minimum is reached between 4 and 7 years of age (the "adiposity rebound"). During late childhood and adolescence, the BMI increases steadily until a maximum is reached in late adolescence or early adulthood.[11] In clinical practice and epidemiological studies, the BMI is nowadays often preferred over weight alone, because it corrects for linear body dimensions, and can therefore be used as an index for excess body weight. Since the BMI is related to the composition, rather than the dimensions of the human body, it will not be further discussed in this chapter.

A third major pattern of growth is seen in the dimensions of the head. Figure 3.3 shows the growth in head circumference from birth to adulthood, as observed in the same girl as studied before. The head grows very rapidly during the first

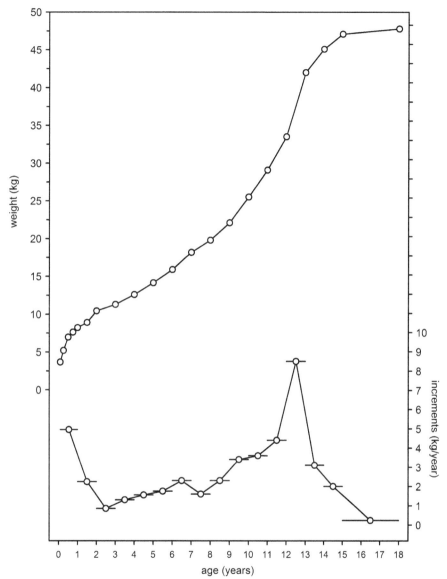

Figure 3.2 Growth in weight of a Belgian girl between birth and 18 years of age. The upper curve is a plot of weight for age (distance curve), and the lower curve shows the corresponding yearly weight increments (velocity curve). Horizontal bars indicate the length of the interval over which the increment was calculated. (Girl no. 29 from the Belgian Growth Study of the Normal Child.[1–3])

postnatal year (the increase in head circumference is on average 12 cm in boys and 10 cm in girls), but the growth velocity in head circumference drops steeply to levels below 1 cm/year by the age of 2 years. Thereafter, the head circumference increases by between a few millimeters and 1 cm per year without a noticeable growth spurt at

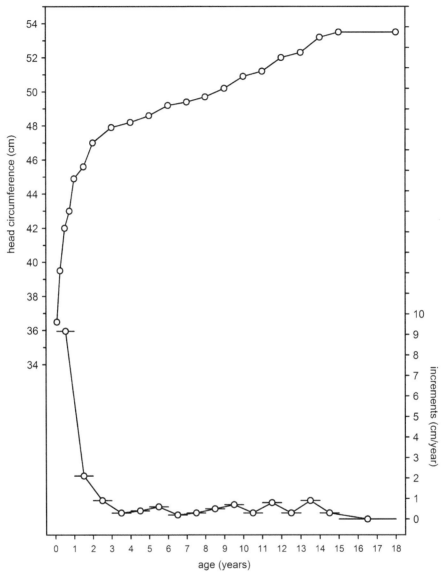

Figure 3.3 Growth in head circumference of a Belgian girl between birth and 18 years of age. The upper curve is a plot of head circumference for age (distance curve), and the lower curve shows the corresponding yearly increments in head circumference (velocity curve). Horizontal bars indicate the length of the interval over which the increment was calculated. (Girl no. 29 from the Belgian Growth Study of the Normal Child.[1–3])

puberty. Approximately half the postnatal growth in head circumference occurs during the first year and, as observed in the given example, almost 90% of the adult head circumference is reached by the age of 3 years. This pattern is typical for all humans, although population differences in size exist. The pattern of growth in head circumference is similar to that of other head dimensions such as head length and head width.[12] Owing to the small growth velocity of head dimensions beyond the age of 3 years, studies of head growth are prone to measurement error and are often limited to the period of infancy.

3.3 TEMPO OF GROWTH

At the beginning of the twentieth century, the famous American anthropologist Franz Boas described that "some children are throughout their childhood further along the road to maturity than others".[13,14] Indeed, individuals vary not only in size, but also in the speed at which they reach mature size, i.e. the tempo of growth. Tempo of growth, also called maturation rate, is correlated with other markers of maturation, such as secondary sexual characteristics and bone age.

The main effects of variations in tempo on the shape of the human growth curve are illustrated in Figure 3.4. The three distance and velocity curves show the growth in stature of typical early-, average- and late-maturing children who have the same size at birth and at adulthood. These three theoretical subjects have the same overall growth potential, but they considerably differ in height at each age along their growth trajectory, and in the shape of their growth pattern. The early maturer clearly reaches final size at an earlier age, and is taller than the average maturer throughout childhood and adolescence. The average maturer reaches adult size at an earlier age and is taller than the late maturer. The effect of differences in tempo of growth on attained height increase with age, and are more apparent in periods where the slope of the growth curve is steeper, i.e. during adolescence.

Differences in tempo thus have an impact on growth distance and velocity during childhood and adolescence, but not on final stature. Longitudinal studies have repeatedly shown that little or no correlation exists between the timing of the pubertal spurt and adult stature, i.e. early-, average- and late-maturing children reach, on average, the same adult height.[15-20] The shorter growth cycle in early maturers is compensated by a slightly but consistently greater growth velocity during childhood and by a more intense pubertal growth spurt. The opposite is seen in late-maturing children, who have a longer childhood growth but a less intense growth spurt. This relationship is reflected in the negative correlation between peak velocity and age at peak velocity in height.[15,21,22] This is also true for other postcranial body dimensions,[16] but not for weight. Early-maturing children have, as adults, on average a higher weight[13] and a higher BMI.[23]

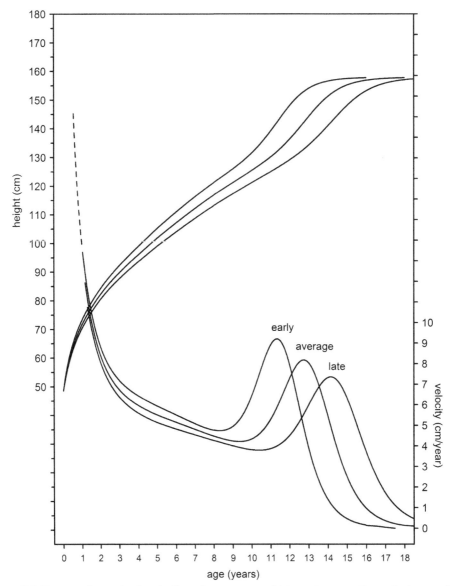

Figure 3.4 Pattern of growth of typically early, average and late maturers: a theoretical example.

Studies on longitudinal growth of twin and family data have shown that tempo of growth is to a great extent genetically determined.[24–27] In a longitudinal study of monozygotic and dizygotic male twins, Hauspie et al.[28] found a strong genetic component in the variance of biological parameters that characterize the shape of the human growth curve, in particular the age at peak velocity, which reflects the tempo of growth. Similar findings were reported by Byard et al.[29] on the basis of familial resemblance in growth curve parameters in the Fels Longitudinal Growth Study and by Hauspie et al.[26] in Bengali children. Tanner[30] suggested that both the growth status and the tempo of growth are under genetic control, but that the genetic factors involved might be quite different. Despite the strong genetic control over tempo of growth, there is also evidence that the human body can adapt to adverse environmental conditions by a slowing down of the developmental growth rate, allowing a child to cope with the physiological and metabolic requirements for a balanced development in suboptimal situations. When the adverse conditions are reversed, a child usually restores its growth deficit by a period of rapid growth to regain his or her original "growth channel", the so-called catch-up growth.[31−33] If, however, the environmental stresses persist for a long period or extend into the adolescent growth cycle, the resulting effect on growth may be a pattern which is typical for late-maturing children. Examples of this can be found in children exposed to chronic mild undernutrition,[34] chronic diseases such as asthma,[35] psychosocial stress[36−38] or socioeconomic deprivation,[39] and in children living at high altitude.[40] These conditions can lead to a small delay in reaching the adolescent growth spurt, achieving sexual maturity and attaining final size. Final stature is usually not affected (i.e. is compatible with the population average) unless the long-lasting adverse conditions are too severe.[33]

3.4 GROWTH MODELING AND BIOLOGICAL PARAMETERS

Serial measurements of size, taken at regular intervals on the same subject, such as shown in Figures 3.1−3.3, form the basis for estimating the expected growth curve of an individual (see Chapter 19). Growth is generally assumed to be a smooth process when based on body measurements taken at intervals varying between several months and a year. To estimate a continuous growth curve, the measurements could be connected linearly (as in Figures 3.1−3.3), or using a smoothing technique such as splines or polynomials. These methods allow one to interpolate or to produce a smooth curve, but they are not useful to summarize or analyze the shape of the growth curve. For the study of growth, various mathematical models have been proposed to estimate a smooth continuous growth curve on the basis of a discrete set of measurements of growth of the same subject over time.[41−43] They all imply that the growth curve increases monotonously with age, and tends towards an upper asymptote when data near adulthood are included. An interesting review of various approaches in modeling human growth has been given by Bogin.[44] More than 200 models have been proposed to describe part or the whole of the human growth process, but only a small number has shown to be of practical use. The possibilities and limitations of commonly used mathematical functions for analyzing human growth have been discussed by Hauspie et al.[45] and Hauspie and Chrzastek-Spruch.[46] This chapter will discuss only the Preece—Baines model 1 (PB1), which has been proven to be very robust to describe the adolescent growth cycle on the basis of growth data covering the period from about 2—5 years of age up to full maturity.[47] The mathematical expression of the PB1 is:

$$y = h_1 - \frac{2(h_1 - h_\theta)}{e^{s_0(t-\theta)} + e^{s_1(t-\theta)}}$$ [3.1]

where y is the height in centimeters, t is the age in years, and h_1, h_θ, s_0, s_1 and θ are the five function parameters. The parameters of PB1 allow a functional interpretation of the growth curve: h_1 is the upper asymptote of the function and thus corresponds to an estimate of mature size; θ is a timing parameter that controls the location of the adolescent growth spurt along the time axis. θ is highly correlated with age at peak velocity. h_θ is the size at age θ, and parameters s_0 and s_1 are rate constants controlling prepubertal and pubertal growth velocity, respectively. Other biological parameters of the growth curve, such as age at take-off, age at PHV, and PHV itself, can be obtained with algebraic expressions that have been derived from the PB1 model.[47]

The parameters of a non-linear growth function like the PB1 curve are usually estimated with non-linear least squares based on numerical minimization techniques. Algorithms for non-linear regression analysis of user-entered functions are offered by most statistical and graphical software packages.

The outcome of modeling serial growth data of an individual is a set of values for the function parameters (five in the case of PB1). Hence, individual growth modeling (or curve fitting) is a technique that summarizes longitudinal growth data in a limited number of parameters. For structural growth models like PB1, these parameters have the same meaning for all subjects, and can be compared between individuals. By feeding the values of the function parameters into the model, one is able to graph the smooth growth curve of an individual (Figure 3.5). Likewise, by entering the parameter values into the first derivative of the function, an estimation of the instantaneous growth velocity is obtained. The formula for the growth velocity of a PB1 model is:

$$y\prime = \frac{2(h_1 - h_\theta)\left(s_0 e^{s_0(t-\theta)} + s_1 e^{s_1(t-\theta)}\right)}{\left(e^{s_0(t-\theta)} + e^{s_1(t-\theta)}\right)}$$ [3.2]

The distance curve in Figure 3.5 shows the PB1 function fitted to the observed height measurements plotted in Figure 3.1. The lower part of Figure 3.5 is a plot of the yearly increments together with the instantaneous velocity curve obtained as the mathematical first derivative of the fitted distance curve. The corresponding values of the function parameters and derived biological parameters are listed in

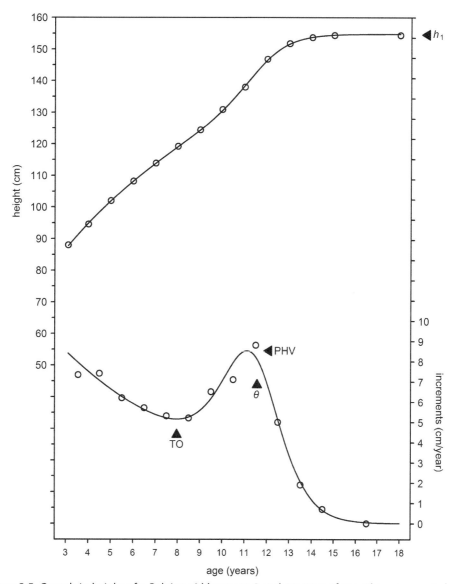

Figure 3.5 Growth in height of a Belgian girl between 3 and 18 years of age. The upper part shows a plot of the height-for-age data together with the Preece–Baines model 1, while the lower part shows the yearly increments in weight with the first derivative of the fitted curve. TO: take-off; PHV: peak height velocity. (Girl no. 29 from the Belgian Growth Study of the Normal Child.[1–3])

Table 3.1. The location of some of these parameters is also indicated on the curves in Figure 3.5.

The precision of the fit of a non-linear model is given by the standard error of estimate (SEE), also called the residual standard deviation (RSD) or root mean square

Table 3.1 Function parameters and biological parameters obtained by fitting Preece—Baines model 1 to the height data of Girl no. 29 from the Belgian Growth Study of the Normal Child[1–3]

h_1	154.4 cm
h_θ	143.7 cm
s_0	0.1374
s_1	1.450
θ	11.63 years
Residual variance	0.144 cm^2
Residual standard deviation	0.380 cm
Age at take-off	8.33 years
Height at take-off	121.2 cm
Velocity at take-off	5.1 cm/year
Age at peak velocity	11.18 years
Height at peak velocity	139.8 cm
Velocity at peak velocity	8.9 cm/year
Adolescent gain	33.2 cm

error (RMSE). The RSD is the square root of the sum of the squared residuals divided by the degrees of freedom of the model:

$$\text{RSD} = \sqrt{\frac{\sum_{i=1}^{N}(y_i - \hat{y}_i)^2}{N - k}} \qquad [3.3]$$

where y_i is the observed height at x_i, $\hat{y}_i$ is the corresponding value of the fitted curve at x_i, N is the number of height measurements, and k is the number of parameters in the model (5 in the case of PB1). It is generally accepted that the growth curve is adequately fitted if the standard error is of the same order as the measurement error of the trait under consideration (typically 0.5 cm for stature during childhood and adolescence). Systematic bias can be estimated with the runs test,[48] and an analysis of autocorrelation in the model residuals. Autocorrelation occurs when consecutive measurements are located on the same side of the curve, and point to a deficiency of the model to describe local patterns that are observed in the data.

Large values of the residual standard deviation, and systematic bias, generally originate from low-precision data or from an inappropriate choice of growth model. Low-precision data are observations that were recorded with a large measurement error, due to inexperienced measurers, large intermeasurer variation, or the use of inadequate measuring techniques or devices. The choice of an appropriate structural growth model depends on the age range under study and the expected shape of the growth curve. The PB1 model was originally designed to study the adolescent growth cycle, and is therefore unsuitable to model growth during infancy or childhood. Whenever longitudinal data from birth are at hand, or the main interest lies in growth before the adolescent spurt, models like the triple logistic function[49] or the JPA-2 function[50] are better alternatives. Structural models explicitly assume a specific shape of the growth curve that includes particular features (e.g. a growth spurt). The fitted curve will always show these features, sometimes prominently, sometimes not, even when they are not present in the data. The PB1 model was designed for postcranial skeletal dimension, and assumes a pubertal spurt with a local minimum in growth velocity at take-off. It is therefore inappropriate for growth in weight, which has a pubertal spurt but no minimum velocity at take-off. Moreover, weight does not increase monotonously with age, and may show a growth pattern that is not compatible with a structural model. The PB1 model is also inappropriate for describing the growth in head circumference, because of the lack of pubertal spurt in this trait. Features that are not structurally incorporated in the model will never be visible in the resulting growth curve, even when clearly present in the data. Examples are prepubertal growth spurts and unusual variations in the growth rate under certain pathological conditions.

One should also be suspicious about estimations of final size by structural growth models when the observed data points do not give a clear indication that the end of the growth phase is nearby. Estimates of final height, for instance, are fairly unreliable if the last observed increment exceeds 2 cm/year. Least-squares techniques are hopelessly weak in fitting parameters beyond the observation range and thus inapt to extrapolate. Analogous problems may arise when the lower bound of the age range does not include the take-off of the adolescent growth spurt in the case of PB1. In such a situation the estimation of the age at take-off and all derived biological parameters by a PB1 fit is not under control of the data and likely to be erroneous. A possible solution to the problem of extrapolation such as the prediction of mature stature (and also to the problem of incomplete data) is the use of a Bayesian approach instead of least-squares techniques for parameter estimation.[49,51]

Growth variables that do not necessarily have a monotonously increasing pattern (e.g. weight, BMI, skinfolds) cannot be successfully described by structural models such as non-linear growth functions. Non-structural approaches, such as polynomials, smoothing splines and kernel estimation are more appropriate for this kind of trait.[15,52]

Besides producing a smooth continuous curve for growth and growth velocity, and summarizing the growth data into a limited number of constants, one of the main goals of mathematical modeling of human growth data is to estimate milestones of the growth process of an individual (biological parameters), such as age, size and velocity at take-off and at peak velocity (Table 3.1). Biological parameters, obtained by fitting a growth model, characterize the shape of the human growth curve and form a basis for studies of genetic and environmental factors that control the dynamics of human growth. On the population level, these growth models can also be used to estimate the "typical average" curve with a mean-constant curve, which will be discussed in the next section.

3.5 INDIVIDUAL VERSUS AVERAGE GROWTH

Our knowledge on the shape of the growth curve comes from longitudinal studies, i.e. based on series of growth measurements of the same subjects over time, which allows part or all of the individual growth process to be established. However, the majority of growth studies are cross-sectional, i.e. based on single measurements taken from individuals who differ in age. Cross-sectional growth data allow one to estimate the central tendency and variation of anthropometric variables at each age in a population, and to construct smooth centile lines showing the "average" and the limits of "normal" variation of a particular trait in that population. These centile lines form the basis of most growth standards and reference curves (see Chapter 21). Despite the immense merits of cross-sectional growth surveys in constructing growth standards, and in epidemiological studies of genetic and environmental factors involved in growth, they only give a static picture of the population variation in growth at different ages, and are very weak in providing information on the dynamics of individual growth patterns over time.

In adolescence, differences in tempo are very pronounced. In early adolescence, some individuals will grow at PHV while others still have to start their growth spurt. In mid-adolescence, early maturers will approach final height (with their growth velocity nearing zero), while others grow at maximum velocity. A consequence of these variations in tempo is that a cross-sectional mean curve, to some extent, smooths out the adolescent growth spurt. This effect is illustrated in Figure 3.6, on the basis of longitudinal growth curves of two boys taken from the Lublin Longitudinal Growth Study.[53] The two subjects differ in timing of their adolescent growth spurt, with a maximum increment in height that occurs at 12.5 years of age for one boy and 14.5 years for the other. Taking the average height at each age, effectively ignoring the difference in timing of the adolescent growth spurt in the two subjects, gives a cross-sectional mean curve that shows a longer but less intense growth spurt. This is reflected in the less steep slope of the mean distance curve, but the effect is even more striking when the yearly increments in height of the subjects are compared with the cross-sectional average of these increments. Both boys have a clear adolescent spurt, with a maximum increment in height of 9.9 and 7.8 cm/year, respectively, while the cross-sectional mean of the increments has a peak that is lower, and a "spurt" that lasts longer than the individual curves. This phenomenon is called the "phase-difference" effect.[9]

The example given in Figure 3.6 illustrates how the pattern of growth of an individual differs from a cross-sectional mean growth curve, especially during adolescence. Although this example is based on averaging two longitudinal growth

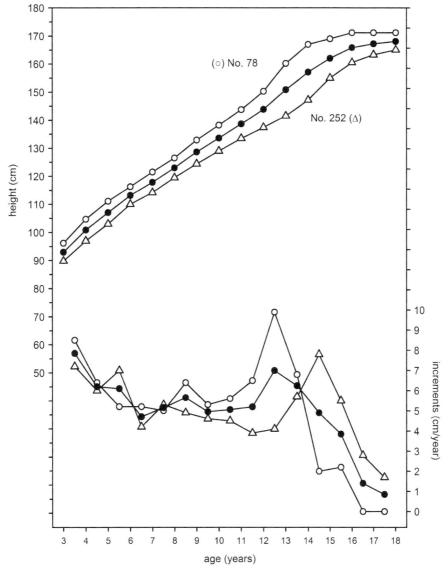

Figure 3.6 Distance and velocity curves of two Polish boys (open markers) with the cross-sectional mean curve (filled markers). Average velocities also correspond to increments of the average distance curve. (Source: *Data from the Lublin Longitudinal Growth Study.*[53])

curves, exactly the same mechanism applies to the means in cross-sectional data, and to the means in longitudinal data when not accounting for differences in tempo. It is the main reason why growth records of an individual over time do not match any of the centile lines shown by cross-sectional growth charts (even when based on longitudinal data) and why these charts are not useful in evaluating the normality of the *pattern* of growth over time. This type of growth standard is also called *unconditioned for tempo*.

The difference between individual and average growth was recognized a long time ago (by Boas in 1892, and by Shuttleworth in 1937; see Tanner[14]), but it was not until the mid-1960s that Tanner et al.[9,10] introduced *tempo-conditioned* growth standards for height, weight, height velocity and weight velocity, based on longitudinal data of British children. These standards show not only the classical cross-sectional centile distribution for attained size, and velocity at each age, but also the "normal" variation in the *shape* of the growth curve. The references for shape were based on an analysis of longitudinal data after centering each individual growth curve around the average age at peak velocity. Chronological age is thus replaced with age corrected for tempo, hence *tempo-conditioned* standards.

Figure 3.7(a) illustrates the effect of this approach on the mean height velocity curves of the two Polish boys from Figure 3.6. When taking averages of height velocity (or height distance), after centering the velocity curves at the mean age at PHV, a mean velocity curve is obtained which can be considered as representative for both individuals, i.e. with an age at peak velocity and a peak velocity that is the average of the two subjects. Later on, Tanner and Davies[6] used the same principle to produce clinical longitudinal standards for height and height velocity in North American children. Wachholder and Hauspie[2,3] achieved the same goal with a technique based on curve fitting, to produce clinical standards for growth and growth velocity in Belgian children. They estimated the typical average pattern of growth with "mean-constant" curves. A mean-constant curve for adolescent growth is obtained by fitting the PB1 to each individual in a sample and by feeding the mean values of the function parameters into the model. The resulting curve represents the average growth pattern in the population, i.e. with a peak velocity and an age at peak velocity that is characteristic or typical for the group.[42] The PB1 velocity curves of the two Polish boys are shown together with their mean-constant curve in Figure 3.7(b). The result is very much like that of the peak velocity centered curves. Note, however, that the PB1 curve slightly underestimates peak velocity; this is a known minor weakness of the PB1 model.[46]

3.6 SEX DIFFERENCES IN GROWTH

It is well known that adult females are on average smaller than adult males for most linear body dimensions, in particular height, sitting height and leg length.[13,54] Although some differences between boys and girls are observed at birth, they remain generally small until the early teens, when the girls start their pubertal growth spurt. Because of the 2-year difference in age of onset of the pubertal spurt, 11–13-year-old European girls are, on average, taller and heavier than boys of the same age.[55] Some studies suggest that the

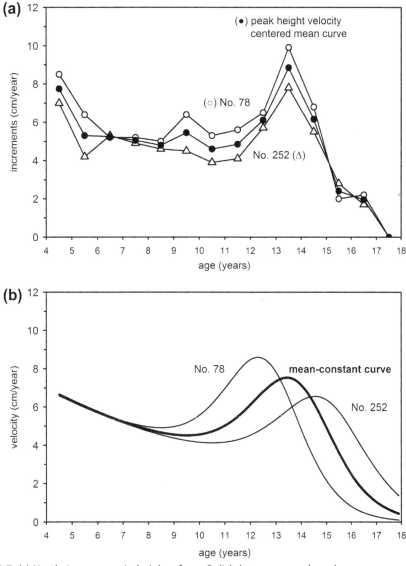

Figure 3.7 (a) Yearly increments in height of two Polish boys, centered on the average age at peak height velocity (open markers) and cross-sectional means of the peak velocity centered curves (filled markers); (b) PB1 velocity curves of these boys with the mean-constant curve. *(Source: Data from the Lublin Longitudinal Growth Study.[53])*

magnitude of sex differences in a population depends on the average size in adults.[56,57] However, Eveleth[58] found a relatively large sex difference in adult stature in Amerindians, a population that has a relatively small adult size. Similar findings were reported for Indians.[54] Eveleth postulated that genetic factors probably play an important role in establishing both the mature size and sex differences, but it is also conceivable that in certain societies boys are more favorably treated, which allows them to better express their genetic growth potential.

Sex differences in growth during infancy, childhood and adulthood, as well as the points of intersection between the male and female average growth curve, can be derived from cross-sectional data in a fairly accurate way. However, for reasons explained above, a study of the manner in which sex differences in size arise during the growth process depends on the availability of longitudinal data and the use of appropriate analytical methods.[21] A suitable approach by which to analyze the dynamics of sexual dimorphism in human growth is to compare the typical average male and female curve in a population, estimated with a mean-constant curve. As an example, the average growth (mean-constant curve) of Belgian boys and girls is shown in Figure 3.8. The total prepubertal growth is considered to be the size achieved up to the age at take-off, and adolescent growth (or adolescent gain) to be the amount of growth between take-off and adulthood. Figure 3.8 illustrates how the sex difference in adult size (D) can be expressed as the sum of three additive components, $D = DP + DT + DA$, where DA is the difference in adolescent gain between boys and girls, DP is the difference in size at take-off in girls, and DT is the amount of growth achieved by the boys between take-off in girls and take-off in boys. DT corresponds to the longer childhood growth that is observed in boys, and $DP + DT$ to the difference in total prepubertal growth between the sexes. Using this technique, Hauspie et al.[54] analyzed the origin of sex differences in height, sitting height, shoulder width and hip width in British children. The results are summarized in Table 3.2.

The largest contribution to the 12.0 cm sex difference in adult height comes from the later onset of the pubertal growth spurt (or longer childhood growth) in boys ($DT = 7.9$ cm). Sex differences at take-off in girls ($DP = 2.1$ cm) and in adolescent gain ($DA = 2.0$ cm) are significantly smaller. The proportional contribution of each of these components may be slightly different in other populations. In West Bengal, sex difference in adult height is larger ($D = 14.2$ cm), mainly due to a more important adolescent gain,[58] but in the Belgian population, the contribution of DT to the sex differences in adult height was comparable to that observed in UK children.[59] The decomposition of the adult sex difference in sitting height is proportional to that of stature, but the adolescent gain in shoulder width is larger in boys than in girls. Adult sex differences in hip width are almost negligible, because the longer childhood growth of 1.5 cm in boys is compensated by a greater adolescent gain in hip width (2.0 cm) in girls (Table 3.2). For all of these body dimensions, differences before take-off in girls are small. In conclusion, a comparison of average growth in boys and girls shows that the differences in linear body dimensions that are observed in adults emerge during adolescence. Within the adolescent period, a longer childhood growth in boys can be identified as an important contributor to these differences.

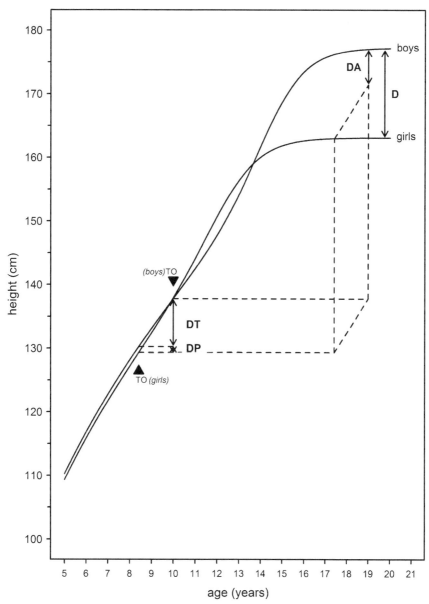

Figure 3.8 Decomposition of sex differences in adult stature (D) into three additive components: differences at take-off in girls (DP), growth in boys between take-off in girls and take-off in boys (DT), and difference in adolescent gain (DA).

Table 3.2 Decomposition of sexual dimorphism in adult size of height, sitting height, shoulder width and hip width into three additive components: a British sample[54]

	Height (cm)	Sitting height (cm)	Shoulder width (cm)	Hip width (cm)
Total difference (D)	12.0	4.5	3.7	−0.6
Difference at take-off in girls (DP)	2.1	0.3	0.3	−0.1
Growth in boys between take-off in girls and take-off in boys (DT)	7.9	3.5	1.7	1.5
Difference in adolescent gain (DA)	2.0	0.7	1.7	−2.0

3.7 SUMMARY

Growth at adolescence is characterized by the presence of a pubertal growth spurt in height and weight, but not in head circumference. Several milestones of the adolescent growth cycle can be identified on the distance and velocity curves. The pubertal spurt starts at take-off and ends when final size is reached. The period of maximum growth is denoted as peak velocity. The timing and magnitude of the pubertal growth spurt are highly variable, as a result of which large transient differences in size may occur between individuals of the same chronological age. Differences in tempo of growth, however, have no impact on final height.

Most of our knowledge on the shape of the growth curve comes from longitudinal studies, in which serial measurements in the same individual are often analyzed with the help of mathematical growth models. These models allow the growth curve to be summarized in a limited number of constants, and important milestones and biological parameters that characterize the pattern of growth to be identified. The Preece−Baines model 1 is an example of a structural model that is particularly useful for adolescent growth in height. The majority of growth studies are, however, based on cross-sectional data. As a consequence of variations in tempo of growth, a cross-sectional mean curve will show a longer but less intense growth spurt, compared to the growth curve of an individual. It is therefore not representative of the pattern of growth of individual children. Methods that account for individual differences in timing and magnitude of peak velocity include centering the individual growth curves around the average age at peak velocity when analyzing the data graphically, and the mean-constant curve when using growth models.

Finally, differences in linear body dimensions that are observed between adult males and females emerge during adolescence. Within the adolescent period, the longer childhood growth in boys is an important contributor to the sexual dimorphism in final height.

REFERENCES

1. Graffar M, Asiel M, Emery-Hauzeur C. La croissance de l'enfant normal jusqu'à trois ans: analyse statistique des données relatives au poids et à la taille. *Acta Paediatr Belg* 1962;**16**:5—23.
2. Wachholder A, Hauspie RC. Clinical standards for growth in height of Belgian boys and girls aged 2 to 18 years. *Int J Anthropol* 1986;**1**:339—48.
3. Hauspie RC, Wachholder A. Clinical standards for growth velocity in height of Belgian boys and girls, aged 2 to 18 years. *Int J Anthropol* 1986;**1**:327—38.
4. Hermanussen M. The analysis of short-term growth. *Horm Res* 1998;**49**:53—64.
5. Lampl M. *Saltation and stasis in human growth: evidence, methods and theory.* Nishimura: Smith-Gordon; 1999.
6. Tanner JM, Davies PSW. Clinical longitudinal growth standards for height and height velocity for North American children. *J Pediatr* 1985;**107**:317—29.
7. Butler GE, McKie M, Ratcliffe SG. An analysis of the phases of mid-childhood growth by synchronisation of growth spurts. In: Tanner JM, editor. *Auxology 88, Perspectives in the science of growth and development.* London: Smith-Gordon; 1989. p. 77—84.
8. Hauspie RC, Chrzastek-Spruch H. The analysis of individual and average growth curves: some methodological aspects. In: Duquet W, Day JAP, editors. *Kinanthropometry IV.* London: E&FN Spon; 1993. p. 68—83.
9. Tanner JM, Whitehouse RH, Takaishi M. Standards from birth to maturity for height, weight, height velocity, and weight velocity: British children. 1965. Part I. *Arch Dis Child* 1966;**41**:454—71.
10. Tanner JM, Whitehouse RH, Takaishi M. Standards from birth to maturity for height, weight, height velocity, and weight velocity: British children. 1965. Part II. *Arch Dis Child* 1966;**41**:613—35.
11. Guo SS, Huang C, Maynard LM, Demerath E, Towne B, Chumlea WC, Siervogel RM. Body mass index during childhood, adolescence and young adulthood in relation to adult overweight and adiposity: the Fels Longitudinal Study. *Int J Obes* 2000;**24**:1628—35.
12. Twiesselmann F. *Développement biométrique de l'enfant à l'adulte.* Paris: Librairie Maloine; 1969.
13. Tanner JM. *Growth at adolescence.* 2nd ed. Oxford: Blackwell Scientific; 1962.
14. Tanner JM. *A history of the study of human growth.* Cambridge: Cambridge University Press; 1981.
15. Largo RH, Gasser Th, Prader A, Stützle W, Huber PJ. Analysis of the adolescent growth spurt using smoothing spline functions. *Ann Hum Biol* 1978;**5**:421—34.
16. Cameron N, Tanner JM, Whitehouse RH. A longitudinal analysis of the growth of limb segments in adolescence. *Ann Hum Biol* 1982;**9**:211—20.
17. Zacharias L, Rand WM. Adolescent growth in height and its relation to menarche in contemporary American girls. *Ann Hum Biol* 1983;**10**:209—22.
18. Marshall WA, Tanner JM. Puberty. In: Falkner F, Tanner JM, editors. *Human growth*, vol. 2. New York: Plenum Press; 1986. p. 171—209.
19. Malina RM, Bouchard C. *Growth, maturation and physical activity.* Champaign, IL: Human Kinetics; 1991.
20. Bielicki T, Hauspie RC. On the independence of adult stature from the timing of the adolescent growth spurt. *Am J Hum Biol* 1994;**6**:245—7.
21. Tanner JM, Whitehouse RH, Marubini E, Resele LF. The adolescent growth spurt of boys and girls of the Harpenden study. *Ann Hum Biol* 1976;**3**:109—26.
22. Hauspie R. Adolescent Growth. In: Johnston FE, Roche AF, Susanne C, editors. *Human physical growth and maturation: methodologies and factors.* New York: Plenum Press; 1980. p. 161—75.
23. Demerath EW, Li J, Sun SS, Chumlea WC, Remsberg KE, Czerwinski SA, et al. Fifty-year trends in serial body mass index during adolescence in girls: the Fels Longitudinal Study. *Am J Clin Nutr* 2004;**80**:441—6.
24. Sharma JC. The genetic contribution to pubertal growth and development studies by longitudinal growth data on twins. *Ann Hum Biol* 1983;**10**:163—71.
25. Mueller WH. The genetics of size and shape in children and adults. In: Falkner F, Tanner JM, editors. *Human growth. Methodology, ecological, genetic, and nutritional effects on growth*, vol. 3. New York: Plenum Press; 1986. p. 145—68.

26. Hauspie RC, Das SR, Preece MA, Tanner JM. Degree of resemblance of the pattern of growth among sibs in families of West Bengal (India). *Ann Hum Biol* 1982;**9**:171–4.

27. Hauspie RC. The genetics of child growth. In: Ulijaszek SJ, Johnston FE, Preece MA, editors. *The Cambridge encyclopedia of human growth and development*. Cambridge: Cambridge University Press; 1998. p. 124–8.

28. Hauspie RC, Bergman P, Bielicki T, Susanne C. Genetic variance in the pattern of the growth curve for height: a longitudinal analysis of male twins. *Ann Hum Biol* 1994;**21**:347–62.

29. Byard PJ, Guo S, Roche AF. Family resemblance for Preece–Baines growth curve parameters in the Fels Longitudinal Growth Study. *Am J Hum Biol* 1993;**5**:151–7.

30. Tanner JM. *Foetus into man: physical growth from conception to maturity.* London: Open Books; 1978.

31. Golden MHN. Catch-up growth in height. In: Ulijaszek SJ, Johnston FE, Preece MA, editors. *The Cambridge encyclopedia of human growth and development*. Cambridge: Cambridge University Press; 1998. p. 346–7.

32. Prader A, Tanner JM, Von Harnack GA. Catch-up growth following illness or starvation. *J Pediatrics* 1963;**62**:646–59.

33. Tanner JM. Growth as a target-seeking function – catch-up and catch-down growth in man. In: Falkner F, Tanner JM, editors. *Human growth: a comprehensive treatise. Developmental biology, prenatal growth*, vol. 1. New York: Plenum Press; 1986. p. 167–79.

34. Hansen JDL, Freesemann C, Moodie AD, Evans DE. What does nutritional growth retardation imply? *Pediatrics* 1971;**47**:299–313.

35. Hauspie RC, Susanne C, Alexander F. Maturational delay and temporal growth retardation in asthmatic boys. *J Allergy Clin Immunol* 1977;**59**:200–6.

36. Skuse DH. Growth and psychosocial stress. In: Ulijaszek SJ, Johnston FE, Preece MA, editors. *The Cambridge encyclopedia of human growth and development*. Cambridge: Cambridge University Press; 1998. p. 341–2.

37. Powell GF, Brasel JA, Blizzard RM. Emotional deprivation and growth retardation simulating idiopathic hypopituitarism. *N Engl J Med* 1967;**276**:1271–83.

38. Widdowson EM. Mental contentment and physical growth. *Lancet* 1951;**i**:1316–8.

39. Bielicki T. Physical growth as a measure of the economic well-being of populations: the twentieth century. In: Falkner F, Tanner JM, editors. *Human growth: a comprehensive treatise. Methodology, ecological, genetic, and nutritional effects on growth*, vol. 3. New York: Plenum Press; 1986. p. 283–305.

40. Malik SL, Hauspie R. Age at menarche among high altitude Bods of Ladakh (India). *Hum Biol* 1986;**58**:541–8.

41. Marubini E, Milani S. Approaches to the analysis of longitudinal data. In: Falkner F, Tanner JM, editors. *Human growth – a comprehensive treatise*. 2nd ed. New York: Plenum Press; 1986. p. 79–94.

42. Hauspie RC. Mathematical models for the study of individual growth patterns. *Rev Epidemiol Sante Publique* 1989;**37**:461–76.

43. Hauspie RC. Curve fitting. In: Ulijaszek SJ, Johnston FE, Preece MA, editors. *The Cambridge encyclopedia of human growth and development*. Cambridge: Cambridge University Press; 1998. p. 114–5.

44. Bogin B. *Patterns of human growth.* 2nd ed. Cambridge: Cambridge University Press; 1999.

45. Hauspie RC, Lindgren G, Tanner JM, Chrzastek-Spruch H. Modelling individual and average human growth data form childhood to adulthood. In: Magnusson D, Bergman LR, Törestad B, editors. *Problems and methods in longitudinal research – stability and change*. Cambridge: Cambridge University Press; 1991. p. 28–46.

46. Hauspie R, Chrzastek-Spruch H. Growth models: possibilities and limitations. In: Johnston FE, Zemel B, Eveleth PB, editors. *Human growth in context*. London: Smith-Gordon; 1999. p. 15–24.

47. Preece MA, Baines MK. A new family of mathematical models describing the human growth curve. *Ann Hum Biol* 1978;**5**:1–24.

48. Siegel S. *Nonparametric statistics for the behavioral sciences.* London: McGraw-Hill; 1956.

49. Bock RD, Thissen DM. Statistical problems of fitting individual growth curves. In: Johnston FE, Roche AF, Susanne C, editors. *Human physical growth and maturation*. New York: Plenum Press; 1980. p. 265–90.

50. Jolicoeur P, Pontier J, Abidi H. Asymptotic models for the longitudinal growth of human stature. *Am J Hum Biol* 1992;**4**:461—8.

51. Bock RD. Predicting the mature stature of preadolescent children. In: Susanne C, editor. *Genetic and environmental factors during the growth period*. New York: Plenum Press; 1984. p. 3—19.

52. Gasser T, Köhler W, Müller HG, Kneip A, Largo R, Molinari L, Prader A. Velocity and acceleration of height growth using kernel estimation. *Ann Hum Biol* 1984;**11**:397—411.

53. Chrzastek-Spruch H, Kozlowska M, Hauspie R, Susanne C. Longitudinal study on growth in height in Polish boys and girls. *Int J Anthropol* 2002;**17**:153—60.

54. Hauspie R, Das SR, Preece MA, Tanner JM, Susanne C. Decomposition of sexual dimorphism in adult size of height, sitting height, shoulder width and hip width in a British and West Bengal sample. In: Ghesquire J, Martin RD, Newcombe F, editors. *Human sexual dimorphism*. London: Taylor & Francis; 1985. p. 207—15.

55. Eveleth PB, Tanner JM. *Worldwide variation in human growth*. Cambridge: Cambridge University Press; 1990.

56. Hall RL. Sexual dimorphism for size in seven nineteenth century northwest coast populations. *Hum Biol* 1978;**50**:159—71.

57. Valenzuela CY, Rothhammer F, Chakraborty R. Sex dimorphism in adult stature in four Chilean populations. *Ann Hum Biol* 1978;**5**:533—8.

58. Eveleth PB. Differences between ethnic groups in sex dimorphism of adult height. *Ann Hum Biol* 1975;**2**:35—9.

59. Koziel SK, Hauspie RC, Susanne C. Sex differences in height and sitting height in the Belgian population. *Int J Anthropol* 1995;**10**:241—7.

60. Hauspie RR, Molinari L. Parametric models for postnatal growth. In: Hauspie RC, Cameron N, Molinari L, editors. *Methods in human growth research*. Cambridge: Cambridge University Press; 2004. p. 205—33.

SUGGESTED READING

Many papers and textbooks give a concise description of the pattern of human growth. When querying databases and repositories, be sure to look for papers based on longitudinal data. Tempo of growth is discussed in Tanner et al.[9,10] A review of various growth models and curve-fitting techniques is given in Hauspie and Molinari.[60] Tanner et al.[9,10] is a citation classic that describes the graphical method to account for individual differences in growth, while Hauspie and Wachholder[3] demonstrate the use of a mean-constant curve. The decomposition of sexual dimorphism in height is based on Hauspie et al.[54] and Koziel et al.[59]

CHAPTER 4

Puberty

Peter T. Ellison, Meredith W. Reiches

Department of Human Evolutionary Biology, Harvard University, Cambridge, MA 02138, USA

Contents

4.1 INTRODUCTION

Puberty refers to the onset of adult reproductive capacity. As a milestone in human development, puberty is quite dramatic, involving a rapid transformation of anatomy, physiology and behavior. Other than pregnancy, it is probably the most abrupt and encompassing developmental transition that human beings undergo between birth and death. It is also a transition of deep cultural significance in most societies around the world, often marked by special rituals and ceremonies.[1]

As dramatic as it is, however, puberty is not really an instantaneous event or a discrete state, but a process that is integrated more or less smoothly with the antecedent and

N. Cameron & B. Bogin (eds): Human Growth and Development, Second edition.
ISBN 978-0-12-383882-7, Doi: 10.1016/B978-0-12-383882-7.00004-0

consequent developmental phases of immaturity and adulthood. At its core, puberty is a neuroendocrine transformation in the processes that regulate reproductive physiology. This transformation arises, however, from physiological mechanisms that are already latent in the prepubertal child, and it initiates a trajectory of change that merges smoothly with other age-dependent changes that continue through adult life. The Rubicon that is crossed in this process is the frontier of reproductive capability. Before puberty, an individual is not capable of producing offspring, however capable he or she may be in every other regard. After puberty, the individual is capable of producing offspring, however deficient in other skills and capacities.

The physiological processes that guide and regulate this pivotal transformation from child into adult have been foci of research for decades, and a great deal of progress has been made in elucidating many of the details. This progress has led to innovative treatments for pathologies of pubertal development, as well as new formulations of the evolution of human life history. But many mysteries and controversies remain. The purpose of this chapter is to introduce students to the current state of knowledge about, and research into, the physiology of human puberty and its relationship to other aspects of human growth and development. The presentation will necessarily be incomplete and will focus on those aspects that the authors think are most important. However, it is also hoped that the chapter will prepare students for further pursuing those issues that may interest them.

The presentation will have three main sections. Section 2 will review the neuro-endocrinology of puberty. A brief sketch is provided of the hormonal axis that controls reproductive physiology and the pulsatile nature of hypothalamic secretion of gonado-tropin-releasing hormone (GnRH) is described. Then the development of GnRH secretion and its central role in puberty is reviewed, and finally some of the major central nervous system controls of GnRH secretion are discussed. At the end of this section, readers should be able to answer the questions: How does the hypothal-amic—pituitary—gonadal (HPG) axis control puberty? What initiates mature HPG axis function?

Section 3 will discuss some of the major upstream factors that influence the timing of puberty, focusing on genetics, energetics and psychosocial factors. The reader may approach this section with the question: Why is the timing of puberty variable across individuals?

Section 4 will discuss some of the major downstream consequences of the activation of the neuroendocrine reproductive axis: the pubertal growth spurt, the development of secondary sexual characteristics and changes in behavior. This section addresses the question: What are the effects of HPG axis activation?

The chapter closes by offering a provisional framework for organizing what is known about human puberty and its timing. In order to describe the function of individual components of the HPG axis, this section will examine, among other evidence, research in animal models, particularly mammals with neuroendocrine function similar to that in humans. In animal models, researchers can experimentally manipulate hormones, components of neural anatomy, gene expression and gonadal function, which is not possible to do systematically or ethically in humans. This evidence allows the function of different organs, hormones and genes to be isolated. In addition, cases of naturally occurring pathology in humans are presented to make inferences about the normal function of receptors, enzymes and organs from the consequences of their malfunction.

The material in this chapter is challenging in two ways. First, it characterizes a wide array of integrated neurological, endocrinological and auxological processes. The reader may wish to organize this material by asking: How does each process contribute to adult reproductive function? Second, the authors have tried to present the major points together with enough supporting detail to make the system comprehensible to general students of human growth and development. Therefore, many details of endocrinology and neurobiology that are necessary to a fuller understanding of reproductive and pubertal physiology may be omitted. But it is hoped that the material in this chapter will prepare interested students for pursuing a deeper understanding using some of the suggested references.

4.2 THE NEUROENDOCRINOLOGY OF PUBERTY

Puberty has its own specific neuroendocrinology, which we will now look at.

4.2.1 Interactions of the Endocrine and Nervous Systems

The endocrine system is one of three major regulatory systems in the body, the two others being the nervous system and the immune system. All three systems utilize chemical messenger molecules to communicate between cells. These molecules interact with specific receptor molecules either on the surface or in the interior of target cells. In the nervous system, the messenger molecules are called neurotransmitters or neuromodulators and ordinarily carry information between immediately adjacent neurons across a tiny synaptic cleft, measured in angstroms. In the endocrine system, the messenger molecules are called hormones and ordinarily carry information through the bloodstream from a site of origin to many dispersed target cells, over distances measured in millimeters to meters. One important exception to these two "ordinary" scenarios that have features of each of them is the secretion of a messenger molecule by a neuron into the blood. When this occurs it is referred to as neuroendocrine secretion. The molecule is produced at the axon terminal of a neuron, like a neurotransmitter, but it is transmitted through the bloodstream, like a hormone.

4.2.2 The Hypothalamic–Pituitary–Gonadal Axis

It is important to understand the mechanisms of interaction between the nervous and endocrine systems because reproduction in humans and other mammals is hormonally controlled by a set of three interacting parts: the hypothalamus of the brain, the pituitary gland suspended at the base of the brain, and the gonads (ovaries in the female, testes in the male), which are endocrine organs. As a unit, these three parts are referred to as the hypothalamic—pituitary—gonadal (HPG) axis. Hormonal signals are passed between the components of this axis and other parts of the body in both sexes. The HPG axis is active in fetal life and again in early infancy, but is only minimally active during childhood.[2,3] The central feature of puberty is now understood as the reactivation the HPG axis after the quiescence of childhood. All the other aspects of puberty, the various transformations of anatomy, physiology and behavior that are externally evident, are downstream consequences of this central, neuroendocrine event.

An understanding of human puberty, therefore, begins with an understanding of the HPG axis and its developmental trajectory. This section will provide a quick sketch of the HPG axis, forgoing much of the detail that would be necessary to understand the full regulation of adult reproduction.

The hypothalamus is a region at the base of the brain where the two hemispheres of the neocortex meet below the third ventricle. The hypothalamus is composed of several separate clusters of cell bodies known as nuclei. These hypothalamic nuclei control a number of different autonomic functions, including the regulation of blood osmolality, circadian rhythms, core body temperature and many others, including reproduction. Inputs to the hypothalamus arrive from many other brain regions, including the limbic system and the cortex. Outputs from the hypothalamus go primarily to the two parts of the pituitary gland.

The pituitary gland is a small organ consisting of two different tissues sitting directly under the hypothalamus in a bony capsule known as the sella turçica. The anterior pituitary is composed of glandular tissue embryologically derived from the roof of the pharynx. It is connected to the hypothalamus by an extremely small vascular connection known as the hypophyseal portal system. This small set of blood vessels has capillary beds both in the hypothalamus and in the anterior pituitary, extends only a few millimeters and contains only about a milliliter of blood. Axons from cell bodies in the hypothalamus terminate in the capillary bed at the hypothalamic end of the hypophyseal portal system.

The posterior pituitary gland is embryologically derived from the same tissue as the hypothalamus and remains physically connected to it by the pituitary stalk. Axons from cell bodies in the hypothalamus extend into the posterior pituitary with terminal buds on the capillaries that supply blood to that part of the gland.

The anterior and posterior pituitary glands differ not only ontogenetically but also functionally. Neuroendocrine products of the hypothalamus, or releasing hormones, enter the hypophyseal portal system of the anterior pituitary and stimulate release of the anterior pituitary's own endocrine products. In contrast, the posterior pituitary stores neurohormones produced in the hypothalamus itself and releases them directly into circulation. The rest of this discussion of HPG axis function in puberty will focus only on the role of the anterior pituitary.

The gonads are the sites of gamete production in both sexes. They are also important endocrine glands, producing steroid and protein hormones. The ovaries of the female are located in the coelomic cavity of the lower abdomen connected by the broad ligament to the fallopian tubes and uterus. The testes of the male are located in the external scrotal sac.

Each part of the HPG axis secretes specific hormones. The hypothalamus secretes a small peptide hormone commonly known as gonadotrophin-releasing hormone (GnRH) (but sometimes referred to as luteinizing hormone-releasing hormone). GnRH is secreted by a diffuse network of less than 2000 neurons located in the mediobasal hypothalamus.[4] Secretion from the axon terminals of these neurons occurs directly into the hypophyseal portal system through which GnRH is carried to the anterior pituitary. GnRH is produced in extremely low amounts and is measurable only in the hypophyseal portal system itself. Once in the systemic circulation the levels become too dilute to measure accurately, and the molecule itself is soon degraded. These facts impose severe limitations on direct observations of GnRH in vivo.

In response to hypothalamic GnRH, the anterior pituitary secretes two protein hormones into the systemic circulation: follicle-stimulating hormone (FSH) and luteinizing hormone (LH). These hormones, together known as gonadotrophins, travel to the gonads where they stimulate gamete production and hormone secretion. Male testes produce two particularly important hormones: testosterone (a steroid) and inhibin (a protein). Female ovaries produce two steroids, estradiol and progesterone, the latter only during the second half of an ovarian cycle after ovulation and during pregnancy. Ovaries also produce inhibin.

Gonadal steroids have potent effects throughout the body, including effects on reproductive organs, body composition and behavior. They also have effects on the hypothalamus and anterior pituitary. Inhibin in both sexes functions to provide feedback control on pituitary activity. In males, these feedback effects generally serve to sustain constant sperm production and to maintain testosterone levels in a broad normal range. In females, the feedback of gonadal steroids and inhibin serves to coordinate monthly waves of gamete maturation and menstrual bleeding.

At first encounter, the student may find the HPG axis unnecessarily complex. Why so many parts and so many hormones? It may help to think of the axis primarily in terms of its two ends: the brain and the gonads. The gonads are the sites of gamete production, clearly essential for reproduction. The brain is the place where all kinds of information about the organism and its environment are integrated, including information that may indicate when conditions (both internal and external) are best for reproduction. The information gathered by the brain can ultimately affect GnRH production by the hypothalamus and thus be passed through the pituitary to the gonads. The pituitary in essence amplifies the tiny neural signal generated by the hypothalamus into a larger, longer lasting hormonal signal that can be received by the gonads.

But why, one might ask, do the gonads secrete hormones in addition to producing gametes? Couldn't the gonads and the rest of the reproductive organs simply respond to the hypothalamic signals relayed through the pituitary? One important part of the answer is that the steroid hormones secreted by the gonads help to coordinate many other aspects of anatomy, physiology and behavior that are important for successful reproduction. They influence the way energy is metabolized and allocated in order to support reproduction in both sexes. They mediate the growth processes that cause individuals to reach adult size and the degree of sexual dimorphism in both the skeleton and soft tissues. They also influence and coordinate reproductive behavior, integrating environmental information with physiology. These functions are particularly evident in many seasonally breeding mammals, deer for instance, where steroids control changes in body composition and antler growth, as well as mating behavior. But similar functions are served, more or less conspicuously, in all mammals.

Thus, the HPG axis can be viewed as the control center for reproduction, where multiple sources of information are integrated and resulting control signals passed on to the gonads. The gonads both produce the gametes that will potentially form a new organism and send signals throughout the body to coordinate changes in anatomy, physiology and behavior, all geared towards securing reproductive success.

4.2.3 Pulsatile Gonadotrophin-Releasing Hormone Secretion

Gonadal and pituitary hormones can be readily measured in various bodily fluids using modern techniques. GnRH, on the other hand, can only be measured in samples of blood taken directly from the hypophyseal portal system. In the 1980s, Knobil and his colleagues published a set of landmark studies based on sampling blood from this tiny portal system in restrained rhesus monkeys.[5-8] They showed that in adult monkeys GnRH is secreted in regular pulses slightly more than an hour apart. Subsequent studies demonstrated that in monkeys where the hypothalamus was surgically ablated, exogenous administration of GnRH at a pulse rate of once every 60—90 minutes resulted in normal gonadotrophin secretion and normal gonadal function.[9] If the pulse rate was slower or faster, however, gonadotrophin secretion was interrupted and gonadal function would eventually stop.[10] They also demonstrated that if a regular, pulsatile pattern of GnRH was administered to an immature female rhesus monkey the monkey would experience puberty and begin normal adult ovarian cycling.[11] Similar results were later obtained for male rhesus monkeys by Plant and colleagues.[2]

Similar experiments are, of course, not possible in humans. However, the knowledge obtained from the monkey experiments has led to effective treatments for various pathological syndromes of precocious and delayed puberty in humans. For example, Kallman's syndrome results from a failure of the GnRH secreting neurons to properly migrate into the hypothalamus during embryogenesis. Affected individuals have no endogenous secretion of GnRH and do not spontaneously go through puberty. However, if GnRH is administered exogenously in 60–90 minute pulses via an intravenous pump system, normal pubertal maturation can be induced. Notably, varying the pulse rate above or below the 60–90 minute frequency interrupts pituitary and gonadal function, just as in the rhesus monkeys.[12]

Variants of GnRH have been synthesized that have much longer half-lives in the blood than the native peptide. Administering these synthetic variants can swamp the normal pulsatile GnRH signal received by the pituitary and instead provide a chronic, non-pulsatile signal. The result is a shutdown of the HPG axis. Administration of these compounds is now used to halt pubertal maturation in various cases of precocious puberty in humans. When the affected individual reaches an appropriate age, the administration of the synthetic compounds can stop and natural GnRH secretion be allowed once again to stimulate pituitary and gonadal function.[13,14] The same compounds can arrest HPG activity in adults and suppress gonadotrophin production in postmenopausal women or women suffering from estrogen-sensitive cancers.[15–17]

Studies such as these have established the role of 60–90 minute pulsatile GnRH as an obligate condition for secretion of adult levels of gonadotrophins and gonadal steroids. Pulsatile GnRH secretion plays the role of an "on/off" switch for the rest of the axis. All other aspects of axis function, including regular ovarian cycling in females, are controlled by feedback from gonadal hormones and do not depend on any variation in GnRH secretory pattern.

4.2.4 Reactivation of Pulsatile Gonadotrophin-Releasing Hormone Secretion

GnRH-secreting neurons arise from the olfactory placode and migrate into position early in embryogenesis.[18,19] They appear to have an innate capacity for pulsatile GnRH secretion and will continue this pattern even when maintained in vitro.[20–22] The pulsatile pattern of secretion is apparently active even as the neurons are completing their embryonic migration, resulting in the in utero activation of the HPG axis.[23] Negative feedback of placental steroids holds fetal pituitary and gonadal activity in check during the latter part of gestation until the withdrawal of placental steroids at birth leads to a second postnatal activation of the axis. Pituitary and gonadal activity recede to low baseline levels, however, during the first years of life. At puberty, the HPG axis is reactivated once more, leading to adult levels of circulating gonadotrophins and gonadal steroids.[3,24]

The pattern of infant and childhood suppression of gonadotrophin secretion and reactivation a decade or more later has been observed in human subjects with gonadal dysgenesis (failure of gonadal development) and thus does not appear to be dependent on feedback of gonadal steroids or inhibin.[25] A similar pattern has been observed in gonadectomized rhesus monkeys.[26,27] In animal models the *GnRH-1* gene is expressed at high levels prior to puberty and GnRH production occurs at adult levels.[28–30] Only the pulsatile pattern of GnRH secretion is lacking. Whether there is a proximate signal that determines the timing of reinitiation of pulsatile GnRH secretion at puberty or whether it occurs on an endogenous timetable has not been established.

4.2.5 Proximate Regulation of Gonadotrophin-Releasing Hormone Secretion

The small population of GnRH-secreting neurons serves as the integrating center for a wide variety of inputs to the control of mammalian reproduction.[2] Some of the signals that converge on these neurons are transmitted by rather generic inhibitory [e.g. gamma-aminobutyric acid (GABA)] and stimulatory (e.g. glutamate) neuroamines. But two recently identified neuropeptides appear to have special roles in the control of GnRH secretion: kisspeptin and neurokinin B.[31–34]

Kisspeptin refers to a family of closely related peptides coded by the *KISS1* gene. Inactivating mutations in the *KISS1* gene or the gene for the kisspeptin receptor, *GPR54*, are associated with idiopathic hypogonadotrophic hypogonadism (IHH, abnormally low gonadotrophin and gonadal steroid levels) and failure to undergo spontaneous puberty in both humans and animal models.[35] In animal models, including both rodents and primates, *KISS1* transcriptional activity increases significantly just before puberty, and *GPR54* expression has been localized on GnRH-secreting neurons.[36,37] Some *KISS1* variants with longer blood half-lives have also been associated with precocious puberty in humans.[35] *KISS1* and *GPR54* knockout models have also demonstrated that kisspeptin signaling is necessary for the normal pubertal activation of pulsatile GnRH secretion in animal models while centrally administered kisspeptin acutely stimulates pulsatile GnRH secretion.[38]

Recently, it has been proposed that neurokinin B (NKB), a distinct peptide that is co-secreted by some kisspeptin-secreting neurons, also provides a necessary signal for normal puberty. Mutations in the gene coding for the peptide (*TAC3*) and its receptor (*TAC3R*) have been associated with IHH in humans, and NKB neuronal projections have been traced to GnRH neurons in ultrastructure studies.[33]

Kisspeptin and NKB neurons in the arcuate nucleus of the hypothalamus express estrogen receptors, and their activity appears to be subject to steroid feedback. Kisspeptin neurons are sexually dimorphic in their abundance in the hypothalami of rodent models, and it has been proposed that the organizing effects that steroids have on the brain early in development may involve organizing effects on kisspeptin and NKB neurons.[34]

Neuropeptide Y (NPY) is another neuropeptide that may help to regulate GnRH secretion. Its effects are primarily inhibitory. Central administration of NPY inhibits pulsatile GnRH secretion in gonadectomized rhesus monkeys.[39,40] But it is unclear whether NPY acts directly on GnRH neurons or through the mediation of other signaling pathways.

Finally, mention must be made here of leptin, a peptide produced by adipose tissue, which may play an important role in the regulation of the HPG axis. While it was proposed early on that leptin might exert direct control of GnRH secretion,[41–43] that now appears unlikely, since GnRH neurons do not express leptin receptor.[44] More will be said about leptin and its putative role in pubertal timing in the next section.

The systems biology of GnRH secretory control is an area of intense research effort at present and much remains to be elucidated. It appears, however, that there is a number of signaling pathways that converge on GnRH neurons with widespread inputs within and beyond the hypothalamus. Rather than positing a single, upstream "trigger" for puberty, a reasonable working hypothesis is that GnRH neurons integrate the various inhibitory and stimulatory signals in ways that ultimately affect pulsatile GnRH secretion. The underlying maturational program may be intrinsic to the GnRH neurons themselves, while upstream factors act principally to advance or delay this program. In the next section, some of the broader genetic and environmental influences that are thought to affect pubertal timing will be considered.

4.3 UPSTREAM FACTORS INFLUENCING PUBERTAL TIMING

There are many upstream factors that influence pubertal timing, as discussed below.

4.3.1 Genetic Factors

The biological factors that influence pubertal timing can be broadly classified as genetic and environmental, although these two domains always interact. Genetic influences on puberty are apparent in a number of clinical pathologies.[45] But evidence for genetic factors influencing normal, non-pathological variation in pubertal timing comes primarily from studies of concordance between different degrees of relatives. Studies comparing monozygotic twins with dizygotic twins and non-twin siblings, for example, indicate that the heritability of menarcheal age in girls is comparable to the heritability of adult height.[46–48] Studies of this kind usually underestimate the potential range of environmental variability, however, and so only provide a relative sense of underlying genetic determination, not a precise measurement. Like height, pubertal timing is likely to be subject to the influence of many genes of individually small effect, an assumption that is consistent with the nearly normal distribution of menarcheal age and other indices of pubertal timing in most populations.[49]

4.3.2 Environmental Factors

Environmental influences on pubertal timing are apparent in well-documented patterns of socioeconomic variation within populations and in the changes in pubertal timing that often accompany migration or other movement between ecological and socio-economic conditions.[49,50] Particular attention has been paid to two types of environmental factor associated with variation in pubertal timing: energetics and psychosocial stress. In addition to correlative evidence, both of these factors are associated with important neuroendocrine pathways regulating GnRH secretion.

Energetics

Frisch and colleagues first called explicit attention to the relationship between energetics and puberty in the 1970s, hypothesizing that a certain level of stored fat was necessary to trigger menarche in girls.[51–53] While this specific hypothesis has been subject to intense criticism,[54–58] the relationship between good nutrition, and hence rapid growth, in childhood and early puberty is more generally accepted. Experimental evidence from animal models including rodents and sheep shows that variation in energy intake and energy expenditure in the prepubertal period can change the timing of puberty.[59–61] Although similar experiments cannot be carried out in humans, both height for age and weight for height in late childhood are significant predictors of pubertal timing, and childhood athletic training and childhood malnutrition are associated with pubertal delay.[62–65]

It is important to keep in mind three distinct metrics when considering the relationship of growth to pubertal timing: rate of growth (usually height velocity), final height and age at puberty. Rate of growth and timing of puberty co-vary; that is, faster growth in childhood is associated with earlier age at puberty. The relationship between rate of growth and final height across populations is more variable, dependent on the genetic potential for adult height within each population.

Pathologies that are associated with short stature are almost invariably also associated with late or absent puberty (Table 4.1). There is a generally inverse relationship between adult height and pubertal timing at the population level as well, despite considerable variability. In contrast, the secular trend towards earlier puberty that has been well documented in many developed and developing countries over the past century has generally also been a secular trend towards greater adult height.[49,50,67] The coincidence of pathologies of growth and pathologies of pubertal maturation suggests that many common physiological mechanisms underlie both processes. The covariance of adult height and pubertal timing across time suggests that many environmental factors also influence both processes.

Several pathways connect information about energy balance with the neuroendocrine mechanisms controlling GnRH secretion. Among the important peripheral signals of energy balance are insulin and leptin. Insulin is secreted by the pancreas to promote glucose uptake from the circulation. Daily rates of insulin production are positively correlated with energy balance.[68–70] Upstream of HPG axis function, insulin receptor (IR) is expressed in the mouse hypothalamus and in immortalized GnRH cell lines.[71,72]

Table 4.1 Pathological causes of slow or stunted growth that are also associated with pubertal delay or absence[66]

Anorexia

Cardiovascular disease (cyanotic heart disease and congestive heart failure)

Chronic anemias

Chronic infection

Constitutional delay of growth

Cushing syndrome

Diabetes mellitus (especially Mauriac syndrome)

GH deficiency

GH insensitivity

IGF-1 deficiency

Hypothyroidism

Intrauterine growth retardation

Malnutrition

Malabsorption

Prader—Willi syndrome

Pulmonary disease

Psychosocial dwarfism

Renal disease

Turner's syndrome

GH: growth hormone; IGF-1: insulin-like growth factor-1.

Suppression of IR in the mouse hypothalamus results in a reduction in GnRH pulsatility.[71] There is thus a potential for insulin to directly regulate pulsatile GnRH release, although the effectiveness of peripheral insulin in exerting this control has not been demonstrated. Downstream, adipose cells are important targets of insulin action, and estrogens facilitate the action of insulin on adipose tissue in women, promoting fat storage.[73,74]

Leptin is a peptide secreted by adipose cells. Levels of leptin secretion vary with fat mass, but they also vary as a consequence of estrogen and insulin stimulation.[75–83] Therefore, females secrete higher levels of leptin per fat mass than males, and individuals of both sexes in positive energy balance secrete more leptin per fat mass than individuals in negative energy balance.

Leptin receptor (LepR) is not expressed by GnRH-secreting neurons,[44] but it is expressed by other hypothalamic cell populations, including kisspeptin-secreting neurons.[32,35] Thus, there is a potential for leptin to indirectly contribute to the regulation of pulsatile GnRH secretion. There is good evidence that this control can be exerted in adult women. For example, in patients with amenorrhea associated with low energy balance, exogenous elevation of peripheral leptin levels results in increases in LH pulse frequency, the number of secondary maturing follicles and circulating estradiol levels.[84,85]

The relationship of insulin and leptin to pubertal timing is more complex. The pubertal period is a period of transient insulin resistance, meaning that circulating insulin levels rise.[86,87] This increase in insulin resistance is caused by increased circulating levels of insulin-like growth factor-1 (IGF-1) during the pubertal growth spurt. IGF-1 reduces the sensitivity of peripheral adipose tissue to insulin, therefore resulting in an increase in baseline insulin necessary to regulate blood glucose levels.[88,89] The increase in peripheral insulin may be correlated with an increase in insulin signaling in the hypothalamus, but this has not been demonstrated. It would seem unlikely that this rise in insulin could contribute to pubertal initiation, however, since it appears to be a consequence of puberty, not an antecedent. Other evidence suggests that prepubertal insulin levels may influence pubertal timing. Women who develop type 2 (adult-onset) diabetes have earlier menarcheal ages than controls, as do first order relatives of type 2 diabetics.[90]

Exogenous administration of leptin to prepubertal mice accelerates vaginal opening, an external sign of reproductive maturation.[91] In primates, including humans, however, increases in circulating leptin, like those of insulin, occur as a consequence of puberty, not as antecedents.[92,93] Some researchers have suggested that leptin is a permissive gate for puberty in humans, meaning that adequate leptin levels are necessary, although not in themselves sufficient, for puberty to occur.[94] Even this hypothesis is in doubt, however, since there are documented cases of individuals with pathological failure of adipose tissue development and extremely low leptin levels who have nevertheless experienced normal puberty and even given birth.[95]

Rather than being a trigger or a gate for puberty, leptin and insulin levels probably contribute to hypothalamic signaling pathways that accelerate or retard pubertal timing without completely determining it. Since both hormones are also important signals of energy balance, a modulating role would be consistent with the documented modulating influence of energy balance on pubertal timing. There is evidence that both insulin and leptin act on kisspeptin neurons in the hypothalamus, which would provide at least one pathway for such a modulatory effect.

Psychosocial Stress

In addition to energetics, there is considerable evidence suggesting that psychosocial stress can influence the timing of puberty, although the evidence can be conflicting. Early studies by Boas documented late maturation among children institutionalized in orphanages.[96] More recent studies find evidence that institutionalization under conditions of chronic maltreatment early in life or a history of childhood abuse can lead to early puberty.[97–99]

Psychosocial stress has also been associated with early puberty in a different context. A number of studies (but not all) have found that girls raised in "father absent" households have earlier ages at menarche than their otherwise similar peers.[100–104] The size of the effect is relatively small, especially when controlled for genetics, and the number of potential confounding factors is large. But several researchers have noted that an acceleration of maturation is predicted to occur when adult mortality risk is high by the branch of evolutionary theory known as life history theory.[103,105–107] It has been suggested that the environment of upbringing might provide important signals about the quality of the environment to be expected in adulthood, and that humans may have evolved to adjust the tempo of maturation in accordance with these signals.

Psychosocial stress is known to affect the HPG axis in animal models.[108–111] Proving that the human HPG axis is sensitive to psychosocial stress independent of other factors is more difficult.[112,113] Most research focuses on interactions between the hypothalamic–pituitary–adrenal (HPA) axis, which is activated under conditions of psychosocial stress, and the HPG axis.[112,114,115] Activation of the HPA axis can be shown to suppress HPG axis function, especially if HPA activation is chronic. High levels of cortisol (the steroid produced by the adrenal gland in response to HPA activation) have also been associated with early pregnancy loss in some studies.[116,117] However, psychosocial stress is only one cause of HPA axis activation. Other physical and metabolic stresses, including energetic stress, can elevate HPA activity.[118,119] Thus, it is not possible to infer from evidence of HPA activation that psychosocial stress is the cause of that activation in any particular situation.

Importantly, the "father absence" hypothesis on the one hand and research on cortisol's impact on HPG axis function on the other generate opposing predictions about the direction of the influence of the HPA axis on the timing of puberty. These findings could be reconciled if chronic psychosocial stress were found to result in diminished activation of the HPA axis, or if psychosocial stress were to act through mechanisms other than the HPA axis.

In summary, a broad range of environmental factors can be shown to influence puberty and reproductive function more generally. Energy balance and psychosocial stress have both attracted attention in humans. Other factors, such as day length and temperature, figure prominently in other mammals.[120] The relatively small population of GnRH neurons in the hypothalamus provides a final common pathway for the integration of information regarding these factors in modulating the timing of puberty.

4.4 DOWNSTREAM CONSEQUENCES OF PUBERTAL ACTIVATION OF THE HYPOTHALAMIC–PITUITARY–GONADAL AXIS

Reactivation of pulsatile GnRH secretion is the central event of puberty. In humans, however, it is an unobservable event. What can be observed are its downstream consequences. These consequences begin with downstream parts of the HPG axis, but soon ramify to include a host of somatic and central nervous system tissues and physiological processes. If the GnRH neurons are the final common pathway for central inputs to the HPG axis affecting puberty, gonadal steroid production can be considered the final common pathway for outputs from the HPG axis to the rest of the body. The extremely broad reach of gonadal steroids in transforming the rest of the body at puberty reflects the fact that puberty is, as noted in the introduction to this chapter, a major life history transition, almost akin to metamorphosis, from an organism whose most basic concerns are staying alive and growing to one concerned with staying alive and reproducing.

Because the downstream consequences of pubertal maturation are so numerous, it will not be possible to review them all here. Some are discussed at greater length in other chapters in this volume. This text will be confined to briefly touching on the downstream consequences for the HPG axis itself, for the pubertal growth spurt and for the development of the most prominent secondary sexual characteristics. These are the downstream consequences that are most often observed in making inferences about puberty, and some of the dangers and assumptions in making those inferences will occasionally be noted.

4.4.1 Downstream Effects on the Hypothalamic–Pituitary–Gonadal Axis

The earliest observable change in HPG axis activity associated with puberty is the appearance of sleep-related pulses of LH that subside during the daytime.[121,122] These are assumed to reflect sleep-related pulsatile release of GnRH, and the circadian pattern suggests some influence of central circadian pacemakers. Over a variable period of time, usually in the order of weeks to months, pulsatile LH secretion extends to the daytime and pulses increase in magnitude. FSH pulses, which occur at much lower levels, may also become observable at this time. The appearance of sleep-related LH pulses occurs earlier in girls than in boys, as is true of virtually all pubertal indicators.

Gonadal steroid production increases as gonadotrophin release patterns become more robust. In girls, rising estradiol levels are assumed to reflect the initial recruitment of primary follicles, the production of testosterone by the outer layer of theca cells in those follicles under the influence of LH, and the aromatization of some of the testosterone into estradiol by the inner granulosa cells of those follicles under the influence of FSH.[123] Circulating levels of gonadal steroids do not reach high levels at first, and there is no evidence of ovulation or menstrual bleeding. Production of testosterone often exceeds the rate of aromatization to estradiol at this stage. In boys, rising testosterone levels are assumed to reflect increasing activity of Leydig cells outside the seminiferous tubules. Proliferation of Sertoli cells inside the tubules occurs as a consequence, although there is no evidence of sperm in urine at this stage.[124,125]

In girls, rising levels of estradiol production from the ovaries cause proliferation of the endometrial lining of the uterus. In the absence of progesterone support, the endometrial lining will eventually slough off.[126,127] The first occurrence of menstrual blood, known as menarche, is often taken as a milestone in female pubertal maturation, but it does not really represent a significant functional boundary in the development of female reproductive potential. Menarche does not, for instance, indicate ovulation and the production of viable gametes. Nor by itself does it signify the ability to successfully initiate gestation. It is, however, a notable external event, often imbued with cultural significance. A similar external marker in boys does not exist. Beard growth and voice change are highly variable in manifestation. Immature sperm may appear in urine at an early stage, but this is rarely observed. Ejaculation and nocturnal emission are extremely variable between individuals and may not occur at early stages of pubertal maturation at all.[128,129]

Ovulation usually first occurs some time after menarche in girls.[130] Ovulation and menstruation become increasingly regular over a period of years. Testosterone to estradiol ratios may remain higher during this period than among fully mature women, and luteal progesterone levels are lower on average.[123,131,132] Peak fecundity, as reflected by the frequency of ovulation and levels of gonadal steroid secretion, usually is not achieved until nearly a decade after menarche. In boys, sperm production, sperm quality and ejaculatory volume reach mature levels more quickly, usually within 5 years of the first signs of pubertal development.[133]

4.4.2 The Pubertal Growth Spurt

Although other hormones, including insulin, growth hormone and IGF-1, play a role in promoting growth at puberty and before, the pubertal growth spurt itself — the relatively rapid acceleration and deceleration of linear growth that normally brings growth in stature to an end — is now understood to be primarily a consequence of the action of estradiol in both sexes.[134,135] In females, the primary source of estradiol is the ovary. In males, testosterone is secondarily converted to estradiol by aromatase in cartilage and bone of the growth plates. Individuals who are chromosomally male but with complete androgen insensitivity syndrome due to disabling mutations of the androgen receptor gene nevertheless experience a normal growth spurt.[135] Although testosterone and other androgens cannot have any direct effect in these individuals, locally converted estradiol still stimulates a normal pubertal growth spurt. In contrast, males who lack functional estradiol receptors or functional aromatase enzyme do not experience a growth spurt — either the acceleration phase or the deceleration. Instead, there are documented cases of such individuals continuing to grow at a prepubertal pace into their thirties[135–138] (Figure 4.1). In contrast, individuals with complete androgen insensitivity due to a defect in the androgen receptor gene undergo normal pubertal growth spurts, reflecting the fact that aromatase conversion of androgens to estrogens is still intact[139,140] (Figure 4.2).

Estradiol receptor is expressed both by active chondrocytes and by osteoblasts and osteoclasts in the growth plates of growing bones.[135] The pubertal acceleration of growth involves the action of estradiol in stimulating chondrocyte proliferation and growth, while the rapid deceleration and eventual closure of the growth plates is due to the

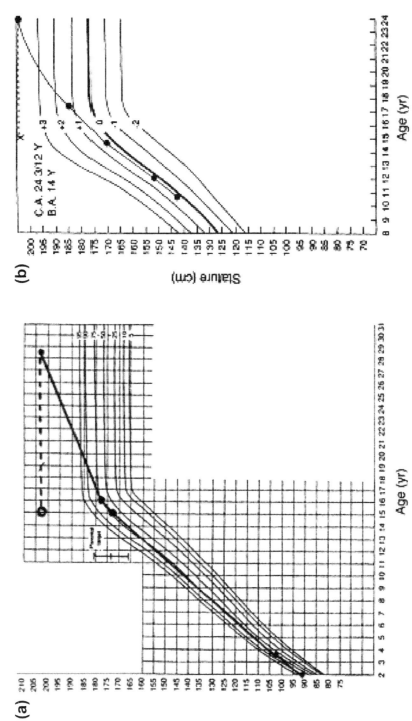

Figure 4.1 (a) Growth chart from a man with a disruptive mutation of the estrogen receptor alpha gene;[135] (b) growth chart of a man with a disruptive mutation in the aromatase gene, *CYP19*.[136] (Source: *Juul et al.*[135])

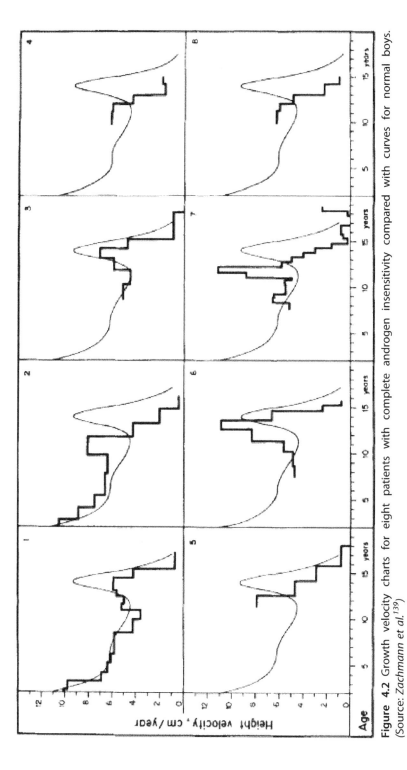

Figure 4.2 Growth velocity charts for eight patients with complete androgen insensitivity compared with curves for normal boys. (Source: Zachmann et al.[139])

actions of estradiol in stimulating mineralization by osteoblasts and opposing deminer-alization by osteoclasts. Both of the major estradiol receptor subtypes are probably involved, so that chondrocyte stimulation predominates at lower estradiol concentrations and osteocyte stimulation predominates at high estradiol concentrations.

4.4.3 Secondary Sexual Characteristics

All the major sexually dimorphic characteristics of adult males and females are conse-quences of the action of gonadal steroids circulating at adult levels. The sequence of appearance of major secondary sexual characteristics tends to follow the sequence of elevation of steroid levels.[141] Pubic and axillary hair is stimulated by androgens in both sexes and thus tends to appear early in the pubertal sequence in girls.[142] These androgens are produced by the adrenal gland. When adrenal androgen production is pathologically elevated, as in cases of untreated congenital adrenal hyperplasia, precocial growth spurts can result.[143,144] Female breast development and pelvic remodeling are stimulated by estradiol and thus tend to occur later.[145,146] Pelvic remodeling, important for parturi-tion, occurs internally and at a rapid rate and is thus difficult to measure; its degree is underestimated by measurements of biiliac breadth.[146] Menstruation in girls occurs only when estradiol levels have reached a high enough level for long enough to stimulate sufficient endometrial growth, and thus menarche tends to occur after the initiation of breast development and usually during the decelerating phase of pubertal growth.[147] Testis growth in boys is a consequence of seminiferous tubule development and the proliferation of Sertoli cells, and thus occurs early in the pubertal sequence in boys. The growth spurt itself, both in height and in shoulder breadth and arm length, occurs later when higher levels of circulating testosterone result in higher levels of local conversion to estradiol. Penis growth is a result of a different secondary local conversion of testosterone to dihydrotestosterone, and occurs more or less in synchrony with the growth spurt.[147]

Changes in body composition are also important downstream consequences of gonadal steroid production.[148,149] Estradiol stimulates adipose tissue growth in girls, particularly in the breasts and gluteofemoral region, acting in synergy with insulin.[149,150] Testosterone, on the other hand, acting through androgen receptors, inhibits adipose tissue growth and stimulates muscle anabolism in boys.[151,152] It is important to realize that these changes in body composition are consequences of pubertal maturation, not antecedents.[92,153] Increases in fat percentage in girls are correlated with pubertal stage, but they are not the primary causes of pubertal maturation.

As with the height spurt, precocial development of secondary sexual characteristics can occur as a result of pathological or idiosyncratic overproduction of adrenal steroids.[144] The adrenal gland normally produces weak androgen hormones whose significance is still debated.[143] These androgens can be locally converted to estrogens in various tissues, including adipose tissue.[154] In postmenopausal women this conversion of adrenal androgens to estrogens (so-called "extragonadal estrogen production") is the principal source of circulating estrogens, levels of which are directly correlated with adiposity.[155,156] Increasing rates of childhood obesity and overweight are thought to be largely responsible for a trend towards early breast development in girls in some populations.[157,158] It is important to realize, however, that extragonadal estrogen production is not a consequence of HPG axis activation and thus does not represent puberty. Nor can it result in ovulation or normal fecundity. Instead, early breast development associated with increasing adiposity should be considered a mild stage along a continuum of pathologically advanced development of secondary sexual characteristics.

4.5 SUMMARY

As complicated as it is, human puberty can be thought of according to a relatively simple model, expressed in Figure 4.3. This model can be thought of as an "hour glass", with a number of factors and pathways converging on pulsatile GnRH secretion as the "on/off switch" of the HPG axis, and a ramifying set of downstream consequences of gonadal steroid production as the final product of the HPG axis.

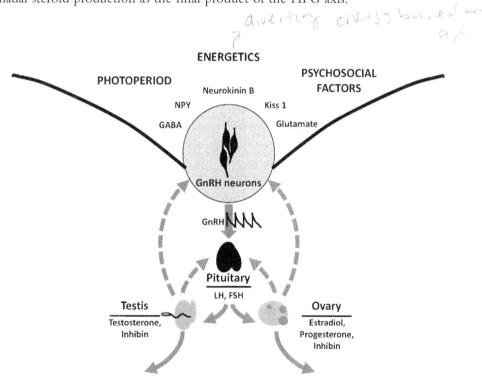

Figure 4.3 A schematic model of human puberty, including upstream modulators and downstream consequences. Upstream modulating pathways converge on the reactivation of pulsatile GnRH secretion. Downstream consequences emanate from changes in circulating levels of gonadal steroids.

At the heart of the pubertal process is the reactivation of pulsatile GnRH secretion by the relatively small population of hypothalamic GnRH neurons. The capacity for this secretion is fully developed years in advance of puberty, and pulsatile secretion probably occurs in both the fetus and the neonate. In gonadectomized rhesus monkeys reactivation of the HPG axis occurs spontaneously at a relatively normal age. If there are simple physiological signals responsible for this spontaneous reactivation they have not yet been discovered.

Activity of the GnRH neurons is subject to control and modulation by a number of upstream neuron populations secreting a variety of neurotransmitters and neuro-modulators. Among those with inhibitory effects on GnRH secretion are GABA and NPY. Among those with stimulatory effects are kisspeptin and glutamate. Other pathways further upstream, such as those involving insulin, leptin and the HPA axis, integrate information about important aspects of the environment and the state of the organism, such as photoperiod (not addressed in this chapter but well documented in many non-human mammals such as rodents), energy balance and psychosocial factors.

At the other end of the HPG axis, production of gonadal steroid hormones results in the maturation of the gonads as producers of mature gametes as well as the mature function of other parts of the reproductive system, such as the endometrium and breasts in girls and accessory glands in males. The pubertal growth spurt is also a downstream consequence of gonadal steroid production along with the development of secondary sexual characteristics and changes in body composition. The aspects of puberty that are normally downstream consequences of gonadal maturation can also occur as pathological consequences of excessive adrenal androgen production or excessive conversion of adrenal androgens to estrogens. The pathological advancement of secondary sexual characteristic development cannot result in activation of the HPG axis, however, and thus should not be confused with true puberty.

The search for "triggers" of human puberty has so far been unsuccessful. It may be more useful to think of normal puberty as a process of innate maturation, the timing of which can be modulated by factors such as nutrition and stress.

REFERENCES

1. Paige KE, Paige JM. *The politics of reproductive ritual*. Berkeley, CA: University of California Press; 1981.
2. Plant TM, Barker-Gibb ML. Neurobiological mechanisms of puberty in higher primates. *Hum Reprod Update* 2004;**10**:67–77.
3. Ebling FJ. The neuroendocrine timing of puberty. *Reproduction* 2005;**129**:675–83.
4. Ebling FJ, Cronin AS. The neurobiology of reproductive development. *Neuroreport* 2000;**11**:R23–33.
5. Krey LC, Butler WR, Knobil E. Surgical disconnection of the medial basal hypothalamus and pituitary function in the rhesus monkey. *Endocrinology* 1975;**96**:1073–87.

6. Knobil E, Plant TM, Wildt L, Belchetz PE, Marshall G. Control of the rhesus monkey menstrual cycle: permissive role of hypothalamic gonadotropin-releasing hormone. *Science* 1980;**207**:1371—3.

7. Knobil E, Plant TM, Wildt L, Belchetz PE, Marshall G. Neuroendocrine control of the rhesus monkey menstrual cycle: permissive role of the hypothalamic gonadotropin-releasing hormone (GnRH). *Science* 1980;**207**:1371—3.

8. Knobil E. Patterns of hypophysiotropic signals and gonadotropin secretion in the rhesus monkey. *Biology* 1981;**24**:44—9.

9. Wildt L, Hausler A, Marshall G, Hutchison JS, Plant TM, Belchetz PE, Knobil E. Frequency and amplitude of gonadotropin-releasing hormone stimulation and gonadotropin secretion in the rhesus monkey. *Endocrinology* 1981;**109**:376—85.

10. Belchetz P, Plant TM, Nakai Y, Keogh EJ, Knobil E. Hypophyseal responses to continuous and intermittent delivery of hypothalamic gonadotropin releasing hormone. *Science* 1978;**202**:631—3.

11. Wildt L, Marshall G, Knobil E. Experimental induction of puberty in the infantile female rhesus monkey. *Science* 1980;**207**:1373—5.

12. Crowley WFJ, McArthur J. Stimulation of the normal menstrual cycle in Kallman's syndrome by pulsatile administration of luteinizing hormone-releasing hormone (LHRH). *J Clin Endocrinol Metab* 1980;**51**:173—5.

13. Santoro N, Filicori M, Crowley WFJ. Hypogonadotropic disorders in men and women: diagnosis and therapy with pulsatile gonadotropin-releasing hormone. *Endocr Rev* 1986;**7**:11—23.

14. Mansfield MJ, Beardsworth DE, Loughlin JS, Crawford JD, Bode HH, Rivier J, et al. Long-term treatment of central precocious puberty with a long-acting analogue of luteinizing hormone-releasing hormone. Effects on somatic growth and skeletal maturation. *N Engl J Metab* 1983;**309**:1286—90.

15. Bryan KJ, Mudd JC, Richardson SL, Chang J, Lee HG, Zhu X, et al. Down-regulation of serum gonadotropins is as effective as estrogen replacement at improving menopause-associated cognitive deficits. *J Neurochem* 2010;**112**:870—81.

16. Torrisi R, Bagnardi V, Rotmensz N, Scarano E, Iorfida M, Veronesi P, et al. Letrozole plus GnRH analogue as preoperative and adjuvant therapy in premenopausal women with ER positive locally advanced breast cancer. *Breast Cancer Res Treat* 2011;**126**:431—41.

17. Torrisi R, Bagnardi V, Pruneri G, Ghisini R, Bottiglieri L, Magni E, et al. Antitumour and biological effects of letrozole and GnRH analogue as primary therapy in premenopausal women with ER and PgR positive locally advanced operable breast cancer. *Br J Cancer* 2007;**97**:802—8.

18. Wray S. From nose to brain: development of gonadotrophin-releasing hormone-1 neurones. *J Neuroendocrinol* 2010;**22**:743—53.

19. Schwanzel-Fukuda M, Crossin KL, Pfaff DW, Bouloux PM, Hardelin JP, Petit C. Migration of luteinizing hormone-releasing hormone (LHRH) neurons in early human embryos. *J Comp Neurol* 1996;**366**:547—57.

20. Wetsel WC, Valenca MM, Merchenthaler I, Liposits Z, López FJ, Weiner RI, et al. Intrinsic pulsatile secretory activity of immortalized luteinizing hormone-releasing hormone-secreting neurons. *Proc Natl Acad Sci USA* 1992;**89**:4149—53.

21. Terasawa E, Fernandez DL. Neurobiological mechanisms of the onset of puberty in primates. *Endocr Rev* 2001;**22**:111—51.

22. Terasawa E, Keen KL, Mogi K, Claude P. Pulsatile release of luteinizing hormone-releasing hormone (LHRH) in cultured LHRH neurons derived from the embryonic olfactory placode of the rhesus monkey. *Endocrinology* 1999;**140**:1432—41.

23. Ronnekleiv OK, Resko JA. Ontogeny of gonadotropin-releasing hormone-containing neurons in early fetal development of rhesus macaques. *Endocrinology* 1990;**126**:498—511.

24. DiVall SA, Radovick S. Pubertal development and menarche. *Ann NY Acad Sci* 2008;**1135**:19—28.

25. Conte FA, Grumbach MM, Kaplan SL. A diphasic pattern of gonadotropin secretion in patients with the syndrome of gonadal dysgenesis. *J Clin Endocrinol Metab* 1975;**40**:670—4.

26. Plant TM. A study of the role of the postnatal testes in determining the ontogeny of gonadotropin secretion in the male rhesus monkey (*Macaca mulatta*). *Endocrinology* 1985;**116**:1341—50.

27. Pohl CR, deRidder CM, Plant TM. Gonadal and nongonadal mechanisms contribute to the prepubertal hiatus in gonadotropin secretion in the female rhesus monkey (*Macaca mulatta*). *J Clin Endocrinol Metab* 1995;**80**:2094—101.

28. El Majdoubi M, Ramaswamy S, Sahu A, Plant TM. Effects of orchidectomy on levels of the mRNAs encoding gonadotropin-releasing hormone and other hypothalamic peptides in the adult male rhesus monkey (*Macaca mulatta*). *J Neuroendocrinol* 2000;**12**:167—76.

29. El Majdoubi M, Sahu A, Plant TM. Changes in hypothalamic gene expression associated with the arrest of pulsatile gonadotropin-releasing hormone release during infancy in the agonadal male rhesus monkey (*Macaca mulatta*). *Endocrinology* 2000;**141**:3273—7.

30. Fraser MO, Pohl CR, Plant TM. The hypogonadotropic state of the prepubertal male rhesus monkey (*Macaca mulatta*) is not associated with a decrease in hypothalamic gonadotropin-releasing hormone content. *Biology* 1989;**40**:972—80.

31. Plant TM. The role of KiSS-1 in the regulation of puberty in higher primates. *Eur J Endocrinol* 2006;**155**(Suppl. 1):S11—6.

32. Castellano JM, Bentsen AH, Mikkelsen JD, Tena-Sempere M. Kisspeptins: bridging energy homeostasis and reproduction. *Brain Res* 2010;**1364**:129—38.

33. Topaloglu AK. Neurokinin B signaling in puberty: human and animal studies. *Mol Cell Endocrinol* 2010;**324**:64—9

34. Navarro VM, Castellano JM, McConkey SM, Pineda R, Ruiz-Pino F, Pinilla L, et al. Interactions between kisspeptin and neurokinin B in the control of GnRH secretion in the female rat. *Am J Physiol* 2011;**300**:E202—10.

35. Oakley AE, Clifton DK, Steiner RA. Kisspeptin signaling in the brain. *Endocr Rev* 2009;**30**:713—43.

36. Gottsch M, Cunningham MJ, Smith JT, Popa SM, Acohido BV, Crowley WF, et al. A role for kisspeptins in the regulation of gonadotropin secretion in the mouse. *Endocrinology* 2004;**145**:4073—7.

37. Plant TM, Ramaswamy S. Kisspeptin and the regulation of the hypothalamic—pituitary—gonadal axis in the rhesus monkey (*Macaca mulatta*). *Peptides* 2009;**30**:67—75.

38. Kauffman A, Clifton DK, Steiner RA. Emerging ideas about kisspeptin-GPR54 signaling in the neuroregulation of reproduction. *Trends Neurosci* 2007;**30**:504—11.

39. Pralong FP. Insulin and NPY pathways and the control of GnRH function and puberty onset. *Mol Cell Endocrinol* 2010;**324**:82—6.

40. Shahab M, Balasubramaniam A, Sahu A, Plant TM. Central nervous system receptors involved in mediating the inhibitory action of neuropeptide Y on luteinizing hormone secretion in the male rhesus monkey (*Macaca mulatta*). *J Neuroendocrinol* 2003;**15**:965—70.

41. Conway GS, Jacobs HS. Leptin: a hormone of reproduction. *Hum Reprod* 1997;**12**:633—5.

42. Mann DR, Plant TM. Leptin and pubertal development. *Semin Reprod Med* 2002;**20**:93—102.

43. Nagatani S, Guthikonda P, Thompson RC, Tsukamura H, Maeda KI, Foster DL. Evidence for GnRH regulation by leptin: leptin administration prevents reduced pulsatile LH secretion during fasting. *Neuroendocrinology* 1998;**67**:370—6.

44. Burcelin R, Thorens B, Glauser M, Gaillard RC, Pralong FP. Gonadotropin-releasing hormone secretion from hypothalamic neurons: stimulation by insulin and potentiation by leptin. *Endocrinology* 2003;**144**:4484—91.

45. Herbison AE. Genetics of puberty. *Horm Res* 2007;**68**(Suppl. 5):75—9.

46. Mustanski BS, Viken RJ, Kaprio J, Pulkkinen L, Rose RJ. Genetic and environmental influences on pubertal development: longitudinal data from Finnish twins at ages 11 and 14. *Dev Psychol* 2004;**40**:1188—98.

47. Loesch DZ, Huggins R, Rogucka E, Hoang NH, Hopper JL. Genetic correlates of menarcheal age: a multivariate twin study. *Ann Hum Biol* 1995;**22**:470—90.

48. Kaprio J, Rimpela A, Winter T, Viken RJ, Rimpela M, Rose RJ. Common genetic influences on BMI and age at menarche. *Hum Biol* 1995;**67**:739—53.

49. Eveleth PB, Tanner JM. *Worldwide variation in human growth*. Cambridge: Cambridge University Press; 1991.

50. Karlberg J. Secular trends in pubertal development. *Horm Res* 2002;**57**(Suppl. 2):19—30.

51. Frisch RE, Revelle R. Height and weight at menarche and a hypothesis of critical body weights and adolescent events. *Science* 1970;**169**:397–9.

52. Frisch RE, Revelle R. Height and weight at menarche and a hypothesis of menarche. *Arch Dis Child* 1971;**46**:695–701.

53. Frisch RE, McArthur JW. Menstrual cycles: fatness as a determinant of minimum weight for height necessary for their maintenance or onset. *Science* 1974;**185**:949–51.

54. Johnston FE, Malina RM, Galbraith MA, Frisch RE, Revelle R, Cook S. Height, weight and age at menarche and the "critical weight' hypothesis". *Science* 1971;**174**:1148–9.

55. Billewicz WZ, Fellowes HM, Hytten CA. Comments on the critical metabolic mass and the age of menarche. *Ann Hum Biol* 1976;**3**:51–9.

56. Trussell J. Statistical flaws in evidence for the Frisch hypothesis that fatness triggers menarche. *Hum Biol* 1980;**52**:711–20.

57. Trussell J. Menarche and fatness: reexamination of the critical body composition hypothesis. *Science* 1978;**200**:1506–13.

58. Ellison PT. Skeletal growth, fatness and menarcheal age: a comparison of two hypotheses. *Hum Biol* 1982;**54**:269–81.

59. Kinoshita M, Moriyama R, Tsukamura H, Maeda KI. A rat model for the energetic regulation of gonadotropin secretion: role of the glucose-sensing mechanism in the brain. *Domest Anim Endocrinol* 2003;**25**:109–20.

60. Prasad BM, Conover CD, Sarkar DK, Rabii J, Advis JP. Feed restriction in prepubertal lambs: effect on puberty onset and on in vivo release of luteinizing-hormone-releasing hormone, neuropeptide Y and beta-endorphin from the posterior-lateral median eminence. *Neuroendocrinology* 1993;**57**:1171–81.

61. Cameron JL. Metabolic cues for the onset of puberty. *Horm Res* 1991;**36**:97–103.

62. Malina RM. Physical activity and training: effects on stature and the adolescent growth spurt. *Med Sci Sports Exerc* 1994;**26**:759–66.

63. Malina RM. Physical growth and biological maturation of young athletes. *Exerc Sport Sci Rev* 1994;**22**:389–433.

64. Martorell R. Physical growth and development of the malnourished child: contributions from 50 years of research at INCAP. *Food Nutr Bull* 2010;**31**:68–82.

65. Morris DH, Jones ME, Schoemaker MJ, Ashworth A, Swerdlow AJ. Determinants of age at menarche in the UK: analyses from the Breakthrough Generations Study. *Br J Cancer* 2010;**103**:1760–4.

66. Lavin N. *Manual of Endocrinology and Metabolism*. 4th ed. New York: Lippincott, Williams & Wilkins; 2009.

67. Cole TJ. Secular trends in growth. *Proc Nutr Soc* 2000;**59**:317–24.

68. Strack AM, Sebastian RJ, Schwartz MW, Dallman MF. Glucocorticoids and insulin: reciprocal signals for energy balance. *Am J Physiol* 1995;**268**:R142–9.

69. Oppert JM, Nadeau A, Tremblay A, Despres JP, Theriault G, Bouchard C. Negative energy balance with exercise in identical twins: plasma glucose and insulin responses. *Am J Physiol* 1997;**272**:E248–54.

70. Obici S, Rossetti L. Minireview: nutrient sensing and the regulation of insulin action and energy balance. *Endocrinology* 2003;**144**:5172–8.

71. Brüning JC, Gautam D, Burks DJ, Gillette J, Schubert M, Orban PC, et al. Role of brain insulin receptor in control of body weight and reproduction. *Science* 2000;**289**:2122–5.

72. Salvi R, Castillo E, Voirol M-J, Glauser M, Rey JP, Gaillard RC, et al. Gonadotropin-releasing hormone-expressing neurons immortalized conditionally are activated by insulin: implication of the mitogen-activated protein kinase pathway. *Endocrinology* 2006;**147**:816–26.

73. Bjorntorp P. The regulation of adipose tissue distribution in humans. *Int J Obes Relat Metab Disord* 1996;**20**:291–302.

74. Jones ME, McInnes KJ, Boon WC, Simpson ER. Estrogen and adiposity – utilizing models of aromatase deficiency to explore the relationship. *J Steroid Biochem Mol Biol* 2007;**106**:3–7.

75. Hilton LK, Loucks AB. Low energy availability, not exercise stress, suppresses the diurnal rhythm of leptin in healthy young women. *Am J Physiol* 2000;**278**:E43–9.

76. Campostano A, Grillo G, Bessarione D, De Grandi R, Adami GF. Relationships of serum leptin to body composition and resting energy expenditure. *Horm Metab Res* 1998;**30**:646−7.

77. Rosenbaum M, Leibel RL. Role of gonadal steroids in the sexual dimorphisms in body composition and circulating concentrations of leptin. *J Endocrinol Metab* 1999;**75**:1784−9.

78. Franks PW, Farooqi IS, Luan J, Wong MY, Halsall I, O'Rahilly S, Wareham NJ. Does physical activity energy expenditure explain the between-individual variation in plasma leptin concentrations after adjusting for differences in body composition? *J Clin Endocrinol Metab* 2003;**88**:3258−63.

79. Franks PW, Loos RJ, Brage S, O'Rahilly S, Wareham NJ, Ekelund U. Physical activity energy expenditure may mediate the relationship between plasma leptin levels and worsening insulin resistance independently of adiposity. *J Appl Physiol* 2007;**102**:1921−6.

80. Rosenbaum M, Leibel RL. Clinical review 107: role of gonadal steroids in the sexual dimorphisms in body composition and circulating concentrations of leptin. *J Clin Endocrinol Metab* 1999;**84**:1784−9.

81. Rosenbaum M, Nicolson M, Hirsch J, Murphy E, Chu F, Leibel RL. Effects of weight change on plasma leptin concentrations and energy expenditure. *J Clin Endocrinol Metab* 1997;**82**:3647−54.

82. Kennedy A, Gettys TW, Watson P, Wallace P, Ganaway E, Pan Q, Garvey WT. The metabolic significance of leptin in humans: gender-based differences in relationship to adiposity, insulin sensitivity, and energy expenditure. *J Clin Endocrinol Metab* 1997;**82**:1293−300.

83. Wadden TA, Considine RV, Foster GD, Anderson DA, Sarwer DB, Caro JS. Short- and long-term changes in serum leptin dieting obese women: effects of caloric restriction and weight loss. *J Clin Endocrinol Metab* 1998;**83**:214−8.

84. Licinio J, Negrao AB, Mantzoros C, Kaklamani V, Wong ML, Bongiorno PB, et al. Synchronicity of frequently sampled, 24-h concentrations of circulating leptin, luteinizing hormone, and estradiol in healthy women. *Proc Natl Acad Sci USA* 1998;**95**:2541−6.

85. Welt CK, Chan JL, Bullen J, Murphy R, Smith P, DePaoli AM, et al. Recombinant human leptin in women with hypothalamic amenorrhea. *N Engl J Metab* 2004;**351**:987−97.

86. Moran A, Jacobs Jr DR, Steinberger J, Cohen P, Hong CP, Prineas R, Sinaiko AR. Association between the insulin resistance of puberty and the insulin-like growth factor-I/growth hormone axis. *J Clin Endocrinol Metab* 2002;**87**:4817−20.

87. Goran MI, Gower BA. Longitudinal study on pubertal insulin resistance. *Diabetes* 2001;**50**:2444−50.

88. Caprio S. Insulin: the other anabolic hormone of puberty. *Acta Paediatr Suppl* 1999;**88**:84−7.

89. Guercio G, Rivarola MA, Chaler E, Maceiras M, Belgorosky A. Relationship between the GH/IGF-I axis, insulin sensitivity, and adrenal androgens in normal prepubertal and pubertal boys. *J Clin Endocrinol Metab* 2002;**87**:1162−9.

90. Goodman MJ, Chung CS. Diabetes mellitus: discrimination between single locus and multifactorial models of inheritance. *Clin Genet* 1975;**8**:66−74.

91. Ahima RS, Dushay J, Flier SN, Prabakaran D, Flier JS. Leptin accelerates the onset of puberty in normal female mice. *J Clin Invest* 1997;**99**:391−5.

92. Demerath EW, Towne B, Wisemandle W, Blangero J, Chumlea WC, Siervogel RM. Serum leptin concentration, body composition, and gonadal hormones during puberty. *Int J Obes Relat Metab Disord* 1999;**23**:678−85.

93. Plant TM, Durrant AR. Circulating leptin does not appear to provide a signal for triggering the initiation of puberty in the male rhesus monkey (*Macaca mulatta*). *Endocrinology* 1997;**138**:4505−8.

94. Rogol AD. Leptin and puberty. *J Clin Endocrinol Metab* 1998;**83**:1089−90.

95. Andreelli F, Hanaire-Broutin H, Laville M, Tauber JP, Riou JP, Thivolet C. Normal reproductive function in leptin-deficient patients with lipoatropic diabetes. *J Clin Endocrinol Metab* 2000;**85**:715−9.

96. Race Boas F. *Language, and culture*. New York: Macmillan; 1940.

97. Wise LA, Palmer JR, Rothman EF, Rosenberg L. Childhood abuse and early menarche: findings from the black women's health study. *Am J Public Health* 2009;**99**(Suppl. 2):S460−6.

98. Charmandari E, Kino T, Souvatzoglou E, Chrousos GP. Pediatric stress: hormonal mediators and human development. *Horm Res* 2003;**59**:161−79.

99. Teilmann G, Pedersen CB, Skakkebaek NE, Jensen TK. Increased risk of precocious puberty in internationally adopted children in Denmark. *Pediatrics* 2006;**118**:e391−9.

100. Bogaert AF. Menarche and father absence in a national probability sample. *J Biosoc Sci* 2008; **40**:623−36.

101. Maestripieri D, Roney JR, DeBias N, Durante KM, Spaepen GM. Father absence, menarche and interest in infants among adolescent girls. *Dev Sci* 2004;**7**:560−6.

102. Tither JM, Ellis BJ. Impact of fathers on daughters' age at menarche: a genetically and environmentally controlled sibling study. *Dev Psychol* 2008;**44**:1409−20.

103. Belsky J, Steinberg L, Draper P. Childhood experience, interpersonal development, and reproductive strategy: and evolutionary theory of socialization. *Child Dev* 1991;**62**:647−70.

104. Grainger S. Female background and female sexual behavior: a test of the father-absence theory in Merseyside. *Hum Nat* 2004;**15**:133−45.

105. Coall DA, Chisholm JS. Evolutionary perspectives on pregnancy: maternal age at menarche and infant birth weight. *Soc Sci Med* 2003;**57**:1771−81.

106. Chisholm JS, Quinlivan JA, Petersen RW, Coall DA. Early stress predicts age at menarche and first birth, adult attachment, and expected lifespan. *Hum Nat* 2005;**16**:233−65.

107. Hoier S. Father absence and age at menarche: a test of four evolutionary models. *Hum Nat* 2002; **14**:209−33.

108. Gotz AA, Wolf M, Stefanski V. Psychosocial maternal stress during pregnancy: effects on reproduction for F0 and F1 generation laboratory rats. *Physiol Behav* 2008;**93**:1055−60.

109. Abbott DH, Saltzman W, Schultz-Darken NJ, Smith TE. Specific neuroendocrine mechanisms not involving generalized stress mediate social regulation of female reproduction in cooperatively breeding marmoset monkeys. *Ann NY Acad Sci* 1997;**807**:219−38.

110. Wise DA, Eldred NL. Maternal social stress disrupts reproduction of hamsters drinking high-calorie fluids. *Pharmacol Biochem Behav* 1986;**25**:449−56.

111. Kinsey-Jones JS, Li XF, Knox AM, Lin YS, Milligan SR, Lightman SL, O'Byrne KT. Corticotrophin-releasing factor alters the timing of puberty in the female rat. *J Neuroendocrinol* 2010; **22**:102−9.

112. Schoofs D, Wolf OT. Are salivary gonadal steroid concentrations influenced by acute psychosocial stress? A study using the Trier Social Stress Test (TSST). *Int J Psychophysiol* 2011;**80**:36−43.

113. Ellison PT, Lipson SF, Jasienska G, Ellison PL. Moderate anxiety, whether acute or chronic, is not associated with ovarian suppression in healthy, well-nourished, western women. *Am J Phys Anthropol* 2007;**134**:513−9.

114. Li XF, Knox AM, O'Byrne KT. Corticotrophin releasing factor and stress-induced inhibition of the gonadotrophin-releasing hormone pulse generator in the female. *Brain Res* 2010;**1364**:153−63.

115. Mastorakos G, Pavlatou MG, Mizamtsidi M. The hypothalamic−pituitary−adrenal and the hypothalamic−pituitary−gonadal axes interplay. *Pediatr Endocrinol Rev* 2006;**3**(Suppl. 1):172−81.

116. Nepomnaschy PA, Welch K, McConnell D, Strassmann BI, England BG. Stress and female reproductive function: a study of daily variations in cortisol, gonadotrophins, and gonadal steroids in a rural Mayan population. *Am J Hum Biol* 2004;**16**:523−32.

117. Nepomnaschy PA, Welch KB, McConnell DS, Low BS, Strassmann BI, England BG. Cortisol levels and very early pregnancy loss in humans. *Proc Natl Acad Sci USA* 2006;**103**:3938−42.

118. Chrousos GP. The role of stress and the hypothalamic−pituitary−adrenal axis in the pathogenesis of the metabolic syndrome: neuro-endocrine and target tissue-related causes. *Int J Obes Relat Metab Disord* 2000;**24**(Suppl. 2):S50−5.

119. Nieuwenhuizen AG, Rutters F. The hypothalamic−pituitary−adrenal-axis in the regulation of energy balance. *Physiol Behav* 2008;**94**:169−77.

120. Bronson FH. *Mammalian reproductive biology*. Chicago, IL: University of Chicago; 1989.

121. Wu FC, Butler GE, Kelnar CJ, Huhtaniemi I, Veldhuis JD. Ontogeny of pulsatile gonadotropin releasing hormone secretion from midchildhood, through puberty, to adulthood in the human male: a study using deconvolution analysis and an ultrasensitive immunofluorometric assay. *J Clin Endocrinol Metab* 1996;**81**:1798−805.

122. Landy H, Boepple PA, Mansfield MJ, Charpie P, Schoenfeld DI, Link K, et al. Sleep modulation of neuroendocrine function: developmental changes in gonadotropin-releasing hormone secretion during sexual maturation. *Pediatr Res* 1990;**28**:213−7.

123. Apter D. Development of the hypothalamic—pituitary—ovarian axis. *Ann NY Acad Sci* 1997; **816**:9—21.

124. Petersen C, Soder O. The sertoli cell — a hormonal target and "super" nurse for germ cells that determines testicular size. *Horm Res* 2006;**66**:153—61.

125. Sharpe RM, McKinnell C, Kivlin C, Fisher JS. Proliferation and functional maturation of Sertoli cells, and their relevance to disorders of testis function in adulthood. *Reproduction* 2003; **125**:769—84.

126. Salamonsen LA. Current concepts of the mechanisms of menstruation: a normal process of tissue destruction. *Trends Endocrinol Metab* 1998;**9**:305—9.

127. Collins J, Crosignani PG. Endometrial bleeding. *Hum Reprod Update* 2007;**13**:421—31.

128. Laron Z. Age at first ejaculation (spermarche) — the overlooked milestone in male development. *Pediatr Endocrinol Rev* 2010;**7**:256—7.

129. Ji CY. Age at spermarche and comparison of growth and performance of pre- and post-spermarcheal Chinese boys. *Am J Hum Biol* 2001;**13**:35—43.

130. Gray SH, Ebe LK, Feldman HA, Emans SJ, Osganian SK, Gordon CM, Laufer MR. Salivary progesterone levels before menarche: a prospective study of adolescent girls. *J Clin Endocrinol Metab* 2010;**95**:3507—11.

131. Apter D, Raisanen I, Ylostalo P, Vihko R. Follicular growth in relation to serum hormonal patterns in adolescent compared with adult menstrual cycles. *Fertil Steril* 1987;**47**:82—8.

132. Apter D, Viinikka L, Vihko R. Hormonal pattern of adolescent menstrual cycles. *J Clin Endocrinol Metab* 1978;**47**:944—54.

133. Matsumoto AM, Bremner WJ. Endocrinology of the hypothalamic—pituitary—testicular axis with particular reference to the hormonal control of spermatogenesis. *Baillieres Clin Endocrinol Metab* 1987; **1**:71—87.

134. Simm PJ, Bajpai A, Russo VC, Werther GA. Estrogens and growth. *Pediatr Endocrinol Rev* 2008; **6**:32—41.

135. Juul A. The effects of oestrogens on linear bone growth. *Hum Reprod Update* 2001;**7**:303—13.

136. Smith EP, Boyd J, Frank GR, Takahashi H, Cohen RM, Specker B, et al. Estrogen resistance caused by a mutation in the estrogen-receptor gene in a man. *N Engl J Metab* 1994;**331**:1056—61.

137. Morishima A, Grumbach MM, Simpson ER, Fisher C, Qin K. Aromatase deficiency in male and female siblings caused by a novel mutation and the physiological role of estrogens. *J Clin Endocrinol Metab* 1995;**80**:3689—98.

138. Carani C, Qin K, Simoni M, Faustini-Fustini M, Serpente S, Boyd J, et al. Effect of testosterone and estradiol in a man with aromatase deficiency. *N Engl J Metab* 1997;**337**:91—5.

139. Zachmann M, Prader A, Sobel EH, Crigler Jr JF, Ritzén EM, Atarés M, et al. Pubertal growth in patients with androgen insensitivity: indirect evidence for the importance of estrogens in pubertal growth of girls. *J Pediatr* 1986;**108**:694—7.

140. Papadimitriou DT, Linglart A, Morel Y, Chaussain JL. Puberty in subjects with complete androgen insensitivity syndrome. *Horm Res* 2006;**65**:126—31.

141. Wheeler MD. Physical changes of puberty. *Endocrinol Metab Clin North Am* 1991;**20**:1—14.

142. Randall VA. Androgens and hair growth. *Dermatol Ther* 2008;**21**:314—28.

143. Nguyen AD, Conley AJ. Adrenal androgens in humans and nonhuman primates: production, zonation and regulation. *Endocr Dev* 2008;**13**:33—54.

144. DiMartino-Nardi J. Pre- and postpubertal findings in premature adrenarche. *J Pediatr Endocrinol Metab* 2000;**13**(Suppl. 5):1265—9.

145. Rosenfield RL. Normal and almost normal precocious variations in pubertal development premature pubarche and premature thelarche revisited. *Horm Res* 1994;**41**(Suppl. 2):7—13.

146. Moerman ML. Growth of the birth canal in adolescent girls. *Am J Obstet Gynecol* 1982;**143**:528—32.

147. Tanner JM, Whitehouse RH. Clinical longitudinal standards for height, weight, height velocity, weight velocity, and stages of puberty. *Arch Dis Child* 1976;**51**:170—9.

148. Siervogel RM, Demerath EW, Schubert C, Remsberg KE, Chumlea WC, Sun S, et al. Puberty and body composition. *Horm Res* 2003;**60**:36—45.

149. Wells JC. Sexual dimorphism of body composition. *Best Pract Res Clin Endocrinol Metab* 2007; **21**:415−30.

150. Pallottini V, Bulzomi P, Galluzzo P, Martini C, Marino M. Estrogen regulation of adipose tissue functions: involvement of estrogen receptor isoforms. *Infect Disord Drug Targets* 2008;**8**:52−60.

151. Bhasin S. Regulation of body composition by androgens. *J Endocrinol Invest* 2003;**26**:814−22.

152. Tipton KD. Gender differences in protein metabolism. *Curr Opin Clin Nutr Metab Care* 2001; **4**:493−8.

153. Demerath EW, Li J, Sun SS, Chumlea WC, Remsberg KE, Czerwinski SA, et al. Fifty-year trends in serial body mass index during adolescence in girls: the Fels Longitudinal Study. *Am J Clin Nutr* 2004; **80**:441−6.

154. Simpson ER. Sources of estrogen and their importance. *J Steroid Biochem Mol Biol* 2003;**86**:225−30.

155. Simpson ER. Aromatization of androgens in women: current concepts and findings. *Fertil Steril* 2002; **77**(Suppl. 4):S6−10.

156. MacDonald PC, Edman CD, Hemsell DL, Porter JC, Siiteri PK. Effect of obesity on conversion of plasma androstenedione to estrone in postmenopausal women with and without endometrial cancer. *Am J Obstet Gynecol* 1978;**130**:448−55.

157. Jasik CB, Lustig RH. Adolescent obesity and puberty: the "perfect storm". *Ann NY Acad Sci* 2008; **1135**:265−79.

158. Walvoord EC. The timing of puberty: is it changing? Does it matter? *J Adolesc Health* 2010;**47**:433−9.

SUGGESTED READING

Castellano JM, Bentsen AH, Mikkelsen JD, Tena-Sempere M. Kisspeptins: bridging energy homeostasis and reproduction. *Brain Res* 2010;**1364**:129−38.

Ebling FJP. The neuroendocrine timing of puberty. *Reproduction* 2005;**129**:675−83.

Ellison PT. *On fertile ground.* Cambridge, MA: Harvard University Press; 2001.

Plant TM, Barker-Gibb ML. Neurobiological mechanisms of puberty in higher primates. *Hum Reprod Update* 2004;**10**:67−77.

INTERNET RESOURCES

Journal of Clinical Endocrinology and Metabolism. Hot Topic: Endocrinology resources on the Internet, http://jcem.endojournals.org/content/86/7/2942.full

A journal article, freely available, on how to access Internet resources related to endocrinology

You and your hormones: http://www.yourhormones.info/

A layperson's guide, produced by the Society for Endocrinology.

The major professional society websites all have links to relevant resources and post recent news items and information:

Society for Endcrinology: http://www.endocrinology.org/index.aspx

Society for Behavioral Neuroendocrinology: http://www.yourhormones.info/

American Neuroendocrine Society: http://www.neuroendocrine.org/

Endocrine Society: http://www.endo-society.org/

Some children's hospitals also provide resources or links on their websites, although these are more often relevant to topics such as diabetes than to puberty per se. One example is:

University of Maryland Children's Hospital. Pediatric Endocrinology Resources, http://www.umm.edu/pediatrics/pediatric_endocrinology_resources.htm

CHAPTER 5

Endocrine Control of Growth

Ron G. Rosenfeld
Oregon Health & Science University, Portland, OR 97239, USA

Contents

5.1 INTRODUCTION

Mammalian growth is an extraordinarily complex biological phenomenon, requiring the coordinated production and integration of multiple hormones and growth factors. Under normal conditions, growth patterns for the size of the skeleton are remarkably consistent from individual to individual within each species. Human stature growth is no exception, underscoring the fundamental biological role of skeletal growth in an individual animal's life. Although growth is characteristic of all mammals, there are a number of distinct features that are characteristic and unique to human growth, including: (1) rapid growth in late gestation; (2) growth deceleration immediately following birth; (3) a prolonged childhood and a mid-childhood growth spurt; (4) an adolescent growth spurt; (5) relatively late attainment of adult height; and (6) minimal sexual dimorphism of adult stature. Each of these stages of growth requires the coordinated integration of specific hormones, growth factors and their receptors. Any secular changes in the height of humans over the evolutionary history of *Homo sapiens* probably reflect nutritional and environmental factors, rather than major genomic changes.

While it is clear that multiple hormones affect growth, both in utero and postnatally, the growth hormone (GH)—insulin-like growth factor (IGF) axis plays a central role throughout fetal and childhood—adolescent growth. GH, after being secreted by the

N. Cameron & B. Bogin (eds): Human Growth and Development, Second edition.
ISBN 978-0-12-383882-7, Doi: 10.1016/B978-0-12-383882-7.00005-2

pituitary, binds to a transmembrane receptor and activates a postreceptor signaling cascade, ultimately leading to phosphorylation of signal transducer and activator of transcription-5b (STAT5b). STAT5b transcriptionally regulates the genes for IGF-1 and for key IGF binding proteins, which transport IGF proteins in serum and other biological fluids, ultimately delivering the IGFs to their target tissues. IGF-1, in turn, binds to the type 1 IGF receptor, resulting in chondrocyte proliferation and statural growth.

Growth failure resulting from IGF-deficient states may be divided into secondary forms, reflecting defects in GH production, and primary forms, characterized by low IGF levels despite normal or even elevated GH production. Molecular defects of the GH—IGF axis have been identified in humans, with phenotypes that correspond to the specific genetic lesions. Therapy with GH or IGF-1 can now be matched to specific defects in the GH—IGF axis.

5.2 WHY DO ANIMALS GROW?

Mammalian growth is an extraordinarily complex biological process. The difference in size between mycoplasma (10^{-13} g) and the blue whale (10^8 g $= 100,000$ kg) is approximately 20 orders of magnitude, and yet both species presumably evolved from similar, if not identical, simple unicellular ancestors. One must infer from this observation that a wide variety of selective pressures directed the growth of each organism over evolutionary time, and that multiple genes and gene products combined to organize and direct the panoply of growth patterns that characterizes life on earth.

Perhaps the most fundamental question is: Why do organisms grow? Why invest so much genetic and ultimately metabolic energy in a growth process, when there may be more urgent needs of an organism? Much of this is dictated, of course, by the simple fact that animals are born small and, generally, helpless. Growth is required to achieve sizes necessary to fulfill the animal's adult functions of gathering, predation, defense, socializing and reproducing. Different societal structures dictate different growth patterns: in gorillas, where the dominant adult male controls a harem, the male gorilla is often twice the size of his female partners; in humans, the adult height differential between adult males and females is only 7%, reflecting the generally monogamous nature of human society.[1] A more detailed view of the human growth pattern over the lifespan demonstrates some additional interesting and unique features that discriminate human growth from that of other mammals, including primates[2] (see Chapter 11):

- Maximal growth occurs during gestation.
- The normal time of gestation for humans is brief relative to maternal body size and adult brain size.
- Growth decelerates immediately following birth.
- Humans undergo relatively late sexual maturation.
- Puberty commences at a time when the growth rate is at the slowest since birth.

- There is a distinct adolescent growth spurt in height.
- A delay occurs between puberty and full reproductive capacity.

Anthropologists and auxologists have debated the selective forces that have shaped human growth patterns. It has been proposed that a relatively short gestation is required to allow accommodation of a relatively large brain/head through the birth canal. A prolonged childhood with a relatively modest growth rate promotes an infantile appearance, which would foster the transmission of knowledge and culture from one generation to the next, as well as prevent juvenile males from presenting a competitive sexual threat to adult males. A rapid and pronounced pubertal growth spurt would then allow adolescents to rapidly attain ideal adult heights at a delayed age. The relative lack of sexually dimorphic growth in human males and females (the 7% difference in adult height is largely explained by different pubertal growth patterns which result in earlier epiphyseal fusion in females) allows for an adult female size that can accommodate the carrying and delivery of a human fetus and newborn; adequate pelvic size is necessary not only to support the size of the fetus, but also to permit delivery of the relatively large fetal cranium characteristic of *H. sapiens*.

This kind of extraordinarily complex pattern of growth requires regulation at multiple, carefully integrated levels. Only a robust system of interacting hormones and growth factors could possibly allow the multiple stages of growth characteristic of complex mammals, as well as the intricate coordination with metabolic and reproductive needs. To attempt to explain growth of mammals in terms of one or two hormones is ludicrously reductive and fails to do justice to the remarkable interspecies differences, as well as the complexity of growth within individual species. Nevertheless, despite this remarkable complexity, it remains clear that the GH—IGF axis plays a predominant role in both intrauterine and postnatal growth (Figure 5.1).[3]

5.3 WHAT CAUSES ANIMALS TO GROW?

Attainment of the complex growth pattern described above requires the interactions of multiple hormones and growth factors, which must be generated in a timely fashion relative to the life cycle and which, in general, interact with one another, rather than working in isolation. Despite this complexity, over the past 50 years, it has become increasingly apparent that the IGFs play a central role in both intrauterine and postnatal growth. The original observation by Salmon and Daughaday[4] in 1957 showed that: (1) the addition of normal serum to rat cartilage stimulated the incorporation of radioactive inorganic sulfate into acid mucopolysaccharides; (2) serum from hypophysectomized rats could not duplicate this effect, unless the rats had first been treated in vivo with GH; and (3) addition of GH to the cartilage medium could not stimulate sulfate incorporation. In subsequent studies, this GH-stimulated factor(s) was also found to enhance the incorporation of leucine into protein—polysaccharide complexes, uridine into RNA and

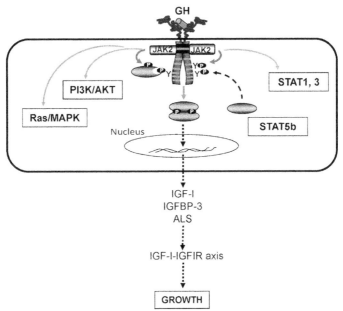

Figure 5.1 Growth hormone (GH) regulation of the insulin-like growth factor (IGF) axis. GH binds to its homodimeric receptor, resulting in association of janus kinase 2 (JAK2) and phosphorylation of key tyrosines on the GH receptor intracellular domain. These tyrosines do not appear necessary for activation of PI3/AKT, MAP kinase (MAPK) or signal transducer and activator of transcription (STAT) 1 and 3, but are obligatory for docking and phosphorylation of STAT5a and STAT5b. After phosphorylation, STAT5b dissociates from the GH receptor, dimerizes, enters into the nucleus, and transcriptionally activates the three critical GH-dependent genes, *IGF1*, *IGFBP3* and *ALS*. This figure is reproduced in the color plate section.

thymidine into DNA.[5] The authors concluded that GH, while not able directly to mediate cellular growth, must stimulate the production of a "sulfation factor", which then enhanced chondrocyte growth and metabolism.

By 1972, a group of proteins that had been initially designated "sulfation factor", "multiplication stimulating activity" and "non-suppressible insulin-like activity" were found to be related and were renamed "somatomedins", signifying their role(s) in mediating growth.[6] With the elucidation of the amino acid sequences of two of these proteins and the discovery of their close structural relationship to insulin, the term somatomedin was replaced by "insulin-like growth factors"-1 and -2 (IGF-1, IGF-2), which is how they are now known.[7]

5.4 STRUCTURE OF INSULIN-LIKE GROWTH FACTOR

IGF-1 is a basic peptide of 70 amino acids, while IGF-2 is a slightly acidic peptide of 67 amino acids.[7] The two peptides share 45 of 73 possible amino acid positions, and have approximately 50% amino acid homology to insulin. Like insulin, both IGFs have A and B chains connected by disulfide bonds. The connecting C-peptide region is 12 amino acids long for IGF-1 and eight amino acids for IGF-2; neither IGF C-peptide bears any homology with the C-peptide region of proinsulin. IGF-1 and -2 also differ from proinsulin in possessing carboxy-terminal extensions, or D-peptides, of eight and six amino acids, respectively. This structural similarity provides a basis for the ability of both IGFs to bind to the insulin receptor and of insulin to bind to the type 1 IGF receptor, thereby explaining the "insulin-like" activity of the IGFs, as well as the growth-promoting ability of insulin. Structural differences between insulin and the IGFs probably explain the failure of insulin to bind with high affinity to the IGF binding proteins.

GH appears to be the primary regulator of IGF-1 gene transcription, at least postnatally. Transcription begins as early as 30 minutes after intraperitoneal injection of GH into hypophysectomized rats. It is critical to note, however, that IGF production in utero is essentially GH independent, and that GH regulation of IGF-1 synthesis does not appear to become a factor until very late gestation, at the earliest. Thus, children with congenital GH deficiency or GH insensitivity are essentially normal size at birth. The factors that regulate IGF-1 synthesis in the fetus remain to be elucidated, but it is likely that placental viability, fetal nutrition and insulin production all play roles.

Investigations over the past decade have demonstrated that STAT5b is the most critical mediator of GH-induced activation of IGF-1 gene transcription, an observation underscored by studies involving target disruption of the STAT5b gene in mouse models[8] and by reports of patients with severe GH insensitivity associated with homozygosity for mutations of the STAT5b gene.[11,12] Two adjacent STAT5 binding sites have been identified in the second intron of the rat IGF-1 gene, within a region previously identified as undergoing acute changes in chromatin structure after GH treatment.[13]

On the other hand, the factors involved in the regulation of IGF-2 gene expression are less clear and, indeed, the role of IGF-2 in growth is still uncertain, especially postnatally. As discussed below, knockout studies have confirmed the importance of IGF-2 in fetal growth, but its role in postnatal life is far less clear. In humans and rats, IGF-2 gene expression is high in fetal life, having been detected as early as the blastocyst stage in mice. In general, fetal tissues have high IGF-2 mRNA levels that decline postnatally, although brain IGF-2 mRNA remains high in the adult rat.

5.5 TARGETED DISRUPTION OF THE INSULIN-LIKE GROWTH FACTOR GENES

Studies involving null mutations for the two IGF peptides and the IGF receptors have greatly enhanced our understanding of the role of the IGF axis in fetal and postnatal growth.[14] Previous investigations had shown that knockouts of either GH or the growth hormone receptor (GHR) genes resulted in little change in birth size, confirming the minimal role of GH in fetal growth. In contrast, mice with knockouts of the gene for either IGF-1 or IGF-2 were found to have birth weights approximately 60% of normal. Mouse mutants lacking both IGF-1 and the GHR are only 17% of normal size. These observations and others indicated that: (1) both IGF-1 and IGF-2 are important embryonic and fetal growth factors; (2) IGF-1 plays a critical role in postnatal growth; and (3) GH does have some modest, apparently IGF-independent role as well. Growth delay began on embryonic day e11 for IGF-2 knockouts and on day e13.5 for IGF-1 knockouts. Those mice with IGF-1 gene disruptions who survived the immediate neonatal period continued to have growth failure postnatally, with weights 30% of normal by 2 months of age. Indeed, postnatal growth was poorer than that observed in mice with GHR mutations, growth hormone-releasing hormone (GHRH) receptor mutations or pit-1 mutations, indicating that both GH-dependent and GH-independent factors are necessary for normal growth. When the genes for both IGF-1 and IGF-2 were disrupted, weight at birth was only 30% of normal, and all animals died shortly after birth, apparently from respiratory insufficiency secondary to muscular hypoplasia.

The results of these experimental observations in mice have been paralleled by human mutational analysis. It had, for example, long been known that children with GH gene deletions or with mutations or deletions of the GHR gene were near normal size at birth, but had severe postnatal growth retardation.[15,16] When the first case of a human IGF-1 gene deletion was reported, the patient was found to exhibit a prenatal and postnatal growth pattern similar to that observed in the mouse knockouts,[17] and this was confirmed in a more recent report of a bioactive IGF-1 molecule, resulting from a missense mutation.[18]

Challenges to the fundamental model of the IGF system and, in particular, the roles of liver-produced, circulating IGF-1, emanated from studies employing specific ablation of hepatic IGF-1 production.[19] These investigations confirmed that the liver is the principal source of circulating IGF-1, but also demonstrated that an 80% lowering of serum IGF-1 levels had no apparent effect on postnatal growth, thereby suggesting that postnatal growth was relatively independent of hepatic IGF-1 production. The authors suggested that either local (paracrine) chondrocyte production of IGF-1, or IGF-1 produced by other tissues (such as adipocytes), was sufficient to maintain adequate production of IGF-1 to account for growth preservation; or, alternatively, that free IGF-1 levels remained within the normal range as a result of the reciprocal increase in GH production, as well as the lowering of serum insulin-like growth factor binding protein (IGFBPs). Additional support for the predominant role in growth of locally produced IGF-1 is provided by the modest decrement of postnatal growth seen in acid-labile subunit (ALS, part of the IGF binding protein system) null mice, who have profound lowering of serum IGF-1 concentrations.[20]

In subsequent studies involving the crossing of liver-derived IGF-1 gene-deleted mice (LID) with ALS gene-deleted mice (ALSKO), an 85–90% reduction in serum IGF-1 was achieved and, in this case, early postnatal growth retardation was observed.[21] These findings suggest that postnatal growth is dependent on both endocrine (i.e. hepatic) and tissue IGF-1, although definite conclusions are problematic in the face of the elevated GH production and perturbations of the IGFBP system observed in these studies. What seems most likely, when the totality of these studies is evaluated, is that both endocrine and autocrine/paracrine IGF play a role in growth.[22,23]

Knockout of the gene for the type 1 IGF receptor, which appears to mediate the growth-promoting actions of both IGF-1 and IGF-2, resulted in birth weights 45% of normal and 100% neonatal lethality. Abuzzahab et al.[24] reported two patients with intrauterine growth retardation and postnatal growth failure, despite elevated serum IGF-1 concentrations. One patient was a compound heterozygote for point mutations in exon 2 of the IGF-1R gene, leading to decreased receptor affinity for IGF-1, while the second had a nonsense mutation of one allele, resulting in reduced numbers of IGF-1 receptors. More recently, fetal and postnatal growth retardation has been observed in a number of patients heterozygous for one mutation of the IGF-1R gene. In mice, concurrent knockout of genes for IGF-1 and the type 1 IGF receptor resulted in no further reduction in birth size (45% of normal), consistent with the concept that all IGF-1 actions in fetal life are mediated through this receptor. On the other hand, simultaneous knockout of the genes for IGF-2 and the type 1 IGF receptor resulted in further reduction of birth size to 30% of normal (as with simultaneous knockouts of IGF-1 and IGF-2); this raises the possibility that some of the fetal anabolic actions of IGF-2 are mediated by a secondary mechanism (perhaps placental growth or IGF-2 interactions with the insulin receptor). Whatever the pathway may prove to be, it does not appear to involve the type 2 IGF receptor, since knockout of this paternally imprinted gene results in an increased birth weight, but death in late gestation or at birth. Since this receptor normally degrades IGF-2, increased growth presumably reflects excess IGF-2 acting through the IGF-1 receptor.

Several conclusions can be drawn from these studies: (1) IGF-1 plays a critical role in both fetal and postnatal growth; (2) IGF-2 is a major fetal growth factor, but has little, if any, role in postnatal growth; (3) the type 1 IGF receptor mediates anabolic actions of both IGF-1 and IGF-2; (4) the type 2 IGF receptor is bifunctional, serving both to target lysosomal enzymes and to enhance IGF-2 turnover; (5) IGF-1 production is involved in normal fertility; (6) placental growth is only impaired with IGF-2 knockouts; (7) GH and the GHR play little role in prenatal growth; and (8) IGF-1 is the major mediator of GH's effects on postnatal growth, although GH and the GHR may have a small IGF-independent effect. Whether these studies in mice are fully applicable to humans is yet unknown, although much has been learned in recent years from rare cases of human mutations of critical genes of the GH–IGF axis.

5.6 GROWTH HORMONE RECEPTOR

The coding and 3′-untranslated regions of the human GHR are encoded by nine exons, numbered 2–10.[25] Exons 3–7 encode the extracellular, GH-binding domain. Examination of the crystal structure of the GH—GHR complex revealed that the complex consists of one molecule of GH bound to two GH-R molecules, initially suggesting that GH-induced receptor dimerization was a necessary step in its action. Recent studies, however, have indicated that the receptor may be constitutively dimerized, and that receptor activation involves a GH-induced conformational change.[26]

It is now clear that the GHR must recruit a cytoplasmic tyrosine kinase, as the receptor lacks intrinsic kinase activity. Janus kinase 2 (JAK2) has been identified as the critical GHR-associated tyrosine kinase; loss of ability of the GHR to bind JAK2 results in loss of GH-induced GHR signaling. Recruitment and/or activation of JAK2 molecules by the GHR promotes their enzymic activity via cross-phosphorylation, and the active kinases then phosphorylate tyrosines on the intracellular portion of the GHR, thereby providing docking sites for critical intermediary proteins, such as the signal transducers and activators of transcription (STATs) (Figure 5.2). There

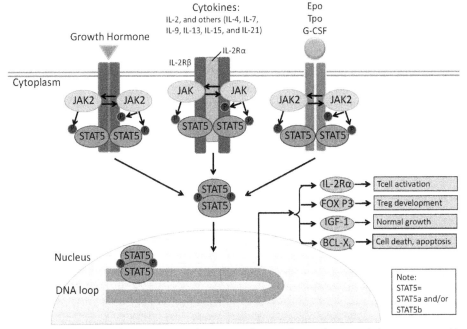

Figure 5.2 A schematic of JAK/STAT5 pathway signaling is shown for growth hormone, cytokines, such as interleukin (IL)-2, IL-4, IL-7, IL-9, IL-13, IL-15 and IL-21, and erythropoietin (Epo), thrombopoieten (Tpo), granulocyte colony-stimulating factor (G-CSF) and their cognate receptors on the cell surface. STAT5 represents both STAT5a and STAT5b. This figure is reproduced in the color plate section. *(Source: Nadeau et al.[25])*

are seven known mammalian STATs; of these, STAT5b appears to be most centrally involved in mediating the growth-promoting actions of the GHR, as indicated by several gene disruption studies in rodent models.[8] Recent investigations have indicated that redundancy exists among the seven intracellular tyrosines, but there are three tyrosines that can each, upon phosphorylation, activate STAT5b.[9,10] The reports of 10 cases of homozygous human STAT5b mutations, all presenting with severe growth failure and GH resistance, have further substantiated the critical intermediary role of STAT5b in GH regulation of IGF-1 gene transcription and growth.[11,12] The STAT proteins dock, via their src-homology-2 (SH2) domain, to phosphotyrosines on ligand-activated receptors, such as the GHR, and are subsequently phosphorylated on single tyrosines at the C-terminus of the protein, dimerize, translocate to the nucleus, bind to DNA through their DNA-binding domain, and in turn regulate gene transcription.

5.7 INSULIN-LIKE GROWTH FACTOR DEFICIENCY

Given the central role of the IGF system in both intrauterine and postnatal growth, assessment of patients with otherwise unexplained growth failure requires an evaluation of the IGF axis.[28] By analogy with other endocrine systems, IGF deficiency has been divided into secondary etiologies (i.e. IGF deficiency resulting from disorders of GH production, at either the hypothalamic or pituitary level) and primary forms (i.e. IGF deficiency despite normal GH production). Primary IGF deficiency was first identified in the 1960s, in patients who ultimately proved to have GH insensitivity resulting from mutations or deletions of the GHR gene.[16] Over the past decade, however, multiple other molecular etiologies (Box 5.1) have been identified, including defects in the post-GHR signaling cascade (STAT5b),[11,12] mutations and deletions of the IGF-1 gene,[17,18] mutations in genes encoding IGF binding proteins[29,30] and mutations in genes for the IGF-1 receptor (the latter actually represents a form of IGF resistance).[24] Molecular analysis of such patients has proven to be invaluable, as specific genotypes predict specific phenotypes. For example, patients with primary IGF deficiency resulting from defects of the GHR or STAT5b genes have normal birth size, but severe postnatal growth failure, while patients with IGF-1 gene defects also have intrauterine growth retardation, microcephaly, developmental delay and variable hearing deficits. These features reflect the relative roles of GH and IGF-1 in prenatal and postnatal growth.

5.8 THERAPEUTIC IMPLICATIONS

For decades, therapy for growth disorders was dominated by GH, first a human cadaver-derived form and then, beginning in the mid-1980s, a recombinant DNA-derived form.

Box 5.1 Molecular defects resulting in primary insulin-like growth factor deficiency

- GHR abnormalities
- Mutations/deletions of *GHR* affecting the extracellular domain of the GHR and resulting in decreased GH binding
- Mutations/deletions of *GHR* affecting the ability of the GHR to dimerize
- Mutations/deletions of *GHR* affecting the transmembrane domain of the receptor and resulting in defective anchoring in the cell membrane
- Mutations/deletions of *GHR* affecting the intracellular domain and signaling
- Post-*GHR* signaling defects
- Mutations of *STAT5b* resulting in defective or absent GH signal transduction
- Mutations/deletions of *IGF-1*
- Deletions of *IGF-1*
- Mutations of *IGF 1* resulting in bioinactive IGF-1
- Defects of IGF-1 transport and/or clearance
- Mutations/deletions of *ALSIGF*, resulting in defective IGF-1 transport and rapid IGF-1 clearance
- IGF-1 resistance
- Mutations of *IGF1R*, resulting in decreased sensitivity to IGF-1

GHR: growth hormone receptor; GH: growth hormone; STAT5b: signal transducer and activator of transcription 5b; IGF-1: insulin-like growth factor-1.

GH remains the treatment of choice for patients with secondary IGF deficiency, as daily dosing allows replacement of deficient pituitary production of GH and adequately stimulates IGF-1 synthesis. The market for GH has greatly expanded, however, to include a wide variety of disorders characterized by short stature but normal GH production, including such conditions as Turner's syndrome, chronic renal failure, Noonan's syndrome, small-for-gestational-age infants with failed catch-up growth, and a heterogeneous group of conditions lumped under the heading of "idiopathic short stature". In general, children with these conditions appear to respond to pharmacological dosages of GH, although growth acceleration generally is not as good as in replacement therapy of GH deficiency. Nevertheless, many such patients experience "catch-up" growth, with eventual improvement of their adult height.

IGF-1 therapy for patients with primary IGF deficiency was first tested in the 1990s, primarily in patients with GHR defects. Growth acceleration was observed and sustained in most of these patients, although, in general, the growth response was not quite as good as observed in patients receiving GH replacement therapy for GH deficiency. The total explanation for this discrepancy remains unclear, but probably involves the failure of systemic IGF-1 administration to fully replace local IGF production, especially at the epiphyseal growth plates. Nevertheless, IGF-1 remains the treatment of choice for patients with defects at the level of the GHR, the JAK-STAT system and the IGF-1 gene.[31]

Both the US Food and Drug Administration and the European authorities have now approved IGF-1 therapy for treatment of children with short stature and "severe primary IGF deficiency", defined as a height below -3 SD, normal GH and a serum IGF-1 below -3 SD (United States) or below -2.5 percentile (Europe). Clinical trials will be necessary to determine whether optimal treatment for such patients and, especially, for less severe forms of IGF deficiency, should be with GH, IGF-1 or, possibly, combination GH plus IGF-1.

5.9 SUMMARY

The combination of animal knockout studies and human mutational analysis has confirmed the primary role of the IGF system in both prenatal and postnatal growth. IGF deficiency may be secondary to defects in GH production, or may be primary, when it exists in the presence of normal or elevated GH production. Established etiologies of primary IGFD include molecular defects of the GHR (extracellular, transmembrane or intracellular), defects of the postreceptor signaling cascade, specifically STAT5b, defects of the IGF-1 gene, and defects of the ALS. To this can be added defects of the IGF-1 receptor, where relative IGF resistance manifests itself as growth failure, despite normal-increased serum IGF-1 concentrations.

REFERENCES

1. Rosenfeld RG. Gender differences in height: an evolutionary perspective. *J Pediatr Endocrinol Metab* 2004;**17**(Suppl. 4):1267–71.
2. Bogin B. *The growth of humanity*. New York: Wiley-Liss; 2001.
3. Rosenfeld RG, Cohen P. Disorders of growth hormone/insulin-like growth factor secretion and action. In: Sperling MA, editor. *Pediatric endocrinology*. 3rd ed. Philadelphia, PA: Saunders; 2008. p. 254–334.
4. Salmon Jr WD, Daughaday WH. A hormonally controlled serum factor which stimulates sulfate incorporation by cartilage in vitro. *J Lab Clin Med* 1957;**49**:825–36.
5. Salmon Jr WD, DuVall MR. A serum fraction with sulfation factor activity stimulates in vitro incorporation of leucine and sulfate into protein–polysaccharide complexes, uridine into RNA, and thymidine into DNA of costal cartilage from hypophysectomized rats. *Endocrinology* 1970;**86**:721–7.
6. Daughaday WH, Hall K, Raben MS, Salmon JD, Van Den Brande L, Van Wyk JJ. Somatomedin: proposed designation for sulphation factor. *Nature* 1972;**235**:107.
7. Rinderknecht W, Humbel RE. Amino-terminal sequences of two polypeptides from human serum with nonsuppressible insulin-like and cell-growth-promoting activities: evidence for structural homology with insulin B chain. *Proc Natl Acad Sci USA* 1976;**73**:4379–81.
8. Udy GB, Towers RP, Snell RG, Wilkins RJ, Park S-H, Ram PA, et al. Requirement of STAT5b for sexual dimorphism of body growth rates and liver gene expression. *Proc Natl Acad Sci USA* 1997;**94**:7239–44.
9. Derr MA, Fang P, Sinha SK, Ten S, Hwa V, Rosenfeld RG. A novel Y332C missense mutation in the intracellular domain of the human growth hormone receptor (GHR) does not alter STAT5b

signaling: redundancy of GHR intracellular tyrosines involved in STAT5b signaling. *Horm Res Pediatr* 2011;**75**:187−99.

10. Derr MA, Aisenberg J, Fang P, Tenenbaum-Rahover Y, Rosenfeld RG, Hwa V. The growth hormone receptor (GHR) c899dupC mutation functions as a dominant negative: insights into the pathophysiology of intracellular GHR defects. *J Clin Endocrinol Metab* 2011;**96**:E1896−904.

11. Kofoed EM, Hwa V, Little B, Woods KA, Buckway CK, Tsubaki J, et al. Growth hormone insensitivity associated with a STAT5b mutation. *N Engl J Med* 2003;**349**:1139−47.

12. Rosenfeld RG, Belgorosky A, Camacho-Hubner C, Savage MO, Wit JM, Hwa V. Defects in growth hormone receptor signaling. *Trends Endocrinol Metab* 2007;**18**:134−41.

13. Woelfle J, Chia DJ, Rotwein P. Mechanisms of growth hormone (GH) action. Identification of conserved STAT5 binding sites that mediate GH-induced insulin-like growth factor-1 gene activation. *J Biol Chem* 2003;**278**:51261−6.

14. Lupu F, Terwilliger JD, Kaechoong L, Segre GV, Efstratiadis A. Roles of growth hormone and insulin-like growth factor 1 in mouse postnatal growth. *Dev Biol* 2001;**229**:141−62.

15. Vnencak-Jones CL, Phillips III JA, Chen EY, Seeburg PH. Molecular basis of human growth hormone gene deletions. *Proc Natl Acad Sci USA* 1988;**85**:5615−9.

16. Rosenfeld RG, Rosenbloom AL, Guevara-Aguirre J. Growth hormone (GH) insensitivity due to primary GH receptor deficiency. *Endocr Rev* 1994;**15**:369−90.

17. Woods KA, Camacho-Hubner C, Savage MO, Clark AJ. Intrauterine growth retardation and postnatal growth failure associated with deletion of the insulin-like growth factor I gene. *N Engl J Med* 1996;**335**:1363−7.

18. Walenkamp MJ, Karperien M, Pereira AM, Hilhorst-Hofstee Y, van Doorn J, Chen JW, et al. Homozygous and heterozygous expression of a novel insulin-like growth factor-1 mutation. *J Clin Endocrinol Metab* 2005;**90**:2855−64.

19. Yakar S, Liu JL, Stannard B, Butler A, Accili D, Sauer B, LeRoith D. Normal growth and development in the absence of hepatic insulin-like growth factor I. *Proc Natl Acad Sci USA* 1999;**96**:7324−9.

20. Ueki I, Ooi GT, Tremblay ML, Hurst KR, Bach LA, Boisclair YR. Inactivation of the acid labile subunit gene in mice results in mild retardation of postnatal growth despite profound disruptions in the circulating insulin-like growth factor system. *Proc Natl Acad Sci USA* 2000;**97**:6868−73.

21. Yakar S, Rosen CJ, Beamer WG, Ackert-Bicknell CL, Wu Y, Liu JL, et al. Circulating levels of IGF-1 directly regulate bone growth and density. *J Clin Invest* 2002;**110**:771−81.

22. LeRoith D, Bondy C, Yakar S, Liu JL, Butler A. The somatomedin hypothesis. *Endocrinol Rev* 2001;**22**:53−74.

23. Kaplan SA, Cohen P. The somatomedin hypothesis 2007: 50 years later. *J Clin Endocrinol Metab* 2007;**92**:4529−35.

24. Abuzzahab MJ, Schneider A, Goddard A, Grigorescu F, Lautier C, Keller E, et al. IGF-1 receptor mutations resulting in intrauterine and postnatal growth failure. *N Engl J Med* 2003;**349**: 2211−22.

25. Leung DW, Spencer SA, Cachianes G, Hammonds RG, Collins C, Henzel WJ, et al. Growth hormone receptor and serum binding protein: purification, cloning and expression. *Nature* 1987; **330**:537−43.

26. Lichanska AM, Waters MJ. New insights into growth hormone receptor function and clinical implications. *Horm Res* 2008;**69**:138−45.

27. Nadeau K, Hwa V, Rosenfeld RG. STAT5b deficiency: an unsuspected cause of growth failure, immunodeficiency and severe pulmonary disease. *J Pediatr* 2011;**158**:701−8.

28. Rosenfeld RG. Molecular mechanisms of IGF-1 deficiency. *Horm Res* 2006;**65**(Suppl. 1):15−20.

29. Domene HH, Bengolea SV, Martinez AS, Ropelato MG, Pennisi P, Scaglia P, et al. Deficiency of the circulating insulin-like growth factor system associated with inactivation of the acid-labile subunit gene. *N Engl J Med* 2004;**350**:570−7.

30. Hwa V, Haeusler G, Pratt KL, Little BM, Frisch H, Koller D, Rosenfeld RG. Total absence of functional acid labile subunit, resulting in severe insulin-like growth factor deficiency and moderate growth failure. *J Clin Endocrinol Metab* 2006;**91**:1826−31.

31. Rosenfeld RG. IGF-1 therapy in growth disorders. *Eur J Endocrinol* 2007;**157**(Suppl. 1):557−60.

SUGGESTED READING

Lupu F, Terwilliger JD, Kaechoong L, Segre GV, Efstratiadis A. Roles of growth hormone and insulin-like growth factor 1 in mouse postnatal growth. *Dev Biol* 2001;**229**:141–62.

Rosenfeld RG, Belgorosky A, Camacho-Hubner C, Savage MO, Wit JM, Hwa V. Defects in growth hormone receptor signaling. *Trends Endocrinol Metab* 2007;**18**:134–41.

Rosenfeld RG, Cohen P. Disorders of growth hormone/insulin-like growth factor secretion and action. In: Sperling MA, editor. *Pediatric endocrinology*. 3rd ed. Philadelphia, PA: Saunders; 2008. p. 254–334.

INTERNET RESOURCES

http://growthgeneticsconsortium.org
   A publicly accessible website which has been established recently for cataloging molecular defects of the GH—IGF.

CHAPTER 6

Nutrition and Growth

Nicholas G. Norgan★, Barry Bogin★★, Noël Cameron★★
★Department of Human Sciences and ★★School of Sport, Exercise & Health Sciences, Loughborough University, Leicestershire LE11 3TU, UK

Contents

The chapter "Nutrition and Growth" in the first edition of this book was written by Nicholas G. Norgan, who died in 2006. This chapter before you is based on Nick's original in the first edition and we have retained his wording as much as possible, while updating purely factual information, websites and the like. We offer this modified chapter as a tribute to Nick's research and his memory as a humane and gracious colleague.

N. Cameron & B. Bogin (eds): Human Growth and Development, Second edition.
ISBN 978-0-12-383882-7, Doi: 10.1016/B978-0-12-383882-7.00006-4

An appreciation of Nick Norgan's achievements in the field of human nutrition may be found in the Proceedings of the Nutrition Society 2006;65:326 (http://journals. cambridge.org/action/displayFullText?type=6&fid=815348&jid=PNS&volumeId=65 &issueId=03&aid=815344&fulltextType=XX&fileId=S0029665106000395).

6.1 INTRODUCTION

It is self-evident that nutrition is essential for growth. Growth is, in this context, an increase in size and mass of the constituents of the body. The only way this can be achieved is from the environment. Nutrition is defined as the process whereby living organisms take in and transform extraneous solid and liquid substances necessary for maintenance of life, growth, the normal functioning of organs and the production of energy.

Essential nutrients are defined as those organic or mineral substances required by the human body that the body cannot manufacture from simpler constituents. The known essential nutrients for human beings are listed in Table 6.1. All human beings require the same list of essential nutrients, but in different amounts according to age, sex, activity levels, health status and, in the case of adult women, pregnancy. The constituents of our bodies, the structure and chemical composition of our cells, tissues and organs, are remarkably similar for people all over the world. Indeed, the similarity extends to much of the animal kingdom. The lean body mass typically consists of 72% water, 21% protein, 7% minerals, and less than 1% of carbohydrate and other nutrients. Yet, we have available and select or tolerate a very wide range of foodstuffs and diet types. This leads to the first truism of nutrition and growth, that *a wide range of diet types is capable of satisfying nutritional needs and promoting optimal growth*. What determines the particular diet we consume involves a myriad of factors. For much of human evolution and for some groups today, the physical environment and climate determined what could be procured or cultivated. However, technology and the economic power to develop and exploit it allow us to inhabit the polar regions, the ocean depths and space — environments that do not naturally support food production. Economic and political systems are the major factors influencing food choice and whether food intake is sufficient to allow adequate growth and health to be achieved.

This chapter has the following aims:

- to illustrate the importance of an understanding of assessment methodology in a consideration of nutrition and growth
- to recognize that, although growth is highly nutrition sensitive, for much of the growth period, the nutritional requirements for growth per se are only a small proportion of the total requirement
- to demonstrate that growth variation, including impairment, is usually of multifac- torial origin

Table 6.1 Essential nutrients of the human diet

Carbohydrate	Minerals	Vitamins
Glucose	*Macronutrient elements*	*Fat-soluble*
	Calcium	A (retinol)
Fat or lipid	Phosphorus	D (cholecalciferol)
Linoleic acid	Sodium	E (tocopherol)
Linolenic acid	Potassium	K
	Sulfur	*Water-soluble*
Protein	Chlorine	Thiamin
Amino acids	Magnesium	Riboflavin
Leucine	*Micronutrient elements*	Niacin
Isoleucine	Iron	Biotin
Lysine	Selenium	Folic acid
Methionine	Zinc	Vitamin B_6 (pyridoxine)
Phenylalanine	Manganese	Vitamin B_{12} (cobalamin)
Threonine	Copper	Pantothenic acid
Tryptophan	Cobalt	Vitamin C (ascorbic acid)
Valine	Molybdenum	
Histidine	Iodine	**Water**
Non-essential amino nitrogen	Chromium	
	Vanadium	
	Tin	
	Nickel	
	Silicon	
	Boron	
	Arsenic	
	Fluorine	

- to understand the need to assess the contribution of nutrition to growth perturbation in the total environment in order to identify the appropriate remedial actions.

6.2 NUTRITION FOR STUDENTS OF GROWTH

6.2.1 Importance of Methodology

Students of growth need an understanding of some of the basic principles of nutrition if they are to understand and critically evaluate issues in nutrition and growth. As in other fields, it is important to be able to assess the strengths and weaknesses of evidence. There are regular surveys of food intake in many countries and frequent research reports on dietary intake and nutritional status. However, a number of issues must be considered. The apparently simple task of collecting accurate, representative data of the food intake of people going about their everyday way of life is notoriously difficult, and that diffi-culty should not be underestimated. Much of the information may have been collected by pragmatic but not necessarily the most accurate techniques.

Similarly, it is not difficult for the student to find in textbooks, on websites or in government and research reports, accounts of what are the nutritional needs of the various members of the population. However, the bases to these figures and their correct use are not straightforward. If the reader is to be informed in order to make critical observation, thought and decisions on nutrition and growth, it is important that these intricacies are known and remembered.

6.2.2 Measurement of Food and Nutrient and Energy Intake

The measurement of habitual food intake is an example of Heisenberg's uncertainty principle: the harder you try to measure it, the more likely you are to affect what you are trying to measure. Nutritionists are able to perform accurate measurements, to less than 1% of the real value, of energy and nutrient intake. The participants are housed in nutrition units and provided with their meals. Duplicate meals and snacks are weighed and analyzed by good physical and chemical analytical methods. (It is true that you can never analyze what someone has eaten because, if they have eaten it, you cannot analyze it, and if you have analyzed it, they cannot eat it.) The problems arise when we become interested in habitual food intake in people leading their everyday lives. It is difficult or inconvenient to weigh the food consumed in many circumstances. The inconvenience of weighing may influence the foods people eat or the number of meals and snacks they consume. A further problem in population studies is the need for statistical consider-ations of adequate numbers. There may be logistical and financial considerations and constraints and these may influence the choice of method. Most dietary surveys adopt the simple methods of questionnaire, interview or other subjective assessment to estimate food intake and translate these to nutrients and energy intakes using tables of food composition. One example is the ongoing National Health And Nutrition Examination Survey (NHANES) of the US population. The tens of thousands of participants are interviewed about food and drink consumed in the past 24 hours or asked about the frequency of foods eaten, but no direct observation of food behavior is done. There are problems with the use of food tables, too. The data on composition in food tables may not match those of the food consumed, especially in the case of homemade recipes or when people do not report the condiments added to the food serving.

There is, thus, a trade-off between ease of use and acceptability to participants and accuracy. In some cases, simple techniques may be appropriate. Some epidemiological investigations may require individuals to be assigned only to a correct tertile of intake: high, medium or low. However, it is crucial when reading the literature to be able to decide whether the nutritional methods used are fit for purpose. The literature abounds with studies purporting to show that children are inadequately nourished, but these are often based on ignorance of either the limitations of the intake data or, more usually, the nature of figures for recommended intake. Many children, for example, may not consume the mean value for a nutrient, such as zinc, but are not zinc deficient because adequate intakes usually fall within two standard deviations ($\pm 2\,SD$) above and below the mean.

6.2.3 Determination of Dietary Requirements and Recommended Intakes

We have nutritional needs, because being in a state of turnover, we have loss of body constituents and because at certain times of the lifespan, such as growth and pregnancy, we have a net gain of tissue. An individual's dietary requirements can be ascertained by measuring these losses and gains. Under most circumstances, such determinations are not practical and our judgments have to be based on the available experimental and other evidence from other individuals. However, individuals vary in their dietary requirements, even after taking into account age, sex and size. Therefore, we can make statements about only the probability of what an individual's requirement for a particular nutrient might be or the probability that a particular intake will be adequate or inadequate. This has led nutritionists to develop a series of values, including their best estimates of the average requirement for a given group of healthy individuals given the circumstances under which they live. The series usually includes a high level, thought to represent the needs of most of the group, and a lower level, below which most individuals would have an inadequate intake. In some cases, a higher level, above which problems of toxicity may be expected, is also identified. As stated above, a range of variation of $\pm 2\,SD$ for intake usually meets the needs of more than 97% of any human population. This range of nutrient intake is often called the recommended daily allowance (RDA), about which more detail will be provided later in this chapter.

Evidence on nutrient requirements has been gained in a variety of ways. Towards the end of the nineteenth century, medical scientists investigated the types and amounts of foods associated with good health or that would lead to the reversal of signs of nutrient deficiencies. They labored under the difficulty of the hegemony of the germ theory of disease following the work of Pasteur and Koch. Around the turn of the twentieth century, physiologists conducted animal and human experiments on artificial and deficient diets with nutrients fed at a variety of levels to ascertain needs. These were essentially balance experiments, with allowances for growth and production. A variation has been the factorial approach, which measures all the avenues of losses from the body — urine, feces, sweat, secretions and other emanations — to calculate the total losses. A further approach has been to determine the level of intake associated with high or maximum levels of the nutrient in the body. This invariably produces higher estimates of requirements than the other approaches.

A problem with the balance approach is that the body can be in balance but in a state of overnutrition or undernutrition that may be hazardous to health owing to changes or adaptations that occur to varying planes of nutrient intakes. Two difficulties emerge. First is the identification of the range of intakes where balance is not associated with risk. The second is whether recommended intakes should be set to maintain the status quo or at levels that are normative and lead to balance at the lower risk range. The problem is most acute for energy. We may calculate an allowance for children to take up aerobic exercise to promote cardiorespiratory fitness and healthy body composition. However, if they are not active, the recommendation would be a prescription for weight gain and obesity. As usual, a decision has to be made according to the context. The Food and Agricultural Organization, World Health Organization and United Nations University (FAO/WHO/UNU)[1] recommendations for energy intended to be applicable to individuals all over the world include components for desirable but discretionary activities; that is, they are normative. The UK recommendations do not.[2] They are values to maintain the status quo.

6.2.4 Nutrient Requirements and Recommended Intake

Many countries have drawn up their own recommendations for nutrient intake. They may have been drawn up and published in slightly different ways, but there is some commonality in their types. These common types are defined and shown in Figure 6.1. The *estimated average requirement* (EAR) is the mean value of any essential nutrient, except in the USA, where it is the median and not the mean value. The EAR refers to

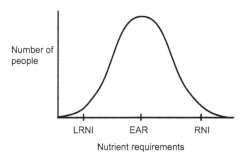

e.g. Vitamin C, adults 19-50 years
LRNI = 10 mg per day, EAR = 25 mg per day, RNI = 40 mg per day

Figure 6.1 Dietary reference values. Estimated average requirement (EAR): similar to the daily recommended value (DRV) of a nutrient; it is an average, so some people will need more than the EAR and some will need less. Reference nutrient intake (RNI): the amount of a nutrient that is enough to cover the needs of almost everyone in a particular group. Lower reference nutrient intake (LRNI): the amount of a nutrient that will meet the needs of only a small number of individuals who have low nutrient needs; intakes lower than this are almost certainly not enough for most individuals *(Source: Department of Health.[4] © Food — a fact of life. British Nutrition Foundation 2008 www.foodafactoflife. org.uk)*

a particular group according to age, sex, possibly body size and composition and, in some cases, lifestyle. It is used as one factor for assessing the adequacy of intake of an individual or of groups and for planning intake of groups. Approximately half the members of the group will need more and half less than the EAR.

The upper level, at or above which daily intake of a nutrient will meet the needs of most individuals in the specified group, is usually set at the average or median requirement + 2 SD. Where distributions of requirements are known to be skewed, the 97.5th percentile may be used. This level is called *reference nutrient intake* (RNI) in the UK and *recommended dietary allowance* (RDA) in the USA. It is intended as a goal for the intake of healthy individuals. It is not intended to be used to assess individual or group intake as most of the population require, and can stay healthy on, an intake lower than this.

One of the most common mistakes or misleading actions in nutrition is to assume that, if an individual or group consumes less than the RNI or RDA, the intake is inadequate. There is no shortage of examples of this occurrence. The US National Institute of Child Health and Human Development (NICHHD) website states, "Unfortunately, fewer than one in ten girls and only one in four boys ages 9 to 13 are at or above their adequate intake of calcium. This lack of calcium has a big impact on bones and teeth".[5] The implication is that up to 90% of girls and 75% of boys in the USA are not getting enough calcium. This website does not define the term "adequate intake", but elsewhere at the NICHHD website it seems to be similar to the EAR. If this is so, then this is not a justifiable implication because many young people consuming less than the EAR are likely to be getting enough! All that can be said is that, as the intake becomes a lower proportion of the EAR or the RNI (RDA), the *risk* of dietary nutrient inadequacy increases. Students of growth should have less of a problem with this point, as they are accustomed not to expect every child to be at the 97.5th percentile for height. Similarly, not every child's nutrient intake needs to be at the 97.5th percentile of requirements. It may be prudent to be at that level, but prudence should not be confused with proof of nutrient inadequacy.

A further point is that, as recommendations vary among countries, given intakes can represent much greater percentages of the reference intake in some countries than others. Such discrepancies do not reflect inherent biological differences but the different approaches and conclusions of the national advisory committees. Table 6.2 compares the UK's RNI and the USA's RDA for a few nutrients for infants and children. The two countries divide age groups differently. All UK RNIs are reported as a single value, but the USA recommends a median value and range of acceptable values for the energy-containing nutrients of protein, carbohydrate and fat. There are only trivial differences between the RNI and the RDA for most of the nutrients shown, except for folate: the US RDA is at least twice that of the UK RNI. Finally, the UK provides a recommendation for salt intake, while the USA provides recommendations for sodium and chloride intake. A major issue with both the UK and US recommendations is that the majority of

Table 6.2 UK daily reference nutrient intakes (RNI)[2] and US recommended dietary allowances (RDA)[3] for infants and children

Nutrient	UK RNI*			US RDA**	
	1—3 years	4—6 years	7—10 years	1—3 years	4—8 years
Protein (g/day)	15	20	28	13 (5—20)	19 (10—30)
Iron (mg/day)	7	6	9	7	10
Zinc (mg/day)	5	6.5	7	3	5
Vitamin A (μg/day)	400	400	500	300	400
Folate (μg/day)	70	100	150	150	200
Vitamin C (mg/day)	30	30	30	15	25
Salt	2	3	5	2.5 (1.0 as sodium)	3.1 (1.2 as sodium)

*The UK advises one value of the RNI.
**For the energy-containing nutrients protein, carbohydrates and fats, the USA advises a median value and an acceptable range for the RDA.

the citizens of either nation have little or no idea whether any of these nutrients are in their food and, probably, no idea of how to interpret the measurement units of gram (g), milligram (mg) and microgram (μg).

The certainty about statements of dietary inadequacy would be increased if a lower level were used. The UK has a lower reference nutrient intake (LRNI), the level below which intakes are likely to be inadequate for the large majority of the specified group. It is taken as the mean $-2\,SD$.

A fourth type of value is published when there is insufficient information to determine the other types. In the UK, *safe intakes* are specified when there is not enough information to estimate RNI, EAR or LRNI. This is the amount that is sufficient for almost everyone in a specified group but not so large as to cause undesirable effects. In the USA, when insufficient evidence exists to determine EAR, an *adequate intake* (AI) is specified. This is intended to cover the needs of most individuals in a specified group but the percentage cannot be stated with certainty. Therefore, these are at a level comparable to RNI and RDA but have even more uncertainty about them.

Finally, the maximum level that is unlikely to pose risks to health to almost all individuals in the specified group is called the *tolerable upper intake level* (UL) in the USA. This does not mean that intakes above RNI or RDA have known nutritional benefits. Also, for many nutrients, there are insufficient data on the levels at which adverse effects occur.

6.2.5 Energy Requirements

According to FAO/WHO/UNU:[1]

The energy requirement of an individual is the level of energy intake from food that will balance energy expenditure when the individual has a body size and composition and level of physical activity consistent with long term good health; and that will allow for the maintenance of economically necessary and socially desirable physical activity. In children and pregnant and lactating women the energy requirement includes the energy needs associated with the deposition of tissues or the secretion of milk at rates consistent with good health.

Not every nutritionist feels comfortable with this definition, because of the problems in establishing what is a state of long-term good health and the possible subjectivity of socially desirable physical activity. However, a more immediate problem is the need to know the energy expenditure. This can be approached at a variety of levels. An estimate can be made knowing the age, sex and weight of the child and assuming a type of lifestyle: inactive, moderately active, and so forth. At the other end of the range of approaches is measurement using stable isotopic doubly labeled water. This has the disadvantage of being expensive and lacking information on the components of the energy expenditure. Somewhere in between is a factorial approach of recording the time and duration of activity and applying energy costs, either measured or taken from the literature, to these to calculate energy expenditure.[6] There are considerable differences in the certainty of estimates of energy expenditure from these different approaches and, hence, in the estimates of energy requirements.

6.2.6 Protein Requirements

The protein requirement of an individual is defined as the lowest level of dietary protein intake that will balance losses of nitrogen in persons maintaining energy balance at modest levels of physical activity. In children and pregnant or lactating women, the protein requirement is taken to include the needs associated with the deposition of tissues or the secretion of milk at rates consistent with good health.[1]

6.2.7 Protein–Energy Ratios

The adequacy of a protein intake is influenced by the adequacy of the energy intake. When energy intake is inadequate, there may be a net negative nitrogen balance that reduces the adequacy of the protein intake. Thus, information on energy and protein intake needs to be considered together. One way of doing this is the protein–energy ratio (PE ratio: protein energy/total energy). When the diet exceeds the safe PE ratio, then any protein nutrition problems will result from inadequate amounts of food rather than low protein content. Most regular diets have PE ratios between 10 and 15%. Human breast milk has a PE ratio of about 7% and is adequate for the rapid growth in the first months of life. In the absence of other detailed information, this figure can be applied to other stages of growth. An allowance has to be made for the efficiency of utilization of the protein, which in most cases is less than that of breast milk.

6.2.8 Vitamin and Mineral Requirements

Current recommendations for a few vitamins and minerals are given in Table 6.2 and more recommendations may be found at the websites referenced with Table 6.2.[2,3] Minimum nutrient intakes to prevent disease were common in the past, for example a recommendation for adults of 60 mg/day for vitamin C will prevent scurvy. In some countries, these minimum requirements have been increased over time, for example to 90 mg/day for vitamin C, to take into account safety factors for variation in food composition, instability resulting from unfavorable storage conditions for food, and the effects of cooking on nutrient availability. These factors, as well as the statistical considerations discussed above in relation to EARs, RDAs and RNIs, are used to establish nutrient intake recommendations for health and performance. Further discussion of specific vitamin and mineral nutrients lies outside the scope of this chapter. Readers are recommended to consult the nutrition textbooks and internet sites listed at the end of this chapter.

6.2.9 Dietary Goals and Guidelines

Dietary goals and guidelines differ from dietary recommended intakes and dietary reference values. Dietary guidelines provide advice on food selection that will help to meet the RDA or RNI and help to reduce the risk of disease, particularly chronic disease. They are thus meant to ensure adequate intake to prevent deficiency states and prevent the inappropriate macronutrient intakes associated with many of the chronic degenerative diseases of affluent societies. The goals set what is to be achieved to reduce the incidence of these diseases in terms that are understood by the professionals; that is, reduce intake of nutrient x to y g per day. The guidelines indicate to the public how the goals are to be achieved. They refer to foods and diets as opposed to nutrients. As most of us base our diet on foods rather than nutrients, they are much more relevant to the population.

The most well-known dietary guideline is to eat a variety of foods. This is usually portrayed as a food block, plate or pyramid of four or five food groups that recommends the kinds and amounts of foods to be eaten each day. The UK's national food guide uses a picture of a plate of food called the Eatwell plate which, "… shows the different types of food we need to eat — and in what proportions — to have a well-balanced and healthy diet" (http://www.nhs.uk/Livewell/Goodfood/Pages/eatwell-plate.aspx). The USA used a food guide pyramid for many years, which indicated that fats, oils and sweets are to be used sparingly, but there can be six to 11 servings of bread, cereal, rice and pasta. In 2011 the pyramid was replaced by the "Choosemyplate" illustration, which is a simplified version of the Eatwell plate (http://www.choosemyplate.gov). The image of the plate comes with advice to, "Make half your plate fruits and vegetables; Make at least half your grains whole grains; Switch to fat-free or low-fat (1%) milk". There are many other pyramids and plates. Walter Willet, a Professor at Harvard University, offers a new "Healthy Eating Plate" which includes advice to consume healthy oils, such as olive oil, stay active and maintain healthy weight. The UK and US plates do not mention healthy oils or exercise as part of a lifestyle approach to health.

Dietary guidelines are regarded as applicable to the whole population. Australia and New Zealand produced guidelines specifically for children of different ages, which allow for fuller consideration of types of infant feeding and the nutritional problems of adolescents. The USA has a food guide pyramid for young children called MyPyramid for Kids, meant to be accessible to 2—6-year-olds.[7]

Values of recommended intakes and dietary goals and guidelines have an important use in addition to assessing or planning diets. This is in the information given on and claims made for food products, particularly on labels. Food manufacturers are interested in dietary recommended values and legislation about food composition and claims. They are important members of the committees that draw up guidelines, often with interests separate to those of nutritionists and clinicians. This may not be counterproductive, as differing views may lead to better evidence on requirements and recommendations in the long term.

6.2.10 Assessment of Nutritional Status

The assessment of the nutritional status of an individual or group involves the collection of information: on diet, biochemical indices, anthropometry, clinical signs, and morbidity and mortality statistics. The value and place of this disparate group of measurements can best be understood by considering the process of becoming malnourished. Figure 6.2 shows the process of moving from a state of good nutrition to malnutrition and eventually death, and shows the place of each type of measurement. The aim should be to correctly describe an individual or group as well nourished, at risk, to be monitored further or in need of remedial action.

The directionality of the process may need to be established by serial measurements, as, for example, poor scores on biochemical, anthropometric and clinical data may persist for some time after the diet has improved. Good dietary assessment is difficult, time consuming and expensive; and its interpretation is rarely clear-cut, given the nature of knowledge of nutritional requirements, and this is often omitted. However, it can be crucial in establishing the true pathogenesis, as many of the other signs are not specific to nutrition but can also arise from other environmental causes such as disease. A wide-ranging ecological assessment may be the only way to ensure that the true causes of low nutritional status are identified and the appropriate remedial action is taken.

The emphasis in assessment is to obtain early warning signs of malnutrition, and the biochemical indices play an important role here. Simple dietary iron deficiency will result in iron-deficiency anemia (IDA) (low levels of hemoglobin, with microcytic and

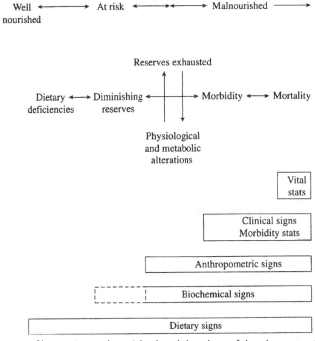

Figure 6.2 The process of becoming malnourished and the place of the elements of nutritional status assessment at different stages of the process. *(Source: Derived from Sabry.[8])*

hypochromic erythrocytes) if left untreated. Before hemoglobin levels fall, body stores diminish. This is described as iron deficiency without anemia. In examining data on nutritional deficiencies, it is important to distinguish between those based on "biochemical" deficiencies and those based on "clinical" signs, in terms of establishing the importance or priority of the problem. Some would have it that assessment of nutritional status should be based more firmly on functionality rather than low levels of body chemicals or size. There is much truth in this, but the cut-off points to identify good and poor nutritional status from these biochemical or clinical indices are based on outcomes and impairments wherever possible. The international cut-off for IDA in pregnant women is a blood level of iron below 11.0 g/dl. Some of the functional consequences of IDA for pregnant women are decreased work capacity, prematurity and low birth weight (LBW) for their newborns, perinatal mortality of the fetus/newborn, maternal mortality, and higher risks for impaired neurocognitive function and mortality of the children later in their life.

Anthropometry plays a major role in nutritional status assessment, particularly in field and clinic studies of children. Growth faltering in infancy (birth to age 36 months) is regarded as a clear sign and symptom of poor nutrition, and nutritionists rely heavily on anthropometry for indices of nutritional status. Not all causes of perturbed or impaired growth are nutritional in origin. Most commonly, growth can be impaired by disease and infection and an associated anorexia or poor appetite. The significance of this is that the first priority may be to treat concurrent infections and eradicate their causes rather than attempt refeeding or development projects in agriculture or subsistence food production. The provision of clean, protected water supplies and sanitation may also be an early priority.

In nutritional status assessment, the "growth" of a population is often described by cross-sectional measurements. There is a risk of circularity in a discussion of nutrition and growth when growth is used as a proxy for nutritional status. The interpretation of a particular size, whether a small child is normally small or growth retarded, is a difficulty, but one outside the scope of this chapter. Height, or length, is a key variable for aux-ologists, but nutritionists are particularly interested in growth of muscle mass, adipose tissue mass and its location, and bone mass because of the greater direct functional implications and the consequences for long-term good health.

6.3 NUTRITION AND GROWTH

The importance of nutrition for growth is well attested by clinical observations of growth reduction in conditions of reduced food intake, for example anorexia nervosa, and in intestinal malabsorption, such as is associated with untreated cystic fibrosis. At the population level, growth faltering has been observed and well documented to be asso-ciated with food shortages in conditions of civil unrest and war. However, nutritional challenges to growth rarely occupy precisely circumscribed epochs and even more rarely do they operate in a vacuum. The secular trends in growth, menarche and skeletal maturation observed in many countries over the past 100 years are a record of the effects of previous living conditions on growth. Most commentators ascribe a key role to improvements in nutrition in these secular changes (see Chapter 9). Nutrition is one of a number of environmental influences on growth. The others include infection, poverty, poor housing and schooling; and it can be difficult to identify and evaluate the precise contribution of nutrition to growth or growth failure. The type, duration and intensity of the nutritional challenge influence the nature of the response in growth, as does the ecological setting. There is thus no quantitative law-like relationship between nutrition and growth, and descriptions of the relationship tend to be either rather general or biosocial case histories.

6.3.1 Normal Nutrition

The aims of this chapter do not include teaching basic nutrition, and it is not covered in any detail. Maternal and fetal nutrition and nutrition in infancy, childhood and adolescence are topics well covered in many textbooks on nutrition. Some representative texts are listed in the Suggested Reading section at the end of this chapter. Most of these are specific to a particular region (e.g. North America or Europe) and few are balanced according to developing country issues.

The infant grows more rapidly in the first year of life than at any subsequent period of life, and breast feeding is recognized as the appropriate method of feeding the newborn and infant in the first months. The advantages expand beyond the provision of a feed nutritionally suited to the human infant that is hygienic and at the correct temperature and with a built-in supply regulator. They extend to better immune competence and more protection against gastroenteritis, ear and chest infections, eczema and childhood diabetes. For the mother, there is a speedier reduction in the size of the uterus and a lower risk of premenopausal breast cancer, ovarian cancer and hip fracture. For a variety of reasons, some mothers choose not to breast feed or may be unable to breast feed. These women should have as much support as breast feeders and should not be made to feel guilty or inadequate.

A basic assumption is that breast milk composition has evolved to meet the nutrient needs of the infant. If the amount produced is sufficient, that is, if energy needs are met, so are nutrient needs. Breast milk intake of 850 ml/day would meet the needs of infants growing along the 50th centile until 4 months old.[9] It would meet the needs of an infant in a developing country growing along the 25th percentile for 6 months. Weaning from the breast and the introduction of appropriate complementary foods should begin at these ages.

Breast-fed babies have in the past been found to grow more slowly in infancy than formula-fed infants in some but not all studies. This meant that breast-fed children often appeared to be growing less satisfactorily than reference growth data as the older growth reference data came from groups of exclusively or mostly formula-fed infants. There is some evidence that this difference has lessened as formula feeds have been "humanized"; that is, modified toward the composition of breast milk. Fears that formula feeding may promote the development of widespread overfeeding and obesity have not been founded. Nevertheless, the World Health Organization (WHO) spent a great deal of time and money to compile data on the growth of approximately 8500 infants and children from widely different ethnic backgrounds and cultural settings (Brazil, Ghana, India, Norway, Oman and the USA). As infants the participants were exclusively, or predominantly, breast fed for six months (http://www.who.int/childgrowth/mgrs/en/). From these data, the WHO published the WHO Growth Standards for infants and children, birth to 60 months of age (http://www.who.int/childgrowth/standards/en/). By calling these growth charts *standards* the WHO maintains that the patterns of growth of their sample are the way infants and children *should* grow, i.e. they are *prescriptive*. Growth *references* reflect how certain groups of people *do* grow, i.e. they are *descriptive* (see Chapter 21). The other major concern of infant nutrition in developed countries, the premature introduction of solid foods, is being addressed by information and education programs.

Once weaned, children and older individuals should eat a wide variety of food, in line with the recommendations of the UK Eatwell Plate or the USA Choosemyplate. Doing so helps to ensure that sufficient nutrients are consumed. The national recommendations of just how much of each nutrient is required for good health are not identical. There is currently much international debate on recommendations for vitamin D and folate (folic acid) intake. One of the biggest differences in recommendation between the UK and the USA is for calcium. The USA currently recommends an RDA of 800 mg/day for children 4—8 years old, rising to 1300 mg/day for 9—18-year-olds. These values are based on the maximal retention of calcium in bone. In the UK the RNI for calcium is 500 mg/day for children 4—10 years old, rising to 600 or 700 mg/day for 11—18-year-olds. The UK recommendations, which are substantially lower figures, are based on a factorial approach to calcium balance in the body, with allowances for gain, loss and absorption. Using the US recommendation will result in more young people being classified as "calcium deficient", and this has implications for public health interventions, governmental expenditures and general worry for parents. Both the RDA and RNI are designed to cover the needs of 97% of the population, so for the majority of people an intake of calcium 50% lower than the RDA or RNI is likely to be sufficient for health.

Nutritional needs in adolescence may be, in absolute terms, greater than at any other time of life. The high rates of proportionate growth may only equal or be less than those of the first few months of life, but they persist for much longer. It is a time when individuals make more of their own choices in food, in some cases using them as part of a relationship struggle with parents and caregivers, but without necessarily too much nutritional knowledge. Independence, or the pursuit of it, may lead to behaviors, such as anorexia, bulimia nervosa and substance abuse, that threaten nutritional integrity and hence growth. However, the growth of teenagers can be remarkably resilient to nutritional challenges, as illustrated later when considering the pubertal growth spurt in poorly nourished Indian adolescents.

6.3.2 Nutritional Demands of Growth

Good nutrition is of fundamental importance to growth. When food becomes limited, one of the earliest responses of the body is to retard growth; indeed, growth assessment by anthropometry is one of the most commonly used indices of nutritional status. Similarly, deficiency of a single nutrient, such as zinc, may cause growth failure. It is easy to move from this to the idea that growth is a costly process that requires most of the energy and nutrient intake. This may be true for some mammals but not for humans, with the exception of the first few months of life. Figure 6.3 shows the energy cost of growth (ECG) in relation to basal metabolic rate (BMR) and activity energy expenditure (AEE) from infancy to adulthood. ECG is always a smaller percentage of

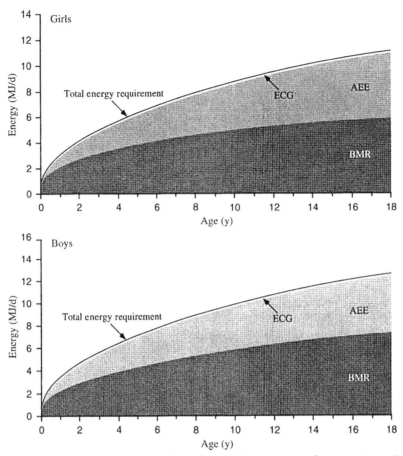

Figure 6.3 Energy requirements of girls and boys from birth to 18 years of age, partitioned into basal metabolic rate (BMR, dark gray), activity energy expenditure (AEE, light gray), and energy cost of growth (ECG, thin white area between AEE and Total energy requirement). *(Source: Butte.[10])*

total energy needs than that required for BMR. As the infant develops and becomes more active, the percentage of energy needed for AEE rises. BMR also rises, but the ECG decreases. Not shown in Figure 6.3 is that at birth the value of ECG is 37–38% of total energy intake and this falls to only 2% of the total energy requirement by the age of 24 months.[9] The ECG increases at puberty and during adolescence as the reproductive system matures and the body undergoes a general growth spurt. However, the total ECG remains below 10% of the total energy intake. Despite these relatively low costs it is important to realize that growth is in the front line of responses to nutritional challenges. If energy intake falls below the needs for BMR, then either AEE or growth, or both, will be curtailed.

6.3.3 Malnutrition

Malnutrition means bad nutrition. The term applies equally to overnutrition as to undernutrition, but it tends to be used more for the latter than the former (as in Figure 6.2). It has been estimated that the proportion of the world's population exhibiting overnutrition now matches that showing undernutrition. The sequelae of overnutrition have traditionally been assumed to take their toll in adulthood [e.g. as non-communicable diseases (NCDs) such as cardiovascular disease or non-insulin-dependent diabetes mellitus]. However, recent evidence demonstrates that increased risk factors for NCDs are present in late childhood and adolescence.[11] In contrast, the sequelae of hunger and undernutrition — that is, increased susceptibility to infectious disease, physical and mental impairment, and possibly death — affect the young most; and undernutrition is given more prominence here.

Growth retardation in developing countries is said to be most common and most marked between 6 and 12 months of age. This timing coincides with the inability of mother's breast milk to sustain infant growth and the introduction of complementary foods. It is also a time of high growth velocities and nutritional needs, a time when the body systems are highly sensitive to stress or perturbation. If the complementary foods given to the infant are not sufficient in energy and other nutrients, or not sanitary, then the growth of the infant will be compromised. There are long-term implications of early life stunting for health in adult life and also for intergenerational effects on growth and health. These later life and intergenerational effects are discussed in detail by Kuzawa in Chapter 12. Kuzawa's perspective focuses on adaptations made by the fetus or infant to survive in a harsh environment. An alternative perspective is presented in Box 6.1;

Box 6.1 Trade-offs in human growth: adaptation or pathology?

It is estimated that more than 80% of the human population lives in the poorer, developing nations of the world. The majority of those people grow up in adverse biocultural environments. By this we mean environments including undernutrition, exposure to infection, economic oppression/poverty, heavy workloads, high altitude, war, racism and religious/ethnic oppression. As a consequence of these biocultural adversities the people may be stunted, have asymmetric body proportions, be wasted, be overweight and be at greater risk for many diseases.

One group of researchers explains this as a consequence of "developmental programming" (DP).[12] In essence, the DP hypothesis states that exposure to adverse environments during gestation results in a body that is smaller at birth and will be unhealthy in adulthood. The smallness at birth is a marker of physiological disruption of the prenatal development of one or more physiological systems, including the cardiocirculatory, neuroendocrine and renal systems. Smaller body size may be in total birth weight, head circumference, body length, organ size or some combination of these.

Soon after the DP hypothesis was first proposed, some researchers found that smaller size at birth plus a greater than expected amount of growth after birth worked synergistically to place people at elevated risk for adult heart disease, glucose intolerance and other metabolic diseases.[13] Gluckman and Hanson[14] and Chapter 12 in this book conceptualize the mismatch between reduced fetal growth and accelerated postnatal growth into the hypothesis of "predictive adaptive responses" (PARs). The developing fetus receives signals from the mother about her nutritional and health environment. The signals may take the form of nutrient levels, biochemicals, neurohormonal products or other substances. The PAR hypothesis is that the fetus uses these signals to adjust its amount and rate of growth to best adapt to its current uterine environment and also the expected postnatal environment. The PAR and the DP hypotheses agree that small size at birth is the result of some type of adversity during fetal development. PAR differs from DP in the expectation of relative health if the postnatal environment is a close match to the prenatal environment. If, however, adversity is ameliorated after birth, then there is the likelihood of a mismatch between the PAR of the fetus and the postnatal environment. Under the relatively good conditions of postnatal life the incorrect PAR of the fetus results in overgrowth and some type of metabolic disease after later in life.

The PAR group considers the alterations at two levels of adaptation: (1) "short-term adaptive responses for immediate survival" and (2) "predictive responses required to ensure postnatal survival to reproductive age".[14]

A life history theory analysis rephrases the DP versus PAR debate from disease or adaptation to the concept of "trade-offs". Life history theory is a branch of biology that studies the evolution and function of life stages and behaviors related to these stages. The life history of a species may be defined as the evolutionary adaptations used to allocate limited resources and energy towards growth, maintenance, reproduction, raising offspring to independence and avoiding death. Life history patterns of any species are often a series of trade-offs between growth versus reproduction, quantity versus quality of offspring, and other possibilities given limited time and resources. Even under good conditions, the stages of human life history (fetal, infant, child, etc.) are replete with trade-offs for survival, productivity and reproduction. Much research shows that under adverse conditions, trade-offs usually result in reduced survival, poor growth, constraints on physical activity and poor reproductive outcomes.[15]

To explain these trade-offs, we offer some examples of human growth and development from research in Guatemala, a small country of Central America. In Guatemala the majority of the population lives under the conditions of adversity described in the opening paragraph of this box. Poverty is the main correlate of this adversity, with 74.5% of the rural population living below the poverty line (based on the 2003 census, the rural population numbers 7.3 million people and the total Guatemala population is 12.3 million; www.ruralpovertyportal.org). An estimated 37.4% of the total Guatemala population lives on less than US $2.00 per capita per day.

The under 5-year-old mortality for the Guatemala population is estimated at 40/1000, but in isolated rural areas the number doubles (UNICEF, http://www.unicef.org/infobycountry/guatemala.html, accessed 30 September 2011). In a narrow sense, the women producing these infants and children are "adapted" because these women live long enough to reproduce. But, far too many of their offspring die and those who live are impaired with poor physical growth, with 60% of all children in Guatemala stunted. Other research shows these children have reduced cognitive development and diminished socioeconomic productivity. Given these outcomes, the trade-off between reproduction versus reduced rates of survival and productivity (i.e. growth and work capacity) indicates a low level of adaptation for the poor in Guatemala.

The INCAP longitudinal study, which is described in the main text of this chapter, provided a nutritional intervention for some poverty-stricken mothers and their infants and children up to the age of 7 years. The PAR hypothesis predicts that a mismatch between the uterine environment and the postnatal environment will result in clinical symptoms of metabolic disease, such as higher blood pressure, and risk for elevated levels of low-density lipoproteins. Follow-up studies of the supplemented infants and children as adolescents and adults find that the group receiving the higher quality protein—energy *atole* supplement (versus the lower quality *fresco* supplement) have lower fasting blood glucose levels, lower systolic blood pressure, a lower triglyceride level and higher high-density lipoprotein cholesterol level. Each of the metabolic measures indicates improved health status. The results do not follow the prediction of the PAR hypothesis, which expects that the mismatch between maternal nutrition and the infant's early life nutrition will result in poorer health. These results are more in accord with the life history trade-off hypothesis, that interventions designed to ameliorate nutritional insufficiency, improve birth weight, and reduce stunting and wasting in the early years of life will have beneficial consequence throughout life.

What happens when poor Guatemalan families live under better biocultural conditions? Beginning in the early 1980s, thousands of Maya families migrated from Guatemala through Mexico and into the USA. The Maya are one of the major ethnic groups of Guatemala. The other major ethnic group are the Ladinos, who are the cultural descendants of the Spanish/Portuguese Conquistadors who arrived in 1500 CE. The contemporary Maya are the cultural descendants of the native people of Guatemala. Since the conquest, the Maya have endured infectious disease epidemics, poor nutrition, forced labor (including slavery), land appropriations and repeated episodes of military attack by national government forces. The Maya population consists of more than 20 linguistically distinct groups of people living in Guatemala, Southern Mexico and Belize. Nevertheless, many commonalities of history, social organization, religion and political economy bind all Maya groups together as an identifiable ethnicity. Together, all the people who collectively self-identify as Maya comprise the largest population of indigenous people in the Americas, with 6—7 million people.

In 1992 and 2000 a research team measured the growth status of the 6—12-year-old children of these immigrants. The Maya—American families live in Indiantown, Florida (a rural community), and Los Angeles, California. Adult Maya work as day laborers in agriculture, landscaping, construction, childcare or in other informal sector jobs. Some Maya work as teacher aids or nursing aids or have opened small businesses such as grocery stores. Still, almost all of the Maya families are of low socioeconomic status by US standards. All of the Maya—American children in the Florida sample, for example, qualify for free breakfast and lunch programs at the schools they attend.

Compared with a sample of Maya of the same ages living in Guatemala and measured in 1998, the Maya—American children are 10.24 cm taller, on average, and most of this height increase is due to longer legs, averaging 7.02 cm longer than the Maya living in Guatemala. These are signs of improved nutrition and health, which are expected as these Maya—Americans now have more food, clean drinking water, better health care, better schooling and a safer social environment than in Guatemala. However, there is evidence of a trade-off, in that the Maya—Americans also are 12—13 kg heavier than Maya of the same ages in Guatemala. A disproportionate amount of this added weight is body fat. The result is that 48% of the Maya—American children are overweight or obese.[15] Excessive body fatness at such a young age is a risk factor for obesity, diabetes, heart disease and other health problems as adults.

This research with Maya—Americans may be interpreted as support for the PAR hypothesis. All of the Maya mothers were born and raised in rural Guatemala and suffered from the adversities of that environment. When pregnant, these mothers may have sent signals to their developing fetus that shaped fetal growth to expect a similarly adverse environment. The excessive fatness for the Maya—American children may be due to the mismatch between the PAR and the reality of the USA.

Another interpretation is based on the intergenerational influences hypothesis (IIH). The IIH was proposed by Irving Emanuel in 1986 and defined as, "… those factors, conditions, exposures and environments experienced by one generation that relate to the health, growth and development of the next generation".[16] Just how these intergenerational influences are passed from parents to offspring is not completely known, and is a very active area of current research. One hypothesis with some experimental support is that uterine/placental nutrient transport is impaired in mothers who experienced malnutrition during their own gestation. The gestational experience of the mother may be passed on to her sons and daughters through gestational imprinting or other epigenetic mechanisms that take place during the fetal development of the offspring (see Chapter 7). Forty years of experimental research with rhesus monkeys has found that it takes at least four generations of good nutritional care to overcome just one generation of undernutrition of pregnant monkeys.[17] Extending this research to human beings may help to explain the so-called secular trend for intergenerational increases in stature and earlier maturation.

A major difference between the PAR hypothesis and the IIH is that the latter removes any notion of "decision making" or adaptive response by the fetus. The IIH posits that there is an epigenetic impact on the fetus over which the fetus has little or no control. The epigenetic marks cause a derangement of optimal genomic expression and development. The result is that the newborn infant has a phenotype with varying degrees of pathology that will result in trade-offs in health in early life, adulthood, or both.

Current research by the co-authors of this chapter (Bogin and Cameron) and their colleagues is investigating the causes and consequences of intergenerational effects on human growth and health. Bogin's teams continue to explore the reasons why the Maya—Americans and Maya in Mexico are so susceptible to childhood fatness and how this relates to adult health.[18] Bogin is also part of a team studying intergeneration influences on the health of Bangladeshi immigrant women and their daughters in the UK.[19] Cameron's teams work in South Africa and in the UK to investigate intergenerational effects on bone health, body composition and health risks.[20,21] The results from this research may lead to better models of human development — models that are able to accommodate a greater range of the biological and cultural sources of adversity as well as their independent and interactive influences on health.

a perspective that a harsh early life environment does not lead to adaptation; rather, it leads to harmful consequences for growth and health.

6.3.4 Protein—Energy Malnutrition

Protein—energy malnutrition (PEM) is the most common and important nutritional deficiency. Half the world's children are said to suffer from undernutrition, mainly as PEM. In its severe form, it exists as kwashiorkor and marasmus. In many African and Asian countries, 1—3% of children under 5 years old suffer from severe malnutrition. In its less severe form, it appears as stunting and wasting, depending on the severity and duration of the insult. In developing countries, 10—50% of children are stunted, that is, with heights more than 2 SD below reference data medians. The model of malnutrition in developing countries is of an iceberg. For every one case of severe malnutrition that is visible at clinics or in hospitals, there may be 15—25 moderate cases and a further 25—35 mild cases in the community. In global terms, World Hunger (http://www.worldhunger.org/) estimates that as of 2010 there were 925 million people in the world suffering from some degree of food hunger. Only 2% (19 million) are people living in the richer, developed countries.

According to conventional methods of classifying causes of death, malnutrition does not appear as a major cause in developing countries and, consequently, receives less of the donor and national health resources. However, results from eight prospective studies in different parts of the world indicate that 42—57% of all deaths of 6—59-month-old children are due to malnutrition's potentiating effects on infectious disease. The relative risk of death for severe malnutrition is eight-fold, for moderate five-fold, and for mild two and a half times that of normally nourished children. However, more than three-quarters of the nutrition-influenced deaths are attributable to mild to moderate malnutrition, because this type of malnutrition is the most common.

We are presented routinely with pictures of malnutrition arising from natural disasters, and it is easy to accept the myth that malnutrition results from scarcity of food. The vast majority of cases of malnutrition have roots more in economic and political considerations, many of which exacerbate any climatic or biological effects. According to the Nobel Prize-winning development economist Amartya Sen, famine has never occurred in a democracy. Inequitable distribution of resources and gender discrimination make a large contribution to malnutrition. It is estimated that 80% of undernourished children live in countries with food surpluses. Suffice to say that malnutrition rarely has its origins in a shortage of food. If we can nourish people at the poles, in the ocean depths and in space, where no human foodstuffs grow naturally, we have the ability to nourish people in less extreme habitats.

The underlying causes of malnutrition are many, and their interrelationships are complex and diverse. There is a well-documented synergism between nutrition and infection in their effects on growth. The undernourished child is more susceptible to infection and the ill child has higher nutritional requirements at a time when they are likely to be anorexic. The interrelationships are too complicated and context specific to allow this synergism to be easily quantified. Effects on the gastrointestinal mucosa causing malabsorption may predominate. Estimates suggest that perhaps one-third of linear growth failure can be ascribed to illness. However, children in rural areas of developing countries who have avoided serious infection are still 10–15 cm shorter than American children at 7 years old. Persistent low-grade infection, parasites, poor water quality, a chronic diet of 90% of energy needs, lack of a single essential nutrient (such as iodine or zinc), and all of the social factors associated with low income and poverty account for this height deficit.

6.3.5 Developed Country Issues

Undernutrition is not a problem confined to developing or transitional societies, to civil unrest and warfare. Poverty, relative and absolute, is found in all countries. Within the UK, a member of the G8 Group of Industrialized Nations, mortality and socioeconomic differentials have widened over the past 30 years and adverse socioeconomic circumstances in childhood influence subsequent mortality that is independent of the continuation of the disadvantage throughout life. An excellent resource on social class differentials in nutrition, growth and health is the website Fair Society, Healthy Lives (http://www.marmotreview.org/).

From the other end of the social spectrum, affluent parents with particular beliefs about food habits have caused growth retardation in their children. Studies have been performed on the growth of children consuming macrobiotic diets in the Netherlands, following observations that these children were lighter and shorter than a control group on an omnivorous diet.[22] Macrobiotic diets resemble those of many children in developing countries in being based on grain cereals, mainly rice, vegetables and pulses, foods high in fiber and starch but relatively low in protein and with intakes of calcium and vitamin D substantially below Dutch RDAs. Growth faltering began in the weaning period, with partial catch-up between 2 and 4 years of age in weight and arm circumference but not in height. Linear growth was associated with protein content but not energy content of the diet. The children of families that subsequently increased their consumption of fatty fish or dairy products grew more rapidly in height than the other children. These observations are significant as they illustrate that growth is sensitive to nutrition even in a good environment. Much of the material concerning developing countries presented above, and more which follows below, may leave the impression that improving nutrition is not very influential, but this is because those environments contain a number of other impediments to growth (such as poverty, poor water and sanitation, poor health care and social unrest).

6.3.6 Catch-Up Growth

A key question in human biology over recent years, and one that has received much attention, is to what extent a growth-retarded child may catch up in stature and weight. Martorell, Keitel Khan and Schroeder hold that stunting (low height for age) arises from events early in life, and once present, it remains for life.[23] The contrary view of Tanner is that the undernourished child slows down and waits for better times.[24] Tanner considers that, "in a world where nutrition is never assured, any species unable to regulate its growth in this way would long since have been eliminated". Obviously, if a child remains in the environment that led to stunting, it is unlikely that this will be conducive to improvements in growth. However, Adair has described catch-up between 2 and 12 years of age in Filipino children staying in the same environment.[25] Severe stunting in the first 2 years was associated with low birth weight, which significantly reduced the likelihood of later catch-up. Positive attributes associated with catch-up were taller mothers, being the first born and having fewer siblings, which illustrates again the actions of intervening non-nutritional factors.

There is a remarkable feature in the growth of children in India, the Gambia, west Africa and elsewhere who were stunted, in some cases severely stunted, before 5 years old and remain stunted at 18 years old. The growth increment between 5 and 18 years may match that of western children.[26] The growth pattern is not the same, as puberty is later and prolonged. It is not clear how this comes about. It is almost as though the last chance for catch-up is taken. Perhaps growth in height is replaced by another factor, growth in weight or physical activity, in the front line of impairments. An example of catch-up with excessive weight gain is presented in Box 6.1.

In contrast to height deficits, weight deficits increase from childhood to adulthood in almost all populations with childhood growth retardation. This may have significant and long-lasting consequences as weight-for-age and weight-for-height may be more important than height per se in terms of functional outcomes such as work capacity, earning power and employment. Linear growth has tended to receive more attention by auxologists than growth in mass, but this bias decreased in recent years.

Improvements in the environment result in catch-up growth, but such improvements are not common. Reports of the long-term effects of treating children with malnutrition provide evidence of catch-up, even though treatment may be for only a short period and the child returns to the same environment. Community-based supplementation studies provide better evidence of what can be achieved in the habitual environment, as described next. Any prediction as to whether catch-up growth will occur in an individual or a population requires a full environmental assessment.

Relocation to better environments in the form of migration (Box 6.1) or adoption involves changes in the environment not restricted to nutrition. Studies of refugees and adoptees from South and Southeast Asia have shown accelerated growth rates, but it is not yet clear whether these translate to increased adult stature. There is some evidence of accelerated puberty, which may have the effect of shortening the growth period and curtailing adult stature.

6.3.7 Supplementation Studies

Supplementation is taken to be an addition to a diet to make up all or part of a deficiency. It can vary according to the type, amount and duration; hence, a dose—response effect should be observed. Thus, it is a classical research design of high internal validity to demonstrate the presence of undernutrition and its effects on growth. This is also a common approach to improving maternal and child health. It might be thought that supplementing the diet of children and mothers in populations with low growth rate would inevitably result in improved growth. However, when this is attempted, the effects are usually much smaller than expected, and growth never achieves the levels of the most affluent groups in the same population or the median values in western reference data.

Several extensive supplementation studies are described in the literature. One such is the Institute of Nutrition of Central America and Panama (INCAP) longitudinal study of the effects of chronic malnutrition on growth and behavioral development that began in 1969. It is unusual, in having a follow-up component several years after supplementation ended.[27] The study was conducted in four villages, which began with small and light mothers on low dietary intakes and with weight gains in pregnancy half those of well-nourished women; 15% of the infants died in the first year of life. Two types of supplement were provided to children up to 7 years old. Two villages selected at random were provided with *atole*, a protein—energy supplement, and the other two with *fresco*, a non-protein supplement with one-third the energy content of *atole*. Both supplements contained minerals and vitamins. Preventive and curative medicine were provided to all villages. Supplement take-up was voluntary and so varied widely, but this allowed a dose—response approach to the analyses using multiple regression, and for confounding factors to be taken into account.

Birth weights increased, but only by some 7 g per 10 MJ of supplement. However, low birth weight and infant mortality fell by a half in those with high supplement intakes. Energy intake was more important than protein for these improvements. In 0—3-year-old children, greater supplementation was associated with better growth in supine length and weight, but not limb circumferences or skinfold thicknesses. In the highest supplementation group, these differences were 1 kg and 4 cm at 3 years old. Supplementation of 420 kJ (100 kcal) per day was associated with additional length gains of 9 mm in the first year, and 5 and 4 mm in years 2 and 3, but had no effect after 3 years of age. Children from supplemented villages continued to weigh more, be taller and have higher lean mass at adolescence, although differences in height were reduced compared to those at 3 years old. Other studies have been analyzed differently but show similar finding when an age-stratified approach is used. Therefore, the effects of supplementation are small and less than expected. The supplemented children did not approach the levels of height and weight seen in more affluent groups or in some other developing countries; and as only some 15% of the energy of the supplement is utilized in growth processes, the fate of most of the supplement is not clear. These data come from research studies that are likely to be more successful than feeding programs because of the enthusiasm of the workers.

There are several possible explanations for this.[28] It may be the result of poorly designed and implemented feeding programs. Alternatively, it may be that the hypotheses and models of the processes involved have been unduly simplistic and inappropriate. The supplement may not reach the intended recipient or it substitutes for rather than supplements the habitual diet. "Leakage" and substitution should not be seen as failures of supplementation, as they may have benefits, albeit away from the intended recipient. The supplement may lead to increased activity, with the benefits of play and fitness and social competence, rather than growth.

Lindsay Allen,[29] in her exemplary review of nutritional influences on linear growth, concluded that there was a lack of clear, consistent evidence and that supplementation of zinc, iron, copper, iodine, vitamin A, or indeed energy or protein alone benefited linear growth in growth faltering in developed countries. In some studies, improvements were seen; in others, weight gain alone was affected; and in still others, there was no effect whatsoever. However, most of these studies have been on children older than the age at which growth faltering is most rapid. Alternatively, multiple deficiencies may be the cause of growth faltering. Except in the case of iodine, low-energy diets are usually low-food diets, which means low-nutrient diets; that is, multiple nutrient deficiencies.

Growth retardation due to zinc deficiency was first described in the Middle East in the 1960s. Zinc deficiency was associated with high-fiber, low-protein diets and with parasitic infections. High-level supplementation of zinc was necessary to achieve improvements in growth, presumably because the high phytate content of the high-fiber diets reduced its bioavailability. Supplemental zinc has been given to low height-for-age well-nourished children in developed countries with variable effects. Linear growth is often improved, but not weight gain. Often, there is no reason for these children to be zinc deficient unless they have unusually high dietary requirements for zinc. The role of zinc in the linear growth retardation of developed countries requires further work.

The most widespread dietary deficiency in the richer nations of North America and Europe is of iron. The incidence of childhood anemia has been falling at a time when iron-fortified formulae and cereals have become widespread and supplemental food programs have been introduced. Linear growth usually improves in response to iron treatment in anemic children. Iron deficiency also has non-hematological effects on behavior and cognitive performance. Non-nutritional factors, such as the social, economic and biological environments, are important in determining the response to supplementation. In each community, different factors may operate, and there may be no circumventing the need for a full description of the ecology of the community. Our hypotheses and models and methods of analyses may need to be refined to identify better what needs to be measured. There are, however, limits to the number of variables that can be studied, determined by the cost and quality of the data obtained. However, these studies provide good examples of the fact that diet does not operate in isolation but in concert with other environmental challenges, such as disease, poor schooling and general deprivation. Some of these issues are described in this book by Schell et al. (Chapter 10) and Steckel (Chapter 9). The answer would then seem to lie in general environmental enrichment. But, in the absence of endless resources, the key deprivations in each particular case need to be established by ecological study and a more targeted approach to entitlement, enrichment and empowerment adopted.

6.3.8 Overnutrition

Malnutrition in the form of overnutrition or obesity is a major nutritional problem in the western world and is becoming common in other parts, too. Here, too, the obvious assumption, that obesity has its origins in gluttony and that the appropriate remedial action is to reduce dietary intake, may not be valid. Obesity can arise only through a positive energy balance, but this does not need a high energy intake to develop. Low levels of physical activity, inactivity with roots in car use and leisure-time television viewing or computer and games console use, is a major etiological factor. Any lowering of energy intake in such children would have an increased risk of precipitating nutrient deficiencies, although changes in the types of macronutrients consumed, such as less fat and a lower energy-dense diet, may be beneficial. Rather, the negative energy balance should be brought about in the main by increased physical activity. From this will flow all the other major health benefits associated with increased activity, fitness and altered body composition, such as increased self-esteem. However, the difficulty of achieving and maintaining weight loss should not be underestimated; indeed, it should be used as a motivation for avoiding overweight and obesity.

One problem in the area of childhood obesity is how to define it. In adults, we have imperfect but widely accepted indices, such as body mass index (BMI) and waist circumference, for which particular values or cut-offs have been associated with a variety of risk factors, of morbidity and mortality. It then is possible to specify "healthy" or "recommended" levels. In children, the median value and other percentiles vary with age and not always in a linear or predictable way; and as yet, few studies have linked these indices to risk factors and very few with outcomes such as morbidity and mortality. Because of this, obesity may then be defined on a statistical basis, such as BMI above the

85th percentile, which is not entirely satisfactory (Chapters 19 and 21 provide more recent information). Bearing in mind the difficulties in assessing the global prevalence of obesity in children and adolescents because of the differing indices employed, longitudinal studies suggest that it has doubled in the last 30 years. The measurement and long-term health risks of childhood and adolescent fatness have been well reviewed by l'Allemand-Jander.[11]

The extent to which obesity in childhood persists into adulthood seems to depend on the time interval between the occurrence in childhood and adulthood. Thus, obesity in adolescence seems more persistent than obesity in childhood. This is the "recency" effect. Until now, most obese adults were not obese as children, but as the prevalence rises, this finding may change.

6.4 TRANSITIONS IN NUTRITION AND GROWTH

It is unlikely that humans would have evolved so successfully if they had selected inappropriate diet types. However, it is equally true that the rate of adoption of new and manufactured foods (decades) is much greater than the rate of human evolution (tens of generations), and we cannot assume that we are well adapted to contemporary foodstuffs and diet types. Some 50% of dietary energy in the world is from cereals or cereal products, but only in evolutionarily recent times has this become the case. The domestication of plants and of animals, begun 10,000 BP, that is, 500 generations ago, altered the diets of humans more than any previous subsistence change. Our societies and cultures have changed, too, and in ways such that technology and globalization have not been uniformly beneficial. Therefore, nutritional problems, albeit different problems from those of the distant past, exist in conditions of affluence.

Nutrition is usually ascribed a key role in the improvements in growth in Europe and other parts of the world over the past centuries. The evidence for this is much less extensive and definitive than we would like, and improvements in growth would not have occurred without concomitant improvements in housing, sanitation and health provision. How these transitions will continue in the future and what new ones will appear in the short or long term is in the realm of prediction but something for which the students of today may have the answer in their lifetime. A major epidemiological interest of the present centers on the hypothesis that maternal and fetal undernutrition program the body to respond to subsequent affluent nutrition in ways that lead to a number of degenerative diseases. If true, and if mothers and fetuses are no longer exposed to undernutrition in developed countries, then we may see a reduction in the incidence and mortality from these degenerative diseases, as is happening already in much of the developed world. However, much of the rest of the world has yet to experience the first stage of this transition. In many parts of Africa, some of which have a host of other health problems such as the prevalence of HIV infection, nutritional health seems set to fall, at least in the short term. This will have an impact on the growth of the next generation of children.

REFERENCES

1. Food and Agriculture Organization. *World Health Organization, and United Nations University (FAO/WHO/ UNU). Human energy requirements. Report of a Joint FAO/WHO/UNU Expert Consultation.* Geneva: WHO, http://www.fao.org/docrep/007/y5686e/y5686e00.htm; 2004.
   Useful as a source of information on the methods adopted to set protein and energy requirements internationally and those requirements.
2. Food Standards Agency. FSA nutrient and food based guidelines for UK institutions. http://www. food.gov.uk/multimedia/pdfs/nutguideuk.pdf
   This report provides some of the current recommendations for the UK. From 1 October 2010 responsibility for nutrition policy transferred from the Food Standards Agency to the Department of Health in England and to the Assembly Government in Wales. These changes will mean that the health departments in these countries will be responsible for nutrition advice, surveys and nutrition research, as well as nutritional labeling, nutrition and health claims, dietetic food and food supplements, calorie information in catering establishments, reformulation to reduce salt, saturated fat and sugar levels in food and reducing portion size (including catering and manufacturing).
3. US Department of Agriculture. Food and Nutrition Information Center, National Agricultural Laboratory. http://fnic.nal.usda.gov/nal_display/index.php?info_center=4&tax_level=1.
   At this site you may find the dietary guidelines for Americans, 2010, and many other publications on nutrition advice and policy.
4. *Department of Health. Dietary reference values for food energy and nutrients for the United Kingdom.* London: HMSO; 1991.
5. http://www.nichd.nih.gov/milk/prob/critical.cfm; 23 September 2011.
6. Norgan NG. Measurement and interpretation issues in laboratory and field studies of energy expenditure. *Am J Hum Biol* 1996;**8**:143−58.
   The methods for measuring energy expenditure are described and critically appraised. Variation and adaptation in energy expenditure are considered in the context of discriminating between these and honest error, particularly in activity where many of the problems arise.
7. MyPyramid for Kids, http://www.cnpp.usda.gov/fgp4children.htm; 26 September 2011. KidsPyralindex.htm.
8. Sabry ZI. Assessing the nutritional status of populations: technical and political considerations. *Food Nutr* 1977;**3**:2−6.
9. Whitehead RG, Paul AA. Long-term adequacy of exclusive breast feeding: how scientific research has led to revised opinions. *Proc Nutr Soc* 2000;**59**:17−23.
   Improved nutritional and anthropometric guidelines are provided for the assessment of lactational adequacy and for when weaning may be initiated. These are based on revised dietary energy requirements.
10. Butte NF. Fat intake of children in relation to energy requirements. *Am J Clin Nutr* 2000;**72**:S1246−52.
11. l'Allemand-Jander D. Clinical diagnosis of metabolic and cardiovascular risks in overweight children: early development of chronic diseases in the obese child. *Int J Obes* 2010;**34**(Suppl. 2): S32−6.
12. Barker DJP, Osmond C, Law CM. The intrauterine and early postnatal origins of cardiovascular disease and chronic bronchitis. *J Epidemiol Community Health* 1989;**43**:237−40.
13. Bogin B, Silva MI, Rios L. Life history trade-offs in human growth: adaptation or pathology? *Am J Hum Biol* 2007;**19**:631−42.
14. Gluckman PD, Hanson MA. *The fetal matrix.* Cambridge: Cambridge University Press; 2005.
15. Smith PK, Bogin B, Varela-Silva MI, Loucky J. Economic and anthropological assessments of the health of children in Maya immigrant families in the US. *Econ Hum Biol* 2003;**1**: 145−60.

16. Emanuel I. Maternal health during childhood and later reproductive performance. *Ann NY Acad Sci* 1986;**477**:27−39.
17. Price KC, Coe CL. Maternal constraint of fetal growth patterns in the rhesus monkey (*Macaca mullata*): the intergenerational link between mothers and daughters. *Hum Reprod* 2000;**15**:452−7.
18. Varela-Silva MI, Azcorra H, Dickinson F, Bogin B, Frisancho AR. Influence of maternal stature, pregnancy age, and infant birth weight on growth during childhood in Yucatan. Mexico: a test of the intergenerational effects hypothesis. *Am J Hum Biol* 2009;**21**:657−63.
19. Bogin B, Thompson JL, Harper D, Merrell J, Chowdhury J, Meier P, et al. Intergenerational and transnational correlates of health for Bangladeshi adult daughters and their mothers: project MINA. *Am J Phys Anthropol* 2011;**144**(Suppl. 52):93.
20. Vidulich L, Norris SA, Cameron N, Pettifor JM. Bone mass and bone size in pre- or early pubertal 10-year-old black and white South African children and their parents. *Calcif Tissue Int* 2011;**88**:281−93.
21. Raynor P. Born in Bradford Collaborative Group. Born in Bradford, a cohort study of babies born in Bradford, and their parents: protocol for the recruitment phase. *BMC Public Health* 2008;**8**:327.
22. Danielle PC, van Dusseldorp M, van Staveren WA, Hautvast JGAJ. Effects of macrobiotic diets on linear growth in infants and children until 10 years of age. *Eur J Clin Nutr* 1994;**48**(Suppl):S103−12.
23. Martorell R, Keitel Khan L, Schroeder DG. Reversibility of stunting: epidemiological findings in children from developing countries. *Eur J Clin Nutr* 1994;**48**(Suppl. 1):S45−57.
24. Tanner JM. *Foetus into man: physical growth from conception to maturity.* 2nd ed. Ware: Castlemead Publications; 1989. p. 130.
25. Adair LS. Filipino children exhibit catch-up growth from age 2 to 12 years. *J Nutr* 1999;**129**:1140−8.
26. Satyanarayana K, Nadamuni Naidu A, Narasinga Rao BS. Adolescent growth spurt among rural Indian boys in relation to their nutritional status in early childhood. *Ann Hum Biol* 1980;**7**:359−65.
27. Martorell R, Scrimshaw NS. The effects of improved nutrition in early childhood: the Institute of Nutrition of Central America and Panama (INCAP) Follow-up Study. *J Nutr* 1995;**125**(Suppl. 4):1027S−138S.
28. Norgan NO. Chronic energy deficiency and the effects of energy supplementation. In: Schürch B, Scrimsaw NS, editors. *Chronic energy deficiency; consequences and related issues.* Lausanne: IDECG; 1987. p. 59−76.
29. Allen LH. Nutritional influences on linear growth: a general review. *Eur J Clin Nutr* 1994; **48**(Suppl. 1):S75−89.
    This review of the literature finds that apart from non-metabolic co-morbidities, namely musculoskeletal complications and attention deficit/hyperactivity disorders, at least one cardiovascular and metabolic risk factor was seen in the majority of the overweight children. Waist circumference, but also to BMI and fat mass, were positively correlated with health risk.

SUGGESTED READING

Nutrition textbooks

There is no shortage of choice of nutrition textbooks, although most are written and published with a US audience in mind. Most are very student friendly, well written and illustrated, with good structure and development of material, end of chapter summaries, assessment activities and, nowadays, lists of Web-based resources. There is some variation in the amount of basic biochemistry and anatomy and physiology that is included. The two books used by Bogin in teaching UK undergraduates are:

Thompson J, Manore N. *Nutrition: an applied approach.* 3rd ed. Pearson. (This book is for the USA market and has a useful companion website:) http://wps.aw.com/bc_thompson_nutrition_3/176/45256/11585553.cw/index.html; 2011.

Webb GP. *Nutrition: a health promotion approach.* 3rd ed. London: Arnold; 2008. (This book is for the UK market.)

Nutrition and growth in bioanthropology, growth and human biology texts

Bogin B. *Patterns of human growth.* 2nd ed. Cambridge: Cambridge University Press; 1999. p. 268—82.

Dr Norgan kindly wrote that, "The biocultural approach to growth and nutrition by this informed and readable author ensures that the relationships of nutrition and growth are considered within the context of the wider environment. However, it would be easy to gain the idea from the section on 'the milk hypothesis' that there is something specific to milk and its effects on growth of undernourished children not shared by other good sources of nutrient. In practice, its advantages are likely to be in non-nutritional terms such as cost and availability". That is a good point; however, more recent research on milk shows it to have some direct beneficial effects on growth and health (see Wiley AS. Consumption of milk, but not other dairy products, is associated with height among US preschool children in NHANES 1999—2002. Ann Hum Biol 2009;36:125—38; Wiley AS. Milk, but not dairy intake is positively associated with BMI among US children in NHANES 1999—2004. Am J Hum Biol 2010;22:517—25).

Eveleth PB, Tanner JM. *Worldwide variation in human growth.* 2nd ed. Cambridge: Cambridge University Press; 1990. p. 191—8, 219—23.

This is a succinct description of nutrition and growth in infancy and adolescence located in a classical account of environmental influences in growth.

Tanner JM. *Foetus into man.* 2nd ed. Ware: Castlemead Publications; 1989. p. 129—40.

There is a short informative account of the effects of nutrition on growth and the tempo of growth which is a useful introduction to the topic. The section on experimental models is unusual in such accounts but illustrates well what can be learned from laboratory-based studies.

Ulijaszek SJ, Strickland SS. *Nutritional anthropology: prospects and perspectives.* London: Smith-Gordon; 1993. p. 119—31.

There is a short section on nutrition and growth but much else of interest and relevance throughout the book, including the consequences of small size, seasonality, and growth and transitions in subsistence.

Ulijaszek SJ, Johnston FE, Preece MA. *The Cambridge encyclopedia of human growth and development.* Cambridge: Cambridge University Press; 1998.

Many entries are relevant to nutrition and growth in the sections on infant feeding and growth (pp. 320—5), nutrition (pp. 325—33), and elsewhere. These provide useful introductions to a number of topics under these headings.

Zerfas AJ, Jelliffe DB, Jelliffe EFP. Epidemiology and nutrition. In: Falkner F, Tanner JM, editors. *Human growth: a comprehensive treatise. Postnatal growth.* 2nd ed. Vol. 2. New York: Plenum; 1986. p. 475—500.

Although much new data has been collected since this chapter was published, its content and structure, with emphases on methods and approaches to the effects of protein—energy malnutrition on human growth, are exemplary and a valuable source of guidance. The three-volume series is widely available and repays consultation.

INTERNET RESOURCES

Many internet resources are given in the main text of this chapter. A few additional sites are:

The International Obesity Task Force, with a section on childhood obesity and links to other obesity sites: http://www.iotf.org

The Human Growth Foundation: http://www.hgfound.org/

British Nutrition Foundation; 2008. Food — a fact of life: www.foodafactoflife.org.uk

CHAPTER 7

Genomics, Epigenetics and Growth

Geert R. Mortier*, Wim Vanden Berghe**

*Department of Medical Genetics, Antwerp University Hospital and University of Antwerp, B-2650 Edegem (Antwerp), Belgium

**Department of Biomedical Sciences, PPES Lab Protein Science, Proteomics and Epigenetic Signaling, Antwerp University, B-2610 Wilrijk, Belgium

Contents

7.1 INTRODUCTION

The major part of the human genome resides in the nucleus of each cell. It is referred to as the nuclear genome. A very small proportion of the genome can be found outside the nucleus, in the mitochondria of the cell. It has the structure of a circular chromosome and is available in many copies in each mitochondrion. This mitochondrial chromosome, the complete DNA sequence of which was unraveled in 1981, is approximately 16.5 kb in size and contains only 37 genes.

The nuclear genome is much more complex than the mitochondrial genome. Over the past 20 years, remarkable progress has been made in our understanding of the structure of the nuclear genome. The completion of the Human Genome Project (started in 1990) and the publication of the first draft of the human genome sequence in 2001 were important milestones. We have known since 1956 that the nuclear genome contains 23 pairs of chromosomes (total 46 chromosomes) with each set of 23 chromosomes inherited from one parent. It is now estimated that these 23 different

N. Cameron & B. Bogin (eds): Human Growth and Development, Second edition.
ISBN 978-0-12-383882-7, Doi: 10.1016/B978-0-12-383882-7.00007-6

chromosomes harbor about 20,000 genes. The exact number of genes is still not precisely known but it is clear that the coding DNA sequences in these genes comprise only approximately 1% of the total DNA.

Each chromosome is a single, continuous double-stranded DNA molecule (double helix) surrounded by chromosomal proteins (histones and non-histone proteins) that are necessary for normal chromosome structure, function and gene expression. The structure of the DNA molecule was elucidated by Watson and Crick in 1953. It consists of two polynucleotide chains that form a double helix. Each nucleotide is composed of a nitrogen-containing base, a phosphate group and a five-carbon sugar (deoxyribose). There are two purine bases (adenine and guanine) and two pyrimidine bases (thymine and cytosine). The genetic information is solely determined by the bases. The phosphate group and sugar moiety serve only as the backbone of the DNA chain and carrier of the bases. The 23 different chromosomes contain altogether 3 billion nucleotides. The average size of a gene is about 10,000—15,000 nucleotides or 10—15 kb (1 kb is 1000 bases or nucleotides).

The function and expression of genes is determined not only by the DNA sequence (sequence of the thousands of nucleotides) within and outside the gene (regulating sequences in the neighborhood of the gene) but also by other molecular mechanisms that modify the DNA without altering the sequence of the nucleotides. The latter mechanisms are referred to as epigenetic mechanisms and are the subject of epigenetic studies (epigenetics). They usually involve the attachment of chemical groups to DNA and its associated proteins. These modifications, including methylation, acetylation, ubiquitination and phosphorylation, often result in alterations of gene expression. Epigenetic mechanisms will be defined and discussed in more detail later in this chapter.

7.2 GENOMIC FACTORS THAT INFLUENCE AND REGULATE GROWTH

Growth is a complex phenomenon influenced by several factors, both genetic and non-genetic. This chapter focuses on height, but all aspects of human growth have some genetic basis. The wide variation of adult height, which approximates a Gaussian distribution in the normal population, suggests that many genes regulate linear growth in humans. There is no doubt that each human chromosome harbors several genes that determine human growth. Chromosomal disorders characterized by short stature illustrate the importance of a correct dosage of genes for normal linear growth. Copy number variations of smaller chromosomal fragments can also cause growth failure. Single nucleotide substitutions in one particular gene may result in more severe forms of dwarfism. Recently, genome-wide association studies have mapped several loci in the human genome that may play an important role in the regulation of linear growth (discussed later in Section 7.5).

7.2.1 Numerical Chromosomal Aberrations

Turner syndrome is the prime example of an abnormal chromosome number causing short stature. This syndrome affects only females. The majority of females with Turner syndrome only have 45 chromosomes. They lack one of the two X chromosomes that females normally have. The absence of this second sex chromosome is responsible for the health problems in girls with Turner syndrome. Typical abnormalities in this syndrome include short stature, gonadal dysgenesis (resulting in abnormal sexual development and infertility) and various congenital anomalies mostly affecting the heart, blood and lymphatic vessels (coarctation of the aorta, lymphedema), and kidneys. Growth hormone (GH) treatment is now standard in Turner syndrome and can result in gains of up to 10 cm to the final height. Some girls with Turner syndrome have 46 chromosomes but lack a portion of one of the X chromosomes. Deletions of the short arm usually result in short stature whereas deletions of the long arm usually cause gonadal dysgenesis. On the tip of the short arm, in the so-called pseudoautosomal region, resides the SHOX gene. This gene codes for a transcription factor that plays an important role in growth regulation.[1] Deletion of this gene alone may also cause Léri—Weill dyschondrosteosis, which is a mild growth disorder.

Lack of one chromosome is very unusual in liveborn babies. Besides Turner syndrome, no other examples of monosomies (i.e. lack of one member of a chromosome pair) are known in newborn babies. The opposite, namely the presence of one extra chromosome (total of 47 chromosomes) is also exceptional and only tolerated for a few autosomes. In newborn babies only trisomy 13 (Patau syndrome), trisomy 18 (Edward syndrome) and trisomy 21 (Down syndrome) are observed. These affected babies usually have multiple congenital anomalies and severe delay in neuromotor development, and may also show signs of growth retardation.

7.2.2 Structural Chromosomal Aberrations

Not only may the lack of an entire chromosome result in a short stature syndrome (such as Turner's syndrome), but also structural rearrangements of one particular chromosome may disturb normal growth. Many structural chromosomal aberrations in humans are known and several of them will cause short stature in addition to other congenital anomalies and mental retardation.[2] One example is the Wolf—Hirschhorn syndrome.[3] This disorder is caused by a deletion of the distal part of the short arm of chromosome 4 (the most common breakpoint is 4p15). The syndrome is characterized by a marked growth retardation of prenatal onset and severe mental retardation. Affected children usually have a distinctive face which allows the experienced clinical geneticist to recognize the condition early in life (Figure 7.1). Congenital malformations of the brain, heart, eye and kidney are very common.

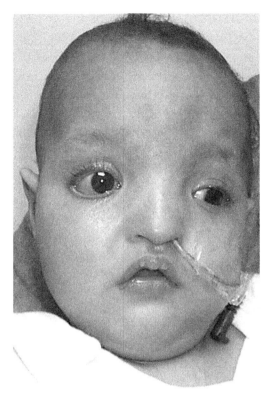

Figure 7.1 Characteristic facial features in an infant with the Wolf—Hirschhorn syndrome. The forehead is high, the nasal root is broad and the eyes are wide set. This figure is reproduced in the color plate section.

The deletion in the Wolf—Hirschhorn syndrome is at least 2 Mb(=2,000,000 nucleotides) in size and contains at least 48 genes. Children with this disorder have for this set of genes only one copy on the normal chromosome 4, thus illustrating the relevance of a normal gene dosage in growth and development.

7.2.3 Copy Number Variations

The development of array-based genomic analysis has enabled small, previously unidentified chromosomal aberrations in the human genome to be detected. With classical cytogenetic analysis numerical chromosomal abnormalities are easily detected, but structural rearrangements that are smaller than 3—5 Mb cannot be picked up with this technique. To overcome this limitation, array-based karyotyping (molecular karyotyping) has been developed over the past decade.[4] This new technology is now widely used in diagnostic genetic laboratories and it can be anticipated that in the near future it will gradually replace classical karyotyping for diagnostic purposes. Molecular karyotyping has shown that each individual may harbor several small deletions or duplications in the genome (copy number variations), many of them being harmless and therefore considered as benign polymorphisms. However, these copy number variations may also cause genetic disorders, of which several have been delineated over the past 5–10 years. These genetic disorders are usually referred to as microdeletion or micro-duplication syndromes because the causal deletion/duplication is too small (sizes ranging from 100 kb to a few Mb) to be detected by classical karyotyping. These disorders were not discovered before the introduction of these array-based techniques. It has become clear that these copy number variations are either benign polymorphisms or rear-rangements with variable effects on health, growth and development. Some of them are considered as predisposing genetic factors for the development of a particular disease (e.g. autism). Others have a more dramatic effect and almost always result in a genetic disorder (e.g. the velocardiofacial syndrome caused by a microdeletion of chromosome band 22q11). Microduplication syndromes are usually milder and more variable than microdeletion syndromes. They usually affect cognitive capabilities, language develop-ment, social skills and behavior of the affected individuals. The more severe conditions result in congenital anomalies. Growth retardation is also a frequently observed feature.

The 12q14 microdeletion syndrome illustrates how a small copy number variation may affect linear growth and development.[5] This microdeletion syndrome was initially identified during a gene mapping study for osteopoikilosis. Osteopoikilosis is a benign genetic disorder characterized by the presence of numerous small and dense spots in the epiphyseal regions of tubular bones (Figure 7.2). It is caused by loss-of-function muta-tions in the LEMD3 gene. To date, at least seven unrelated patients with the 12q14 microdeletion syndrome have been reported in the literature. These affected individuals have as common features learning disabilities, failure to thrive, proportionate short stature and variable degrees of osteopoikilosis on skeletal radiographs. They all have a small deletion at chromosome 12q14, ranging in size from 1.35 Mb to 9 Mb with a common deleted region of about 2.5 Mb, containing 10 RefSeq genes (Figure 7.3). Patients with a deletion encompassing the LEMD3 gene usually develop osteopoikilosis lesions in the bones.

In the common deleted region resides the HMGA2 gene, which is a strong candidate for the growth failure in these patients. The HMGA2 gene codes for an architectural protein that regulates the transcription of a number of genes. Several lines of evidence indicate that HMGA2 plays an important role in regulating linear growth in humans. First, inactivating mutations in mice result in small-sized animals (the pygmy mouse phenotype). Second, a chromosomal rearrangement disrupting the HMGA2 gene has been reported in a boy with extremely tall stature and advanced growth. In addition, a small intragenic deletion has been identified in a Belgian family with short stature.[5] Third, whole genome association studies have revealed an association between a nucle-otide change in the HMGA2 gene and adult height in the general population. The

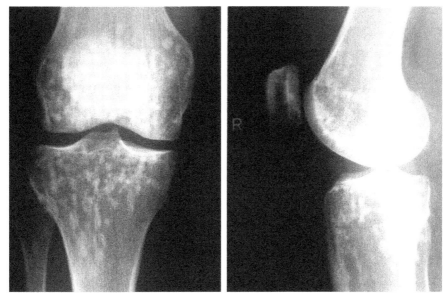

Figure 7.2 Anteroposterior and lateral views of the right knee showing multiple osteopoikilosis lesions in the lower part of the femur and upper part of the tibia.

characterization and delineation of the 12q14 microdeletion syndrome show how advances in technology and better insights into the human genome can result in better understanding of the function of individual genes in complex processes such as growth and development.

7.3 MUTATIONS IN GENES

Not only do small deletions and duplications in the genome affect linear growth; single nucleotide changes in individual genes can also impair growth and development. Whereas numerical and structural chromosomal aberrations usually exert their effect through the loss of one or more genes (gene dosage effect), mutations in individual genes can have different mechanisms in the pathophysiology of the growth disorder they are causing. They can result in either a loss of function or a gain of function of the mutated gene. In other words, they can either inactivate or enhance the function of the mutated protein.

Genes affecting growth and development have different functions. Some of them regulate cell division; others affect intracellular signaling pathways. Some genes code for growth factors or hormones; others for proteins that control bone formation and growth.

The presumably most frequent and best known mutation in the human genome that causes a serious growth disorder is the mutation responsible for achondroplasia

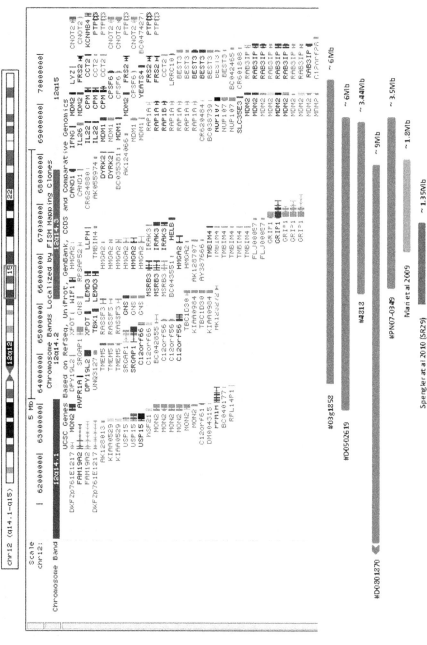

Figure 7.3 Overview of seven patients with the 12q14 microdeletion syndrome. The microdeletion in each patient is shown as a horizontal bar at the bottom of the figure. Red bars represent patients investigated by the authors. The microdeletions are plotted against the corresponding genomic region on chromosome 12q14–q15 (UCSC Genome Browser) showing the RefSeq genes within this region. With the exception of the patient reported by Mari et al., all deletions encompass the LEMD3 gene. The HMGA2 gene is deleted in all patients. This figure is reproduced in the color plate section.

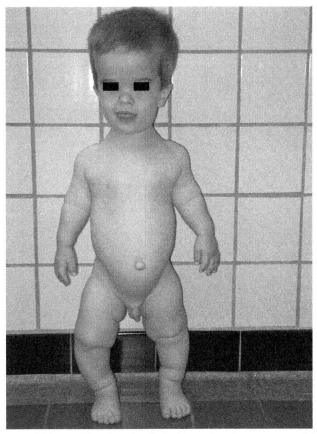

Figure 7.4 Three-year-old boy with achondroplasia. Note the short stature with short limbs and bowing of the legs. This figure is reproduced in the color plate section.

(Figure 7.4). Achondroplasia is an autosomal dominant disorder affecting approximately one in 10,000 individuals.[6] It is the most common form of dwarfism in humans. Achondroplasia is most frequently caused by a single nucleotide mutation. This mutation will result in the change of one amino acid (glycine in position 380 is changed to arginine) in the transmembrane domain of the fibroblast growth factor receptor 3 gene (FGFR3). The FGFR3 gene encodes a membrane receptor that belongs to the family of fibroblast growth factor receptors. Upon binding of the ligand (fibroblast growth factors), the FGFR3 receptors become activated and will dimerize, which results in increased intracellular signaling of the activated cell. It has been shown that the mutation that causes achondroplasia constitutively activates the receptor without binding of the ligand. The mutation has a gain–of–function effect which will result in increased intracellular signaling and ultimately decreased chondrocyte proliferation in the growth plate. FGFR3 is a negative regulator of longitudinal bone growth. FGFR3 knockout mice are longer than the wild-type animals. Heterozygous mutations in FGFR3 not only result in achondroplasia but may also cause a lethal disorder with severe dwarfism (thanatophoric dysplasia) and a milder phenotype such as hypochondroplasia. The net effect on growth mainly depends on the nature and location of the mutation within the FGFR3 gene and its capability to activate the receptor protein in the absence of ligand binding.

Many genetic disorders with short stature have been defined at the nucleotide level. The causal genes have been unraveled and this new knowledge has provided more insights into the genetic regulation of growth. However, many genes that regulate growth are still unknown. For many growth disorders the genetic cause is still obscure.

7.4 ASSOCIATION STUDIES

Until recently, there has been little success in identifying the genetic factors influencing normal variation in human height. Before 2007, the hunt for genes had followed a familiar path of genome-wide linkage studies and candidate-gene approaches. As was the case for most common traits, these approaches had limited success. The most likely reason for the lack of success of the linkage approach is that height variation is explained by many variants, each of which explains only a small proportion of the heritability of the trait. In this scenario, association has been shown to be a much more powerful approach to gene identification than linkage. Until recently, it was only feasible to genotype a small subset (tens or hundreds) of the more than 10 million common variants that occur across the human genome in association studies, so only a small number of candidate genes could be assessed. Now, however, with technological advances giving researchers the ability to assay efficiently and cost-effectively a large proportion (over two-thirds) of the common variation across the human genome, progress has been rapid and genome-wide association studies (GWAS) have become possible. These GWAS have been designed to discover the genetic factors underlying complex genetic disorders. They have also been used to identify genetic variation associated with quantitative traits such as lipid levels, body mass index and height.

Researchers have found common variation in over 180 loci which are associated with height.[7,8] Some of these loci were replicated in several independent studies. For some of them there is biological evidence that these loci (genes) are involved in height regulation. The involved genes fall into known pathways and processes such as cell division, cellular growth and apoptosis, energy metabolism, GH signaling, endochondral ossification and growth plate functioning. For many of these genes the final proof that they regulate human height will come with the identification of mutations causing growth disorders. However, the common variants within these genetic loci seem to explain less than 10% of the population variation in height. It is not yet clear where the "missing heritability" comes from. Possible explanations include: (1) the heritability has been overestimated in the past; (2) GWAS fail in detecting rare and new genetic variants; and (3) gene—gene and gene—environment interactions have not been taken into account.

There is considerable evidence for the induction of phenotypical variations in human growth by variation in the quality of the early life environment.[9] Studies show consistent associations between prenatal famine and adult body size, diabetes and schizophrenia.[10] Although the mechanisms behind these relationships are unclear, an involvement of epigenetic regulation has been hypothesized and will be discussed in the following section. Epigenetics suggests that the genome can "remember" certain influences to which it is exposed, particularly early in life, which cause modifications to DNA that in turn alter the way it operates later in life. On occasion, these changes may even be passed on from one generation to the next.[11]

7.5 EPIGENETIC FACTORS THAT INFLUENCE AND REGULATE GROWTH

Not only genetic defects but also epigenetic changes may disturb normal growth and development. The major neuroendocrine axis that regulates growth [the GH—insulin-like growth factor-1 (IGF-1) axis] responds to epigenetic control of DNA expression. Epigenetic mechanisms usually involve the attachment of chemical groups to DNA or its associated chromatin proteins, which affects the accessibility for gene transcription. These modifications, including methylation, acetylation, ubiquitination and phosphorylation, often result in alterations of gene expression (Figure 7.5).

The term "epigenetics" refers to the heritable changes in gene expression and cell phenotype in the absence of alterations to the DNA sequence. Structural modifications to genes that do not change the nucleotide sequence itself but instead control and regulate gene expression include DNA methylation, histone modification and non-coding RNA regulation. Epigenetic changes are believed to be a result of changes in an organism's environment that result in fixed and permanent changes in most differentiated cells. Hence, while the genome generally remains uniform in all the different cells of a complex organism, the epigenome controls the differential gene expression in most cell types, silencing or activating genes, and defining when and where they are expressed. It is well established in a number of animal and human studies that environmental exposures during critical periods of development may result in permanent alterations in the structure and function of affected organs (the "developmental programming" concept).[12]

Most human physiological systems and organs begin to develop early in gestation but become fully mature only after birth. A relatively long gestation and period of postnatal maturation allows for prolonged prenatal and postnatal interactions with the environment.[9] The mechanisms underlying developmental programming remain still unclear. However, recent findings suggest that epigenetic modifications induced during gestation and development can play a fundamental role.[13] Such "non-genomic tuning" of phenotype through developmental plasticity is adaptive because it attempts to match an

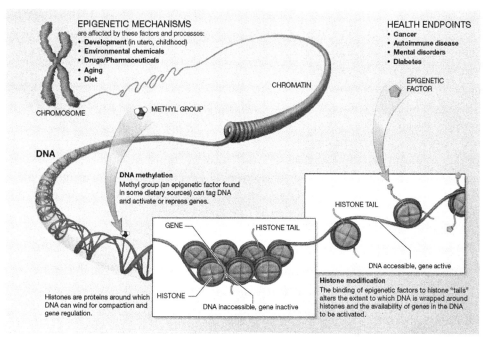

Figure 7.5 DNA methylation and histone acetylation control DNA accessibility and gene activation/ inactivation. This figure is reproduced in the color plate section. (Source: *Image from: http:// nihroadmap.nih.gov/EPIGENOMICS/epigeneticmechanisms.asp)*

individual's responses to the predicted environment. Most epidemiological studies of developmental origins of health and disease have used birth weight as an indicator of fetal nutrition and intrauterine growth. A growing body of literature shows an association between low birth weight and other measures of retarded fetal growth in many populations.[14] A number of studies have shown that prenatal exposure to famine is associated with various adverse metabolic and mental phenotypes later in life, including a higher body mass index and elevated plasma lipids, as well as increased risks of obesity, cardiovascular disease and schizophrenia. Most of these associations are dependent on the timing of the exposure during gestation and lactation periods. Early gestation seems to be an especially vulnerable period.[15]

The biological mechanisms contributing to associations between the prenatal exposure to famine and adult health outcomes are unknown but may involve persistent epigenetic changes. To answer the question whether prenatal exposure to famine can lead to persistent changes in the epigenome, the level of methylation of the IGF-2 gene, a key factor in human growth and development, was determined. Individuals who were exposed to famine during early gestation had a much lower level of methylation of IGF-2 than controls six decades after the exposure.[16] More recently, this observation was extended by testing a set of 15 additional candidate loci implicated in growth, metabolic and cardiovascular phenotypes.[17] Methylation of six of these loci (IL10, GNASAS, INSIGF, LEP, ABCA1 and MEG3) was associated with exposure to famine in utero. Collectively, these data indicate that DNA methylation differences after exposure to prenatal famine may persist during the human life course.

Furthermore, in a number of recent studies, it has been shown that a variety of environmental conditions can result in epigenetic alterations that can be passed to the next generations without continued exposure.[18,19] One of the most significant features of developmental programming is that adverse consequences of harsh environmental conditions during fetal or neonatal life can be passed from the first generation to the next.[6] For example, children born to mothers who were in their first trimester during the Dutch famine were heavier and those born to mothers who were in their third trimester were lighter than the children of women born before the famine.[20] Later, it was found that famine exposure in utero was associated with increased neonatal adiposity and poor health in later life of the offspring.[21] These findings suggest that inadequate feeding during the human developmental stages can result in transgenerational epigenetic effects and that these effects can be passed to the subsequent generations. Of special note, it is well known that immigrants from developing countries take two to three generations to reach an upper plateau in average stature when they have migrated to a more favorable environment.[22] This transgenerational increase in stature may well be due to changes in the epigenetic markers that regulate gene expression and body growth.

Another epigenetic mechanism was identified following pronuclear transfer experiments in mice in the early 1980s, which showed that maternal and paternal genetic contributions were non-equivalent and that both were indispensable for normal development. The introduction of reciprocal translocations into mice showed that discrete areas of the mouse genome were subject to differential parental regulation. In parallel with this fascinating mouse work, it was observed that several non-Mendelian human syndromes showed similar inheritance to phenotypes seen in the disomic mice. The mapping of deletions causative in Prader–Willi syndrome (PWS) and Angelman syndromes, for example, permitted localization of parentally non-equivalent genomic regions in humans. This parent-of-origin, monoallelic gene expression, with its associated differential DNA methylation (first shown in 1993), became defined as genomic imprinting. Research since the early 1990s has established the profound significance of imprinting for correct mammalian ontogenesis.[23]

Defined more simply, genomic imprinting refers to an epigenetic marking of genes that results in monoallelic expression depending on their parental origin, i.e. genes that are silent when maternally inherited but expressed when paternally inherited, or vice versa.[23,24] The regulation of imprinted loci is orchestrated by epigenetic factors including regulatory non-coding (ncRNAs), CpG DNA methylation of imprinting control regions (ICRs), post-translational modifications (PTMs) of histone proteins and changes in higher order chromatin structure.[24]

The reasons for the existence of imprinting are not completely known. A widely accepted hypothesis for the evolution of genomic imprinting is the parental conflict hypothesis.[25] Also known as the kinship theory of genomic imprinting, this hypothesis states that the inequality between parental genomes due to imprinting is a result of the differing interests of each parent in terms of the evolutionary fitness of their genes. The mother's evolutionary imperative is often to conserve resources for her own survival while providing sufficient nourishment to current and subsequent litters. In contrast, the father's evolutionary interest is to favor his own progeny at the expense of the mother and any of her future offspring. Accordingly, paternally expressed genes tend to be growth promoting whereas maternally expressed genes tend to be growth limiting.

In support of this hypothesis, genomic imprinting has been found in all placental mammals, where postfertilization offspring resource consumption at the expense of the mother is high; it has not been found in oviparous birds or monotremes (a class of oviparous mammals) where there is relatively little postfertilization resource transfer and therefore less parental conflict. The conflict between genes of different parental origin leads to the silencing of one copy but the expression of the other.[26] Parental-specific monoallelic expression thus balances fetal growth to the equal benefit of both parental genomes, in spite of the resulting potentially damaging haploinsufficiency. The latter occurs when a diploid organism only has a single functional copy of a gene and the single functional copy of the gene does not produce enough of a gene product (typically a protein) to bring about a wild-type condition, leading to an abnormal or diseased state. Haploinsufficiency is therefore an example of incomplete or partial dominance, as a heterozygote (with one mutant or silenced allele and one normal allele) displays a phenotypic effect.

There are two critical periods in epigenetic reprogramming: gametogenesis and early preimplantation development. Major reprogramming takes place in primordial germ cells in which parental imprints are erased and totipotency is restored. Imprint marks are then re-established during spermatogenesis or oogenesis, depending on sex. Upon fertilization, genome-wide demethylation occurs followed by a wave of new methylation. Early embryogenesis is a critical time for epigenetic regulation, and this process is sensitive to environmental exposures such as heavy metals, bioflavonoids and endocrine disruptors (e.g. bisphenol A, phthalates).[27,28] Considerable epidemiological, experimental and clinical data have amassed showing that the risk of developing disease in later life is dependent on early life conditions, mainly operating within the normative range of developmental exposures. This relationship reflects plastic responses made by the developing organism as an evolved strategy to cope with immediate or predicted circumstances, to maximize fitness in the context of the range of environments that it may potentially face. There is increasing evidence, in both animals and humans, that such developmental plasticity is mediated in part by epigenetic mechanisms. Epigenetic alterations are being linked to several important reproductive outcomes, including early pregnancy loss, intrauterine growth restriction, congenital syndromes, preterm birth and pre-eclampsia. The diversity of environmental exposures linked to adverse reproductive effects continues to grow.[18,19]

7.5.1 Imprinting Defects and Human Growth Disorders

The human chromosome 11p15 encompasses two imprinted domains important in the control of fetal and postnatal growth (Figure 7.6). Each domain is differentially methylated and regulated by its own ICR. ICR1 is paternally methylated and located in the telomeric region of the 11p15 locus. It regulates the H19 and IGF2 genes. ICR2 is maternally regulated and located in the centromeric region. It regulates the KCNQ1 and CDKN1C genes. Imprinting defects in these two domains are implicated in two clinically opposite growth disorders, the Silver—Russell syndrome (SRS) and the Beckwith—Wiedemann syndrome (BWS).[26,29–31]

SRS is a genetic disorder characterized by prenatal and postnatal growth retardation. Mean birth weight at term is around 1900 g. Affected children have a proportionate short stature with normal head circumference. The average adult height of males is 151 cm and that of females is 139 cm. Despite the normal head size, developmental delay and learning disabilities are not uncommon in children with SRS. Hypoglycemia is

Insulator model – Igf2 cluster

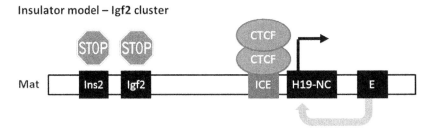

Binding of the insulator protein CTCF to imprint control element (ICE) blocks Ins2 and Igf2 mRNA transcription, enhancers activate noncoding RNA H19

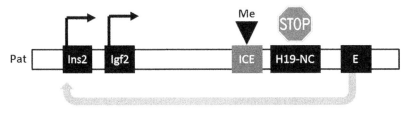

Paternal DNA methylation imprinting silences ICE and transcription of ncRNA H19 , enhancers activate Ins2 and Igf2 mRNA transcription

Figure 7.6 Schematic overview of imprinting at the Igf2/H19 loci on the human chromosome 11p15. This figure is reproduced in the color plate section.

a well-known complication in the neonatal period. Food aversion and recurrent vomiting result in severe failure to thrive during infancy and early childhood.

SRS is a genetically heterogeneous disorder. Different genetic defects have been identified in children with SRS.[29,30] In over 50% of the children with SRS a loss of DNA methylation is found at ICR1 of the 11p15 locus. Hypomethylation at ICR1 causes epigenetic dysregulation of the H19 and IGF-2 genes, ultimately resulting in reduced expression of IGF2 with growth restriction as a consequence. H19 and IGF2 are both imprinted genes, with H19 only expressed on the maternal allele (paternal imprinting) and IGF2 only expressed on the paternal allele (maternal imprinting). H19 is a growth-suppressing gene, whereas IGF2 is a growth-stimulating gene. The degree of hypomethylation at ICR1 seems to correlate with the severity of the growth phenotype as well as additional SRS diagnostic features. Other genetic defects identified in children with SRS include uniparental disomy (UPD) for chromosome 7 (see below) and structural chromosomal aberrations affecting genes that code for growth factors or GH (e.g. rearrangements at chromosome locus 17q25).

BWS is a genetic disorder characterized by prenatal and postnatal overgrowth.[31] Accelerated growth may start prenatally, usually in the last trimester, with increased birth weight and length as a result, or only manifest later on, after birth. At birth, polyhydramnios and large placenta are frequently seen. Growth velocity and bone age are advanced in the first 4—6 years of life, after which they may return to normal. Adult height tends to be at the upper end of the normal range. Characteristic clinical findings in infants and children with BWS include macroglossia, umbilical hernia and hemihypertrophy. Intelligence is usually normal but learning difficulties have been reported particularly if neonatal hypoglycemia was present and untreated. Children with BWS have an increased risk for developing embryonal tumors, particularly before the age of 7 years. Wilms' tumor (tumor of the kidneys) is most commonly observed. The risk of tumors is greater in children with hemihypertrophy, where one side of the body is more developed than the other side. The genetic basis of BWS is also complex.[31] In 10% of children with BWS a gain of DNA methylation is found at the ICR1 locus. Loss of methylation at ICR2 is found in 60% of BWS cases. There is also evidence that in both SRS and BWS patients multiple loci in the genome can show imprinting defects.[32]

Although not yet completely understood, SRS and BWS can be used as models to decipher the functional link between the observed (epi)genetic mutations and the clinical features in individuals with disturbed growth. Thus, fetal and postnatal growth is epigenetically controlled by different ICRs at 11p15 and other chromosomal regions. Although DNA methylation defects in imprinting control regions in SRS and BWS are clearly established, little information is available regarding the mechanism responsible for defective ICR DNA methylation. Despite mutation screening of several factors involved in the establishment and maintenance of methylation marks, including ZFP57, MBD3,

DNMT1 and DNMT3L, the molecular clue for the ICR1/ICR2 hypomethylation remains unclear. Genetic analysis of H19 and the IGF2/H19 imprinting control region has uncovered various new genetic defects, including mutations and duplications that may be relevant in (epi)genetic (re)organization of the locus and etiology of SRS.[33]

The use of assisted reproductive technology (ART) has been shown to induce epigenetic alterations and to affect fetal growth and development. In humans, several imprinting disorders, including BWS, occur at significantly higher frequencies in children conceived with the use of ART than in children conceived spontaneously.[34,35]

The cause of these epigenetic imprinting disorders associated with ART, including hormonal hyperstimulation, in vitro fertilization (IVF), intracytoplasmic sperm injection (ICSI), micromanipulation of gametes, exposure to culture medium, in vitro ovocyte maturation and time of transfer, remains unclear. However, recent data have shown that in patients with BWS or SRS, including those born following the use of ART, the DNA methylation defect involves imprinted loci other than 11p15 [11p15 region: CTCF binding sites at ICR1, H19 and IGF2 DMRs, KCNQ1OT1 (ICR2); SNRPN (chromosome 15 q11−13), PEG/MEST1 (chromosome 7q31), IGF-2 receptor and ZAC1 (chromosomes 6q26 and 6q24, respectively), DLK1/GTL2-IG-DMR (chromosome 14q32) and GNAS locus (chromosome 20q13.3)].[34]

7.5.2 Uniparental Disomy

Some rare genetic disorders illustrate that it is not enough to have the correct number and structure of chromosomes. For some chromosomes it is essential to have one copy from each parent (and not both copies from the same parent). Zygotes in which both sets of 23 chromosomes originate from the same parent (uniparental diploidy) will not develop normally. The same holds true for some chromosomes: having both homologs derived from the same parent [uniparental disomy (UPD)] may cause a genetic abnormality. UPD may go undiagnosed if it causes no obvious phenotypic effects. If UPD for a particular chromosome causes an abnormality, it usually indicates the presence of imprinted genes on that chromosome. UPD may involve a whole chromosome or only part of a chromosome. UPD for the maternal chromosome 7 is found in 10% of children with SRS. Although this observation suggests the presence of at least one imprinted gene on chromosome 7 that is responsible for SRS, studies have so far failed to identify this gene (or genes). UPD for the 11p15 region may cause BWS. UPD for other chromosomes or chromosomal portions has been documented in other genetic syndromes such as PWS and Angelman syndrome. Both disorders can be caused by UPD of the 15q11−q13 region, with maternal UPD resulting in PWS and paternal UPD in Angelman syndrome. These disorders have mental retardation as a common feature. In addition, children with PWS have obesity, short stature and abnormal sexual development. Growth retardation is also seen in children with UPD for the paternal

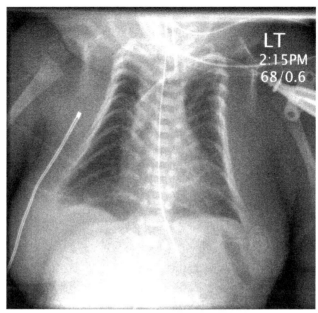

Figure 7.7 Chest radiograph of an infant with uniparental disomy of chromosome 14. The configuration of the ribs resembles that of a coat-hanger.

chromosome 14. A characteristic and diagnostic helpful feature in these children is the narrow thorax with so-called "coat-hanger" ribs on radiographs (Figure 7.7).

7.6 CONCLUSION

Growth is a complex phenomenon which is influenced by many factors in the genome and environment. This chapter has presented a discussion of genetic (numerical and structural chromosomal aberrations, copy number variations, gene mutations) and epigenetic (imprinting disorders, UPD, prenatal famine exposure) anomalies that result in growth failure. For many years, disease susceptibility was believed to be determined predominantly by the genetic information carried in the DNA sequence. In recent years, however, it has become clear that epigenetic modifications play an equally essential role in growth development and that this process, especially at key developmental periods, is very susceptible to environmental modulation. The theory behind the developmental origins of adult height suggests that there are mammalian epigenetic fetal survival mechanisms that down-regulate fetal growth, both for the fetus to survive until birth and to prepare it for a restricted extrauterine environment. Insights into the interplay of genetic and epigenetic growth mechanisms should provide avenues to develop possible therapies for short stature syndromes and common adult diseases associated with intra-uterine growth restriction.

REFERENCES

1. Marchini A, Rappold G, Schneider KU. SHOX at a glance: from gene to protein. *Arch Physiol Biochem* 2007;**113**:116—23.
2. Schinzel A. *Catalogue of unbalanced chromosome aberrations in man.* 2nd ed. Berlin: Walter de Gruyter; 2001.
3. Zollino M, Murdolo M, Marangi G, Pecile V, Galasso C, Mazzanti L, et al. On the nosology and pathogenesis of Wolf—Hirschhorn syndrome: genotype—phenotype correlation analysis of 80 patients and literature review. *Am J Med Genet C Semin Med Genet* 2008;**148C**:257—69.
4. Buysse K, Delle Chiaie B, Van Coster R, Loeys B, De Paepe A, Mortier G, et al. Challenges for CNV interpretation in clinical molecular karyotyping: lessons learned from a 1001 sample experience. *Eur J Med Genet* 2009;**52**:398—403.
5. Buysse K, Reardon W, Mehta L, Costa T, Fagerstrom C, Kingsbury DJ, et al. The 12q14 microdeletion syndrome: additional patients and further evidence that HMGA2 is an important genetic determinant for human height. *Eur J Med Genet* 2009;**52**:101—7.
6. Horton WA, Hall JG, Hecht JT. Achondroplasia. *Lancet* 2007;**370**:162—72.
7. Lettre G. Genetic regulation of adult stature. *Curr Opin Pediatr* 2009;**21**:515—22.
8. Lanktree MB, Guo Y, Murtaza M, Glessner JT, Bailey SD, Onland-Moret NC, et al. Meta-analysis of dense genecentric association studies reveals common and uncommon variants associated with height. *Am J Hum Genet* 2011;**88**:6—18.
9. Gordon L, Joo JH, Andronikos R, Ollikainen M, Wallace EM, Umstad MP, et al. Expression discordance of monozygotic twins at birth: effect of intrauterine environment and a possible mechanism for fetal programming. *Epigenetics* 2011;**6**(5):579—92.
10. Lumey LH, Stein AD, Susser E. Prenatal famine and adult health. *Annu Rev Public Health* 2011;**32**:237—62.
11. Lumey LH, Stein AD. Transgenerational effects of prenatal exposure to the Dutch famine. *BJOG* 2009;**116**:868.
12. Gluckman PD, Hanson MA, Mitchell MD. Developmental origins of health and disease: reducing the burden of chronic disease in the next generation. *Genome Med* 2010;**2**:14.
13. Godfrey KM, Lillycrop KA, Burdge GC. Epigenetic mechanisms and the mismatch concept of the developmental origins of health and disease. *Pediatr Res* 2007;**61**:5—10R.
14. Hochberg Z, Feil R, Constancia M, Fraga M, Junien C, Carel JC, et al. Child health, developmental plasticity, and epigenetic programming. *Endocr Rev* 2011;**32**:159—224.
15. Heijmans BT, Tobi EW, Lumey LH, Slagboom PE. The epigenome: archive of the prenatal environment. *Epigenetics* 2009;**4**:526—31.
16. Heijmans BT, Tobi EW, Stein AD, Putter H, Blauw GJ, Susser ES, et al. Persistent epigenetic differences associated with prenatal exposure to famine in humans. *Proc Natl Acad Sci USA* 2008;**105**:17046—9.
17. Tobi EW, Lumey LH, Talens RP, Kremer D, Putter H, Stein AD, et al. DNA methylation differences after exposure to prenatal famine are common and timing- and sex-specific. *Hum Mol Genet* 2009;**18**:4046—53.
18. Gluckman PD, Hanson MA, Low FM. The role of developmental plasticity and epigenetics in human health. *Birth Defects Res C Embryo Today* 2011;**93**:12—8.
19. Robins JC, Marsit CJ, Padbury JF, Sharma SS. Endocrine disruptors, environmental oxygen, epigenetics and pregnancy. *Front Biosci Elite Ed* 2011;**3**:690—700.
20. Stein AD, Lumey LH. The relationship between maternal and offspring birth weights after maternal prenatal famine exposure: the Dutch Famine Birth Cohort Study. *Hum Biol* 2000;**72**:641—54.
21. Painter RC, Osmond C, Gluckman P, Hanson M, Phillips DIW, Roseboom TJ. Transgenerational effects of prenatal exposure to the Dutch famine on neonatal adiposity and health in later life. *BJOG* 2008;**115**:1243—9.
22. Hall JG. Review and hypothesis: syndromes with severe intrauterine growth restriction and very short stature — are they related to the epigenetic mechanism(s) of fetal survival involved in the developmental origins of adult health and disease? *Am J Med Genet A* 2010;**152A**:512—27.

23. Frost JM, Moore GE. The importance of imprinting in the human placenta. *PLoS Genet* 2010;**6**. e1001015.
24. Wu HA, Bernstein E. Partners in imprinting: noncoding RNA and polycomb group proteins. *Dev Cell* 2008;**15**:637–8.
25. Haig D. The kinship theory of genomic imprinting. *Annu Rev Ecol Syst* 2000;**31**:9–32.
26. Ubeda F. Evolution of genomic imprinting with biparental care: implications for Prader–Willi and Angelman syndromes. *PLoS Biol* 2008;**6**:e208.
27. Jirtle RL, Skinner MK. Environmental epigenomics and disease susceptibility. *Nat Rev Genet* 2007;**8**:253–62.
28. Time Magazine. *How the first nine months shape the rest of your life.* 22 September 2010. http://www.time.com/time/health/article/0,8599,2020815,00.html#ixzz1NLGx9ndv, September 2010;
29. Eggermann T, Begemann M, Binder G, Spengler S. Silver–Russell syndrome: genetic basis and molecular genetic testing. *Orphanet J Rare Dis* 2010;**5**:19.
30. Eggermann T, Begemann M, Spengler S, Schröder C, Kordass U, Binder G. Genetic and epigenetic findings in Silver–Russell syndrome. *Pediatr Endocrinol Rev* 2010;**8**:86–93.
31. Choufani S, Shuman C, Weksberg R. Beckwith–Wiedemann syndrome. *Am J Med Genet C Semin Med Genet* 2010;**154C**:343–54.
32. Azzi S, Rossignol S, Steunou V, Sas T, Thibaud N, Danton F, et al. Multilocus methylation analysis in a large cohort of 11p15-related foetal growth disorders reveals simultaneous loss of methylation at paternal and maternal imprinted loci. *Hum Mol Genet* 2009;**18**:4724–33.
33. Demars J, Shmela ME, Rossignol S, Okabe J, Netchine I, Azzi S, et al. Analysis of the IGF2/H19 imprinting control region uncovers new genetic defects, including mutations of OCT-binding sequences, in patients with 11p15 fetal growth disorders. *Hum Mol Genet* 2010;**19**:803–14.
34. Le Bouc Y, Rossignol S, Azzi S, Steunou V, Netchine I, Gicquel C. Epigenetics, genomic imprinting and assisted reproductive technology. *Ann Endocrinol Paris* 2010;**71**:237–8.
35. Pozharny Y, Lambertini L, Clunie G, Ferrara L, Lee MJ. Epigenetics in women's health care. *Mt Sinai J Med* 2010;**77**:225–35.

INTERNET RESOURCES

Genomic website of US Department of Energy: http://genomics.energy.gov/

Online Mendelian Inheritance in Man: http://www.ncbi.nlm.nih.gov/sites/entrez?db=omim

Database of submicroscopic chromosomal imbalances (DECIPHER): http://decipher.sanger.ac.uk/

ORPHANET: the portal for rare diseases and orphan drugs: http://www.orpha.net/consor/cgi-bin/index.php

Epigenome: http://epigenome.eu/

Learn Genetics – Epigenetics: http://learn.genetics.utah.edu/content/epigenetics/

Raine Study in Australia: http://www.rainestudy.org.au/

The International Society for Developmental Origins of Health and Disease (DOHaD): http://www.mrc.soton.ac.uk/dohad/index.asp?page=1

Epigenetics: the ghost in your genes (BBC): http://video.google.com/videoplay?docid=4942166965081178368

Epigenome at a glance: http://www.youtube.com/watch?v=s7dDd1bvNfA

CHAPTER 8

The Genetic Epidemiology of Growth and Development

Bradford Towne*, Ellen W. Demerath**, Stefan A. Czerwinski*

*Division of Epidemiology, Lifespan Health Research Center, Department of Community Health, Boonshoft School of Medicine, Wright State University, Dayton, OH 45420, USA

**Division of Epidemiology and Community Health, School of Public Health, University of Minnesota, Minneapolis, MN 55454, USA

Contents

N. Cameron & B. Bogin (eds): Human Growth and Development, Second edition.
ISBN 978-0-12-383882-7, Doi: 10.1016/B978-0-12-383882-7.00008-8

8.1 INTRODUCTION

In spite of the predominant role of genetic variation in causing the observed variability among children in their growth and development, studies of genetic influences on growth and development are few in comparison to the plethora of descriptive studies, population comparisons and studies of the impact of specific environmental factors. There are two main reasons for this. First, the courses of study under which many investigators of growth and development are trained (e.g. physical anthropology or human biology) usually provide little formal training in human genetics and statistical genetic analysis. Second, in order to best study genetic influences on growth and development, data from related children are needed. Preferably those data from related children are longitudinal, and ideally they are longitudinal data from large numbers of related children reared under different household environments. Unfortunately, such data are very rare.

The purpose of this chapter is to provide an overview of the genetic epidemiology of normal human growth and development. Although a treatise on quantitative genetic approaches to the study of growth and development is beyond the scope of this chapter, as is a complete review of the existing literature on the genetics of growth and development, the references, suggested readings and internet resources will provide a good starting point for the interested student to pursue further study. This chapter is meant to serve as an introduction to how auxologists can most profitably study the genetics of growth and development with the methods and approaches available today.

Almost half a century ago, Neel and Schull[1] proposed that the epidemiological approach can be extended to the study of non-diseased states, and argued that, "… genetic concepts must be an integral part of the armamentarium of the modern epidemiologist" (p. 302). The "epidemiological genetics" that Neel and Schull[1] envisioned has become known as "genetic epidemiology". Upon the establishment of the International Genetic Epidemiology Society (IGES) in 1992, its founding president, James V. Neel, succinctly defined genetic epidemiology as, "The study of genetic components in complex biological phenomena".[2] From this perspective, the genetic epidemiology of growth and development may be considered as the study of the genetic underpinnings of the size, conformation and maturity status of individuals over the course of childhood. This includes characterizing the magnitude of genetic influences on growth and development phenotypes, examining how those genetic influences operate over time, identifying and localizing genes and specific genetic polymorphisms in those genes that contribute to variation in growth and development, and elucidating how genetic and environmental factors interact during growth and development. The advances made over the past few decades in both molecular and statistical genetics have led to current highly sophisticated analyses used to increasingly better elucidate the roles of genes and environment in the complex biological phenomena that comprise growth and development.

This chapter is divided into four sections that follow below. Section 8.2 provides an introduction to basic statistical genetic terminology. Section 8.3 discusses different study designs used to examine genetic influences on primarily quantitative traits. Section 8.4 summarizes published findings from various studies of the genetics of growth and development. Section 8.5 presents some example findings from current genetic epidemiological studies of the growth and development of US children in the well-established Fels Longitudinal Study. Throughout the chapter, important terms or concepts are italicized the first time they are mentioned. Those that are not defined in the text of the chapter are briefly defined in the Glossary.

8.2 STATISTICAL GENETIC TERMS AND CONCEPTS

Statistical genetics refers to a variety of methods for analyzing *phenotypic* variation among related individuals. These methods include those tailored for the study of both discrete and continuous traits. Most growth and development phenotypes exhibit a continuous distribution over a delimited range, and because the growth and development status of a child can usually be measured in some way, most growth and development phenotypes are quantitative traits. Growth and development phenotypes also are referred to as being complex traits, meaning that genes at a few and perhaps several loci contribute to the variation observed in the trait, as do environmental factors, possibly through interaction with those genes. The field of quantitative genetics deals with the analysis of complex traits. As with any specialized field of study, it contains a number of specific terms and concepts. This section provides a brief discussion of those quantitative genetic terms and concepts most important for an understanding of the genetic epidemiology of normal growth and development. Thorough discussion of quantitative genetic methods can be found in books listed in the Further Reading section at the end of this chapter.

8.2.1 Relatedness of Individuals

To start with, because related individuals are not independent, but share some of their genes by virtue of sharing common ancestry, it is necessary to consider their degree of relatedness in assessing the extent of their resemblance for a trait. The *coefficient of kinship* between two individuals is the probability that an *allele* taken at random from the two alleles at a locus in one individual is identical to an allele taken at random from the two alleles at the same *locus* in another individual. The coefficient of kinship between first degree relatives is 0.25, meaning that, for example, between a pair of full siblings there is a 25% chance that they each have at a locus the very same allele that they each inherited from a common ancestor. Most of what we know about the genetic control of growth and development comes from family-based studies in which the correlations between relatives and between unrelated individuals for a trait such as stature or weight are calculated. The basic premise underlying these investigations is straightforward: if the variation in a trait is largely under genetic control, then related individuals will be more similar for the trait than will unrelated individuals (i.e. the intrafamily variance of the trait is low compared to the interfamily variance). Conversely, if the variation in a trait is only partly determined by genes, then related individuals may only resemble each other a little bit more so than do unrelated individuals (i.e. the intrafamily variance of the trait is a little smaller than the interfamily variance).

8.2.2 Heritability

Through examination of correlations between different pairs of relatives, heritabilities can be calculated. The concept of *heritability* (h^2) is central to understanding the nature of genetic control for any trait. The h^2 of a trait is a measure of the degree of genetic control of a phenotype, ranging from 0 (no genetic control) to 100% (complete genetic control). Heritabilities are population-level estimates, specific to a particular population in a given environment, and this can sometimes be an important consideration when comparing h^2 estimates across populations.

According to classical quantitative genetics theory (e.g. see texts by Falconer and Mackay, 1996, Lynch and Walsh, 1998)[3,4] the observed phenotypic variation (σ^2_P) in a trait can be expressed as the sum of both genetic (σ^2_G) and random environmental effects (σ^2_E). This is written as:

$$\sigma^2_P = \sigma^2_G + \sigma^2_E \qquad [8.1]$$

In its simplest form, this model provides a starting point for understanding the quantitative genetics of complex traits. For example, σ^2_P can be decomposed further into components representing the variance due to *additive effects* of genes at several loci (σ^2_A), *dominance effects* (σ^2_D) and *epistasis* (σ^2_I), while σ^2_E can be decomposed into the variance due to specific measured environmental factors ($\sigma^2_{E\ factor\ \#1}$) and that due to random, unmeasured environmental factors ($\sigma^2_{E\ random}$). *Broad-sense heritability* refers to the proportion of the phenotypic variance attributable to all sources of genetic variance, and is written as:

$$h^2 = \sigma^2_G / \sigma^2_P \qquad [8.1]$$

Narrow-sense heritability refers to the proportion of the phenotypic variance attributable only to the additive genetic variance, and is written as:

$$h^2 = \sigma^2_A / \sigma^2_P \qquad [8.3]$$

Generally speaking, at least initially, the narrow-sense heritability is the most useful in characterizing the genetic effects of continuously distributed traits such as stature or weight. Inheritance of such quantitative traits is likely to be influenced by a number of genes exerting mostly small to moderate effects. For that reason, quantitative traits are often referred to as being *polygenic traits*. However, not all genes influencing a trait are likely to make the same contribution to the phenotypic variance of the trait. Also, since it is typically very difficult (e.g. because of sample size constraints) to identify genes explaining only a small proportion of the phenotypic variance of a trait (e.g. 5% or less), it is perhaps more practical to refer to most quantitative traits as being *oligogenic traits*, meaning that it is likely that a few genes with pronounced and identifiable effects of varying degrees are together responsible for most of the genetic contribution to the phenotypic variance of a trait. In most instances, h^2 estimates refer to narrow-sense heritabilities. The variance components approach to decomposing the phenotypic variation exhibited in a quantitative trait, briefly described here, has its roots in the seminal work by Fisher,[5] and is an elegant and powerful method for evaluating the different sources of variation contributing to the overall variance of a complex trait.

8.2.3 Genetic and Environmental Correlations

Quantitative genetics is much more than simply calculating h^2 estimates. Since it is well established that measures of growth and development have substantial and significant heritable components, intellectual focus turns to the nature of the genetic regulation of growth and development. For example, significant phenotypic correlations often exist between different measures of growth and development. These phenotypic correlations may be due to *pleiotropy*, the joint effects of a gene or genes on different traits, or to shared environmental factors. In most cases, significant phenotypic correlations between two traits are due to both pleiotropy and shared environmental effects.

Just as the phenotypic variance of one trait can be decomposed into genetic and environmental variance components, so too can the phenotypic correlation between two traits be decomposed into genetic and environmental covariance components. Thus, the phenotypic correlation between two traits is a function of the h^2 of each trait and the genetic and environmental correlations between them. This is written as:

$$\rho_P = \sqrt{h_1^2}\sqrt{h_2^2}\,\rho_G + \sqrt{(1-h_1^2)}\sqrt{(1-h_2^2)}\,\rho_E \qquad [8.4]$$

where ρ_p is the phenotypic correlation, ρ_G is the genetic correlation, ρ_E is the environmental correlation, h_1^2 is the heritability of trait 1 and h_2^2 is the heritability of trait 2. If both traits have low heritabilities, the phenotypic correlation between them is due largely to the environmental correlation, whereas if both traits have high heritabilities, the phenotypic correlation between them is due largely to the genetic correlation.

As with phenotypic correlations, additive genetic and random environmental correlations range from -1.0 to 1.0. A genetic correlation of 1.0, for example, indicates complete positive pleiotropy between two traits. That is, there are genes that affect in the same manner both of the traits being examined. A genetic correlation significantly less than one indicates incomplete pleiotropy, meaning that the two traits are influenced to some extent by the same set of genes, but that other genes also are influencing the value of one or the other of the two traits. A genetic correlation of zero between two traits indicates that the two traits have different genes controlling them. Finally, a negative genetic correlation indicates that the same set of genes operates in an opposite manner on the two traits. Similarly, the random environmental correlation is a measure of the direction and strength of the correlated response of two traits to non-genetic factors. If specific non-genetic factors have been identified and measured that influence the covariance of the two traits, however, then the environmental correlation can be decomposed into non-random and random components.

Multivariate quantitative genetic analyses, in which the heritabilities of two (or more) traits are estimated along with the genetic and environmental covariances between them, are powerful tools for investigating the nature of relationships between different aspects or measures of growth and development.

8.2.4 Applications of Genetic and Environmental Correlations to Longitudinal Data

Another topic of particular interest in the field of growth and development is the nature of the genetic control of a trait over time. For these types of analysis it is necessary to have serial measurements of the trait or traits of interest. Serial measurements of traits separated by time are normally correlated to some degree, with higher phenotypic correlations often found over short intervals and lower phenotypic correlations found over longer intervals. *Canalization* is a familiar term to auxologists, referring to the tendency of a trait to follow a certain course or trajectory over time. The more highly canalized a trait, the higher the phenotypic correlations between repeated measurements. From a genetic perspective, traits that are highly canalized, and that are relatively insensitive to changes in environmental conditions, are likely to have relatively high heritabilities. The same genes, however, may or may not be influencing the trait to the same extent over the entire course of growth and development.

To test hypotheses concerning the genetic control of growth at different ages, the same approach discussed above for the examination of two traits at one point in time is taken. In its simplest form, however, the "two traits" are now the same trait measured at two points in time. The genetic and environmental correlations between repeated measures of the trait at different ages are then calculated. This approach allows for disentangling shared genetic effects from shared environmental effects on a trait measured over the course of childhood.

The strength of a genetic correlation for a single trait with repeated measures is indicative of the degree of consistency or uniformity in the genetic control of the trait over time. For example, if a genetic correlation of 1.0 is found between stature measured at age 8 years and measured again at age 18, then it can be inferred that the genes influencing stature during the middle of childhood are the same as those that influence height in early adulthood. If a genetic correlation is obtained that is significantly lower than 1.0, however, then there is evidence that a different suite of genes controls stature at ages 8 and 18 years. Similarly, the environmental correlation is a measure of the consistency or uniformity of the response of the trait to non-genetic factors over time. The discussion will return to genotype by age interaction after first discussing genotype by environment and genotype by sex interactions.

8.2.5 Genotype by Environment Interaction

Understanding how genes interact with aspects of the physical and internal biological environments is essential for better understanding the genetic architecture of complex traits. In studies where relatives live in different environments, genotype by environment (G × E) interactions can be examined using extensions of variance components methods for studying quantitative trait variation.

G × E interaction is likely to be an important influence on the variation observed among children in their growth and development, particularly in populations with high prevalences of environmental factors known to negatively impact growth and development. The key to G × E interaction, however, is that not all children may respond the same to such environmental factors, and a portion of that differential response or susceptibility at the phenotypic level may be due to genetic variation among individuals.

The simplest approach to modeling G × E interaction is to make the genetic variance in a trait a function of a dichotomous environmental variable. Examples of this could be the presence or absence of a particular disease in a child, high or low protein intake, etc. Figure 8.1 shows a simple hypothetical depiction of the response of three genotypes at a locus to two different environments. In the presence of G × E interaction, the relationship between trait levels and specific genotypes will vary as a function of the environment. In this case, trait levels in Environment 1 are substantially less variable than trait levels in Environment 2. For genotypes AA and AB, trait levels remain stable or decrease from Environment 1 to Environment 2. For genotype BB, trait levels increase from Environment 1 to 2. This example demonstrates how gene expression may vary under different environmental conditions.

In G × E analyses of the response of a quantitative trait, the variance components method is expanded to include environment-specific additive genetic variances that are then estimated. For example, a large number of related children might be measured for a trait (e.g. stature) at a specific age, and also tested for the presence of a particular infection at that age. If the additive genetic variances of the measured trait are not significantly different between infected and non-infected children, then that would be an indication that there is no G × E interaction between that trait and that infection at that age. If, on the other hand, the additive genetic variances of the measured trait are significantly different between infected and non-infected children, then that would indicate a genetic basis to the differential response of the growth status of children to infection at that age. G × E

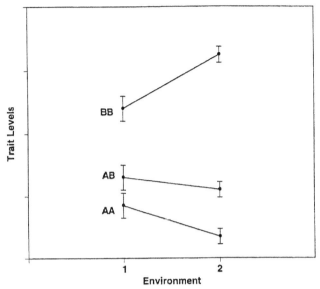

Figure 8.1 Hypothetical depiction of gene by environment interaction with the response of three genotypes at a locus to two different environments.

interaction is also tested by examining the genetic correlation between the trait measured in different environments. A genetic correlation significantly different from 1.0 is another indication of G × E interaction. In the example here, a genetic correlation significantly less than 1.0 would indicate that the G × E interaction is due to an incompletely correlated genetic response of the trait in infected and non-infected children.

8.2.6 Genotype by Sex Interaction

Sexual dimorphism in the growth and development of children is well known, but the genetic basis of this sexual dimorphism is poorly understood. The approach for studying G × E interaction using related individuals living in different environments described above can be used to study genotype by sex (G × S) interactions. The rationale here is that the hormonal environments of males and females differ considerably, and the expression of autosomal genes controlling a quantitative trait may be influenced by the sex—environment encountered.

In analyses of G × S interaction, the variance components method is again expanded. Additional parameters are estimated, the most important being sex-specific variance components and the genetic correlation between the sexes for the trait. G × S interaction is indicated by significantly different additive genetic variances for males and females and/or a genetic correlation between the sexes significantly less than 1.0.

G × S interaction analyses can be used to examine the genetic basis to the sexual dimorphism in measures of growth and development. The aim of G × S interaction analyses is to determine whether the sexual dimorphism evidenced in a trait during childhood is itself a heritable trait. In some families, for example, male and female children might not be very different in a measure of growth or development at a particular age, while in other families there might be significant differences between male and female relatives in that measure of growth or development at that age.

8.2.7 Genotype by Age Interaction

The nature of genetic influences on measures of growth and development may change over the course of childhood. As initially discussed earlier, the genetic correlation between a trait measured at two points in time can provide insight into the genetic control of a trait over time. If extensive longitudinal data from related children are available, genotype by age (G × A) interactions can be more rigorously examined. Like G × S interactions, G × A interactions are a type of G × E interaction. In this case, the "environment" is the age of the child at the time of the measurement of a trait. In these analyses, the additive genetic variance of a trait is modeled as a function of age. From these age-specific additive genetic variances, age-specific heritabilities of the trait can be determined. Also estimated are the additive genetic and environmental correlations between the trait measured across time.

G × A interaction is indicated by an additive genetic variance of a trait changing over a span of ages. This suggests that the genetic expression of a trait is dependent upon the age of the child. G × A interaction also is indicated by a change in the genetic correlation between a trait measured over time. For example, a genetic correlation between time-points of a serially measured trait that decreases significantly from 1.0 over a span of ages indicates G × A interaction. And, if sufficient serial data are available, the function or shape of a genetic correlation curve can provide further insights into dynamic genetically mediated biological processes underlying such G × A interactions.

8.2.8 Identifying Genes Influencing Growth and Development

Once it has been determined that a trait has a significant heritability, interest quickly turns to locating and identifying the actual genes that influence variation in the trait. Advances in molecular and statistical genetic methods make it possible to search for genes and specific genetic polymorphisms influencing complex traits. Unlike monogenic traits that are influenced by a single gene with large effects, most complex traits are largely (but not exclusively) influenced by genes at a number of loci whose individual effects can be of small to moderate size. While understanding of monogenic growth disorders has significantly increased over the last several decades, understanding the genetics of normal variation in quantitative measures of growth and development has continued to be a daunting task. Technological advances in molecular biology, however, including relatively inexpensive high-throughput genotyping of upwards of millions of *single-nucleotide polymorphisms*

(*SNPs*) and increasingly lower cost exome and whole-genome sequencing, along with attendant methodological advances in statistical genetics, have made it possible to identify genes exerting small or moderate effects and even to identify rare polymorphisms influencing a trait in only some populations or pedigrees. There are two basic strategies to follow in the search for genes involved in the regulation of growth and development: population-based association studies or family-based quantitative trait linkage studies.

8.2.9 Population-Based Association Studies

The first approach is the candidate gene association approach. Here, genes suspected to be physiologically involved in the trait are examined. For example, a sample of unrelated individuals is selected and genotyped for a specific polymorphism in or near the candidate gene. Simple statistical tests are then used to evaluate associations between marker genotype status and the value of a trait. Carriers of a particular allele, for example, may have a mean value for the trait that is significantly different from the mean value of the trait in those who do not have a copy of that allele. Population-based association studies have obvious appeal, in that they are computationally straightforward compared to the analysis of marker genotype and quantitative trait data from family members.

There is a significant problem with population-based association studies, however, that has become evident as greater knowledge has been gained regarding *linkage disequilibrium*. Two loci are in *equilibrium* when alleles at the two loci are randomly associated with each other. If the relationship between the loci is not random, then linkage disequilibrium is present. Unfortunately, linkage disequilibrium can occur for a number of reasons including new mutations and genetic drift, and in the presence of selection. The main problem with association studies, however, is that disequilibrium is difficult to predict. Two loci may be very close to each other and yet be in equilibrium. Conversely, two loci may be relatively far apart from each other and yet be in disequilibrium. There is no sure way to know that the marker that has been typed is in disequilibrium with the true functional variant. Given that within a given locus, numerous genetic variants may be in high disequilibrium with one another, when a significant association of a variant and trait is found, this only points to a general region, and does not mean that the variant with the strongest association is the functional variant. If it is known a priori, however, that the typed marker is in fact a functional polymorphism (that is, there is a measurable difference among marker genotypes in gene expression; e.g. one genotype results in much lower levels of a particular protein compared to the other genotypes), then association studies become a more viable strategy to pursue.

8.2.10 Genome-Wide Association Studies

In recent years, *genome-wide association* (*GWA*) studies of complex traits have proliferated. GWA analysis is an extension of the candidate gene association approach, and is made possible by relatively low-cost genotyping of now typically from 500,000 to 1,000,000

SNPs in each subject in a study sample. SNPs are biallelic genetic markers that are coded either "0" or "1", and are most often separated by only fairly small intervals across the entire genome. Associations between genotypes and phenotypes are evaluated over every marker. Given the large number of markers, these analyses are computationally intensive and stringent strategies to control for multiple testing must be followed.

GWA study designs typically take one of two forms. One is the case–control approach where individuals with a certain disease or condition are compared to unaffected individuals with regard to genotype status at every genotyped SNP across the genome. SNPs significantly more prevalent in either cases or controls are identified for follow-up to assess possible causative or protective roles of nearby functional polymorphisms. A second GWA study design focuses on quantitative traits where different genotypes at a single locus are examined for differences in trait levels. Associations are denoted when individuals with certain SNP genotypes have consistently and significantly higher or lower trait values than individuals with the other SNP genotypes.

There are several strengths to the GWA approach. First, the high density of SNP markers helps to better localize association signals. GWA signals can be typically reduced to approximately a 500 kb interval, compared to a much broader interval obtained from quantitative trait linkage discussed below. Another strength, alluded to above, is that data can be obtained from unrelated individuals, potentially making data collection more efficient. Indeed, numerous GWA studies have been based on already existing epidemiological study samples. Potential problems with population stratification can be ameliorated by using principal component-based (or ancestral marker-based) adjustments obtained from the SNP marker set. Replication of findings is critically important for GWA studies, however, given the large number of comparisons made across the genome in a single study. Such replication of significant findings provides confirmation of the role that at least common polymorphisms play in contributing to phenotypic variation. All GWA studies should include plans for some form of replication in independent samples.

Despite the popularity of GWA studies over the past decade, important criticisms of the approach have emerged in recent years. To date, GWA studies have collectively had somewhat limited success, despite numerous published studies and several large meta-analyses of various complex traits that have sometimes included samples sizes of over 100,000 subjects. In most cases, reported associations have accounted only for a very small proportion of the overall heritability of the traits examined. Some researchers speculate that the reason for this is that GWA studies are only useful for identifying common disease variants. Unfortunately, it now appears that many common disease or quantitative risk factor variants explain only a relatively small proportion of the total phenotypic variance in the disease or trait examined. The population-based association approach is therefore not likely to be able to identify rare disease variants that are most likely to have larger effect sizes.

In addition, significant associations can be due to heterogeneity in the population sampled. This occurs when population subgroups differ systematically in both allele frequencies and levels of the quantitative trait of interest. Even among seemingly fairly homogeneous families (e.g. similar ethnic background) there can be significant differences in specific allele frequencies; and all families carry unique "private" polymorphisms, some of which may affect a trait in some families but not in others.

8.2.11 Quantitative Trait Linkage Analysis

The second predominant approach for discovering genes influencing the types of measurable traits of most interest to auxologists is quantitative trait linkage mapping. Linkage studies require a good deal of planning before their initiation in order to obtain maximal statistical power to detect genes of modest to moderate effect. The premise behind linkage analysis is that if two loci are physically located close to each other, then alleles at these loci will be more likely to be inherited together. In this case, the loci are said to be linked. As the distance between loci increases, the probability that alleles at these loci will *cross over* or *recombine* during meiosis increases. Through investigation of the frequency of recombination events among genetic markers one can identify chromosomal regions harboring genes that influence variation observed in a trait. Once a region has been identified, molecular mapping techniques such as high-density SNP typing or sequencing can be used to better delineate chromosomal regions of interest and to identify functional polymorphisms.

Over the past two decades there have been many advances in quantitative trait linkage analysis as applied to complex traits. Over that time, allele-sharing methods have gained prominence for the analysis of quantitative traits. The key premise behind allele-sharing methods is the concept of *identity by descent* (IBD). In comparisons between relatives, two alleles that are structurally identical are said to be *identical by state* (IBS); alleles that are structurally identical and inherited from a common ancestor (e.g. two siblings getting the same allele from their mother) are further classified as IBD. A pair of relatives can share zero, one or two alleles IBD at any given marker locus. The likelihood of their sharing zero, one or two alleles IBD is contingent upon their coefficient of kinship. Linkage between a quantitative trait locus (QTL) and a marker exists in chromosomal regions when pairs of relatives who are more phenotypically similar share more alleles at a marker locus than pairs of relatives who are less phenotypically similar.

The power to detect and localize QTLs is a function of several factors, the most important being the strength of the genetic effect. Traits that are highly heritable will tend to have a higher probability of being mapped compared to those with low to modest heritability, but this is not always the case. Also, as in any statistical analysis, sample size is of importance, but in linkage studies other aspects of the study sample are also important, most especially the family structure of the study sample. Having many families is good, but having fewer more complex extended pedigrees, preferably with several generations represented, will yield increased statistical power because of the greater number and variety of relationships between relatives.

Linkage analysis has several strengths and some weaknesses. One strength of linkage analysis is the ability to identify rare genetic variants in family-based samples. Because genes segregate in families, the ability to identify rare genes of moderate effect is possible. Identification of rare variants may help to explain what has been termed the "missing heritability" observed from population-based GWA studies of common traits (i.e. the portion of the heritability not explained by the common variants).[6] Rare variants are likely to have larger effect sizes and could contribute to the unexplained heritability. Further, some researchers suggest that these rare variants are likely to have obvious functional consequences.[6,7] A benefit of pedigree-based studies for the identification of rare variation is that rarer variants, if present, will be present at a much higher frequency than in the general population. Thus, pedigree-based studies inherently have greater power to detect the effects of such rare variants. However, traditional linkage-based studies have limited resolution (owing to typically having fewer genetic markers typed, although this is somewhat ameliorated with SNP-based linkage analysis compared to earlier STR-based linkage analysis; see also below) and are only able to localize QTL to approximately 10–15 Mb of sequence, a much broader region than GWA studies.

8.2.12 Quantitative Trait Linkage and Association

In recent years there has been an effort to combine linkage and association approaches. This approach effectively utilizes the strength of both genetic paradigms. Combined linkage and association analysis can only be accomplished with family-based data, however. As a first step, a family-based approach to association analysis (e.g. measured genotype) can be implemented to test for associations assuming additive genetic effects on each available SNP in the panel.[8,9] Since data from family members cannot be treated as independent observations, family-based methods such as variance-component analysis are able to use a polygenic component to absorb any non-independence among individuals by incorporating a residual heritability parameter. In this context, quantitative trait linkage provides an additional, independent source of information that when used in conjunction with GWA can augment power to detect loci influencing growth-related traits. SNPs used for linkage can be selected from among the typed SNP panel to maximize *heterozygosity* and minimize linkage disequilibrium among the selected markers. Approximately 10,000 SNPs are required for adequate genomic coverage in SNP-based linkage analysis. A joint test of linkage and association can performed by comparing the likelihood of a model in which both the SNP-specific association parameter and the linkage variance component are estimated to the likelihood of a model in which both are constrained to be zero. The power of this test depends on the underlying trait model and on how many functional variants there are within a gene or region; however, under certain circumstances combined linkage and association can be more powerful than association alone. In the coming years, the field of genetics will continue to move towards the use of more advanced technology and the new wave of studies will focus on the sequencing of entire exomic regions and ultimately whole-genome sequencing.

8.3 STUDY DESIGNS

Various family-based study designs can be used to examine the genetics of complex traits. Each study design has certain advantages and disadvantages. This section describes some of the major types of study design used by genetic epidemiologists to study complex quantitative traits.

8.3.1 Twin Studies

Over the years, studies of twins have been useful in establishing the familial aggregation of many complex traits. In its basic form, the twin model compares phenotypic differences between two classes of twins, monozygotic (MZ) and dizygotic (DZ). MZ twins share 100% of their genetic make-up, while DZ twins share on average only half of their genetic make-up (i.e. on the genetic level they are the same as any other pair of full sibs). Because of this, phenotypic differences observed between MZ twins are assumed to be the result of environmental factors only, while phenotypic differences between DZ twins are considered to be due to differences in both genes and environmental exposure. Thus, by calculating phenotypic correlations in groups of MZ and DZ twins and comparing them, assumptions can be made about the degree of genetic control of different traits.

One important assumption in the classical twin study design is that both MZ and DZ twin pairs are equally likely to share a common environment. This assumption may not necessarily be valid, however, because MZ twins are often more likely to share common activities, foods and other aspects of the environment to a greater extent than DZ twins. Because there is no fully satisfactory way to separate shared genetic and environmental effects, studies of twins often yield inflated h^2 estimates.

The twin study design is especially problematic if the focus of the study is a growth-related outcome. Twin births are physiologically different from singleton births owing to competition over maternal resources during pregnancy. Fetal growth rates among twins may therefore be considerably discordant, and the postnatal growth of twins is often different from that of siblings from singleton births (e.g. early catch-up growth in twins).

8.3.2 Nuclear Families

Another commonly used study design is that of nuclear families. In this study design, correlations between the various classes of first degree relatives in a nuclear family are estimated. These include parent–offspring, sibling–sibling and spouse–spouse correlations. Heritabilities can be estimated from these different familial correlations. Heritability estimates calculated from nuclear family data, however, are subject to inflation owing to the effects of shared environmental factors such as diet and lifestyle among family members living in a single residence. Given this, heritabilities are often adjusted by taking into account the degree of spousal correlation in the family. It is assumed that any correlation found between spouses is the result of shared environmental factors. Such spousal correlations may depend upon the length of time that the couple has been married. However, such spousal correlations may also be the result of assortative mating.

There are practical considerations to be taken into account in studies of nuclear family members apart from those just mentioned. For example, it is sometimes difficult to obtain information about certain life events because they are often separated in time by a generation: it may take 20 to 30 years of waiting to collect growth measures of the children of parents who were measured when they were children. Also, generational differences in growth may be due to secular trends. This may effectively reduce the heritability of certain traits by diminishing the degree of phenotypic correlation observed. These two problems can be eliminated by examining only sibling correlations, but the problem of shared environment remains.

8.3.3 Extended Pedigrees

The study design that offers considerable promise for elucidating the genetic architecture of complex traits is the extended family approach. This approach involves collecting information from all available family members and estimating phenotypic correlations between all relatives of varying degrees of relationship. By sampling members outside the immediate nuclear family, many of the problems encountered with immediate shared environmental effects in other study designs are minimized because family members come from a number of different households. This results in more accurate and reliable h^2 estimates. In addition, sampling family members in different households (who thereby live in potentially different environmental circumstances) provides the opportunity to investigate G × E interactions. With regard to the study of growth and development, within large extended pedigrees there will be several related children of approximately the same age. This will enable analyses to proceed very quickly after the initiation of data collection.

There are a few practical drawbacks to this approach, however. The single most important consideration is that the methods involved in calculating statistical genetic parameters can be computationally intensive. This, however, is much less of an obstacle as computer technologies continue to progress. Indeed, advances in computer technology over the past three decades have made the statistical genetic analyses of data from large pedigrees tractable. In addition, collecting data from large numbers of related individuals of varying ages who may live some distance from each other requires a great deal of planning, effort and research funding.

8.4 STUDIES OF THE GENETICS OF GROWTH AND DEVELOPMENT

The preceding sections introduced several basic terms and concepts necessary for discussing genetic epidemiological approaches to growth and development. This section provides a brief overview of numerous studies of genetic influences on growth and development that have been conducted over the past century. These studies fall into two general categories: those that infer genetic determination of growth and development through comparison of different human populations, and those that examine growth and development traits in families. The review presented here provides a sampling of a considerable part of this literature, focusing on studies of height, birth weight, menarche and skeletal development.

8.4.1 Population Differences in Growth and Development

There is considerable variation across populations in growth in height, weight, and other body dimensions, as well as in the tempo and timing of maturation.[10] For example, mean adult height varies from approximately 150 cm for males in the shortest populations on earth (e.g. Mbuti pygmies of central Africa) to 180 cm for males in northern European populations. These long-standing observations of racial or ethnic differences in growth and development rendered support for the notion that genetic factors are likely involved. The degree to which genetic factors influence growth and development cannot be addressed, however, by the simple comparison of measures of growth and development traits across populations. The populations compared often are exposed to vastly different environments, and the shortest and smallest populations also tend to have the poorest economic status, while the tallest populations tend to be from industrialized nations. Between-population differences may be due to differences in both genetic and environmental factors, whose relative importance of is often confounded. For example, evidence of secular trends in stature and pubertal maturation,[10] and the degree of similarity for stature in high socioeconomic status groups from various parts of the world (e.g. Martorell, 1988)[11] argue that a significant part of interpopulation variation in growth and development is due to environmental factors.

8.4.2 Family Studies of Growth and Development

Population comparisons provide only indirect evidence of a connection between genetic factors and phenotypic variation in growth and development. Only family studies within populations can clearly define the relationships between genes and growth, because it is with these designs that environmental and genetic sources of variation can be explicitly modeled.

As an initial overview of genetic influences on growth and development, Table 8.1 summarizes published familial correlations and/or the heritability estimates for birth weight, height, weight and other anthropometrics, as well as age at menarche in females, from a large selection of family studies from diverse populations. Table 8.1 does not contain an exhaustive listing of all published findings, but provides a starting point; the studies listed in Table 8.1 were published in widely circulated journals and represent the range of findings typically reported in the literature.

Several general comments can be made regarding these investigations. First, most studies have been based on first degree familial correlations. That is, they are based on either nuclear family or twin pair designs. As discussed above, there are important concerns when studying only first degree relatives, particularly when studying growth and development. These concerns include secular trends that may reduce correlations between parents and offspring, and the shared environments of siblings, especially twins, that may inflate correlations between them. Second, specific environmental sources of variation such as diet and disease usually have not been incorporated into the analyses. Not accounting for the variance in a trait attributable to such environmental factors can lead to underestimation of the h^2 of the trait. Third, the majority of studies have focused solely on height at a given point in time (mostly adult height). A smaller number of studies have examined other anthropometrics. Fourth, the majority of studies are based on cross-sectional data. Only a few studies have longitudinal growth and development data from related individuals that permit examination of genetic influences on patterns of change in height, weight and other measures over time. And fifth, almost all of the studies have focused solely on heritability estimation. There are very few multivariate quantitative genetic analyses of measures of growth and development, or analyses of genotype by environment, sex or age interactions.

Table 8.2 summarizes recently published results from GWA or linkage analyses of birth weight, height, body mass and age at menarche. Although relatively few studies of measures of growth and development have been conducted, this is an expanding area of auxological genetics research as primary interest has shifted to identifying specific genes and genetic polymorphisms that influence such measures. Again, Table 8.2 does not contain an exhaustive listing of all published findings, but provides a starting point for entry into the literature.

Birth Weight

The genetics of prenatal growth has largely been approached by examining the heritability of birth weight. Initially, genetic influences on birth weight were deduced from the known effects of quantitative changes in chromosomes. For example, supernumerary autosomes (trisomy 21, 18 and 13) and abnormal numbers of X chromosomes

Table 8.1 Heritability estimates of anthropometrics during childhood and adolescence

Trait	Reference	Population	Design and family structure	Sample size	Familial correlations	Heritability or genetic variance estimate	Age range
Birth weight							
	Penrose, 1954[12]	UK	Cross-sectional, nuclear			Fetal genetic factors: 18% Maternal genetic factors: 20% Environmental factors: 62%	Birth
	Morton, 1955[13]	Japan	Cross-sectional, nuclear/twins		$r_{twins} = 0.56$ $r_{sibs} = 0.52$ $r_{halfsibs-mo} = 0.58$ $r_{halfsibs-fa} = 0.10$		Birth
	Nance et al., 1983[14]	USA	Cross-sectional, nuclear/twins	Offspring of 385 twin pairs	$r_{sibs} = 0.48$ $r_{halfsibs-mo} = 0.31$ $r_{halfsibs-fa} = -0.03$		Birth
	Clausson et al., 2000[15]	Sweden	Cross-sectional, twins	868 MZ, 1141 DZ		$h^2 = 0.25-0.40$	Birth
	Magnus et al., 2001[16]	Norway	Cross-sectional, trios (Fa, Mo, first born)	67,795 trios	$r_{fa-mo} = 0.02$ $r_{fa-off} = 0.129$ $r_{fa-son} = 0.126$ $r_{fa-da} = 0.133$ $r_{mo-off} = 0.226$ $r_{mo-son} = 0.222$ $r_{mo-da} = 0.231$	$h^2 = 0.25$	Birth
	van Dommelen et al., 2004[17]	Netherlands	Longitudinal, twins	4649 twin pairs		$h^2 = 0.14$ $h^2 = 0.24$	Birth, females Birth, males
	Arya et al., 2006[18]	USA (Mexican Americans)	Cross-sectional, nuclear/extended	840 subjects		$h^2 = 0.72$	Birth

Reference	Population	Study design	Sample	Results	Heritability	Age
Grunnet et al., 2007[19]	Denmark	Cross-sectional, twins	138 MZ, 214 DZ	$r_{MZ} = 0.75$ $r_{DZ} = 0.56$	$h^2 = 0.38$	Birth
Choh et al., 2011[20]	USA	Longitudinal, nuclear/extended	917 subjects		$h^2 = 0.67$	Birth
Height/recumbent length						
Vandenberg & Falkner, 1965[21]	USA	Longitudinal, twins (stature curve parameters)	29 MZ, 31 DZ	Concordance between MZ and DZ twins: MZ = DZ initial value (birth) MZ < DZ (velocity) MZ < DZ (acceleration)		Birth–6 y
Welon & Bielicki, 1971[22]	Warsaw, Poland	Longitudinal, nuclear	496 parent–child pairs	$r_{parent-son} = 0.36$ $r_{parent-son} = 0.43$ $r_{parent-da} = 0.54$ $r_{parent-da} = 0.59$		8 y, male 18 y, male 8 y, female 18 y, female
Garn et al., 1976[23]	USA	Cross-sectional, (adopted/biological siblings)	6726 biological, 504 adoptive parent–offspring, pairs	$r_{adopted\ sibs} = 0.29$ $r_{adoptive-biological\ sibs} = 0.35$		Birth–18 y
Malina et al., 1976[24]	USA, white and black	Cross-sectional, nuclear	422 black families, 384 white families		$h^2 = 0.49$ (white) $h^2 = 0.37$ (black)	6–12 y
Mueller, 1976[25]	Colombia, Africa, Peru, New Guinea, Japan	Cross-sectional, nuclear		$r_{pc} = 0.29$ (average)		
Mueller, 1976[25]	USA, UK, West Europe, East Europe	Cross-sectional, nuclear		$r_{pc} = 0.37$ (average)		

(Continued)

Table 8.1 Heritability estimates of anthropometrics during childhood and adolescence—cont'd

Trait	Reference	Population	Design and family structure	Sample size	Familial correlations	Heritability or genetic variance estimate	Age range
	Wilson, 1976[26]	USA	Longitudinal, twins	159 MZ, 195 DZ	$r_{MZ} = 0.58$ $r_{MZ} = 0.94$ $r_{DZ} = 0.69$ $r_{DZ} = 0.61$		Birth 4 y Birth 8 y
	Fischbein, 1977[27]	Sweden	Longitudinal, twins	94 MZ, 233 DZ	$r_{MZ} = 0.90$ $r_{DZ} = 0.60{-}0.70$		10–16 y
	Mueller & Titcomb, 1977[28]	Colombia	Cross-sectional, nuclear	403 families	$r_{mo-child} = 0.28$ $r_{fa-child} = 0.27$	$h^2 = 0.49$ (males) $h^2 = 0.47$ (females)	7–12 y
	Susanne, 1977[29]	Belgium	Cross-sectional, nuclear	125 families	$r_{pc} = 0.51$	$h^2 = 0.82$	17–35 y
	Roberts et al., 1978[30]	West Africa	Cross-sectional, nuclear, full and half siblings	276 sibships		Fa–child: $h^2 = 0.61$ Mo–child: $h^2 = 0.85$ Mid-parent–child: $h^2 = 0.65$ Full siblings: $h^2 = 0.81$ Paternal half-siblings: $h^2 = 0.56$	
	Fischbein & Nordqvist, 1978[31]	Sweden	Longitudinal, twins	94 MZ, 133 DZ	Average growth profile similarity within twin pair: $r_{MZ} = 0.85$ $r_{DZ} = 0.54$		10–16 y (growth curve concordance)
	Kaur & Singh, 1981[32]	India	Cross-sectional, nuclear	82 families	$r_{pc} = 0.48$	$h^2 = 0.92$	18–59 y

Reference	Country	Study type	Sample	Results	Heritability	Age
Solomon et al., 1983[33]	Finland	Cross-sectional, nuclear	2869 subjects		$h^2 = 0.58$	<55 y
Devi & Reddi, 1983[34]	India	Cross-sectional, nuclear	436 families	$r_{pc} = 0.34$, $r_{sibs} = 0.33$	$h^2 = 0.65$	6–13 y
Sharma et al., 1984[35]	India	Cross-sectional, nuclear/twins	610 subjects	$r_{sibs} = 0.30$, $r_{DZ} = 0.59$, $r_{MZ} = 0.98$		3–26 y
Byard et al., 1993[36]	USA	Longitudinal, nuclear (height growth curve parameters)	228 families	Age at TO: $r_{pc} = 0.17$, $r_{sibs} = 0.32$ TOV: $r_{pc} = 0.26$, $r_{sibs} = 0.35$ Age at PHV: $r_{pc} = 0.22$, $r_{sibs} = 0.35$ PHV: $r_{pc} =$ ns, $r_{sibs} = 0.32$		2–18 y
Towne et al., 1993[37]	USA	Longitudinal, nuclear/extended (height curve parameters)	569 subjects		Recumbent length at birth: $h^2 = 0.83$ velocity 0–2 y: $h^2 = 0.67$ acceleration change 0–2 y: $h^2 = 0.78$	0–2 y
Hauspie et al., 1994[38]	Poland	Longitudinal, twins (stature curve parameters)	44 MZ, 42 DZ		Age at TO: $h^2 = 0.49$ age at PHV: $h^2 = 0.74$ PHV: $h^2 = 0.76$	8.5 y–adulthood
Beunen et al., 1998[39]	Belgium	Longitudinal, twins	99 twin pairs		age at TO: $h^2 = 0.93$ TOV: $h^2 = 0.90$ age at PHV: $h^2 = 0.92$	10–18 y

(Continued)

Table 8.1 Heritability estimates of anthropometrics during childhood and adolescence—cont'd

Trait	Reference	Population	Design and family structure	Sample size	Familial correlations	Heritability or genetic variance estimate	Age range
	Price et al., 2000[40]	USA (African-Americans)	Cross-sectional, extended families	1185 families	$r_{po} = 0.26$ $r_{sib} = 0.27$		18–92 y
	Price et al., 2000[40]	USA (Caucasians)	Cross-sectional, extended families	1185 families	$r_{po} = 0.37$ $r_{sib} = 0.37$		18–92 y
	Silventoinen et al., 2000[41]	Finland	Longitudinal, twins	3466 MZ, 7450 DZ		$h^2 = 0.66–0.82$	Birth cohorts 1928—earlier through birth cohort 1947—1957
	Luke et al., 2001[42]	Jamaicans	Cross-sectional, nuclear, extended	623 subjects		$h^2 = 0.74$ $h^2 = 0.44$ $h^2 = 0.84$	Mean 39.5 y Mean 38.8 y Mean 37.5 y
	Silventoinen et al., 2001[43]	Finland	Longitudinal, twins	4873 twin pairs		$h^2 = 0.78–0.87$	Birth cohort 1938—49
	Silventoinen et al., 2001[43]	Finland	Longitudinal, twins	2374 twin pairs		$h^2 = 0.67–0.82$	Birth cohort 1975—79
	Arya et al., 2002[44]	India	Cross-sectional, nuclear	1918 subjects (342 families)		$h^2 = 0.36$	6—72 y
	Brown et al., 2003[45]	USA	Longitudinal, nuclear	2885 subjects		$h^2 = 0.88$ $h^2 = 0.88$ $h^2 = 0.88$	>40 y >55 y >70 y
	Silventoinen et al., 2003[46]	Multiple European nationalities	Longitudinal, twins	30,111 twin pairs		$h^2 = 0.84–0.93$	20—40 y
	Li et al., 2004[47]	China	Nuclear	1169 subjects (385 families)		$h^2 = 0.65$	Mean Fa = 62.3 y, Mo = 59 y, Da = 31 y

Study	Country	Design	Sample	Correlations	Heritability	Age
Schousboe et al., 2004[48]	Denmark	Longitudinal, twins	299 male twin pairs, 325 female twin pairs		$h^2 = 0.69$ $h^2 = 0.81$	18–67 y 18–67 y
van Dommelen et al., 2004[17]	Netherlands	Longitudinal, twins	4649 twin pairs		$h^2 = 0.10$ $h^2 = 0.44$ $h^2 = 0.52$ $h^2 = 0.15$ $h^2 = 0.74$ $h^2 = 0.58$	Birth females 1 y females 2 y females Birth males 1 y males 2 y males
Malkin et al., 2006[49]	Chuvashes (Russia)	Cross-sectional, nuclear	743 subjects		$h^2 = 0.87$	18–89 y males
Macgregor et al., 2006[50]	Australia	Longitudinal, twins	618 MZ females, 239 MZ males, 338 DZ females, 143 DZ males, 334 DZ OS	$r_{MZ} = 0.92$ $r_{MZ} = 0.92$ $r_{DZ} = 0.44$ $r_{DZ} = 0.39$ $r_{DZ} = 0.42$	$h^2 = 0.911$	17–90 y females 32–44 y
Saunders & Gulliford, 2006[51]	UK	Longitudinal, extended families	22,297 subjects		$h^2 = 0.49$	Standardized for age
Bayoumi et al., 2007[52]	Arab	Cross-sectional, consanguineous	1277 subjects		$h^2 = 0.68$	16–80 y
Czerwinski et al., 2007[53]	USA	Longitudinal, nuclear/extended	403 subjects		$h^2 = 0.98$	Mean 38.5 y
Dubois et al., 2007[54]	Canada (Quebec)	Longitudinal, twins	85 MZ, 92 DZ		$h^2 = 0.445$ $h^2 = 0.223$ $h^2 = 0.241$ $h^2 = 0.54$	Birth 5 mo males 5 mo females 60 mo
Pan et al., 2007[55]	Hutterites	Longitudinal, extended, multiple lines of descent	806 subjects		$h^2 = 0.90$	6–89 y
Reis et al., 2007[56]	Brazil	Cross-sectional, twins	5 MZ, 9 DZ		$h^2 = 0.95$	Mean 13 y

(*Continued*)

Table 8.1 Heritability estimates of anthropometrics during childhood and adolescence—cont'd

Trait	Reference	Population	Design and family structure	Sample size	Familial correlations	Heritability or genetic variance estimate	Age range
	Silventoinen et al., 2007[57]	Netherlands	Longitudinal, twins	7753 pairs (at age 3)		$h^2 = 0.71–0.79$ $h^2 = 0.58–0.71$	3–12 y males 3–12 y females
	Silventoinen et al., 2008[58]	Sweden	Longitudinal, twins	99 MZ, 76 DZ, male twin pairs		0.97	17.5–20 y
	Silventoinen et al., 2008[59]	Sweden	Multiple, twins/siblings	1582 MZ, 1864 DZ, 154,970 full brother pairs		0.81	16–25 y
	Axenovich et al., 2009[60]	Netherlands	Cross-sectional, extended pedigrees	2940 subjects		0.86	Mean 48.26 y
	Jowett et al., 2009[61]	Mauritius	Cross-sectional, extended pedigrees	400 subjects		$h^2 = 0.84$	Mean 50 y
	Mathias et al., 2009[62]	Chennai (South India)	Cross-sectional, extended families	498 subjects from 26 pedigrees		$h^2 = 0.72$	Mean 42.65 y
	Poveda et al., 2010[63]	Belgium	Cross-sectional, nuclear	460 subjects		0.84	17–72 y
	Choh et al., 2011[20]	USA	Longitudinal, nuclear/extended	917 subjects		$h^2 = 0.95–0.96$ $h^2 = 0.74–0.95$	30–36 mo 0–24 mo
Weight							
	Garn et al., 1976[23]	USA	Cross-sectional (adopted/biological siblings)	6726 biological, 504 adoptive parent–offspring pairs	$r_{\text{adopted sibs}} = 0.18$ $r_{\text{biological sibs}} = 0.27$		Birth–18 y

Reference	Country	Design	Sample	Results	Heritability	Age
Wilson, 1976[26]	USA	Longitudinal, twins	159 MZ, 195 DZ	$r_{MZ} = 0.61$ $r_{MZ} = 0.86$ $r_{DZ} = 0.68$ $r_{DZ} = 0.55$		Birth 4 y Birth 8 y
Fischbein, 1977[27]	Sweden	Longitudinal, twins	94 MZ, 233 DZ	$r_{MZ} = 0.80–0.90$ $r_{DZ-males} = 0.60–0.70$ $r_{DZ-females} = 0.70–0.20$		10–16 y
Mueller & Titcomb, 1977[28]	Colombia	Cross-sectional, nuclear	403 families	$r_{mo-child} = 0.36$ $r_{fa-child} = 0.31$	$h^2 = 0.16$ (males) $h^2 = 0.21$ (females)	7–12 y
Susanne, 1977[29]	Belgium	Cross-sectional nuclear	125 families	$r_{pc} = 0.34$	$h^2 = 0.64$	17–35 y
Fischbein & Nordqvist, 1978[31]	Sweden	Longitudinal, twins	94 MZ, 133 DZ	Average growth profile similarity within twin pair: $r_{MZ} = 0.79$ $r_{DZ} = 0.22$ (females) $r_{DZ} = 0.53$ (males) $r_{pc} = 0.34$		10–16 y Growth curve concordance
Kaur & Singh, 1981[32]	India	Cross-sectional nuclear	82 families		$h^2 = 0.39$	18–59 y
Arya et al., 2002[44]	India	Cross-sectional, nuclear	1918 subjects (342 families)		$h^2 = 0.314$	6–72 y
van Dommelen et al., 2004[17]	Netherlands	Longitudinal, twins	4649 twin pairs		$h^2 = 0.64$ $h^2 = 0.58$ $h^2 = 0.55$ $h^2 = 0.59$	1 y females 2 y females 1 y males 2 y males
Estourgie-van Burk et al., 2006[64]	Netherlands	Cross-sectional, nuclear/twins	478 MZ males, 517 DZ males, 561 MZ females, 478 DZ females, 962 DZ opposite sex		$h^2 = 0.59$ $h^2 = 0.78$	5 y males 5 y females
Dubois et al., 2007[54]	Canada (Quebec)	Longitudinal, twins	85 MZ, 92 DZ		$h^2 = 0.399$ $h^2 = 0.871$ $h^2 = 0.9$ $h^2 = 0.877$	Birth 5 mo males 5 mo females 60 mo

(Continued)

Table 8.1 Heritability estimates of anthropometrics during childhood and adolescence—cont'd

Trait	Reference	Population	Design and family structure	Sample size	Familial correlations	Heritability or genetic variance estimate	Age range
	Silventoinen et al., 2008[59]	Sweden	Multiple, twins, siblings	1582 MZ pairs, 1864 DZ pairs, 154,970 full brother pairs		$h^2 = 0.64$	16–25 y
	Choh et al., 2011[20]	USA	Longitudinal, nuclear/extended	917 subjects		$h^2 = 0.74$–0.85	1–36 mo
Biacromial breadth							
	Mueller & Titcomb, 1977[28]	Colombia	Cross-sectional, nuclear	403 families	$r_{mo-child} = 0.33$ $r_{fa-child} = 0.32$	$h^2 = 0.63$ (males) $h^2 = 0.40$ (females)	7–12 y
	Susanne, 1977[29]	Belgium	Cross-sectional, nuclear	125 families	$r_{pc} = 0.33$	$h^2 = 0.58$	17–35 y
	Kaur & Singh, 1981[32]	India	Cross-sectional, nuclear	82 families	$r_{pc} = 0.38$	$h^2 = 0.75$	18–59 y
	Devi & Reddi, 1983[34]	India	Cross-sectional, nuclear	436 families	$r_{pc} = 0.30$ $r_{sibs} = 0.37$	$h^2 = 0.49$	6–13 y
	Sharma et al., 1984[35]	India	Cross-sectional, nuclear/twins	610 subjects	$r_{sibs} = 0.32$ $r_{DZ} = 0.56$ $r_{MZ} = 0.95$		3–26 y
	Arya et al., 2002[44]	India	Cross-sectional, nuclear	1918 subjects (342 families)		$h^2 = 0.44$	6–72 y

Reference	Country	Design	Sample	Correlation	h^2	Age
Salces et al., 2007[65]	India	Mixed-longitudinal, nuclear	238 brothers, 214 sisters (134 families)		$h^2 = 0.30–1.0$	4–19 y

Biiliac breadth

Reference	Country	Design	Sample	Correlation	h^2	Age
Susanne, 1977[29]	Belgium	Cross-sectional, nuclear	125 families	$r_{pc} = 0.49$	$h^2 = 0.73$	17–35 y
Devi & Reddi, 1983[34]	India	Cross-sectional, nuclear	436 families	$r_{pc} = 0.18$ $r_{sibs} = 0.18$	$h^2 = 0.34$	6–13 y
Ikoma et al., 1988[66]	Japan	Cross-sectional, nuclear	3632 subjects	$r_{sibs} = 0.30$	$h^2 = 0.54–0.55$	>14 y
Salces et al., 2007[65]	India	Mixed-longitudinal, nuclear	238 brothers, 214 sisters (134 families)	$r_{pc} = 0.27$	$h^2 = 0.47–1.0$	4–19 y

Upper arm circumference

Reference	Country	Design	Sample	Correlation	h^2	Age
Mueller & Titcomb, 1977[28]	Colombia	Cross-sectional, nuclear	403 families	$r_{mo-child} = 0.37$ $r_{fa-child} = 0.32$	$h^2 = 0.20$ (males) $h^2 = 0.34$ (females)	7–12 y
Susanne, 1977[29]	Belgium	Cross-sectional, nuclear	125 families	$r_{pc} = 0.30$	$h^2 = 0.50$	17–35 y
Kaur & Singh, 1981[32]	India	Cross-sectional, nuclear	82 families	$r_{pc} = 0.23$	$h^2 = 0.24$	18–59 y
Devi & Reddi, 1983[34]	India	Cross-sectional, nuclear	44 MZ, 436 families	$r_{pc} = 0.26$ $r_{sibs} = 0.24$	$h^2 = 0.46$	6–13 y
Sharma et al., 1984[35]	India	Cross-sectional, nuclear/twins	610 subjects	$r_{sib} = 0.26$ $r_{DZ} = 0.52$ $r_{MZ} = 0.95$		3–26 y
Arya et al., 2002[44]	India	Cross-sectional, nuclear	1918 subjects (342 families)		$h^2 = 0.301$	6–72 y
Poveda et al., 2010[63]	Belgium	Cross-sectional, nuclear	460 subjects		$h^2 = 0.57$	17–72 y

Age at menarche

Reference	Country	Design	Sample	Correlation	h^2	Age
Damon et al., 1969[67]	USA	Retrospective, nuclear	78 Mo–Da pairs	$r_{mo-da} = 0.24$		

(Continued)

Table 8.1 Heritability estimates of anthropometrics during childhood and adolescence—cont'd

Trait	Reference	Population	Design and family structure	Sample size	Familial correlations	Heritability or genetic variance estimate	Age range
	Orley, 1977[68]	Hungary	Retrospective, nuclear	550 Mo–Da pairs	$r_{mo-da} = 0.25$		
	Kaur & Singh, 1981[32]	India	Retrospective, nuclear	72 Mo–Da pairs	$r_{mo-da} = 0.39$		
	Brooks-Gunn & Warren, 1988[69]	USA	Retrospective, nuclear (daughters)	307 Mo–Da pairs	$r_{mo-da} = 0.26$ (non-dancers) $r_{mo-da} = 0.32$ (ballet dancers)		14–17 y
	Meyer et al., 1991[70]	Australia	Retrospective, twins	1178 MZ	$r_{MZ} = 0.71$ $r_{DZ} = 0.22$	$h^2 = 0.17$ (additive effects) $d^2 = 0.54$ (dominance effects)	
	Malina et al., 1994[71]	USA	Retrospective, nuclear (university athletes)	109 mo–da pairs, 77 sib pairs	$r_{mo-da} = 0.25$ $r_{sib} = 0.44$		
	Loesch et al., 1995[72]	Poland	Longitudinal, twins (examined genetic correlations among maturity traits)	95 MZ female, 97 DZ female		h^2 (raw) = 0.95 $h^2 = 0.44$ (unique genetic effects) $h^2 = 0.53$ (shared genetic effects with skeletal maturity)	0–18 y
	Kirk et al., 2001[73]	Australia	Longitudinal, twins	1001 pairs, 708 subjects	$r_{MZ} = 0.51$ $r_{DZ} = 0.17$	$h^2 = 0.5$	Mean 13 y

Author, year	Country	Study design	Sample	Correlations	Heritability	Age
Sharma, 2002[74]	India	Cross-sectional, twins	60 female twin pairs (30 MZ, 30 DZ)	$r_{MZ}=0.93$ $r_{DZ}=0.55$	$h^2=0.78$	Mean 17.5 y
Towne et al., 2005[75]	USA	Longitudinal, nuclear/extended	371 subjects		$h^2=0.46$	9–16 y
Pan et al., 2007[55]	Hutterites	Longitudinal, extended, multiple lines of descent	806 subjects		$h^2=0.46$	

BMI

Author, year	Country	Study design	Sample	Correlations	Heritability	Age
Magnusson & Rasmussen, 2002[76]	Sweden	Cross-sectional, extended/nuclear	196,743 sons, 19,972 fathers	Full bro = 0.36 mat half bro = 0.21 pat half bro = 0.11 father–son = 0.28		18–19 y
Silventoinen, et al., 2007[57]	Netherlands	Longitudinal, twins	7753 pairs (at age 3)		$h^2=0.60-0.78$ $h^2=0.57-0.82$	3–12 y males 3–12 y females
Haworth et al., 2008[77]	UK	Longitudinal, twins/nuclear	3582 twin pairs (at age 3)		$h^2=0.48$ $h^2=0.65$ $h^2=0.82$ $h^2=0.78$	4 y 7 y 10 y 11 y
Silventoinen et al., 2008[59]	Sweden	Multiple, twins/siblings	1582 MZ pairs, 1864 DZ pairs, 154,970 full brother pairs		$h^2=0.59$	16–25 y
Wardle et al., 2008[78]	UK	Longitudinal, twins	5092 twin pairs		$h^2=0.77$	8–11 y
Lajunen et al., 2009[79]	Finland	Longitudinal, twins	2413 twin pairs		$h^2=0.69$ $h^2=0.58$ $h^2=0.66$ $h^2=0.58$ $h^2=0.83$ $h^2=0.74$	11–12 y males 11–12 y females 14 y males 14 y females 17 y males 17 y females

(Continued)

202 Bradford Towne, Ellen W. Demerath, Stefan A. Czerwinski

Table 8.1 Heritability estimates of anthropometrics during childhood and adolescence—cont'd

Trait	Reference	Population	Design and family structure	Sample size	Familial correlations	Heritability or genetic variance estimate	Age range
	Martin et al., 2010[80]	USA	Longitudinal, nuclear	821 subjects		$h^2 = 0.70$	Mean 12.6 y
	Salsberry & Reagan, 2010[81]	USA	Longitudinal, mother–offspring	5453 subjects, 4994 subjects,		$h^2 = 0.29$ $h^2 = 0.20$ $h^2 = 0.61$ $h^2 = 0.56$	6–8 y males 6–8 y females 12–14 males 12–14 females
	Choh et al., 2011[20]	USA	Longitudinal, nuclear/ extended	917 subjects		$h^2 = 0.43–0.78$	0–36 mo
Growth pattern parameters							
	Beunen et al., 2000[82]	Belgium	Longitudinal, twins	99 twin pairs		Adolescent stature growth curve parameters: $h^2 = 0.89–0.96$	10–18 y
	van Dommelen et al., 2004[17]	Netherlands	Longitudinal, twins	4649 twin pairs		Stature at different ages: $h^2 = 0.12–0.44$ $h^2 = 0.33–0.74$	Birth–2.5 y females Birth–2.5 y males
	Czerwinski et al., 2007[53]	USA	Longitudinal, nuclear/ extended	403 subjects		Adolescent stature growth curve parameters: age at PHV: $h^2 = 0.72$ PHV: $h^2 = 0.65$ height at PHV: $h^2 = 0.98$	2–18 y
	Silventoinen et al., 2008[58]	Sweden	Longitudinal, twins	99 MZ males, 76 DZ males	$r_{MZ} = 0.92$ $r_{DZ} = 0.41$	$h^2 = 0.93$	17.5–20 y

Table 8.2 Recent genome-wide association and large-scale genetic association studies of growth and development traits

Trait	Reference	Population(s)	Study design	N (discovery and replication)	Trait age range (years)	No. of significant loci identified	% of variance explained by identified loci	Findings (loci/pathways identified)
Age at menarche								
	Liu et al., 2009[84]	EU, Chinese	GWA	3,480	~9–17	1	—	SPOCK-7 (proteoglycan; inhibits MMP-2 which mediates endometrial menstrual breakdown)
	Perry et al., 2009[85]	EU	GWA	17,510	9–17	2	—	The two loci identified were LIN28B (expressed in placental, fetal liver, testis; previously associated with normal variation in adult ht) and intergenic locus (9q31.2)
	Elks, et al., 2010[86]	EU	GWA	102,533	9–17	42	3.6–6.1%	Data indicate enrichment for gene pathways involved in (1) cellular growth, proliferation, function and maintenance; and (2) lipid metabolism, small molecule biochemistry and molecular transport, including fatty acid biosynthesis; specific loci include TAC3R, ESR1 (estrogen receptor) and four obesity-susceptibility loci (FTO, TMEM18, SEC16B, TRA2B)

(*Continued*)

Study	Ethnicity	Study type	N	Age	Loci	%	Comments
Sulem et al., 2010[87]	EU, Icelandic	GWA	20,954	7–19	1	—	The locus identified was LIN28B
Birth weight							
Freathy et al., 2010[88]	EU		38,214	Birth	2	0.3–0.1%	Loci identified were (1) ADCY5 (encodes enzymes responsible for synthesis of cAMP, may influence on insulin secretion); and (2) CCNL1 (encodes protein associated with cyclin-dependent kinases)
Kilpeläinen et al., 2011[89]	EU	Association study of 12 obesity-susceptibility loci	28,219	Birth	2	—	Loci identified were (1) MTCH2 (encodes mitochondrial membrane protein critical for apoptosis); and (2) FTO (an obesity-susceptibility locus linked to food intake, energy expenditure and adiposity)
BMI							
den Hoed et al., 2010[90]	EU	Association study of 16 obesity-susceptibility loci	13,071	9–16	9	1%	Nine of 16 loci examined replicated in children, and were primarily those that are expressed in the hypothalamus, suggesting importance of neuronal control of energy balance
Kang et al., 2010[91]	African	GWA	1,931	18–74	0	—	No SNPs reached genome-wide significance

Fat mass

Loos et al., 2008[92]	EU	GWA	5,988 Children	0–11	1	0.24%	In children aged 7–11 years, each additional copy of rs17782313 was associated with BMI changes. rs17782313 mapped 188 kb downstream of MC4R, a known obesity gene

Height

Soranzo, et al 2009[93]	EU	GWA	19,798	16–99	17	< 0.20%	Numerous, including CATSPER4 (associated with male fertility), TMED10 (TMP21), NPR3 (encodes natriuretic peptide (NCP) involved in blood pressure regulation, JAZF1 (implicated in type 2 diabetes and prostate cancer susceptibility)
Kang et al., 2010[91]	African	GWA	1,931	18–74	14	0.20%	No SNPs reached genome-wide significance
Lango et al., 2010[94]	EU	GWA	183,727	Adults	180	10%	Numerous, including hedgehog signaling, TGF-β signaling, many skeletal growth and skeletal dysplasia genes and pathways identified

(Continued)

Lanktree et al., 2011[95]	6 Ethnicities	Association study of 2000 cardiovascular disease susceptibility loci	114,223	21–80	64	—	Numerous; results indicate enrichment of the following pathways: energy metabolism, insulin and growth hormone signaling, heart morphogenesis, cellular growth and apoptosis circadian rhythm, and collagen formation and remodeling

PHV, age at PHV

| Sovio et al., 2009[96] | EU | Association study of 43 height-related loci | 3,538 | 0–31 | 24 | — | Seven SNPs [SF3B4/SV2A, LCORL (×2), UQCC, DLEU7, HHIP, HIST1H1D] were associated with PHV (infancy); five SNPS (SF3B4/SV2A, SOCS2, C17orf67, CABLES1, DOT1L) were associated with PHV (puberty); no significant associations with age at PHV (puberty). HHIP is a component of the hedgehog signal transduction pathway which is involved with embryogenesis and development. SOCS2 is a negative regulator of cytokine/cytokine hormone pathway JAK/STAT which influences growth and development |

BMI: body mass index; PHV: peak height velocity; GWA: genome-wide association; SNP: single-nucleotide polymorphism; TGF: transforming growth factor.

(as in Turner's syndrome) all result in growth retardation. Formal quantitative genetic analyses of birth weight find somewhat lower heritability estimates than for body weight and length in postnatal life, which are both highly heritable (see below). Assessment of genetic influences on birth weight is complicated, however, by the fact that prenatal growth (at least as measured by birth weight) is influenced by both the genetic make-up of the fetus and the maternal intrauterine environment. There is no fully satisfactory way to partition these two sources of variation. Therefore, not surprisingly, estimates of the influences of fetal genes, maternal genes, non-genetic maternal factors and random environmental effects on fetal growth vary considerably across studies. The role of fetal genes varies from 0 to 50%, maternal factors from 27 to 50%, and random environmental factors from 8 to 43% in the variation in birth weight.[97]

For example, a classic study by Penrose[12] attempted to partition the variance in birth weight among fetal genes, maternal genes, non-genetic maternal factors and random environmental effects. He concluded that fetal genes accounted for approximately 18% of the phenotypic variance, while "maternal factors" (a combination of both genetic and uterine environment) explained approximately 40% of the phenotypic variance. The importance of uterine environment in the control of prenatal growth is also demonstrated by the changes in twin correlations from birth onwards (e.g. Wilson, 1976).[26] Intrapair differences in the birth weight of MZ twins are often significant at birth (tending to be larger than differences between DZ twins) because MZ twins compete for placental resources. Differences in weight between MZ twins decrease over time. By 3 years of age, the MZ twin correlation is about 0.80—0.90 and the DZ twin correlation is about 0.40—0.50.

A problem with the use of birth weight as a measure of prenatal growth is that it represents growth status at a variety of maturational ages depending on gestational age. Most studies of the genetics of birth weight have not controlled for gestational age. This flaw has probably led to underestimates of genetic influences. Indeed, using a variance components method for pedigree data, and modeling a gestational age covariate effect, the present authors found a high heritability of birth weight in the Fels Longitudinal Study population ($h^2 = 0.67$;[20]). Continued work along these lines will help to identify specific factors influencing fetal growth and development. However, progress depends on measurement strategies that better capture the process of fetal development (e.g. serial ultrasound biometry).

Height

Data from nearly 4000 individuals in 1100 nuclear families in England analyzed by Pearson and Lee[98] provide perhaps the earliest evidence for the inheritance of height. In this landmark study, Pearson and Lee found a significant correlation between spouses (0.28), showing positive assortative mating for height, but higher correlations between siblings (0.54) and between parents and offspring (0.50). Since the expected correlation between full siblings and between parents and offspring would be 0.50 if the h^2 of the trait was 1.0, they concluded that the population variation in height was highly determined by genetic factors. These early results have been corroborated by hundreds of subsequent family studies. In populations around the world, the estimates of the h^2 of height range from 0.60 to above 0.90, clearly showing that height is a highly heritable trait.

In a review of 24 studies of parent–child correlations of height and weight, however, Mueller[25] indicated that population estimates of heritability tend to be systematically lower in developing countries than in affluent countries. There are several reasons why this might be so. As mentioned earlier, according to classic quantitative genetic theory, the heritability of height or any trait is a function of the population in which the estimate is made, as well as of the trait itself. Heritability estimates will tend to be higher if there is positive assortative mating (i.e. a significant phenotypic correlation between parents). And indeed, assortative mating for height has been found in European or European-derived populations more frequently than in non-European populations. Also, non-European populations in the developing world tend to live under more nutritional and disease stress than European populations. In these populations such environmental factors have the potential to affect a given trait more than in affluent populations. Since heritability is the proportion of variance due to genetic influences, a larger proportion of environmentally induced variation will reduce the heritability. In addition, many non-European populations are experiencing rapid economic change, resulting in the growth environment of children differing quite markedly from that of their parents, thus decreasing parent–offspring correlations and the estimate of total variation attributable to genes.

Weight, Circumferences and Skinfolds

Whereas the heritability of skeletal lengths (e.g. height, sitting height) tends to be high, the h^2 of skeletal breadths (e.g. biiliac and biacromial diameters) tends to be somewhat lower, averaging between 0.40 and 0.80. In turn, skeletal breadths tend to have higher heritabilities than circumferences and skinfolds. It has been assumed that soft-tissue traits are more easily altered by the changing nutritional environment of individuals than are skeletal tissues, which respond less quickly to changes in nutritional status, and as a result have a greater proportion of their variance explained by environmental, rather than genetic, factors.

Longitudinal Studies

As mentioned earlier, the vast majority of family studies of growth and development are cross-sectional. Only a few studies have longitudinal growth and development data from related children that permit genetic analyses of the processes of growth and development.

Some of these longitudinal studies of the genetics of growth, for example, examined changes in parent—child or sibling correlations from age to age. Reports from the Fels Longitudinal Study,[99] Poland[22] and elsewhere[100,101] found that parent—child correlations for height increased during the first 4 years of life, decreased during adolescence (when heterogeneity of the maturational tempo disrupted familial similarity) and subsequently rose above the prepubertal level.

Modern longitudinal genetic epidemiological studies of growth and development use growth curve-fitting methods to pinpoint growth and maturational events, particularly of changes in the tempo of growth in a measure, and then examine growth curve parameters in genetic analyses. For example, Beunen et al.[82] report high h^2 estimates for the ages at take-off and at peak height velocities, and the heights at those ages. Van Dommelen et al.[17] fitted curves to serial infant height (length) and weight data from a large sample of Dutch twin infants and found significant heritabilities of various growth curve parameters. Similar analyses of Fels Longitudinal Study data are discussed in more detail in Section 8.5.

Maturation

Not only is physical size heritable, but the timing and tempo of maturation are also significantly controlled by genes. A number of early studies of dental development found that radiographic measures of the timing of tooth formation (calcification) and dental emergence were more highly correlated within MZ twin pairs than DZ twin pairs, suggesting a heritability of 0.85—0.90.[102] The number and pattern of dental cusps were also found to be under genetic control. The rate of skeletal maturation has been compared in siblings over time in several reports, with the general finding being that there is a great deal of similarity between siblings in the age of ossification onset of bones in the hand and foot. The general pattern of skeletal maturation (i.e. the tendency to be an "early" or "late" maturing individual) also suggests that the tempo of development is highly heritable, with sib—sib correlations of 0.45.[103]

The process of maturation is commonly believed to be controlled, at least partially, by genes independent from those controlling final size. This conjecture stems from the observation that siblings may reach identical height even though they differed in the timing of maturational events.[104] Further and more widespread use of the multivariate quantitative approaches discussed in Section 8.2, in which genetic and environmental correlations between different traits may be calculated, will allow for greater understanding of the extent of shared genetic and non-genetic factors underlying growth and development traits.

Age at menarche is one of the most studied developmental traits. A number of early studies suggested that age at menarche has a genetic basis (e.g. Boas, 1932).[105] The mother—daughter and sister—sister correlations in the age at menarche were close to 0.50, indicating a high degree of genetic determination of age at menarche. These and later studies, however, have relied primarily on recalled ages at menarche, and thus recall bias (greater in mothers than in daughters) is introduced into these estimates. Later studies have confirmed a strong genetic influence on age at menarche,[37,106] although the familial correlations were lower than in the early studies (~0.25—0.45). In a sample of 371 female Fels Longitudinal Study subjects of varying degrees of relationship to each other, and from whom age at menarche data had been collected during their participation in the growth and development aspect of the study, Towne et al.[75] found a substantial and significant heritability of 0.49 for age at menarche.

8.5 EXAMPLES FROM THE FELS LONGITUDINAL STUDY

This final section highlights some of the topics discussed in the preceding sections through examples of published and ongoing genetic analyses conducted over the years in the Fels Longitudinal Study.

The Fels Longitudinal Study began in 1929 in Yellow Springs, Ohio. It was one of several longitudinal studies of child growth and development initiated in the USA between the end of World War I and the start of the Great Depression, and it is the only one that has survived to today. Although the Fels Longitudinal Study did not begin with an interest in genetics, familial data began to be collected soon after the study began. Most of the mothers who enrolled their children in the early years of the study had more children later, and many of those children were subsequently enrolled. A set of MZ, dichorionic triplets was recruited early in the study specifically to examine their similarities in growth and development. Another set of triplets and a few twin pairs were also recruited in later years. Over time, other relatives were incorporated into the study, the first of these being siblings of original subjects as mentioned above, and then offspring of original study subjects. The Fels Longitudinal Study today has more than 1000 active research subjects with various serial data from infancy, and mixed cross-sectional and longitudinal data from more than 1000 of their relatives. These individuals represent about 200 kindreds consisting of both nuclear and extended families.

The description of the "Genetics Program" of the Fels Longitudinal Study written by its first director, Lester W. Sontag,[107] is remarkable for its modern-sounding tone. Sontag noted that many aspects of growth and development were likely to have significant genetic determination, but are influenced by environmental factors as well. He noted that the study included many families with two or more children, and that these "… constitute the material for the study of inheritance of growth patterns as well as of metabolic characteristics".

For example, the set of MZ, dichorionic triplets mentioned above were the subject of three early reports that described their similarities in physical and mental traits as young children, striae in their bones and the onset of ossification from infancy through pubescence.[108–110] Soon after the triplets' eighteenth birthday, Reynolds and Schoen[111] published a description of their growth patterns. A paper by Reynolds[112] is especially noteworthy because it used familial data from different types of relatives to examine the effects of degree of kinship on patterns of ossification. Included in this analysis were the set of identical triplets, as well as three pairs of identical twins, 22 pairs of siblings, eight pairs of first cousins and 18 unrelated children. Reynolds found that close relatives were very similar in pattern of ossification, distant relatives less so, and unrelated participants even less similar.

A series of studies from the late 1950s to the late 1960s by Garn and colleagues used data from siblings, parents and offspring to examine patterns of familial correlations in traits pertaining especially to dental and skeletal maturation. An example of the analyses and sample sizes from this period is provided by Garn et al.,[113] who examined ossification data from radiographs of the hand–wrist and chest for 72 parent–child pairs, 318 sibling pairs, four pairs of DZ twins and four pairs of MZ twins. Since these were serial data taken at half-yearly intervals from the ages of 1 to 7 years, there were 1211 pairings of parent–child data, 6690 pairings of sibling data, 102 pairings of data from DZ twins and 176 pairings of data from MZ twins. Garn et al.[113] concluded that, "In these well-nourished … Ohio-born white children, genes appear to account for a major proportion of ossification variance during growth". These investigators also examined the genetics of various dental traits, including the timing of stages of dental development,[102] tooth morphology[114] and the appearance of discrete dental traits.[115] The influence of familial factors on growth in body size was also examined.[99,102]

Genetic analyses of growth and development data from the Fels Longitudinal Study have had a resurgence in the last 20 years. This is largely due to advances in statistical genetic methods that maximize the amount of information available in longitudinal data from large numbers of relatives of varying degrees of relationship to one another, as well as advances in molecular genetic methodology that allow for relatively low-cost genotyping. For example, Towne et al.[37] fitted a three-parameter function to serial recumbent lengths from 569 infants in order to characterize each individual's unique pattern of growth during infancy. Figure 8.2 shows the growth curves of two infant boys who differ in their patterns of growth. Boy #1 started out in life shorter than boy #2, but had a rate of increase in recumbent length that was much greater than that of boy #2. Both boys, however, experienced about the same amount of growth (~42 cm) from birth to the age of 2 years. In this study, substantial h^2 estimates of 0.83 for recumbent length at birth, 0.67 for rate of increase in length and 0.78 for a parameter describing the curvilinear shape of growth in recumbent length from birth to 2 years were found.

Fels Longitudinal Study investigators have used the triple logistic model of Bock et al.[116] as implemented in the AUXAL program[117] to fit growth curves to extensive serial stature data from some 600 study subjects in order to examine individual

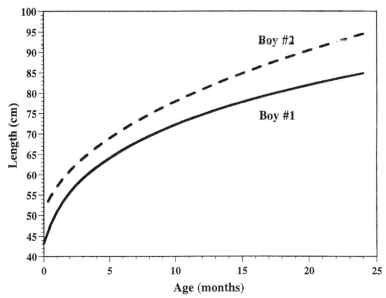

Figure 8.2 Height distance curves for two boys with differing growth patterns between birth and 24 months old.

differences in patterns of growth and to conduct multivariate quantitative genetic analysis of different parameters of the pubertal growth spurt. For example, Figure 8.3 shows the growth and velocity curves of two girls with visibly different growth patterns. Girl #1 was only 9.37 years old when she was at the peak of her pubertal growth spurt, whereas the age at peak height velocity of girl #2 was 13.09 years. At the time of peak height velocity, girl #1 was shorter than girl #2 (141.1 vs 156.5 cm), which is expected given her younger age at peak height velocity, but at the age at peak height velocity, girl #1 had a higher rate of growth than girl #2 (8.8 vs 6.4 cm/year). By the end of their growth, girl #1 was a petite woman (158.5 cm) while girl #2 was somewhat taller than average (170.6 cm). Highly significant h^2 estimates in the order of 0.85 for age at peak height velocity, 0.61 for growth rate at peak height velocity and 0.96 for stature at the age of peak height velocity were found. Especially interesting was the finding of additive genetic correlations between these pubertal growth spurt parameters that were significantly lower than 1.0, suggesting incomplete pleiotropic effects of genes on different aspects of growth. That is, these three different growth curve parameters may have, to some extent, unique genetic underpinnings.

In an association study, Towne et al.[118] found evidence of the effects of a functional polymorphism in the β-subunit of the luteinizing hormone gene (LH-β) on stature during childhood. A total of 736 individuals, from 137 nuclear and extended families, measured a total of 13,300 times between the ages of 2 and 18 years, were genotyped for the LH-β polymorphism. Individuals with the less common LH-β allele were shorter

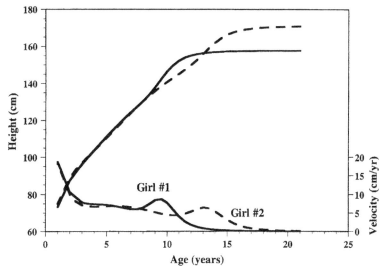

Figure 8.3 Height distance and velocity curves for two girls with different growth patterns between 12 months and 20 years old.

than those homozygous for the common LH-β allele at all ages, with this difference steadily increasing with age (e.g. on average, heterozygotes were 0.5 cm shorter than homozygotes at age 2, and they were 2.0 cm shorter at age 18). These results suggest that the LH-β polymorphism is associated with at least modest differences in the stature of children at all childhood ages.

Towne et al.[83] recently examined the mid-childhood growth spurt in Fels Longitudinal Study subjects. Presence of a mid-childhood growth spurt was found to have a significant heritability of 0.37 ± 0.14 ($p = 0.003$). From linkage analysis, two QTL with suggestive LOD scores were found: one at 12 cM on chromosome 17p13.2 (LOD = 2.13) and one at 85 cM on chromosome 12q14 (LOD = 2.06). While the chromosome 17 finding was novel, other investigators have found evidence of linkage or association of growth measures to markers in the same region of chromosome 12.

Fels Longitudinal Study investigators have used multivariate variance components methods incorporating parametric correlation functions to model the heritability and genetic and correlational structures of skeletal maturity throughout childhood. For example, a total of 6893 annual hand–wrist radiograph skeletal age assessments from a sample of 807 children aged 3–15 years representing 192 nuclear and extended families were simultaneously analyzed. The best fitting model had 65 parameters and allowed for an exponential decay in genetic and environmental correlations as a function of chronological age differences. From this model, the h^2 estimates of skeletal age at each chronological age were: $3 = 0.71$, $4 = 0.73$, $5 = 0.77$, $6 = 0.93$, $7 = 0.78$, $8 = 0.77$, $9 = 0.73$, $10 = 0.63$, $11 = 0.45$, $12 = 0.39$, $13 = 0.34$, $14 = 0.23$ and

15 = 0.11. The genetic correlation matrix showed a pattern of decreasing correlations between skeletal age at different chronological ages as age differences increased (e.g. ρ_G between skeletal age at age 3 and skeletal age at age 4 was 0.96, but between skeletal age at age 3 and skeletal age at age 15 ρ_G was 0.56). The random environmental correlation matrix showed an even more pronounced pattern of decreasing correlations between skeletal age at different chronological ages as age differences increased (e.g. ρ_E between skeletal age at age 3 and skeletal age at age 4 was 0.77, but ρ_E between skeletal age at age 3 and skeletal age at age 15 was only 0.12). These results show a high heritability of skeletal age through early puberty, and suggest that skeletal maturation at different stages of development is influenced by different sets of genes and environmental factors.

8.6 SUMMARY

For over a century there has been scientific interest in the genetic underpinnings of growth and development. But, as with any area of scientific inquiry, to one degree or another all of these studies were limited by the methods and technologies available to them at the time. For that reason, most of the literature on the genetics of growth and development until relatively recently is limited to h^2 estimates of measures of growth and development gathered once from first degree relatives. The opportunities exist today, however, for far more sophisticated genetic epidemiological studies of growth and development.

One major problem, though, is that modern genetic epidemiological studies of growth and development can be expensive undertakings. Such studies are readily justified, however, on very practical and applied grounds. Foremost among these is that the growth and development of children can have health consequences later in life. Thus, to a large extent, genetic epidemiological studies of growth and development are inherently of biomedical interest. Indeed, much of the current research emphasis in the Fels Longitudinal Study pertains to studies of the relationships between age-related changes in body composition (including those that occur during childhood) and the development and progression of cardiovascular disease (CVD) and type 2 diabetes mellitus risks in later life, an area of active research today.

For example, work by Barker and colleagues in the UK,[119] as well as others, suggests that early growth variation in size and body composition (during prenatal as well as postnatal periods) influences the risk of a number of disorders including hypertension, obesity, heart disease and diabetes. Longitudinal family studies of growth and development are needed to fully evaluate these hypotheses. A current Fels Longitudinal Study project is aimed at evaluating the role of birth weight and infant growth in predisposing to adult-onset disease risk factors, taking into account the significant heritable components of both early growth and various adult disease risk factors.

Demerath et al.,[120] for example, found that birth weight was negatively associated with fasting insulin concentration in adulthood after adjusting for body mass index and age, but after taking into account the significant heritability of insulin concentration, birth weight accounted for only 1–2% of the phenotypic variance of fasting insulin concentration.

Demerath et al.[121] found significant heritabilities of measures of infant weight and weight change, and Demerath et al.[122] found significant associations between measures of postnatal weight gain (which had earlier been found to be significantly heritable) and direct measures of adiposity in adulthood (which also are significantly heritable). In a similar vein, another ongoing Fels Longitudinal Study project is examining changes in traditional CVD risk factors during growth and development, and associations between growth-related events and CVD and type 2 diabetes risk in adulthood. For example, although serum lipid and lipoprotein levels track from childhood to adulthood, Czerwinski et al.[123] found differences in the heritabilities of lipid and lipoprotein measures in children sampled before and after puberty. In general, heritability estimates were higher after puberty, suggesting that the genetic control of lipid and lipoprotein levels may be influenced by maturational factors. Czerwinski et al.[53] found highly heritable measures of patterns of growth (e.g. timing of the pubertal growth spurt) to be associated, at least on the phenotypic level, with heritable measures of adult disease risk such as blood pressure and body mass index. Future studies based on these findings will explicitly explore pleiotropic effects of genes on measures of growth and measures of disease risk in adulthood.

Placing studies of growth and development more squarely in the context of biomedical research will allow auxological investigations to move beyond being descriptive studies, and will help to open the door to resources needed to conduct modern genetic epidemiological studies of growth and development.

GLOSSARY

Allele: A variant of the DNA sequence at a particular locus. Typically, individuals possess two allelic variants at each locus, one each derived from the maternal and paternal chromosomes. The two alleles may be identical or different, making the individual homozygous or heterozygous, respectively, at that locus.

Canalization: The tendency of a growth-related trait to follow a certain course or trajectory over time.

Complex trait/phenotype: Any phenotype whose expression is influenced by multiple genes, or by one or more genes and one or more environmental factors. Complex traits can be quantitative or discrete.

Epistasis: Interactions between alleles at different loci. Also known as gene × gene interaction.

Gene: A segment of DNA that codes for a specific protein or enzyme.

Genotype: The group of genes making up an organism. The genotype at a particular locus consists of the two alleles present at that locus.

Genome-wide association (GWA) study: Use of high-density SNPs to test for associations between genomic variation and trait variation across the genome.

Heritability: A measure that expresses the extent to which phenotypes are determined by genes transmitted from parents to their offspring. Heritability (in the narrow sense) is defined as the proportion of the total phenotypic variance that is attributable to the additive effects of genes.

Identity by descent (IBD): Identical alleles at the same locus found in two related individuals that are identical because they originated from a common ancestor.

Identity by state (IBS): Identical alleles found within two individuals. If the two individuals are related, the two alleles may also be identical by descent if they are replicates of the same ancestral allele from a previous generation.

Kinship coefficient: The probability that two genes from two individuals for a given locus are identical by descent. A general measure of relatedness.

Linkage analysis: A test of co-segregation of traits and genomic variants used to localize a trait to a chromosomal region.

Linkage disequilibrium: Non-random association within a population of alleles at two or more linked loci. Linkage disequilibrium decays with increasing genetic (recombination) distance between loci.

Locus: The particular position on a chromosome where a gene resides.

Monogenic: A trait is monogenic if that trait is influenced primarily or entirely by alleles at only one genetic locus.

Mutation: Specific sequence variants in the nucleotide sequence of a gene. These variants may or not be inherited.

Oligogenic: A trait is oligogenic if it is influenced by a few loci of significant, individually detectable effects.

Phenotype: The observable characteristics of an organism, or a specific trait produced by the genotype in conjunction with the environment.

Polygenic: A phenotype is polygenic if it is influenced by many genes of relatively small individual effects, such that the influence of any single locus is very difficult to detect on its own.

Polymorphism: The joint occurrence in a population of two or more genetically determined alternative phenotypes, each occurring at an appreciable frequency (arbitrarily, 1% or higher). A polymorphism may be defined at either the protein level (e.g. Rh+ and Rh− red blood cell groups) or at the DNA level (alternative alleles at a locus).

Quantitative trait locus (QTL): Any locus harboring genetic variants that influence variation in a complex phenotype.

Recombination (crossover): The exchange of segments of homologous chromosomes following chromosomal duplication and synapse formation during meiosis. Recombination is responsible for the production of offspring with combinations of alleles at linked loci that differ from those possessed by the two parents.

Single-nucleotide polymorphism (SNP): A type of DNA sequence variant where a single nucleotide (A, T, C or G) at a particular location in the genome differs between individuals.

REFERENCES

1. Neel J, Schull W. *Human heredity*. Chicago, IL: University of Chicago Press; 1954.
2. International Genetic Epidemiology Society. Available at: http://www.geneticepi.org/ (accessed 14 October 2011).
3. Falconer DS, Mackay TFC. *Introduction to quantitative genetics*. 4th ed. Harlow: Longman; 1996.
4. Lynch M, Walsh B. *Genetic analysis of quantitative traits*. Sunderland, MA: Sinauer Associates; 1998.
5. Fisher RA. The correlation between relatives on the supposition of Mendelian inheritance. *Trans R Soc Edinb* 1918;**52**:399−433.
6. Manolio TA, Collins FS, Cox NJ, Goldstein DB, Hindorff LA, Hunter DJ, et al. Finding the missing heritability of complex diseases. *Nature* 2009;**461**:747−53.
7. Cirulli ET, Goldstein DB. Uncovering the roles of rare variants in common disease through whole-genome sequencing. *Nat Rev Genet* 2010;**11**:415−25.
8. Almasy L, Blangero J. Exploring positional candidate genes: linkage conditional on measured genotype. *Behav Genet* 2004;**34**:173−7.
9. Boerwinkle E, Chakraborty R, Sing CF. The use of measured genotype information in the analysis of quantitative phenotypes in man. I. Models and analytical methods. *Ann Hum Genet* 1986;**50**:181−94.
10. Eveleth PB, Tanner JM. *Worldwide variation in human growth*. 2nd ed. Cambridge: Cambridge University Press; 1990.
11. Martorell R, Mendoza F, Caastillo R. Poverty and stature in children. In: Waterlow JC, editor. *Linear growth retardation in less developed countries*, Vol. 14. New York: Raven Press; 1988. p. 57−73.
12. Penrose L. Some recent trends in human genetics. *Caryologia* 1954;**6**(Suppl):521−30.
13. Morton N. The inheritance of human birth weight. *Ann Hum Genet* 1955;**20**:125−34.
14. Nance W, Kramer A, Corey L, Winter PM, Eaves LJ. A causal analysis of birth weight in the offspring of monozygotic twins. *Am J Hum Genet* 1983;**35**:1211−23.
15. Clausson B, Lichtenstein P, Cnattingius S. Genetic influence on birthweight and gestational length determined by studies in offspring of twins. *BJOG* 2000;**107**:375−81.
16. Magnus P, Gjessing HK, Skrondal A, Skjaerven R. Paternal contribution to birth weight. *J Epidemiol Community Health* 2001;**55**:873−7.
17. van Dommelen P, de Gunst MC, van der Vaart AW, Boomsma DI. Genetic study of the height and weight process during infancy. *Twin Res* 2004;**7**:607−16.
18. Arya R, Demerath E, Jenkinson CP, Göring HH, Puppala S, Farook V, et al. A quantitative trait locus (QTL) on chromosome 6q influences birth weight in two independent family studies. *Hum Mol Genet* 2006;**15**:1569−79.
19. Grunnet L, Vielwerth S, Vaag A, Poulsen P. Birth weight is nongenetically associated with glucose intolerance in elderly twins, independent of adult obesity. *J Intern Med* 2007;**262**:96−103.
20. Choh AC, Curran JE, Odegaard AO, Nahhas RW, Czerwinski SA, Blangero J, et al. Differences in the heritability of growth and growth velocity during infancy and associations with FTO variants. *Obesity (Silver Spring)* 2011;**19**:1847−54.
21. Vandenberg S, Falkner F. Hereditary factors in human growth. *Hum Biol* 1965;**37**:357−65.

22. Welon Z, Bielicki T. Further investigations of parent—child similarity in stature as assessed from longitudinal data. *Hum Biol* 1971;**43**:517—25.

23. Garn S, Bailey S, Cole P. Similarities between parents and their adopted children. *Am J Phys Anthropol* 1976;**45**:539—43.

24. Malina R, Mueller W, Holman J. Parent—child correlations and heritability of stature in Philadelphia black and white children 6 to 12 years of age. *Hum Biol* 1976;**48**:475—86.

25. Mueller WH. Parent—child correlations for stature and weight among school aged children: a review of 24 studies. *Hum Biol* 1976;**48**:379—97.

26. Wilson R. Concordance in physical growth for monozygotic and dizygotic twins. *Ann Hum Biol* 1976;**3**:1—10.

27. Fischbein S. Intra-pair similarity in physical growth of monozygotic and of dizygotic twins during puberty. *Ann Hum Biol* 1977;**4**:417—30.

28. Mueller W, Titcomb M. Genetic and environmental determinants of growth of school-aged children in a rural Colombian population. *Ann Hum Biol* 1977;**4**:1—15.

29. Susanne C. Heritability of anthropological characters. *Hum Biol* 1977;**49**:573—80.

30. Roberts D, Billewicz W, McGregor I. Heritability of stature in a West African population. *Ann Hum Genet* 1978;**42**:15—24.

31. Fischbein S, Nordqvist T. Profile comparisons of physical growth for monozygotic and dizygotic twin pairs. *Ann Hum Biol* 1978;**5**:321—8.

32. Kaur D, Singh R. Parent—adult offspring correlations and heritability of body measurements in a rural Indian population. *Ann Hum Biol* 1981;**8**:333—9.

33. Solomon P, Thompson E, Rissanen A. The inheritance of height in a Finnish population. *Ann Hum Biol* 1983;**10**:247—56.

34. Devi M, Reddi G. Heritability of body measurements among the Jalari population of Visakhapatnam. *Ann Hum Biol* 1983;**10**:483—5.

35. Sharma K, Byard P, Russell J, Rao DC. A family study of anthopometric traits in a Punjabi community: I. Introduction and familial correlations. *Am J Phys Anthropol* 1984;**63**:389—95.

36. Byard P, Guo S, Roche A. Family resemblance for Preece—Baines growth curve parameters in the Fels Longitudinal Study. *Am J Hum Biol* 1993;**5**:151—7.

37. Towne B, Guo SS, Roche AF, Siervogel RM. Genetic analysis of patterns of growth in infant recumbent length. *Hum Biol* 1993;**65**:977—89.

38. Hauspie RC, Bergman P, Bielicki T, Susanne C. Genetic variance in the pattern of the growth curve for height: a longitudinal analysis of male twins. *Ann Hum Biol* 1994;**21**:347—62.

39. Beunen G, Maes H, Vlietinck R, Malina RM, Thomis M, Feys E, et al. Univariate and multivariate genetic analysis of subcutaneous fatness and fat distribution in early adolescence. *Behav Genet* 1998;**28**:279—88.

40. Price RA, Reed DR, Guido NJ. Resemblance for body mass index in families of obese African American and European American women. *Obes Res* 2000;**8**:360—6.

41. Silventoinen K, Kaprio J, Lahelma E, Koskenvuo M. Relative effect of genetic and environmental factors on body height: differences across birth cohorts among Finnish men and women. *Am J Public Health* 2000;**90**:627—30.

42. Luke A, Guo X, Adeyemo AA, Wilks R, Forrester T, Lowe Jr W, et al. Heritability of obesity-related traits among Nigerians, Jamaicans and US black people. *Int J Obes Relat Metab Disord* 2001;**25**:1034—41.

43. Silventoinen K, Kaprio J, Lahelma E, Viken RJ, Rose RJ. Sex differences in genetic and environmental factors contributing to body-height. *Twin Res* 2001;**4**:25—9.

44. Arya R, Duggirala R, Comuzzie AG, Puppala S, Modem S, Busi BR, Crawford MH. Heritability of anthropometric phenotypes in caste populations of Visakhapatnam, India. *Hum Biol* 2002;**74**:325—44.

45. Brown WM, Beck S, Lange E, Davis CC, Kay CM, Langefeld CD, et al. Age-stratified heritability estimation in the Framingham Heart Study families. *BMC Genet* 2003;**4**(Suppl. 1):S32.

46. Silventoinen K, Sammalisto S, Perola M, Boomsma DI, Cornes BK, Davis C, et al. Heritability of adult body height: a comparative study of twin cohorts in eight countries. *Twin Res* 2003;**6**:399—408.

47. Li M-X, Liu P-Y, Li Y-M, Qin Y-J, Liu Y-Z, Deng H-W. A major gene model of adult height is suggested in Chinese. *J Hum Genet* 2004;**49**:148—53.

48. Schousboe K, Visscher PM, Erbas B, Kyvik KO, Hopper JL, Henriksen JE, et al. Twin study of genetic and environmental influences on adult body size, shape, and composition. *Int J Obes Relat Metab Disord* 2004;**28**:39—48.

49. Malkin I, Ermakov S, Kobyliansky E, Livshits G. Strong association between polymorphisms in ANKH locus and skeletal size traits. *Hum Genet* 2006;**120**:42—51.

50. Macgregor S, Cornes B, Martin N, Visscher PM. Bias, precision and heritability of self-reported and clinically measured height in Australian twins. *Hum Genet* 2006;**120**:571—80.

51. Saunders CL, Gulliford MC. Heritabilities and shared environmental effects were estimated from household clustering in national health survey data. *J Clin Epidemiol* 2006;**59**:1191—8.

52. Bayoumi RA, Al-Yahyaee SA, Albarwani SA, Rizvi SG, Al-Hadabi S, Al-Ubaidi FF, et al. Heritability of determinants of the metabolic syndrome among healthy Arabs of the Oman family study. *Obesity (Silver Spring)* 2007;**15**:551—6.

53. Czerwinski SA, Lee M, Choh AC, Wurzbacher K, Demerath EW, Towne B, Siervogel RM. Genetic factors in physical growth and development and their relationship to subsequent health outcomes. *Am J Hum Biol* 2007;**19**:684—91.

54. Dubois L, Girard M, Girard A, Tremblay R, Boivin M, Pérusse D. Genetic and environmental influences on body size in early childhood: a twin birth-cohort study. *Twin Res Hum Genet* 2007;**10**:479—85.

55. Pan L, Ober C, Abney M. Heritability estimation of sex-specific effects on human quantitative traits. *Genet Epidemiol* 2007;**31**:338—47.

56. Reis VM, Machado JV, Fortes MS, Fernandes PR, Silva AJ, Dantas PS, et al. Evidence for higher heritability of somatotype compared to body mass index in female twins. *J Physiol Anthropol* 2007;**26**:9—14.

57. Silventoinen K, Bartels M, Posthuma D, Estourgie-van Burk GF, Willemsen G, van Beijsterveldt TC, et al. Genetic regulation of growth in height and weight from 3 to 12 years of age: a longitudinal study of Dutch twin children. *Twin Res Hum Genet* 2007;**10**:354—63.

58. Silventoinen K, Haukka J, Dunkel L, Tynelius P, Rasmussen F. Genetics of pubertal timing and its associations with relative weight in childhood and adult height: the Swedish Young Male Twins Study. *Pediatrics* 2008;**121**:e885—91.

59. Silventoinen K, Magnusson PKE, Tynelius P, Kaprio J, Rasmussen F. Heritability of body size and muscle strength in young adulthood: a study of one million Swedish men. *Genet Epidemiol* 2008;**32**:341—9.

60. Axenovich T, Zorkoltseva I, Belonogova N, Struchalin MV, Kirichenko AV, Kayser M, et al. Linkage analysis of adult height in a large pedigree from a Dutch genetically isolated population. *Hum Genet* 2009;**126**:457—71.

61. Jowett JB, Diego VP, Kotea N, Kowlessur S, Chitson P, Dyer TD, et al. Genetic influences on type 2 diabetes and metabolic syndrome related quantitative traits in Mauritius. *Twin Res Hum Genet* 2009;**12**:44—52.

62. Mathias RA, Deepa M, Deepa R, Wilson AF, Mohan V. Heritability of quantitative traits associated with type 2 diabetes mellitus in large multiplex families from South India. *Metabolism* 2009;**58**:1439—45.

63. Poveda A, Jelenkovic A, Susanne C, Rebato E. Genetic contribution to variation in body configuration in Belgian nuclear families: a closer look at body lengths and circumferences. *Coll Antropol* 2010;**34**:515—23.

64. Estourgie-van Burk GF, Bartels M, van Beijsterveldt TC, Delemarre-van de Waal HA, Boomsma DI. Body size in five-year-old twins: heritability and comparison to singleton standards. *Twin Res Hum Genet* 2006;**9**:646—55.

65. Salces I, Rebato E, Susanne C, Hauspie RC, Saha R, Dasgupta P. Heritability variations of morphometric traits in West Bengal (India) children aged 4—19 years: a mixed-longitudinal growth study. *Ann Hum Biol* 2007;**34**:226—39.

66. Ikoma E, Kanda S, Nakata S, Wada Y, Yamazaki K. Quantitative genetic analysis of bi-iliac breadth. *Am J Phys Anthropol* 1988;**77**:295—301.

67. Damon A, Damon S, Reed R, Valadian I. Age at menarche of mothers and daughters with a note on accuracy of recall. *Hum Biol* 1969;**41**:160–75.

68. Orley J. Analysis of menarche and gynecological welfare of Budapest school girls. In: Eiben O, editor. *Growth and development. Physique.* Budapest: Adademiai Klado; 1977.

69. Brooks-Gunn J, Warren MP. Mother–daughter differences in menarcheal age in adolescent girls attending national dance company schools and non-dancers. *Ann Hum Biol* 1988;**15**:35–43.

70. Meyer J, Eaves L, Heath A, Martin N. Estimating genetic influences on the age-at-menarche: a survival analysis approach. *Am J Med Genet* 1991;**39**:148–54.

71. Malina R, Ryan R, Bonci C. Age at menarche in athletes and their mothers and sisters. *Ann Hum Biol* 1994;**21**:417–22.

72. Loesch D, Huggins R, Rogucka E, Hoang N, Hopper J. Genetic correlates of menarcheal age: a multivariate twin study. *Ann Hum Biol* 1995;**22**:479–90.

73. Kirk KM, Blomberg SP, Duffy DL, Heath AC, Owens IP, Martin NG. Natural selection and quantitative genetics of life-history traits in western women: a twin study. *Evolution* 2001; **55**:423–35.

74. Sharma K. Genetic basis of human female pelvic morphology: a twin study. *Am J Phys Anthropol* 2002;**117**:327–33.

75. Towne B, Czerwinski SA, Demerath EW, Blangero J, Roche AF, Siervogel RM. Heritability of age at menarche in girls from the Fels Longitudinal Study. *Am J Phys Anthropol* 2005;**128**:210–9.

76. Magnusson PK, Rasmussen F. Familial resemblance of body mass index and familial risk of high and low body mass index. A study of young men in Sweden. *Int J Obes Relat Metab Disord* 2002;**26**: 1225–31.

77. Haworth CMA, Carnell S, Meaburn EL, Davis OS, Plomin R, Wardle J. Increasing heritability of BMI and stronger associations with the FTO gene over childhood. *Obesity* 2008;**16**:2663–8.

78. Wardle J, Carnell S, Haworth CM, Plomin R. Evidence for a strong genetic influence on childhood adiposity despite the force of the obesogenic environment. *Am J Clin Nutr* 2008;**87**:398–404.

79. Lajunen HR, Kaprio J, Keski-Rahkonen A, Rose RJ, Pulkkinen L, Rissanen A, Silventoinen K. Genetic and environmental effects on body mass index during adolescence: a prospective study among Finnish twins. *Int J Obes* 2009;**33**:559–67.

80. Martin LJ, Woo JG, Morrison JA. Evidence of shared genetic effects between pre- and postobesity epidemic BMI levels. *Obesity (Silver Spring)* 2010;**18**:1378–82.

81. Salsberry PJ, Reagan PB. Effects of heritability, shared environment, and nonshared intrauterine conditions on child and adolescent BMI. *Obesity (Silver Spring)* 2010;**18**:1775–80.

82. Beunen G, Thomis M, Maes HH, Loos R, Malina RM, Claessens AL, et al. Genetic variance of adolescent growth in stature. *Ann Hum Biol* 2000;**27**:173–86.

83. Towne B, Williams KD, Blangero J, Czerwinski SA, Demerath EW, Nahhas RW, et al. Presentation, heritability, and genome-wide linkage analysis of the mid-childhood growth spurt in healthy children from the Fels Longitudinal Study. *Hum Biol* 2008;**80**:623–36.

84. Liu Y-Z, Guo Y-F, Wang L, Tan L-J, Liu X-G, Pei Y-F, et al. Genome-wide association analyses identify SPOCK as a key novel gene underlying age at menarche. *PLoS Genet* 2009; **5**(3):e1000420.

85. Perry JRB, Stolk L, Franceschini N, Lunetta KL, Zhai G, McArdle PF, et al. Meta-analysis of genome-wide association data identifies two loci influencing age at menarche. *Nat Genet* 2009;**41**: 648–50.

86. Elks CE, Perry JRB, Sulem P, Chasman DI, Franceschini N, He C, et al. Thirty new loci for age at menarche identified by a meta-analysis of genome-wide association studies. *Nat Genet* 2010;**42**: 1077–85.

87. Sulem P, Gudbjartsson DF, Rafnar T, Holm H, Olafsdottir EJ, Olafsdottir GH, et al. Genome-wide association study identifies sequence variants on 6q21 associated with age at menarche. *Nat Genet* 2009;**41**:734–8.

88. Freathy RM, Mook-Kanamori DO, Sovio U, Prokopenko I, Timpson NJ, Berry DJ, et al. Variants in ADCY5 and near CCNL1 are associated with fetal growth and birth weight. *Nat Genet* 2010;**42**: 430–5.

89. Kilpeläinen TO, den Hoed M, Ong KK, Grøntved A, Brage S. Obesity-susceptibility loci have a limited influence on birth weight: a meta-analysis of up to 28,219 individuals. *Am J Clin Nutr* 2011;**93**:851—60.

90. den Hoed M, Ekelund U, Brage S, Grontved A, Zhao JH, Sharp SJ, et al. Genetic susceptibility to obesity and related traits in childhood and adolescence: influence of loci identified by genome-wide association studies. *Diabetes* 2010;**59**:2980—8.

91. Kang SJ, Chiang CWK, Palmer CD, Tayo BO, Lettre G, Butler JL, et al. Genome-wide association of anthropometric traits in African and African-derived populations. *Hum Mol Genet* 2010;**19**:2725—38.

92. Loos RJF, Lindgren CM, Li S, Wheeler E, Zhao JH, Prokopenko I, et al. Common variants near MC4R are associated with fat mass, weight and risk of obesity. *Nat Genet* 2008;**40**:768—75.

93. Soranzo N, Rivadeneira F, Chinappen-Horsley U, Malkina I, Richards JB, Hammond N, et al. Meta-analysis of genome-wide scans for human adult stature identifies novel loci and associations with measures of skeletal frame size. *PLoS Genet* 2009;**5**(4):e1000445.

94. Lango AH, Estrada K, Lettre G, Berndt SI, Weedon MN, Rivadeneira F, et al. Hundreds of variants clustered in genomic loci and biological pathways affect human height. *Nature* 2010;**467**:832—8.

95. Lanktree MB, Guo Y, Murtaza M, Glessner T, Bailey SD, Onland-Moret NC, et al. Meta-analysis of dense genecentric association studies reveals common and uncommon variants associated with height. *Am J Hum Genet* 2011;**88**:6—18.

96. Sovio U, Bennett AJ, Millwood IY, Molitor J, O'Reilly PF, Timpson NJ, et al. Genetic determinants of height growth assessed longitudinally from infancy to adulthood in the northern Finland birth cohort 1966. *PLoS Genet* 2009;**5**(3):e1000409.

97. Mueller WH. Genetic and environmental influences on fetal growth. In: Ulijaszek SJ, Johnston FE, Preece MA, editors. *The Cambridge encyclopedia of human growth and development*. Cambridge: Cambridge University Press; 1998. p. 133—6.

98. Pearson K, Lee A. On the laws of inheritance in man. I. Inheritance of physical characteristics. *Biometrika* 1903;**2**:357—462.

99. Garn SM, Rohmann CG. Interaction of nutrition and genetics in the timing of growth and development. *Pediatr Clin North Am* 1966;**13**:353—79.

100. Tanner J, Goldstein H, Whitehouse R. Standards for children's height at ages 2—9 years allowing for height of the parents. *Arch Dis Child* 1970;**45**:755—62.

101. Furusho T. On the manifestations of the genotypes responsible for stature. *Hum Biol* 1968;**40**:437—55.

102. Garn S, Lewis A, Polacheck D. Sibling similarities in dental development. *J Dent Res* 1960;**39**:170—5.

103. Hewitt D. Some familial correlations in height, weight and skeletal maturity. *Ann Hum Genet* 1957;**22**:26—35.

104. Tanner JM. *Growth at adolescence*. Oxford: Blackwell; 1962.

105. Boas F. Studies in growth. *Hum Biol* 1932;**4**:307—50.

106. Zacharias L, Rand W, Wurtman R. A prospective study of sexual development and growth in American girls: the statistics of menarche. *Obstet Gynecol Surv* 1976;**31**:325—37.

107. Sontag LW. Biological and medical studies at the Samuel S. Fels Research Institute. *Child Dev* 1946;**17**:81—4.

108. Sontag L, Comstock G. Striae in bones of a set of monozygotic triplets. *Am J Dis Child* 1938;**56**:301—8.

109. Sontag L, Nelson V. A study of identical triplets. Part I. Comparison of the physical and mental traits of a set of monozygotic dichorionic triplets. *J Hered* 1933;**24**:473—80.

110. Sontag L, Reynolds E. Ossification sequences in identical triplets. A longitudinal study of resemblences and differences in the ossification patterns of a set of monozygotic triplets. *J Hered* 1944;**35**:57—64.

111. Reynolds E, Schoen G. Growth patterns of identical triplets from 8 through 18 years. *Child Dev* 1947;**18**:130—51.

112. Reynolds E. Degree of kinship and pattern of ossification. *Am J Phys Anthropol* 1943;**1**:405—16.

113. Garn S, Lewis A, Kerewsky R. Third molar agenesis and size reduction of the remaining teeth. *Nature* 1963;**200**:488–9.
114. Garn S, Lewis A, Walenga A. The genetic basis of the crown-size profile pattern. *J Dent Res* 1968;**47**:1190.
115. Garn S, Lewis A, Kerewsky RS, Dahlberg A. Genetic independence of Carabelli's trait from tooth size or crown morphology. *Arch Oral Biol* 1966;**11**:745–7.
116. Bock RD, du Toit SHC, Thissen D. *AUXAL: auxological analysis of longitudinal measurements of human stature*. Chicago, IL: Scientific Software International; 1994.
117. Bock R, du Toit S, Thissen D. AUXAL: auxological analysis of longitudinal measurements of human stature, *Version 3*. Chicago, IL: Scientific Software International; 2003.
118. Towne B, Parks JS, Brown MR, Siervogel RM, Blangero J. Effect of a luteinizing hormone b-subunit polymorphism on growth in stature. *Acta Med Auxol* (abstract) 2000;**32**:43–4.
119. Barker DJ. Childhood causes of adult diseases. *Arch Dis Child* 1988;**63**:867–9.
120. Demerath EW, Towne B, Czerwinski SA, Siervogel R. Covariate effect of birth weight in a genetic analysis of fasting insulin and glucose concentrations in adulthood. *Diabetes* 2000; **49**(Suppl. 1):A183.
121. Demerath EW, Choh AC, Czerwinski SA, Lee M, Sun SS, Chumlea WC, et al. Genetic and environmental influences on infant weight and weight change: the Fels Longitudinal Study. *Am J Hum Biol* 2007;**19**:692–702.
122. Demerath E, Reed D, Choh A, Soloway L, Lee M, Czerwinski SA, et al. Rapid postnatal weight gain and visceral adiposity in adulthood: the Fels Longitudinal Study. *Obesity (Silver Spring)* 2009;**17**:2060–6.
123. Czerwinski SA, Towne B, Guo SS, Chumlea WC, Roche AF, Siervogel RM. Genetic and environmental influences on lipid and lipoprotein levels in pre- and post-pubertal children. *Am J Hum Biol* 2000;**12**:289.

SUGGESTED READING

Hartl DL. *A primer of population genetics*. 3rd ed. Sunderland, MA: Sinauer Associates; 2000.
Hartl DL, Clark AG. *Principles of population genetics*. 4th ed. Sunderland, MA: Sinauer Associates; 2007.
Khoury MJ, Cohen BH, Beaty TH. *Fundamentals of genetic epidemiology*. Oxford: Oxford University Press; 1993.
Khoury MJ, Bedrosian S, Gwin M, Higgins J, Ioannidis J, Little J. *Human genome epidemiology: building the evidence for using genetic information to improve health and prevent disease*. 2nd ed. New York: Oxford University Press; 2010.
Lynch M, Walsh B. *Genetic analysis of quantitative traits*. Sunderland, MA: Sinauer Associates; 1998.
Ott J. *Analysis of human genetic linkage*. Baltimore, MD: Johns Hopkins University Press; 1999.
Palmer LJ, Burton P, Smith GD. *An introduction to genetic epidemiology*. Bristol: Policy Press; 2011.
Terwilliger JD, Ott J. *Handbook of human genetic linkage*. Baltimore, MD: Johns Hopkins University Press; 1994.
Thomas DC. *Statistical methods in genetic epidemiology*. New York: Oxford University Press; 2004.
Weiss KM. *Genetic variation and human disease: principles and evolutionary approaches*. New York: Cambridge University Press; 1995.

INTERNET RESOURCES

Analytical resources

Columbia University: Terwilliger laboratory (various analytical software): http://linkage.cpmc.columbia.edu/index_files/Page434.htm
Department of Genetics, Texas Biomedical Research Institute (various analysis programs including SOLAR): http://txbiomed.org/departments/genetics.aspx
Division of Statistical Genetics, University of Pittsburgh (various analytical resources): http://watson.hgen.pitt.edu/

Eigenstrat (software for population stratification adjustment): http://genepath.med.harvard.edu/~reich/Software.htm

Laboratory of Statistical Genetics: Rockefeller Univ. (a comprehensive analytical resource): http://linkage.rockefeller.edu/

Merlin (linkage analysis software): http://www.sph.umich.edu/csg/abecasis/Merlin

The Human Genetic Analysis Resource: http://darwin.cwru.edu/

UCLA Human Genetics (various analytical resources): http://www.biomath.medsch.ucla.edu/faculty/klange/software.html

University of Michigan, Center for Statistical Genetics: Abecesis Lab (various analytical software programs): http://www.sph.umich.edu/csg/abecasis/

General resources

GENATLAS (Database): http://www.dsi.univ-paris5.fr/genatlas/

National Center for Biotechnology Information: http://www.ncbi.nlm.nih.gov/

National Human Genome Research Institute (NHGRI): Genome - wide Association Database: http://www.genome.gov/gwastudies/

National Human Genome Research Institute (NHGRI): http://www.genome.gov/

Office of Genomics and Disease Prevention of the Centers for Disease Control and Prevention (CDC): http://www.cdc.gov/genomics/default.htm

Online Mendelian Inheritance (Database): http://www.ncbi.nlm.nih.gov/omim

SNAP (database): http://www.broadinstitute.org/mpg/snap/

The Center for Human Genetics (Marshfield Clinic Research Foundation): http://www.marshfieldclinic.org/chg/pages/default.aspx

The Genome Database: http://www.ncbi.nlm.nih.gov/sites/genome

U.S. Department of Energy Genomics Site: http://genomics.energy.gov/

Wikipedia (general concepts in genetics): http://en.wikipedia.org/wiki/Statistical_genetics

CHAPTER 9

Social and Economic Effects on Growth

Richard H. Steckel

Economics, Anthropology and History Departments, Ohio State University, Columbus, OH 43210, USA

Contents

9.1 INTRODUCTION

Although human biologists and physical anthropologists have known for some time that socioeconomic factors such as social class impinge on child growth and therefore adult height, a richer understanding of the relationship began to emerge when economists, historians and other social scientists joined the conversation in the 1970s.[1] Economic historians introduced new sources of data, added several useful concepts, and discovered numerous puzzles or apparent anomalies in the past that elucidated the contribution of socioeconomic factors to growth. Because the historical record encompasses a rich variety of human experience, their efforts helped to illuminate intergenerational influences on body size, measure human capacity for growth following extreme deprivation, and expand knowledge of cultural conditions that are ultimately expressed through proximate influences on growth.

N. Cameron & B. Bogin (eds): Human Growth and Development, Second edition.
ISBN 978-0-12-383882-7, Doi: 10.1016/B978-0-12-383882-7.00009-X

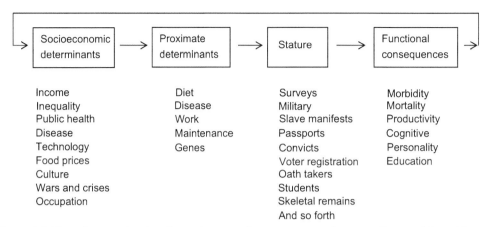

Figure 9.1 Flow diagram showing determinants and consequences of stature. *(Source: Adapted from Steckel.[8])*

Figure 9.1 illustrates the ideas of relationships involving stature. The arrows show that human growth is directly affected by diet, which is a blend of protein, calories and micronutrients, and by work, disease and nutritional resources required for basal metabolism. In turn, these variables respond to a host of environmental conditions such as income, inequality, public health measures, food prices, technology and the like, as shown in the figure. Social scientists are also interested in ways that stature influences demographic and socioeconomic outcomes such as mortality, morbidity, labor productivity, cognitive development and educational attainment. Figure 9.1 shows that these variables shape various social outcomes such as income, inequality and technology used in the production process. This chapter explains and illustrates the links from socioeconomic conditions to stature via proximate determinants of growth, and discusses ways in which adult stature affects socioeconomic outcomes.

9.1.1 Methodology

Some 30 years ago, economic historians formulated the concept of net nutrition (similar in meaning to nutritional status as used by nutritionists), which can be explained metaphorically by viewing the human body as a biological machine. Our machine operates on food as fuel, which it expends at idle (resting in bed) and while fighting infection or engaging in physical activity. Diseases may stunt growth if they divert nutritional intake to mobilize the immune system to combat infection or cause incomplete absorption of food that is eaten. Similarly, arduous physical activity or work places a substantial claim on the diet, which makes it possible to lose weight or even starve on 4000 calories per day. For these reasons, average height reflects a population's history of net nutrition; growth occurs only if enough fuel is available after other expenditures, needed for survival, have been met. If better times follow a period of deprivation, growth may exceed that ordinarily found under good conditions. Catch-up (or compensatory) growth is an adaptive biological mechanism that complicates the study of child health using adult height because it can partially or completely erase the effects of deprivation. Between birth and maturity, a person may undergo several episodes of deprivation and recovery, thereby obscuring important fluctuations in the quality of life. Chronically poor net nutrition during the first 15—20 years of life inevitably results in slower and less growth, which may reduce adult height by several inches depending upon the severity.

Diet, disease, and work or physical activity are proximate or immediate determinants of growth that have been widely studied by human biologists. They also study genetic influences on growth, but the distinction between individual and population-level outcomes is crucial. The vast majority of individual differences in stature are genetically or biologically regulated.[2] Whether growth occurs under poor or good environmental conditions, how tall we become depends heavily on our inheritance from biological parents. At the population level these individual differences tend to cancel out, such that average heights reflect primarily environmental conditions. This raises the question of whether systematic differences in growth potential exist around the world, i.e. if populations of different ancestry lived under the same environmental conditions for several generations, would their average heights differ? DNA analysis may ultimately prove the most effective way to address this question but the scientific community has yet to provide an answer. We do know, however, that considerable convergence in average height has occurred around the globe where economic development has unfolded. In 1950 Japanese men were the shortest of any industrial nation (about 160 cm) but young adult men today are approximately 173 cm, or roughly 5 cm below modern American height standards, and may well catch up within a generation.[3–5] While modest ethnic or ancestral differences in the potential for growth may exist, perhaps as much as a few centimeters, it is safe to say that environmental factors are known to have had a much larger effect on height differences around the world over the past two centuries. For example, in the early to mid-nineteenth century the average heights of young Dutch men were short for Europe (about 164—165 cm), but today they reach 183 cm and are the tallest in the world.[6,7]

9.2 SOCIOECONOMIC INFLUENCES

Several socioeconomic influences on growth are interrelated. Urban areas often exhibit high inequality, technology affects food prices, gross domestic product (GDP) limits expenditure on public health, and so forth. Therefore, it can be difficult to isolate or otherwise measure the "pure" effect of a particular influence on stature. This chapter discusses several of these variables in sequence, but readers should understand that correlations may exist among them.

9.2.1 Income or Gross Domestic Product

Early research on height and income was motivated by a desire to link the familiar with the unfamiliar. GDP equals the value of goods and services produced by a country in a particular year. Economic historians knew GDP and its determinants quite well but understood little about stature and its proximate causes. Indeed, they were highly skeptical that height was a useful measure of human welfare and the standard of living; most believed that human growth overwhelmingly reflected genetic inheritance and/or that height differences and fluctuations were caused by genetic drift.

Therefore, Richard Steckel gathered data on both measures, stature from national height studies found in the appendix of Eveleth and Tanner's *Worldwide Variation in Human Growth* and GDP information from the World Bank in 1985 dollars, i.e. converting values to what they would have been at the price level of 1985.[8,9] The latter was taken for the year the height study was published, but a more elaborate model would consider lagged relationships between GDP per capita and average height because human growth depends not only on current but also on past conditions, including those affecting the mother as a child. In principle, the relationship could be quite complex.

Steckel argued that the two measures should be positively correlated at the national level because many people who lived in poor countries could not afford basic necessities. Their physical growth was constrained by lack of food, hard work and poor medical care in an environment plentiful with debilitating pathogens. As income or GDP increased, more people had access to nutritious food and good medical care, and work was typically less arduous. At the highest levels of GDP per capita, the vast majority of the population had nutritious diets, good medical care, adequate housing, clean water and so forth. Therefore, much or most of the potential for growth by the population was realized, and they were tall by international comparisons.

One could reasonably expect that the height–income relationship would be non-linear. If it was linear then height would continue to increase with income such that people became physical giants; Bill Gates' kids and those of his neighbors might be 10 feet tall! A scatter diagram (Figure 9.2) makes clear that average height continues to increase with per capita GDP but at a decreasing rate. The figure gives results for boys, but a similar (natural log) relationship was found for girls, men and women, as shown in Table 9.1.

9.2.2 Inequality

Readers will note considerable scatter around the regression line in Figure 9.2. It is clear that per capita GDP is highly correlated with average height but it is obvious that other

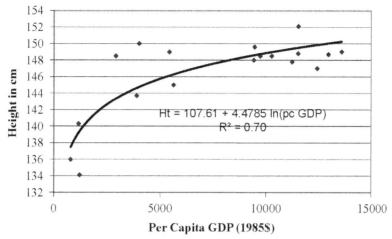

Figure 9.2 Per capita GDP and height at age 12 (boys). This figure is reproduced in the color plate section.

Table 9.1 Correlations between average height and the log of per capita income

Group	Correlation	Number of countries	Significance level
Boys aged 12★	0.87	22	0.0001
Girls aged 12★	0.82	21	0.0001
Adult men★★	0.82	15	0.0002
Adult women★★	0.88	16	0.0001

★The countries represented for boys and girls are Argentina, Australia, Czechoslovakia, Denmark, Egypt, Hungary, Italy, Ghana, India, Japan, Netherlands, New Zealand, Republic of Korea, Soviet Union, Taiwan, USA and Uruguay; the boys also include Mozambique. The USA, Czechoslovakia, Netherlands and Japan have two height studies in the sample.
★★The countries represented for adults are Czechoslovakia, India, Indonesia, Netherlands, Paraguay, Soviet Union, Taiwan and USA. The adult men sample also includes Denmark and Zaire, and the adult women sample also includes France, New Zealand, Republic of Korea and Ireland. The USA, India and Zaire have multiple height studies in the sample.
Source: Steckel (1995).[8]

conditions matter as well. One of these is the extent of income inequality within a country. To illustrate, consider the thought experiment of taking $1000 per year from the richest family in a country and giving it to the poorest. Average height should increase. Why? The children of the rich are so well off that all their basic necessities important for health and physical growth will still be met. Yet the poorest family is so impoverished (even in rich countries) that they will spend some and perhaps most of their additional income on food, clothing, shelter, medical care and the like. Their children will therefore grow more than they would have otherwise.

Statistical analysis requires an empirical measure of inequality, which economists measure by a Gini coefficient that by definition has a range of 0 to 1, and the lower the number the greater the degree of equality. Data for calculating a Gini are usually obtained from surveys, tax records and so forth, and the process begins by ranking households from poorest to richest. A line depicting the share of income received by the bottom x per cent of households (as x varies from 0 to 1) is called a Lorenz curve (Figure 9.3). Compare this curve with the line of perfect equality (a 45 degree line from the origin) where the bottom x per cent of households always has x per cent of total income. The Lorenz curve and the 45 degree line delimit two areas, A and B, used to define inequality:

$$\text{Gini} = A/(A + B)\qquad\qquad [9.1]$$

In the case of perfect equality (every household has the same income), the Lorenz curve coincides with the 45 degree line and the area of A equals 0, and hence the Gini equals 0. In a situation of perfect inequality one household has all the income and the area of B shrinks to 0. Of course, neither extreme exists in reality; studies generally report values of 0.3 at the low end and 0.6 at the high end for inequality at the household level. Conspicuous among the former are rich European countries under governments of democratic socialism, and at the other end are developing countries with non-transparent governments having command over some valuable natural resources whose rewards are highly concentrated among a few families.

With this motivation, Steckel added a Gini coefficient to the regression of height on per capita GDP. To expand the sample, subgroups of the population and appropriate dummy variables were added for urban, rural, rich, poor, student, military, female, groups of various continental ancestries, adolescents and age. The Gini coefficient was not only statistically significant but powerful. Holding per capita GDP and other

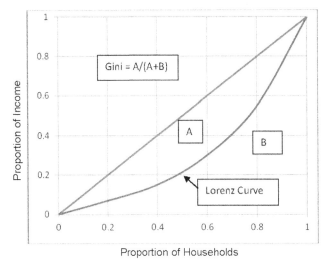

Figure 9.3 Lorenz curve and Gini coefficient. This figure is reproduced in the color plate section.

explanatory variables constant, increasing the Gini by 0.1 reduced average stature by 3.3 cm for adults and 1.4 cm for adolescents.

The coefficients on dummy variables show that a large gap in height (about 11.7 cm) existed between rich and poor adolescents but few systematic differences existed across other groups. Students and military populations were slightly taller (but not significantly so) and no systematic (statistically significant) difference in height existed between urban and rural populations. After controlling for income and the Gini, no systematic height differences existed across groups with the exception of adolescents from Asian and Indo-Mediterranean countries. These results suggest that a large share of height differences across countries can be explained by environmental conditions correlated with income and inequality (see Table 9.2).

9.2.3 Life Expectancy

The two most widely used measures of social performance are per capita GDP and life expectancy, which are highly interrelated. While economists debate the strength of causation between the variables, it seems clear that each influences the other. Rich countries have greater ability than poor ones to provide public health services, medical care, adequate housing and nutritious diets. They also have machines that alleviate much hard physical labor in the production process. Of course, even rich governments may give low priority to public health and/or the challenges of public health may be exceptionally difficult and expensive.

Going in the other direction, healthy populations have greater physical vigor and energy for work and fewer days lost from work, which enhances labor productivity and per capita GDP. Citizens live longer and have a greater incentive to acquire education and on-the-job training because the payback period of these investments is longer than in high-mortality societies.

There are also good biological reasons why stature and life expectancy should be positively correlated. There is a well-known synergy between diet and disease.[10] Poor nutrition that stunts growth also weakens the immune system and makes the body vulnerable to illness and death.

It is therefore unsurprising that life expectancy rises with average stature. Figure 9.4 plots the relationship for countries with national height studies (the same ones used in Figure 9.2). Similar results are obtained for girls aged 12, and for adult men and adult women. One can see that a linear functional form provides a tight fit with the data. A 10-year increase in life expectancy is associated with an increase of 6 cm in stature for boys aged 12.

9.2.4 Public Health

The relationship between stature and socioeconomic variables changed dramatically with the rise of the public health movement in the late nineteenth century.[11,12] Public

Table 9.2 Explaining average height by per capita income, Gini coefficient, place of residence, gender, ethnic group and age

Variable	Adolescents			Adults		
	Coeff.	t-Value	Sample mean	Coeff.	t-Value	Sample mean
Intercept	100.56	24.24		151.14	12.26	
Log per capita GDP	4.90	11.99	8.04	3.97	2.75	7.67
Gini coefficient	−14.34	−2.84	0.41	−32.60	−4.00	0.43
Urban	0.81	1.16	0.21	−0.44	−0.31	0.13
Rural	−1.35	−1.95	0.11	−2.82	−1.49	0.03
Poor	−6.29	−5.17	0.05			
Rich	4.42	6.37	0.10			
Student				1.22	1.23	0.13
Military				2.02	1.56	0.13
Female	0.89	2.44	0.49	−11.41	−16.51	0.47
European ancestry	−2.26	−2.32	0.29	−1.26	−0.70	0.09
African	2.83	1.89	0.05			
African ancestry	−2.14	−1.69	0.05	−1.36	−0.74	0.13
Asian	−4.46	−4.39	0.25	−2.09	−0.98	0.19
Indo-Mediterranean	2.51	2.04	0.25	4.09	1.25	0.47
Age 11	5.37	9.64	0.21			
Age 12	11.14	19.99	0.21			
Age 13	16.80	29.47	0.19			
Age 14	21.62	36.96	0.18			
R^2	0.93			0.96		
Sample size	191			32		
Method		OLS			2SLS	

Definition of variables: Dependent variable = average height in centimeters. Income is measured in 1985 US dollars at international prices for the year that the height study was published. The mean of the dependent variable is 144.05 cm for adolescents and 163.69 cm for adults. The Gini coefficients are for households.

The omitted class refers to a national height study of Europeans. Age 10 is an excluded variable in the regression on adolescent height. Observations on "poor" and "rich" groups do not exist for the adults.

The countries represented for adolescents are Argentina, Australia, Egypt, France, Hong Kong, India, Japan, Republic of Korea, Malaysia, New Zealand, Spain, Sudan, Taiwan, Turkey, USA, Uruguay and Yugoslavia. The countries represented for adults are Egypt, France, Hong Kong, India, Republic of Korea, New Zealand, Taiwan, Thailand, Turkey, UK and USA. Several countries have more than one height study.

GDP: gross domestic product; OLS: ordinary least squares; SLS: second least-squared regression

Source: Steckel (1995).[8]

health measures helped to separate pathogens from people, who prior to the movement had little effective knowledge of the origins of disease. Up to that time many disease-causing organisms were found in water, human waste, garbage and filthy living conditions found in the home. Inspired by of the germ theory of disease, various public works projects provided cleaner water, garbage removal and sewage remediation, the last initially being privies, decentralized cesspools or sanitary sewers whereby waste water was diluted by rivers and oceans. These sewers existed in ancient Rome but the rapid urbanization during the nineteenth century posed significant problems for waste removal

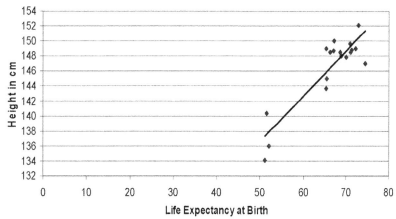

Figure 9.4 Height of boys age 12 and life expectancy at birth. This figure is reproduced in the color plate section.

and treatment. The emergence of cheap steel in the 1870s and the creation of elevators soon permitted the construction of high-rise buildings that demanded a unified system of waste removal. Progress in secondary (biological) treatment of waste water, however, was slow and continued well into the twentieth century, with centralized systems gaining the upper hand.

At the household level, soon families were cleaning their kitchens and washing their clothes in a belief that "cleanliness was next to godliness". The process and the technology unfolded over many decades and in some parts of the industrial world significant progress continued into the middle of the twentieth century.

One benefit of the public health movement is clear from Figure 9.5, which shows the heights of native-born American men from the early eighteenth to the late twentieth

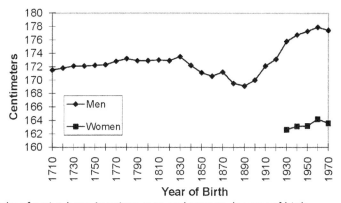

Figure 9.5 Height of native-born American men and women, by year of birth.

century, arranged by birth cohorts. The most rapid increase in American history occurred in the four decades after 1890, during which the average stature of men grew by 7 cm (from 169 to 176 cm).

9.2.5 Migration, Trade and Urbanization

These items are discussed as a cluster because they often occur together. Cities historically have grown through in-migration and they are usually centers of trade and commerce. Migrants did go elsewhere, such as the American west during the nineteenth and early twentieth centuries, but many settlers were moving near the frontier. Mortality rates were higher on the frontier than in the rural areas of the east and excess mortality might have originated with the difficulties of living where housing was widely dispersed, the food supply was uneven, and the community safety net was weak or had yet to be created through churches and small commercial centers.[13]

Transcontinental migration stunted growth, or at least contributed to high mortality. Slaves from Africa and indentured servants from Europe experienced a well-known phenomenon called "seasoning", whereby death rates and morbidity were two to three times higher than normal for the first couple of years after arrival.[14] Newcomers born on a different continent had immune systems adapted to their places of origin and the new disease environment posed challenges for health and physical growth. Over time, the disease environment has become more homogeneous around the globe, but movement from the industrial world to the developing world and even within the developing world may still pose difficulties for health.

The impact of city living was greatest prior to the public health movement, when the vast majority of people had little accurate understanding of the origins of disease. Therefore, as the transportation revolution unfolded in the nineteenth century, with steamboats, interior canals, railroads and transocean canals (e.g. the Panama and Suez canals), as people traveled they also unwittingly transmitted pathogens that exposed other people. It is no accident that cholera, for example, spread rapidly through the USA in the 1830s and the 1850s along transportation routes.[15]

As focal points of in-migration and trade and concentration points for waste, cities were particularly unhealthy. Their reputations for bad health also descended from another common feature, inequality and poverty. Between 1850 and 1925 large numbers of poor immigrants came to America. Eventually, some moved up but the inflow was so great that at any one time a significant share of urban populations lived in impoverished ghettos.

The impact on human growth was apparent from height differences between rural and urban populations. Among Civil War soldiers of the Union Army, who were measured in the 1860s, those from rural areas had an advantage of 1−1.5 inches.[16] By World War II, however, the advantage had vanished and troops from urban areas were about 1 inch taller than the rural troops.[17] The public health revolution, improved diets and higher incomes had transformed life and well-being in the cities.

The long-term trend in heights in the USA, depicted in Figure 9.5, shows a substantial decline for people born after 1830. Economic historians have put forth several possible explanations, including urbanization. Between 1830 and 1880 the share of the American population living in urban areas increased by 19.6 percentage points and if urban men were 1.5 inches shorter than rural men, this shift would explain that about 20% of the height decline from 1830 to 1880. Of course, other factors also contributed to the height decline, including the rise of public schools, which spread pathogens among children that caused diphtheria, whooping cough and scarlet fever.[8] Higher food prices, discussed below, and the Civil War itself were also relevant. The latter disrupted food production and the widespread deployment of troops spread diseases.

9.2.6 Technology and Labor Organization

Everyone familiar with the concept of net nutrition knows that work or physical activity drains fuel or dietary intake that can constrain physical growth. Therefore, diets cannot be evaluated in isolation but must be considered in relation to demands placed upon them. Machines exist to measure energy expenditure, and while useful they are often applied to synthetic situations rather than used to gauge actual energy expenditure in the wide variety of jobs that people do, from working on an assembly line to cutting trees as a lumberjack. Still, it is safe to say that physical activity is probably a significant drain on net nutrition for many children, particularly if they train for sports teams, walk numerous blocks to school or engage in vigorous play on a regular basis (the equivalent of aerobic exercise discussed below).

Physiologists have estimated energy expenditures for a wide variety of common activities.[18] Expenditures are expressed as MET units, which is energy expended in the activity relative to that experienced by sitting quietly at rest. Some activities in which the MET exceeds 5.0 are farm chores such as bailing hay and cleaning the barn (8.0); bicycling at 12–15 mph (19–24 kph) (8.0); swimming (7.0); jogging (7.0); aerobic exercise (6.5); and construction work (5.5). Low MET activities include moderate walking (2.8); driving (2.3); work as a store clerk (2.0); office work (1.5); and watching television (1.0).

Over the past century the work of people living in industrial countries has become far less demanding. In the USA, for example, the share of the labor force working on farms declined from 30.7% to 2.5% from 1910 to 1990, and the pre-tractor machinery of the earlier period was far more arduous to operate.[19] Child labor was common in part because few children went to school beyond the eighth grade. These conditions still exist in the world but mainly in developing nations. So the constraints on growth imposed by physical labor are largely concentrated in poor countries where work is performed with little power equipment.

9.2.7 Food Prices

All human biologists know that the price of food affects the quantity and quality that can be purchased within the family budget. But not all prices are equally important for nutrition and growth. Protein is not only crucial for vigorous physical development but typically the most expensive food item, especially in the form of meat. The type of meat also varies considerably in price, with pork and poultry being cheaper than beef per pound.

Therefore, it is useful to consider trends in prices of various types of foods, many of which have fallen and then risen around the globe over the past half century: fallen as a result of improved agricultural technology, such as mechanization, irrigation, fertilizers and higher yielding seed varieties, but risen from the diversion of grain supplies such as corn to ethanol production, extreme weather fluctuations and bad harvests. Figure 9.6 depicts price indices of internationally traded food items since 1960. The overall trend is U-shaped since 1975, downward until the mid-1990s and upward in the past 15 years. The price declines were good news in overcoming chronic malnutrition in low-income countries but the recent increases have had the opposite effect. These prices do not tell the whole story, however, because many foods important in diets are locally traded, and some localities have suffered poor harvests that are disguised by the aggregate prices.

9.2.8 Crises

The physical growth of children quickly declines or ceases with the arrival of adverse socioeconomic conditions, and if bad times persist, short stature registers in adult heights. The growth of children is therefore best for pinpointing the time and duration of

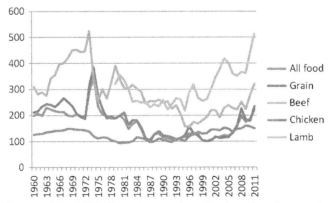

Figure 9.6 World prices of food. This figure is reproduced in the color plate section.

short-term adversity, but adult heights decline or fall below trend if adverse conditions are particularly severe, continue for a substantial period and/or occur during sensitive periods of growth, which are early childhood and adolescence.

The expansion of research on crises marks one of the most interesting policy-related developments in the anthropometric literature. In contrast with heights, traditional measures such as unemployment rates and GDP do not monitor or reflect resource allocation within the household. By traditional measures, children are largely invisible. The extent to which children are nutritionally vulnerable in times of crisis is an important mark of a society's priorities and its willingness to invest in the future.

Two strands of literature have emerged on the health of children during crises, the most numerous focusing on developing countries, where considerable height data have been collected since the mid-1980s. In this vein, papers have appeared on Cameroon,[20] Kazakhstan,[21] North Korea,[22] Novi Sad,[23] South Africa[24] and Zimbabwe.[25-27] The growth of adolescent boys fluctuated seasonally with the supply of food in Czechoslovakia.[28]

One might expect that seismic shifts or differences in political systems would adversely affect the health and welfare of children. The stature of children in Nazi Germany stagnated from 1933 to 1938 following autarchy and market disintegration.[29] North Koreans declined in health relative to the South in the late twentieth century,[30] and were poorly equipped to deal with a famine that appeared in the 1990s.[22] Delivery of United Nations food aid, however, seems to have improved anthropometric outcomes.[31] Eastern Europe and Russia during the transition in the 1990s were different, whereby life expectancy deteriorated, especially in Russia, but anthropometric data suggest that children were protected.[32] Apparently adults bore the brunt of declining socioeconomic conditions during and immediately after these regime changes.

9.2.9 Culture

This term has many meanings and it is used here with regard to practices of food preparation and distribution. Several cultures have food taboos which may be permanent or temporary or apply only to men or women or certain occupations such as priests.[33] Hindus, for example, do not eat beef and Arabs do not eat pork. Many Jews do not consume shellfish. Various beliefs and rituals may complicate dietary choices, making it more difficult or expensive to obtain a balanced, nutritious diet that promotes child growth.

A related point concerns the distribution of food within the family in times of extreme hardship. Under famine conditions, for example, the working adults are often privileged because their work is essential to purchase or otherwise find food needed for family survival. Several researchers have noted, for example, the higher mortality rates of the young and the old in times of famine and a rise in the incidence of nutrition-related diseases among children.[34]

In the past two decades the "missing women of Asia" have been discussed in the press. Amartya Sen noted that more than 100 million women were missing through sex-selective abortion, infanticide or differential nutrition of boys and girls.[35] Critics such as Emily Oster questioned the explanation, suggesting that a higher prevalence of hepatitis B, which selectively aborts females, in Asia than in Europe could account for the missing women.[36] Research later cast doubt on the hepatitis B explanation by showing that the sex ratio was roughly 50–50 for firstborn children and that the male–female ratio was higher among poor than rich families, suggesting that poor but not rich families practiced selective control over the sex ratio.

In the wealthier nations, until recently, men dominated or controlled labor market activity, political decision making and property. In poorer nations this was and is still the case. In constructing and analyzing traditional measures of social performance such as GDP, earnings and wealth, which are (or were) largely produced or owned by men, many economists tacitly assume that material goods and services are (or were) equitably distributed within the household. But is (or was) this the case? It is desirable to have direct information on the fates of children relative to adults and girls relative to boys. How did these groups fare during industrialization? Is (or was) there gender discrimination in some societies? Numerous methodologies have been used to address the issue,[37] and heights can provide some answers. To make proper comparisons, the heights of men and women must be standardized for biological differences in the amount of growth achieved, with results translated into Z-scores or percentiles of modern height references or standards.

In the developing world considerable research has gone forward using the Living Standards Measurement Surveys (LSMS) and similar data sources.[38] Contrary to predictions of the unitary model of intrahousehold resource allocation, which maintains that both parents have the same preferences for the well-being of sons and daughters, some studies find preferences in allocation of resources to boys or to girls that correspond to the resources or human capital available to the parent of the same sex. For example, pensions received by women in South Africa had a greater positive impact on the physical growth of girls than of boys.[39] A study by Sahn and Stifel using the Demographic Health Surveys reports that this result holds more generally in Africa.[40] Similarly, mothers' education had a greater positive impact on the height of daughters within an indigenous population of Brazil.[41,42]

If mothers and fathers have preferences for children of their sex, one may suspect that the sex composition of the household affects the allocation of dietary resources. A study in rural India suggests that both boys and girls had poorer outcomes (height) if they were born after multiple same-sex siblings; boys born after multiple daughters, however, had the best outcomes.[43] In the 1990s economic growth improved living standards in India, and height gains occurred for both sexes of young children in the cities, but girls lagged behind boys in rural areas, especially in the northern part of the country, where research documents a preference for sons.[44] Deaton reports that an excess of Indian males (consistent with relatively high female mortality) was positively correlated with the percentage difference in their heights relative to the average height of males and females.[45] In the Dominican Republic, boys and girls fared similarly in female-versus male-headed households, but if the household was poor, children fared somewhat better with a female head.[46] According to evidence gathered upon admission to the London House of Correction in the Victorian era, anthropometric indicators disproportionately declined for women if the family became larger and/or their economic resources declined.[47] Some believe that strong son preferences exist in Vietnam, but a study on data from the early 1990s found no differences in standardized height or weight for age achieved by preteen boys and girls.[48]

Much of the debate over the distributional consequences of industrialization has focused on differences by class or region, but recent studies now shed light on gender. The general expansion of anthropometric history has led researchers to develop new sources of data such as prison records, which are skewed towards males but otherwise represent reasonably well the teenagers and younger adults of the working class. Although women are a distinct minority in prison admissions, in absolute numbers the female heights are abundant in many states and countries.

England has garnered the largest share of research on the relative health of women during industrialization. Johnson and Nicolas show that heights of men and women reported by London's Scotland Yard declined from the late 1820s to the 1850s, with the hungry forties being the worst period in which to have been born.[49] Note, however, that the negative coefficients on their time period dummies are generally larger for women than for men, a result that understates the stress on women because they were shorter (hence their percentage decline in height was larger) and for biological reasons, women are ordinarily more resilient than men. In subsequent work on heights of convicts sent to New South Wales from England and Ireland covering birth years from 1785 to 1815, Nicolas and Oxley report that women's heights fell relative to men's and that women living in rural as opposed to urban areas underwent more physiological stress,[50] but Jackson disputes their result of a downward trend on grounds of statistical procedure.[51] In a final effort to settle the question, Johnson and Nicholas analyze 30,000 heights on female prisoners from three sources, which they compare with male heights assembled by other researchers.[52] They reiterate the presence of a growing height gap between men and women up to the end of the Napoleonic Wars, but thereafter their heights moved in parallel, which suggests that factors endogenous to the household economy led to deterioration for women during the early phase of industrialization. Moreover, heights in England declined relative to those in Ireland, which suggests that disamenities of industrialization imposed health penalties. Unlike the rest of pre-famine Ireland where heights declined, Ulster's nascent industrial economy competed favorably with that of England after the Napoleonic Wars, which translated into improving heights for women.[53]

Both men and women in nineteenth century Bavaria benefited from a diet rich in milk and potatoes. Heights declined for women relative to men, however, during the hungry 1840s and in contrast to what is known about biological resiliency, women's heights were more sensitive to economic conditions in early childhood.[54] On the other hand, research on heights gathered from passports issued by the USA shows that female heights increased more rapidly than those for men after the Civil War of the 1860s.[55]

9.2.10 Mother's Education

Several studies report the importance of mother's education for child health and physical growth, with a recent case study for Russia.[56] The mechanism operates in two ways on the health and height of the child, via cognitive ability[57] and informed child-care practices.[58,59] Consistent with this interpretation, a study on Vietnamese children found that the effect of parental education remains powerful when controlling for income, which suggests that the effect of education operates directly on height rather than via income alone.[48] In Jamaica, where females in the household often share child-care responsibilities, increased education of any woman had a positive impact on child health.[60] Children of the household head were nearly half a standard deviation taller than other children in the household, which suggests that the head was able to direct resources to their own children. One may wonder whether the head was also able to enhance the education of her own children. In these same households, the presence of the father had a large positive impact on children's height, even when controlling for household income.

9.3 CONSEQUENCES OF STATURE

Physical growth and development in childhood influence several measures of social performance such as longevity, capacity for work, productivity and skills of the labor force. Research continues to elucidate these relationships and the understanding of some connections, such as nutrition and personality, are very preliminary.

9.3.1 Wages, Cognition and Labor Productivity

In recent years numerous papers have appeared on heights as a determinant of wages or income, the majority appearing in economics journals. Development economists have known for some time that taller men earn more in occupations for which physical strength is an asset. This research continues with studies on coalminers in India,[61] manual workers in Brazil[62] and farmers in Ethiopia.[63] In line with this work, Schultz finds an

8—10% increase in wages for every centimeter increase in stature of men or women in Ghana and Brazil.[64]

In a new direction for this type of research, the majority of papers since 1995 consider the implications of stature and other anthropometric measures for income or wages in developed countries.[65—67] Some papers now investigate the implications not only of height but of weight and beauty for labor market remuneration.[68,69] One analysis finds that, possibly, employers discriminate against obese people for higher paying positions of leadership and decision making in nine European countries.[70] In Denmark's private sector, obesity was a disadvantage for women but actually beneficial for men.[71]

A few years ago, Persico et al. suggested that boys who were taller in adolescence gain confidence, social skills and related forms of human capital that serve them well in adult labor markets.[72] The most recent strand of this research argues that taller people earn more because they have greater cognitive ability. Case and Paxson challenge the Persico et al. mechanism linking stature to higher wages, showing that people who were tall as adolescents and were also tall and did well on cognitive tests in early childhood, before schooling could have selectively boosted their human capital.[73] The benefits of robust physical growth in childhood extend into adulthood as reflected in better cognitive function, improved mental health and ability to perform activities of daily living.[74] This finding has several implications, including an explanation for why economic returns to height continue to be observed in wealthy countries. It also demonstrates the long reach into adulthood of early childhood conditions, suggesting that societies probably underinvest in prenatal and early childhood health and nutrition.

If confirmed by other studies, the mechanism outlined by Case and Paxson will have considerable influence on economic history, economic development and our understanding of the sources of economic growth. It will bring nutritional status and stature to the fore in comprehending rates of return to schooling, socioeconomic inequality that stems from cognitive deficits and conceivably rates of technological change. One could imagine dissertations on whether nutritional improvements helped to trigger the Enlightenment or whether increases in agricultural productivity, which preceded or accompanied industrialization in many countries, also contributed to economic growth via more effective decision making.

REFERENCES

1. Steckel RH. Strategic ideas in the rise of the new anthropometric history and their implications for interdisciplinary research. *J Econ Hist* 1998;**58**:803—21.
2. Silventoinen K. Determinants of variation in adult body height. *J Biosoc Sci* 2003;**35**:263—85.
3. Honda G. Differential structure, differential health: industrialization in Japan, 1868—1940. In: Steckel RH, Floud R, editors. *Health and welfare during industrialization.* Chicago, IL: University of Chicago Press; 1997. p. 251—84.
4. Mosk C. *Making health work: human growth in modern Japan.* Berkeley, CA: University of California Press; 1996.

5. Health Service Bureau. *National health and nutrition survey in Japan. Nutrition survey report 2007*. Tokyo: Ministry of Health Labour and Welfare; 2009.

6. Drukker JW, Tassenaar V. Paradoxes of modernization and material well-being in the Netherlands during the nineteenth century. In: Steckel RH, Floud R, editors. *Health and welfare during industrialization*. Chicago, IL: University of Chicago Press; 1997. p. 331–77.

7. Statline. Reported height. In: Health, lifestyle, use of medical facilities. *The Hague: Centraal Bureau voor de Statistiek*; 2011.

8. Steckel RH. Stature and the standard of living. *J Econ Liter* 1995;**33**:1903–40.

9. Steckel RH. Height and per capita income. *Hist Methods* 1983;**16**:1–7.

10. Scrimshaw NS, Taylor CE, Gordon JE. *Interactions of nutrition and infection*. Geneva: World Health Organization; 1968.

11. Rosen GA. *History of public health*. Baltimore, MD: Johns Hopkins University Press; 1993.

12. Duffy J. *The sanitarians: a history of American public health*. Urbana, IL: University of Illinois Press; 1992.

13. Steckel RH. The health and mortality of women and children, 1850–1860. *J Econ History* 1988;**48**:333–45.

14. Curtin PD. *Death by migration: Europe's encounter with the tropical world in the nineteenth century*. Cambridge. Cambridge University Press; 1989.

15. Rosenberg CE. *The cholera years: the United States in 1832, 1849, and 1866*. Chicago, IL: University of Chicago Press; 1962.

16. Margo R, Steckel RH. Heights of native born whites during the antebellum period. *J Econ History* 1983;**43**:167–74.

17. Karpinos BD. Height and weight of selective service registrants processed for military service during World War II. *Hum Biol* 1958;**30**:292–321.

18. Ainsworth BE, Haskell WL, Herrmann SD, Meckes N, Bassett Jr DR, Tudor-Locke C, et al. Compendium of physical activities: a second update of codes and MET values. *Med Sci Sports Exerc* 2011;**43**:1575–81.

19. Carter SB. Labor Force. In: Carter SB, Gartner SS, Haines MR, Olmstead AL, Sutch R, Wright G, editors. *Historical statistics of the United States. Millennial edition*. New York: Cambridge University Press; 2006.

20. Pongou R, Salomon J, Majid E. Health impacts of macroeconomic crises and policies: determinants of variation in childhood malnutrition trends in Cameroon. *Int J Epidemiol* 2006;**35**:648–56.

21. Dangour AD, Farmer A, Hill HL, Ismail SJ. Anthropometric status of Kazakh children in the 1990s. *Econ Hum Biol* 2003;**1**:43–53.

22. Schwekendiek D. The North Korean standard of living during the famine. *Soc Sci Med* 2008;**66**:596–608.

23. Bozić-Krstić VS, Pavlica TM, Rakić RS. Body height and weight of children in Novi Sad. *Ann Hum Biol* 2004;**31**:356–63.

24. Hendriks S. The challenges facing empirical estimation of household food (in)security in South Africa. *Dev South Africa* 2005;**22**:103–23.

25. Alderman H, Hoddinott J, Kinsey B. Long term consequences of early childhood malnutrition. *Oxford Econ Papers* 2006;**58**:450–74.

26. Alderman H, Hoogeveen H, Rossi M. Reducing child malnutrition in Tanzania: combined effects of income growth and program interventions. *Econ Hum Biol* 2006;**4**:1–23.

27. Hoddinott J. Shocks and their consequences across and within households in rural Zimbabwe. *J Dev Stud* 2006;**42**:301–21.

28. Cvrcek T. Seasonal anthropometric cycles in a command economy: the case of Czechoslovakia, 1946–1966. *Econ Hum Biol* 2006;**4**:317–41.

29. Baten J, Wagner A. Autarky, market disintegration, and health: the mortality and nutritional crisis in Nazi Germany, 1933–37. *Econ Hum Biol* 2003;**1**:1–28.

30. Pak S. The biological standard of living in the two Koreas. *Econ Hum Biol* 2004;**2**:511–21.

31. Schwekendiek D. Determinants of well-being in North Korea: evidence from the post-famine period. *Econ Hum Biol* 2008;**6**:446–54.

32. Stillman S. Health and nutrition in Eastern Europe and the former Soviet Union during the decade of transition: a review of the literature. *Econ Hum Biol* 2006;**4**:104−46.
33. Meyer-Rochow VB. Food taboos: their origins and purposes. *J Ethnobiol Ethnomed* 2009;**5**:18.
34. Ó Gráda C. *Famine: a short history*. Princeton, NJ: Princeton University Press; 2009.
35. Sen A. More than 100 million women are missing. *NY Rev Books*; 1990. p. 37.
36. Oster E, Chen G. *Hepatitis B does not explain male-biased sex ratios in China. NBER working paper series no. 13971*. Cambridge, MA: National Bureau of Economic Research; 2008. p. 1v.
37. Lundberg S. Sons, daughters, and parental behaviour. *Oxford Rev Econ Policy* 2005;**21**(3).
38. Strauss J, Thomas D. Human resources: empirical modeling of household and family decisions. In: Chenery H, Srinivasan TN, editors. *Handbook of development economics*. Amsterdam: North-Holland; 1995. p. 1883−2023.
39. Duflo E. Grandmothers and granddaughters: old age pensions and intra-household allocations in South Africa. *World Bank Econ Rev* 2003;**17**:1−25.
40. Sahn DE, Stifel DC. Parental preferences for nutrition of boys and girls: evidence from Africa. *J Dev Stud* 2002;**39**:21−45.
41. Godoy RA, Leonard WR, Reyes-Garcia V, Goodman E, McDade T, Huanca T. Physical stature of adult Tsimane' Amerindians, Bolivian Amazon in the 20th century. *Econ Hum Biol* 2006;**4**:184−205.
42. Godoy RA, Reyes-Garcia V, Vincent V, Leonard WR, Huanca T, Bauchet J. Human capital, wealth, and nutrition in the Bolivian Amazon. *Econ Hum Biol* 2005;**3**(1).
43. Pande RP. Selective gender differences in childhood nutrition and immunization in rural India: the role of siblings. *Demography* 2003;**40**:395−418.
44. Tarozzi A, Mahajan A. child nutrition in India in the nineties. *Econ Dev Cult Change* 2007;**55**:441−86.
45. Deaton A. Height, health, and inequality: the distribution of adults heights in India. *Am Econ Rev Papers Proc* 2008;**98**:468−74.
46. Rogers BL. The implications of female household headship for food consumption and nutritional status in the Dominican Republic. *World Dev* 1996;**24**:113−28.
47. Horrell SH, Meredith DG, Oxley DJ. Measuring misery: body mass, ageing and gender inequality in Victorian London. *Explor Econ Hist* 2009;**46**(1):93−119.
48. Haughton D, Haughton J. Explaining child nutrition in Vietnam. *Econ Dev Cult Change* 1997;**45**: 541−56.
49. Johnson P, Nicholas S. Male and female living standards in England and Wales, 1812−1857: evidence from criminal height records. *Econ Hist Rev* 1995;**48**:470−81.
50. Nicholas S, Oxley D. Living standards of women in England and Wales, 1785−1815: new evidence from Newgate Prison records. *Econ Hist Rev* 1996;**49**:591−9.
51. Jackson RV. The heights of rural-born English female convicts transported to New South Wales. *Econ Hist Rev* 1996;**49**:584−90.
52. Johnson P, Nicholas S. Health and welfare of women in the United Kingdom, 1785−1920. In: Steckel RH, Floud R, editors. *Health and welfare during industrialization*. Chicago, IL: University of Chicago Press; 1997. p. 201−49.
53. Oxley D. Living standards of women in prefamine Ireland. *Soc Sci Hist* 2004;**28**:271−95.
54. Baten J, Murray JE. Heights of men and women in 19th-century Bavaria: economic, nutritional, and disease influences. *Explor Econ Hist* 2000;**37**:351−69.
55. Sunder M. *Passports and economic development: an anthropometric history of the US elite in the nineteenth century*. Munich: Ludwig-Maximilians-Universität; 2007.
56. Fedorov L, Sahn DE. Socioeconomic determinants of children's health in Russia: a longitudinal study. *Econ Dev Cult Change* 2005;**53**:479−500.
57. Rubalcava LN, Teruel GM. The role of maternal cognitive ability on child health. *Econ Hum Biol* 2004;**2**:439−55.
58. Shariff A, Ahn N. Mother's education effect on child health: an econometric analysis of child anthropometry in Uganda. *Indian Econ Rev* 1995;**30**:203−22.
59. Christiaensen L, Alderman H. Child malnutrition in Ethiopia: can maternal knowledge augment the role of income? *Econ Dev Cult Change* 2004;**52**:287−312.
60. Handa S. Maternal education and child height. *Econ Dev Cult Change* 1999;**47**:421−39.

61. Dinda S, Gangopadhyay PK, Chattopadhyay BP, Saiyed HN, Pal M, Bharati P. Height, weight and earnings among coalminers in India. *Econ Hum Biol* 2006;**4**:342−50.

62. Thomas D, Strauss J. Health and wages: evidence on men and women in urban Brazil (analysis of data on health). *J Econometrics* 1997;**77**:159−86.

63. Croppenstedt A, Muller C. The impact of farmers' health and nutritional status on their productivity and efficiency: evidence from Ethiopia. *Econ Dev Cult Change* 2000;**48**:475−502.

64. Schultz TP. Wage gains associated with height as a form of health human capital. *Am Econ Rev* 2002;**92**:349−53.

65. Meyer HE, Selmer R. Income, educational level and body height. *Ann Hum Biol* 1999;**26**:219−27.

66. Heineck G. Up in the skies? The relationship between body height and earnings in Germany. *Labour* 2005;**19**:469−89.

67. Rashad I. Height, health, and income in the US, 1984−2005. *Econ Hum Biol* 2008;**6**:108−26.

68. Harper B. Beauty, stature and the labour market: a British cohort study. *Oxford Bull Econ Stat* 2000;**62**(s1):771−800.

69. Wada R, Tekin E. *Body composition and wages. NBER Working Paper No. 13595.* Cambridge, MA: National Bureau of Economic Research; 2007.

70. Atella V, Pace N, Vuri D. Are employers discriminating with respect to weight? European evidence using quantile regression. *Econ Hum Biol* 2008;**6**:305−29.

71. Greve J. Obesity and labor market outcomes in Denmark. *Econ Hum Biol* 2008;**6**:350−62.

72. Persico N, Postlewaite A, Silverman D. The effect of adolescent experience on labor market outcomes: the case of height. *J Polit Econ* 2004;**112**:1019−53.

73. Case A, Paxson C. Stature and status: height, ability, and labor market outcomes. *J Polit Econo* 2008;**116**:499−532.

74. Case A, Paxson C. Height, health, and cognitive function at older ages. *Am Econ Rev Papers Proc* 2008;**98**:463−7.

SUGGESTED READING

Bogin B. *The growth of humanity.* New York: Wiley-Liss; 2001.

Floud R, Fogel RW, Harris B, Hong SC. *The changing body: health, nutrition, and human development in the western world since 1700.* New York: Cambridge University Press; 2011.

Komlos JH, editor. *The biological standard of living on three continents: further explorations in anthropometric history.* Boulder, CO: Westview Press; 1995.

Komlos J, Baten J, editors. *The biological standard of living in comparative perspective.* Stuttgart: Franz Steiner; 1998.

Steckel RH. Stature and the standard of living. *J Econ Liter* 1995;**33**:1903−40.

Steckel RH. Height and human welfare: recent developments and new directions. *Explor Econ Hist* 2009;**46**:1−23.

Steckel RH, Floud R, editors. *Health and welfare during industrialization.* Chicago, IL: University of Chicago Press; 1997.

INTERNET RESOURCES

Official web page of Economics & Human Biology: http://www.elsevier.com/wps/find/journaldescription.cws_home/622964/description

CHAPTER 10

Environmental Effects on Growth

Lawrence M. Schell*, Kristen L. Knutson**, Stephen Bailey[†]

*Department of Anthropology and Department of Epidemiology and Biostatistics, University at Albany, State University of New York, Albany, NY 12222, USA
**Department of Medicine, University of Chicago, Chicago, IL 60637, USA
[†]Department of Anthropology, Tufts University, Medford, MA 02155, USA

Contents

10.1 INTRODUCTION

A chief characteristic of human growth and development is that it is "eco-sensitive"; it is sensitive to a wide variety of features of the environment. Among the most often studied are features of the natural environment, and usually these are studied as extremes (extreme cold or heat, aridity, high altitude). To these we must add anthropogenic features such as air pollution, metals (mercury, lead), pesticides and herbicides such as DDT, and energy (radiation and noise). Most anthropogenic factors are recent

N. Cameron & B. Bogin (eds): Human Growth and Development, Second edition.
ISBN 978-0-12-383882-7, Doi: 10.1016/B978-0-12-383882-7.00010-6

developments, and may pose adaptive challenges that are reflected in altered patterns of growth.

The study of human growth in relation to the natural environment has been one of the fundamental research areas in the study of human variation and adaptation.[1] By the mid-twentieth century, patterns of growth that were responses to environmental extremes, including slower maturation and reduced growth, were interpreted as adaptations, that is, relatively beneficial to the individual by providing some benefit in terms of function, survival and/or reproduction. While these benefits have rarely been measured, the theory that growth is a way for individuals to adapt to their immediate physical environment has been around since the 1960s.[2] Thus, the idea that growth responses are part of the adaptive potentialities of *Homo sapiens* is found in virtually all texts on human biological adaptation.[3-5]

Another interpretation of reduced growth and slowed maturation is that it is a direct result of adverse circumstances. James Tanner, who led the field of human growth and development for decades, noted the relationship between adverse circumstances in childhood and poor growth.[6] He championed the use of child growth as an index of community well-being, of health, and even of the moral status of a society as inequality of growth among different social groups reflected the unequal distribution of health resources. He called this study auxological epidemiology. In this view, slow or less growth indicates poorer health and the lack of adaptation in the face of nutritional or social disadvantage and adversity. Thus, researchers use two general and somewhat contradictory interpretations of environmentally influenced growth patterns (see Bailey and Schell for a review and discussion of the applications of these interpretations).[7]

This chapter focuses on environmental influences on growth including aspects of the natural environment and anthropogenic factors. Because of this dual focus, the contradictory interpretations of growth will be considered after a review of the relevant data on growth and the environment.

10.1.1 Research Design Issues

Studies of growth patterns in relation to environmental factors demonstrate several issues in the design of growth studies. Foremost of these is the issue of balanced precision, the idea that the independent variable (the "cause") and the dependent variable (the "effect") should be measured with equal precision. The earliest growth studies examined size (the dependent variable) in relation to age (the independent variable). Today, studies of growth and the environment require accurate and reliable measurements of both individual growth and the environmental factors, but this is not always achieved. Measuring the environment is straightforward when the environmental factor is not modified by behavior or culture and everyone living in one community has basically the same exposure throughout their lives (e.g. high-altitude studies). However, it is more difficult to measure pollutant exposures because individuals in a single community can vary greatly in level of exposure. Some pollutants leave long-term residues in the body that can be measured retrospectively to estimate past exposure, for example lead measured in blood and bone, while others leave little trace of past exposure. Exposure to energy, such as radiation or noise, does not leave a residue at all and this makes retrospective studies very difficult. This fact explains much of the difficulty in determining the effects of mobile phones that emanate microwave radiation, since past radiation exposure is extremely difficult to reconstruct. At present, the study of environmental influences on growth is limited by our ability to measure environmental factors, and the information reviewed below should be understood as a limited picture wrested from substantial difficulties measuring the environmental factors of greatest concern to human well-being.

10.2 TEMPERATURE AND CLIMATE

Climate appears to influence growth and development, helping to determine body size and proportions. According to Bergmann's and Allen's rules, body size and proportions of warm-blooded, polytypic animals are related to temperature. In humans, Allen's rule predicts longer extremities and appendages relative to body size in warmer climates, and shorter ones in colder climates. Bergmann's rule predicts larger body sizes in colder versus warmer climates.

There is ample statistical evidence for a relationship between adult size and shape consonant with Bergmann's and Allen's rules. Roberts[8] examined published data on body dimensions of multiple samples of males from around the world and correlated the sample means with measures of local temperature. There is a significant negative correlation between body weight and mean annual temperature, as well as a negative relationship between sitting height as a proportion of total height and temperature (Figure 10.1). Newman[9] tested Bergmann's and Allen's rules through examination of aboriginal males in North and South America spanning 1000 years: smaller statures are observed near the equator consonant with Bergmann's rule, while the shorter legs among the Inuit are consonant with Allen's rule. In addition, the amount of body surface area tends to increase from cold to hot climates.[10] An analysis of samples measured since Roberts' landmark paper of 1953 showed that temperature was still related to body size, weight to height proportions and height to torso proportions, although the relationships were weaker than in Roberts' analysis. The effect moderation was due to greater weight among samples from more tropical regions, which could well be due to a nutritional transition in these lands.[11] The combined effect of nutrition and temperature demonstrates how phenotypic development is the product of multiple environmental influences.

The relationships observed between body proportion and environmental temperature can be explained in terms of the body's thermoregulatory process. In hot, dry environments, a body that has greater surface area relative to total body size or volume will more efficiently dissipate heat produced by the body's metabolism and activity.

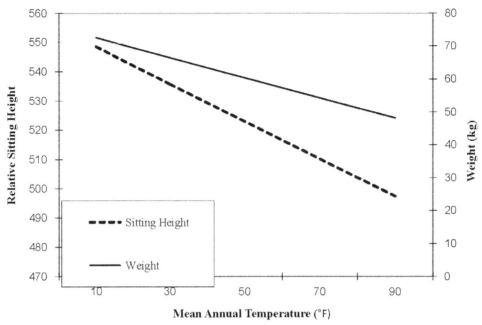

Figure 10.1 Relationship between mean annual temperature and body weight. *(Source: Adapted from Roberts.[8])*

The reverse is true for cold environments where heat retention is important to avoid hypothermia; thus less surface area through which heat would be lost is more adaptive. The small stature and low body mass of populations in tropical rainforests, for example the pygmies in Africa, is an adaptive body shape because the high humidity in these environments limits the effectiveness of sweating, which dissipates heat through evaporation. The smaller body mass of these populations minimizes heat retention.[10]

The focus on the effects of temperature has primarily been on adult form, and only a few studies have examined the relationships of growth parameters to temperature. Malina and Bouchard[12] suggest that the typical body shapes associated with extremes in temperature have implications for development. For example, studies of the mean age at menarche demonstrate a negative correlation with annual mean temperature, indicating earlier maturation among females in hotter climates.[8,12]

Eveleth[13] conducted a longitudinal study of well-off, well-nourished American children in Brazil to determine the effects of the hot climate on growth. She observed that the Rio children weighed less than well-nourished, middle-class US children from Iowa, and had less weight for height, indicating a more linear body form in Brazil. Limb growth also was more linear and less stocky, with more surface area to volume. This growth of size and shape is consistent with expectations from Bergmann's and Allen's rules. Age at menarche, however, did not differ between the Brazilian and US populations, indicating that the populations were maturing at similar rates.

In humans, there appears to be a general relationship between climate and body size that roughly adheres to Bergmann's and Allen's rules. How this relationship develops through growth patterns has not been intensively studied and is not well understood.

10.3 SEASON

Seasonal variation in growth rates has been observed in healthy children. A classic study by Palmer[14] showed that growth rates for height are greater during the spring and summer months, while rates of weight gain are greater during the fall (autumn) and winter. The greatest increases in weight are often in September to November, and can be up to five times the weight gain in the minimal months from March to May.[15] Approximately two-thirds of the annual weight gain occurs between September and February. This seasonal rhythm in weight gain is not established in children until about 2 years of age. These observations were based on a temperate latitude population living in the northern hemisphere.

Height growth, on the other hand, reaches its maximum from March to May in the northern hemisphere. The average velocity is two to two-and-a-half times the average velocity during September to November, the period of minimal height growth.[15] Finally, one study has suggested that minimum weight gains and maximum height gains occur simultaneously.[16] It is worth noting that these are seasonal trends in the average growth velocities. Individual patterns of growth do not necessarily conform to these seasonal peaks in growth; in fact, the timing of individual peaks in growth can vary significantly.[15]

Seasonal variation in growth is not limited to temperate zones, and variation in growth rates between dry and rainy seasons has been observed in tropical climates. In Guatemala City male and female preadolescent and male postadolescent children follow a seasonal pattern, but adolescent children do not.[17] The absence of an effect during adolescence may be due to the pubertal growth spurt, which is quite large and its occurrence is highly variable between individuals.

A possible explanation for all seasonal variation in growth in height may be variation in sunlight, which influences the hormones involved in growth regulation.[10] Swedish boys exposed to sunlamps during the winter averaged 1.5 cm more growth in height than unexposed controls (Nylin, 1929, cited in Refs 10 and 15). However, during the summer, the control group grew more rapidly than the exposed boys, resulting in no overall difference in mean annual growth in height. Among blind children, the months of maximal growth were evenly distributed throughout the year while for normally sighted children the months of maximal growth occurred between January and June.[18] The seasonal variation in day length was posited as an explanation for the consolidation of maximal growth into the 6 months for children with normal vision. A test of this hypothesis examined growth rates of children living on the Orkney Islands, where there is very large seasonal variation in day length. Growth was poorly correlated with climatic variables, suggesting that day length is the critical variable.[19]

What is critical, day length or sunlight exposure? A study of children in Zaire, Africa,[20] found that the growth in height was more rapid during the dry season than during the rainy season. Even though length of day was longer in the rainy season, actual exposure to bright sunlight was greater for the children during the dry season, supporting the role of sunlight in growth regulation.[10] The study by Bogin[17] of children in Guatemala also demonstrated that children aged 5—7 years grew more rapidly during the dry than the rainy season, and similarly to the African sample, exposure to sunlight was greater during the dry season.

Exposure to sunlight has long been recognized as an important factor for skeletal development. Ultraviolet light stimulates the production of cholecalciferol, vitamin D_3, in human skin, and vitamin D_3 increases intestinal absorption of calcium and regulates the rate of skeletal remodeling and mineralization of new bone tissue.[21] Despite the fortification of milk in some countries with vitamin D_2, the major source of vitamin D for humans is the body's synthesis of the D_3 form under stimulation from sunlight.[22] In addition, the marked seasonal variation of vitamin D derivatives in plasma coincide with variation in sunlight.[23] Thus, exposure to sunlight and the production of vitamin D_3 may account for the observation that children in temperate climates have a faster rate of growth in height during the spring and summer, and that children in the tropics grow more during the dry season, when sunlight exposure is greatest.

The seasonal variation of weight gain can be explained in some populations by seasonal differences in resource availability, nutritional variation and rates of disease, but for other, more well-nourished populations this explanation is not sufficient and there may exist an endogenous seasonal rhythm of weight gain in children.[10] Further studies are required to better understand this phenomenon.

There appears also to be seasonal variation in menarche, with higher incidences of menarche in spring and summer possibly due to variation in light, nutrition and/or disease.[24]

10.4 HIGH-ALTITUDE HYPOXIA

As we climb above 3000 m (about 10,000 feet), environments present reduced biomass, aridity, shortened growing seasons, diurnal cold stress, high solar radiant energy loads and reduced oxygen availability. Because this last challenge, called hypoxia, cannot be ameliorated by culture or technology, it should allow us a window on growth responses to a single, uniquely constant environmental stressor. In practice, however, the view is obscured. Growth outcomes at altitude are a cumulative history of myriad adaptive successes or failures. Not all moments in that history have equal impact. Our physiological adaptations are themselves mediated by genes facing natural selection. Pathways between those genes and growth outcomes are poorly mapped. Lastly, growth responses of the approximately 140 million people who live worldwide at high altitude are constrained by disparate gene pools filtered through differing histories of occupation. The windows may not be comparable.

Altitude's effects on growth begin with how cells respond to hypoxia. Normal cells require partial oxygen pressures (PO_2) averaging about 23 mm in the cytoplasm, which dips below 5 mm in the mitochondria.[25,26] Since rates of oxygen exchange are determined by simple diffusion, cells face inevitable stress as we climb and the gradient between atmospheric and cellular oxygen declines. At sea level, the difference is 136 mm. At 4572 m (15,000 feet), this gradient is halved. Moreover, because of reduced carbon dioxide at high altitude, initial respiratory alkalosis also compromises hemoglobin's release of oxygen. Cellular effects of reduced PO_2 in tissue can be detected as low as at 600 m (1968 feet), but become clinically obvious by 3000 m (9842 feet).[25,27,28] At this point, cells begin to trade growth metabolism for maintenance of oxygen tension. Our bodies respond to these local trade-offs through physiological adaptations. Over time, cardiovascular and respiratory morphology also change.[25,26] Growth thus can be seen as a snapshot of how successfully these adaptations buffer our cells from hypoxia.

Adaptation has its limits, however. The highest permanent settlements are La Rinconada in Peru and Ma Gu in Tibet, both at roughly 5050 m (16,570 feet).[29] When the gradient between capillary and cell PO_2 disappears altogether, at approximately 9000 m (29,528 feet), cellular metabolism fails and death ensues.[25,26] Not surprisingly, then, occupation of high-altitude regions has been comparatively recent. While the Tibetan Plateau may have been sporadically exploited since 25,000 BP, and the Andean altiplano for half that long, sustained settlement occurred only about 6000 and 3000 BP, respectively, while the Ethiopian Semien highlands were occupied after 6000 BP.[30]

10.4.1 Prenatal and Early Postnatal Adaptation

The most dramatic effects of hypoxia on growth occur before birth. Placentas of indigenous Andean women at high altitude are up to 15% heavier relative to birth weight, and have three times more non-symmetrical shapes; both changes enhance surface area for oxygen transport.[29,31−33] Some studies show increased rates of blood flow in placental tissue of indigenous women[34−37] associated with higher birth weights. Flow rates in Tibetans are associated with markedly higher levels of nitric oxide (NO) metabolites, which are powerful vasodilators.[36,38] Bolivian Aymara show lower NO expression than Tibetans at comparable altitudes, but higher than lowlanders initially exposed to hypoxia. However, different flow rates or oxygen saturations were not found to influence net oxygen delivery to fetuses of Aymara or European ancestry.[37] In this context, early tissue growth, particularly skeletogenesis, is already mildly hypoxic, with well-described molecular compensations.[39] At high altitude, theory predicts that such existing pathways will be utilized for adaptation.

Hypoxia also drives changes in fetal size and body proportions. Ultrasound measurements of second trimester fetuses deprived of oxygen due to maternal smoking or diabetes showed reduced lower limb growth, changes in head shape and reduced length.[40,41] Models of fetal development have been developed that variously emphasize maternal oxygen transport, metabolic control mechanisms in the placenta or trade-offs between oxygen tension and glucose utilization.[32,34,37,42–46]

At birth, reductions of 50–100 g per 1000 m of altitude are the rule. Lean tissue comprises most of the loss.[31,32,42,47,48] Interpopulation variance can be pronounced. Neonates of Han immigrants in Lhasa, for instance, had mean birth weights over 800 g lower than normal-term sea-level white neonates in the USA, and about 250–300 g below European or Aymara Bolivians, while Tibetan neonates in Lhasa ranged from 450 g below the US norms to slightly above.[35]

Such disparate outcomes to a single constant stress may reflect different adaptive mechanisms. Higher Tibetan birth weight, compared to Peruvian Indians or residents of Leadville, Colorado (3026 m), were significantly correlated only with increased maternal uterine artery flow rates during pregnancy, while Peruvian and Leadville birth weights correlated only with maternal blood oxygen saturation.[42]

Significantly, the roughly normal distribution of birth weights characteristic of sea level shifts left at high altitude. Low birth weight (LBW) can become up to four times more prevalent, approaching half of all births. Thus, while mean change between sea level and high-altitude birth weight may have minor clinical significance, the left shift of the new distribution, already populated by children at or near risk of LBW, will extend many more neonates into high-risk categories.[48]

Peruvian neonates at high altitude also demonstrated reduced muscle mass relative to sea-level controls, but similar levels of body fat.[49] This could indicate gestational constraints, or prenatal adaptation to energy or oxidative stress through reduction of actively metabolizing lean tissue and associated mitochondria. For the latter reason, optimal birth weight at high altitude has been estimated to be 170 g lower than at sea level.[28]

Parental ancestry influences birth outcomes at altitude. Despite their socioeconomic advantages, one-third of all neonates of pure European ancestry in La Paz, Bolivia, were small for gestational age (SGA), compared to 16% of lower socioeconomic status Mestizo neonates and 13% of pure indigenous Aymara mothers.[34,35] Unlike lowland populations, maternal weight, parity or level of prenatal care were not significant predictors of birth weight, suggesting that hypoxic stress is the limiting factor. Birth weight response also may be associated with the sex of the parent of high-altitude ancestry. Controlled for gestational age, European fathers contributed more to birth weight increases than mothers. The effect, consistent with parental competition over epigenetic imprinting, dropped out when both parents were Aymara. The selective impact of hypoxia thus can override mechanisms that operate at sea level.[50]

In sum, the period from second trimester to birth reflects foundational responses to intrauterine hypoxia that produce reduced birth weights and lower body growth. The size of this effect varies significantly. Isolated from nutritional status, however, birth weight reduction appears directly correlated with length of residency at high altitude. Among indigenous groups, reduced birth weight is greatest in the Americas.

10.4.2 Childhood and Adolescence

After birth, children at high altitude tend to remain small for age. Prepubescent Nepalese Sherpa children, living between 3400 and 3800 m, are the shortest and lightest indigenous populations, while Sichuanese Tibetan children living at 3100 m are the tallest and heaviest. European Bolivian children living at 3600 m are the largest non-indigenous. While a portion of these differences reflect nutritional status, all high-altitude samples of children, regardless of ancestry and socioeconomic or nutritional status, are markedly shorter and lighter than US reference samples, with reduced skeletal muscle. Similar trends have been found in comparisons of genetically similar populations at high and low altitude, which argues for an independent effect of hypoxia.

Differences in attained growth partly reflect developmental delay. By puberty, skeletal maturation lags by 20% in both Tibetan and Andean children. Both groups approach US standards by 20 years of age.[29,51,52] Ethiopian children, by contrast, show smaller delays, reaching about 1 year at 12 years of age, with a velocity pattern comparable to Quechua children through most of childhood.[53] Both Quechua and Ethiopians were more advanced in hand—wrist skeletal age than Tibetans or Nepalese—Tibetans.[54]

Uniquely, Ethiopian children's maturational status catches up to chronological age in their sixteenth year. This has been variously attributed to reduction in disease stress in the Ethiopian highlands, to the moderate altitude (3000 m) at which these children live or to genetically unique physiological adaptations.[25,54,55] Delayed maturation followed by "catch-up" has also been viewed as part of an adaptive complex to inhibit growth of actively metabolizing lean tissue characteristic of adolescence.[29]

Estimates of the impact of high altitude on sexual maturation vary from little or no effect in Ethiopia, to small delays in the order of 6—9 months in Chinese or Tibetan populations, to over 1 year in the Andes and Nepal.[25,53,56] One study found moderate to profound delays among all high-altitude groups, with age at menarche ranging from 14.6 years for Ethiopian girls to 18.1 years for Sherpa girls and 16.1 years for Tibetan girls.[54] The only study to compare sexual maturation of genetically similar groups at low and high altitude found that Bolivian girls of European ancestry had a 0.8-year delay in menarche attributable to hypoxia relative to indigenous girls.[56]

Such wide variation in children's growth and maturation, given a constant hypoxic load, has led many investigators to cite nutritional stress as the larger causal factor, particularly in differences between indigenous and immigrant populations.[11,57—64]

In the clearest example of this, privileged French schoolchildren living at 3200–3600 m in La Paz, Bolivia, were 6 cm shorter than US reference standards, but 13 cm taller than their poorer Bolivian peers. The French contrast with US standards argues for an independent hypoxic effect, while the larger French advantage over Bolivian peers demonstrates the independent contribution of nutrition or factors such as access to health care. Importantly, growth status of the French children was related to their length of residence at high altitude; those with the shortest exposure were nearly 4 cm taller than those with the longest.[62]

In Central Asia, the picture is more complicated. Stunting, or reduced height for age by World Health Organization (WHO) standards, of Tibetan children increases with altitude, after control for nutritional and socioeconomic variables. Above 4000 m, over a quarter of all children were found to be stunted. By contrast, wasting, or reduced weight for age, was attributed to nutritional status, rather than hypoxia, except for the youngest children living above 4000 m.[65] Linear growth, then, appears particularly sensitive to hypoxic stress.

Differences between Tibetan and immigrant Han children are not systematic. In some studies,[64] Tibetans had more height and muscularity only at higher altitudes. Compared to lowland peers, Han were lighter and shorter. While they did not have larger chests, their lung volumes were greater, and they had elevated hemoglobin only at the highest altitudes.[66] In other samples,[61,62] Tibetan boys in mid-childhood were significantly taller, heavier and fatter than Han schoolmates of either sex, with higher vital capacity and percentage blood oxygen.

Urban–rural or sex variation in access to health care or food, rather than income per se, may account for some of the observed differences in nutritional status.[67] While suburban Tibetan boys in Sichuan were the largest, heaviest and most muscular among both sexes of four ethnic groups, urban Tibetan boys in Lhasa (3670 m) were systematically shorter and leaner, with lower body mass indices (BMIs) and skinfold thicknesses, than their female peers, and comparable to Han children. Unexpectedly, they also were smaller than either sex at a higher but less urbanized Tibetan site.[68,69]

The impact of undernutrition at high altitude can become dramatic when associated with broader political upheaval or socioeconomic neglect. Despite substantial growth improvements over 35 years in neighboring regions of the same altitude, there was no evidence of a secular trend in size among cohorts of Quechua Peruvian children living at 4250 m in an area torn by civil war.[70] Stunting and wasting were less common, but stunting still reached nearly 60%. Differences between these children and genetically similar peers elsewhere at the same altitude patently are nutritional, rather than hypoxic.

Data are rarer from high-altitude regions outside the Tibetan Plateau or Andean altiplano. Saudi children born and raised at 3000 m were lighter and shorter than National Center for Health Statistics (NCHS) medians, but their BMIs were similar before late childhood,[71] indicating a balanced impact on size and weight.

Indian–Tibetans born and raised at moderate to high altitudes were taller and heavier, and had thicker skinfolds than their peers in Tibet, but had similar elevated NO production. At 3521 m, they showed a pattern of leg growth retardation reported elsewhere.[72] Ethiopian children at 3000 m were closest to US sea-level references for height or weight. After mid-childhood, boys also were systematically taller than lowland peers, while girls and younger boys showed no significant differences. Compared to Peruvian or Nepalese Tibetan children, however, Ethiopians had reduced summed skinfolds and markedly lower chest circumferences, reflecting a more linear build.[54]

Overall, then, reductions of growth in size and weight at high altitude reflect independent contributions from hypoxia and undernutrition. How these are apportioned varies according to the parental gene pool, length of residency at high altitude, and balance between availability of oxygen and calories. Where the effects of nutrition can be controlled for, children's linear growth in Africa and Central Asia appears to be somewhat less affected by hypoxia per se than in the Americas.

10.4.3 Changes in Shape and Proportion

Finally, children's body proportions and shape appear to be influenced by high altitude. Across several Tibetan samples at 8 to 12 years of age, from 3000 to 4100 m, axial proportions including stature and lower leg length relative to stature were independently determined by oxygen availability, while stature, sitting height, chest circumference and arm proportions were independently influenced by caloric status.[67] Compared to Han peers, Tibetan children had longer lower legs,[60,61] although Han with the highest forced vital capacity (FVC) had leg proportions like those of average Tibetans[60] (see also refs 73 and 74). This suggests some overlap in adaptive efficiency that may reflect length of residence.

Andean children's axial proportions respond differently. Rural Aymara children averaged longer trunks relative to their stature, that is, shorter legs, compared to urban European ancestry peers. However, there was significant overlap in trunk to leg proportions between the two groups of children, and across ages, which could reflect maturational timing or adaptation.[75]

Larger chests are a consistent feature in high-altitude children, although the magnitude of the effect varies across regions. Prepubescent Ethiopian and Qinghai Tibetan boys have the most slender trunks relative to height, while Quechua boys have uniquely large chest circumferences for their size.[27,57,71,72]

Chest growth reflects functional adaptations. Children at high altitude typically have up to 1000 cm^3 greater FVC, and 500 cm^3 greater residual volume. Andean children show the most dramatic chest growth. At 4259 m, Peruvian children's and adolescents' chest circumference and FVC runs over 2 cm above US reference standards. Their residual volume is nearly 80% more than US standards.[76] Often mischaracterized as "dead air space", the residual volume actually represents a buffer of oxygenated air to

buffer against localized hypertension.[26] It has a higher heritability, and appears to be under active selection. Vital capacity is more representative of phenotypic plasticity; it will show dramatic increases in immigrants. Both compartments drive growth in chest dimensions.

Changes in chest size and function are less dramatic in Asia. Chest circumference is larger in Tibetan than Han children.[69] Chest length, correlated with FVC, increases with altitude in Tibetan children and adolescents, as does chest circumference.[71,72,77,78] At similar altitudes, immigrant Han children show significant changes in lung volume, but not chest dimensions.[64,66] Tibetans demonstrate significantly greater residual volumes than Han, while having more similar vital capacities.[25] However, despite their larger chests, blood oxygen saturation was only slightly greater among Tibetan boys. This may reflect Tibetan reliance on greater blood flow rates, rather than just on oxygen per unit blood, to combat hypoxia.[77,79]

Populations also differ in how altitude affects the timing of chest and lung growth. Andean populations showed increased velocity of trunk growth immediately upon birth, while European children in the same environment did not.[29,58,80] Moreover, chest circumference in Andean populations appears to expand into the third decade of life.[76] This latter effect has also been shown in Bod children of India at 3514 m,[25] and Ethiopian children at 3000 m,[53] but not Russian children from the Tien Shan region of Kirghistan (cited in Frisancho[25]).

Thus, while both indigenous and immigrant children ultimately produce larger chests and higher lung function than lowland peers, indigenous growth responses begin earlier and remain more profound. These increases, taken with enhanced lower leg growth, appear to be signature adaptive responses in body shape and proportion to hypoxia alone.

10.4.4 Molecular Control of Adaptation and Growth

Overall, variation in morphological growth within and between high-altitude regions suggests disparate gene pools confronting the same environmental stress. While Tibetan and Andean populations achieve basal metabolic rates and other measures of oxygen demand similar to lowlanders, they achieve them quite differently. From birth, Tibetans have significantly higher resting ventilation rates, reaching 15 liters per minute in adults, compared to 10.5 l/minute for Andeans, and adults show double the hypoxic ventilatory response to experimental stress.[77] Surprisingly, Tibetans do not develop local pulmonary hypertension, an otherwise universal mammalian response to hypoxia, nor do they develop enlarged chests as a result of obstructed pulmonary blood flow.[25,77] Blood flow is higher for Tibetans than immigrants from the second trimester of gestation through adulthood because of system-wide vasodilatation triggered by higher NO secretion by the arterial epithelium.[38] Finally, Tibetans show a greater density of muscle capillary beds, which enhances diffusion to cells.

By contrast, Andeans support morphological growth through elevated hemoglobin levels, enhanced lung volume compartments and total blood oxygen saturation relative to Tibetans. Andean populations show only mildly enhanced NO delivery, and demonstrate local pulmonary hypertension from birth; by adulthood clinical pulmonary hypertension and various cardiac myopathies are common.[28,76]

A third pattern of adaptation to hypoxia is shown by Ethiopians on the Semien Plateau. From 2 years of age onward, their linear build and some physiology, such as NO responsiveness, resemble Tibetan children. But unlike Tibetans, they also manage oxygen saturation that attains typical lowland US values. In the words of one investigator, it is "as if the Ethiopian sample were not living at high altitude".[55]

If we track growth outcomes to their sources, molecular mechanisms underlie the populational differences described above. From embryogenesis through adulthood, cellular trade-offs between oxygen tension and glucose metabolism are mediated by hypoxia inducible factor-1 (HIF-1) and HIF-2. HIF-1 is an evolutionarily conserved transcription factor independently triggered by hypoxia and heightened NO production. The body's primary angiogenic mediator, HIF-1 triggers early circulatory development in the naturally hypoxic environments of embryonic and fetal tissue. Postnatally, it orchestrates new vessel growth in the hypoxic conditions of a wound, organ lesion or bone fracture.[81]

HIF-2 is structurally similar, with some overlapping functions,[82] but is uniquely involved with maturation of the lung epithelium and blood cell production, and is thus linked to NO production via the epithelium.

Both factors collaborate in promoting gene expression for cartilage matrix deposition in bone and meniscus[73,78] via vascular endothelial growth factor (VEGF), which up-regulates blood vessel growth factors. By mediating how much blood is available, VEGF paces the maturation of all bone cells. Osteocytes and NO, in turn, can up-regulate HIF. The feedback has been termed angiogenic−osteogenic coupling.[74] HIF factors are particularly active in cartilage, where normal oxygen tensions are between 7% at the outer surface and 1% at the inner core.[39,78,83] Thus, skeletal growth under hypoxic conditions is HIF dependent.

A cellular trade-off between calories and oxygen occurs because HIF-1 also promotes glucose transporter-1 (GLUT-1), which increases uptake of glucose for anaerobic glycolysis.[84] Because anaerobic glycolysis is less efficient, there will be reduced energy available for bone growth.[46] Under normoxic conditions, HIF systems are damped throughout the body. Skeletal growth will be primarily limited by glucose availability. However, even comparatively mild levels of hypoxic stress to fetuses may trigger direct HIF inhibition of glucose metabolism in their mitochondria, which will limit growth.[44] For a fetus, then, high altitude may have a significantly lower stress threshold. Lower limb vulnerability to hypoxia, which begins in utero, may reflect an overlay of hydraulic compromises, arising out of bipedalism, set upon basic mammalian predispositions towards enhanced perfusion of the forebody to defend brain metabolism.[40,41,68,69]

The genes controlling these molecular growth trade-offs differ among populations. Quantitative genetic approaches have not identified a major gene complex that differentiates between high- and low-altitude Andean Indian populations' responses to hypoxia. Their growth patterns may represent lowland developmental plasticity extended as far as a lifetime at high altitude permits. Tibetans, by contrast, have two rapidly evolving gene variants in the EPAS1 (HIF-2) system that now characterize over 90% of all Tibetans.[85] These variants may reduce the need for accelerated lung or chest growth, or right ventricular hypertrophy, characteristic of Andean and other populations at high altitude by 3 months of age. Tibetan maturational timing is slowed, but the etiology — hypoxic or caloric — remains unclear. Unfortunately, we have no comparable molecular studies of Ethiopians. Their growth patterns suggest adaptive mechanisms that have achieved high equilibrium frequencies. Ethiopian growth patterns could reflect an African stamp on original central Asian adaptive complexes reflected in their 20% Asian mitochondrial DNA.[25]

We are confronted, then, with an embarrassment of windows into dissimilar populations of growing children. Central Asian and African adaptive responses to hypoxia may be more evolutionarily effective, if measured by comparatively better growth outcomes and lower morbidity and mortality attributable to hypoxia. However, all children pay a cost for living in thin air. Calculating that cost, and apportioning it among various environmental stressors, remains daunting.

10.5 SLEEP

Sleep is at the intersection of environmental and internal influences on growth and human biology. Inasmuch as society and its behavioral norms influence sleep characteristics, it can be considered an environmental influence and so is included here.

Sleep duration and quality have been associated with a variety of physiological systems, including immune function, glucose metabolism, neurobehavioral performance and hormonal profiles, which indicates that sleep plays an important role in human health. Total sleep duration per day declines from birth to adulthood ranging from an average of 14 hours at 6 months of age to an average of 8 hours (SD 0.8 hours) at 16 years of age.[86] Sleep is comprised of two major sleep stages, rapid-eye-movement (REM) sleep and non-rapid-eye-movement (NREM) sleep. Sleep stages develop in utero, at approximately 28–32 weeks' gestational age.[87] The amount of REM is greater earlier in development, and increases from 30–32 weeks' gestational age until 1–2 postnatal weeks.[87] By 2 years of age, REM sleep is established at 20–25% of the total sleep time, and this proportion remains stable throughout adulthood.[88]

Many hormones related to growth are affected by the sleep–wake cycle, including growth hormone (GH), luteinizing hormone (LH), testosterone, follicle-stimulating hormone (FSH) and prolactin (PRL). In adult men, the largest and often the only pulse of GH occurs shortly after sleep onset. In women, smaller daytime GH pulses are more frequent, but the sleep-onset pulse, while reduced in amplitude relative to that observed in men, is usually present.[89] In children, GH levels are typically higher during sleep than during wake, with a peak occurring shortly after sleep onset.[90–92] When sleep is interrupted by waking, GH secretion is abruptly suppressed.[89] An early study observed that among prepubertal children, In children, GH was secreted only during sleep while among pubertal adolescents and young adults GH was secreted during both wake and sleep, but the amount secreted during sleep was approximately double the amount secreted during wake.[93] A study in infants ranging in age from 1 week to 12 months found that GH levels were higher during sleep than wake only after the age of 3 months.[94] Thus, the secretion of GH is tightly coupled to the onset of sleep in children as young as 3 months up through adulthood. LH levels also appear to increase during sleep in pubertal children but not prepubertal children or young adults.[95,96] This sleep-induced increase in LH will occur even if sleep occurs during the daytime, indicating that it is not simply a circadian effect.[97] Furthermore, in pubertal boys, there is a marked increase in testosterone secretion during sleep, which appears to be dependent on the increased LH secretion.[96,98] One study also demonstrated a sleep-related increase in FSH in boys and girls in late puberty.[95] In adults, levels of PRL are normally lowest at midday and increase slightly throughout the afternoon, with a major nocturnal elevation shortly after sleep onset.[89] Regardless of time of day, sleep onset stimulates release of PRL; however, maximal simulation occurs only when sleep occurs at night.[89] A sleep-dependent release of PRL in both prepubertal and pubertal children has also been observed.[89] Given the invasive nature of frequent blood sampling, only a few studies have examined the relationship between sleep and hormonal secretion, but they have generally found that sleep is associated with the release of many hormones involved in growth and development.

Few studies have examined the relationship between sleep and linear growth or sexual maturation prospectively. Two studies examined the relationship between sleep duration and linear growth but found no association.[99,100] A small study of four children aged 1–3 years did observe increases in both sleep and growth after treatment for psychosocial dwarfism,[101] but whether there is a causal link between the changes in sleep and changes in growth cannot be determined. Prospective studies need to use objective measures of sleep and anthropometry among children of varying ages to determine whether sleep duration or quality can affect growth.

Several changes in sleep behavior typically occur during pubertal development. For example, total amount of sleep obtained per night decreases, which is associated with an increase in daytime sleepiness.[99,102–104] The decrease in amount of sleep obtained in adolescence, however, does not reflect a decrease in sleep need. In fact, sleep need appears similar across childhood and adolescence and is estimated to be approximately 9 hours.[102,105] The timing of sleep becomes delayed during pubertal development, which means that propensity to sleep occurs later in the evening and spontaneous awakening occurs later in the morning.[102,106] Owing to social commitments, such as school, adolescents must often wake earlier than spontaneous awakening and this probably explains both the shorter sleep durations and increased daytime sleepiness associated with greater sexual maturation. Thus, sleep changes during sexual maturation, but whether the timing or tempo of maturation is affected by sleep is not known.

The duration and quality of sleep have also been associated with body weight and BMI. Over 65 observational studies have found significant cross-sectional associations between short sleep duration and increased BMI in both adults and children (see Refs 107—109 for reviews). Some studies have also found that poor subjective sleep quality was associated with higher BMI.[110,111] Two meta-analyses analyzed data from cross-sectional studies in children and both found significant associations between short sleep duration and increased odds of being obese.[112,113] There have also been a few prospective studies of sleep and weight gain in children, which reported that shorter sleep durations or greater sleep problems were significantly associated with increased weight gain or risk of obesity.[114—120] Experimental studies of sleep restriction in young adults have suggested that one potential mechanism for a link between sleep and weight gain is dysregulation of hormones that contribute to appetite regulation, including a reduction in leptin, an appetite suppressant, and a concomitant increase in ghrelin, an appetite stimulant.[121,122] Similar studies have not been conducted in children. Nonetheless, the prospective studies suggest that short sleep duration is associated with increased body weight in children, which could increase the risk of developing obesity.

In summary, several hormones that are important for normal, healthy growth and development are associated with sleep. Furthermore, insufficient sleep may be a risk factor for the development of obesity, the rates of which are increasing dramatically for children and adolescents worldwide. Thus, more research is required to understand better the impact of chronic sleep restriction and impaired sleep quality in children and adolescents. Finally, since bedtimes are a volitional behavior, researchers need to examine cultural variation in sleep practices, particularly with respect to children.

10.6 POLLUTANTS

Pollution is usually defined as unwanted materials (e.g. lead, mercury, particulate matter) or energy (e.g. noise and radiation) produced by human activity or natural processes such as volcanic action. Anthropogenic pollutants are produced from power plants that generate energy, manufacturing industries, transportation, the construction of homes and factories, and even agriculture. Once created, pollutants are dispersed globally to virtually all populations by wind and water currents, and through the food chain.

In the past most of our knowledge of biological effects of pollutants came from occupational studies, but the information was not very generalizable as it usually concerned effects of large exposures on adult males. Developments in measurement technology have made it possible to make accurate measurements of low levels of pollutants using very small biosamples. Pollutants are now routinely detected in pregnant women, newborn babies and children, and we need to understand their effects on the developing organism when environmental insults can have irreparable, long-lasting effects. Fetal programming (see Chapter 12), the impact of environmental factors on the fetus that affect its functioning postpartum and its health in later life, can be thought of as a reformulation of reproductive toxicology that includes nutritional insults as well as chemical ones.

The study of human development and toxicants is based on observation without modifying exposures, since an experiment in which exposure is randomized to subjects is obviously unethical. Purely observational studies yield statistical associations and these must be judged in terms of the likelihood that the association is based on biological cause. There are six commonly used criteria for judging the causal basis of statistical associations (Box 10.1), and studies of growth and environmental factors should be designed to meet these criteria as much as possible.

All of the listed criteria depend on the accurate and reliable measurement of exposure. The best way to assess exposure is to measure the pollutant of interest in the person. For example, in a study of lead, it is best to measure lead in the blood or bone. An inexpensive but far less accurate method of assessing exposure is the substitution of a measurement made in a geographical zone, such as a postal zone, for the exposure of every child living in the zone. However, people in one zone are likely to experience different amounts of true exposure and grouping them together and using an average value leads to misclassification. This produces large errors in the independent (exposure) variable and less statistical power to detect effects. Too often studies of growth and pollution are forced for economical reasons to use this latter method, but effects on growth are more likely to be accurately determined if we can employ the most accurate

Box 10.1 Criteria for judging the causal basis of statistical associations

- A strong association
- Biological credibility to the association
- Consistency with other studies
- Compatible sequence of cause and effect
- Evidence of a dose—response relationship

measures of exposure. Despite the challenges in studying pollutants, there is now considerable evidence that human physical growth and development are sensitive to several pollutants including lead, the components of air pollution, organic compounds such as polychlorinated biphenyls, as well as some forms of energy such as radiation and noise.

10.6.1 Cigarette Smoking

Cigarette smoking is a perfect example of an anthropogenic influence on growth and development. Exposure is a function of human behavior, the exposure composition is complex, and human experience with smoke is fairly recent, although some could argue that smoke is the oldest pollutant.[123]

Cigarette smoke contains a large variety of compounds including carbon monoxide and cyanide. These compounds can cross the placenta and affect the fetus, and second-hand cigarette smoke may affect children in households with smokers.[124] Postnatal exposure to cigarette smoke may also affect growth, but this problem has not been studied sufficiently.

After gestational age, maternal cigarette smoking is the single greatest influence on birth weight in well-off countries.[125] In most populations suffering from nutritional stress, very few women smoke during pregnancy, so the effect of smoking is minimal or absent. Women who smoke during pregnancy have babies weighing on average 200 g less than babies of non-smokers and the reduction in birth weight is related to the number of cigarettes smoked. This dose—response relationship is good evidence for a causal relationship between smoking and prenatal growth. Gestation length is reduced by only 2 days or less, which cannot account for the birth-weight decrement. When birth weights of smokers' and non-smokers' infants are compared at each week of gestation from weeks 36 to 43, smokers' babies consistently have lower mean birth weights. Just living with a smoker may affect birth weight, as women whose husbands smoked had lower birth-weight babies.[126,127]

The reduction in mean birth weight is part of a downward shift of the entire distribution of birth weights. Thus, the frequency of LBW (less than 2500 g) is more common among smokers, again irrespective of gestational age, and it is approximately doubled among smokers.

Maternal smoking also is significantly associated with shorter body lengths (about 1 cm), reduced arm circumference and, in some studies, slightly reduced head circumference.[128–131] The sizes of the decrements depend on the amount and timing of cigarette consumption by the mothers in the sample. Weight growth is strongly affected by smoking in the last trimester. In one longitudinal study using repeated ultrasound imaging, biparietal diameter of the head increased significantly more rapidly among fetuses of non-smokers from the 28th week of gestation onwards, i.e. starting near the beginning of the last trimester of pregnancy.[40]

The effect of quitting smoking after conception also informs us of when smoking acts to reduce prenatal growth. Quitting before the fourth month of pregnancy is thought to reduce or remove the effects of smoking. However, quitting is more common among light smokers than heavy smokers. When both the amount of smoking and the quitting are considered, very heavy smokers who quit may not fully lower their risk of LBW.[132] However, from a practical point of view, quitting or reducing smoking is advised for all women who smoke and who are pregnant or who may become pregnant, because smoking has such a strong, detrimental effect on the fetus.

The primary constituent of tobacco smoke is carbon monoxide. Carbon monoxide, with an affinity for adult hemoglobin 200 times that of oxygen, has an even greater affinity for fetal hemoglobin. It is estimated that if a mother smokes 40 cigarettes per day there is a 10% concentration of carboxyhemoglobin equivalent to a 60% reduction in blood flow to the fetus. Thus, cigarette smoking exacerbates fetal hypoxia, which has some similarities to high-altitude hypoxia. In fact, placenta ratios are larger among smokers largely owing to the reduction in birth weight, as they are among high-altitude births. Some studies have noted that the placentas of heavy smokers are heavier than non-smokers' placentas,[133,134] a finding consistent with effects seen at high altitude.[33] However, other studies have found no difference in placenta size associated with heavy smoking and the nicotine content of cigarettes smoked.[130,135] Smokers' placentas also are thinner, with larger minimum diameters.[133] Some of these changes in placental morphology may be adaptive given the reduced oxygen-carrying capacity of the blood, but other changes, such as calcification or ones indicative of aging or chronic ischemia (lack of blood flow) in the placenta do not appear to be adaptive.

Cigarette smoke also contains nicotine, which stimulates adrenal production of epinephrine, norepinephrine and acetylcholine, and this results in less uteroplacental perfusion (blood flow through the uterus and placenta). It also can act on the fetus directly to increase fetal blood pressure and respiratory rate. In addition, cyanide, lead and cadmium are contained in cigarette smoke and are all toxic.[136] Smoking can also affect hormone levels[137,138] and this, in turn, could affect prenatal growth.

It is tempting to think that smoking may not act through any of these means, but indirectly by reducing maternal appetite and weight gain, but this is not the case. Studies have compared weight gains of smoking and non-smoking pregnant women and found that weight gains are similar. Other studies have controlled for weight gain by matching for weight or through statistical procedures, and the effect of smoking on size at birth is still present.

Postnatal effects of cigarette smoking are less well studied and less clear. Follow-up studies of smokers' offspring have difficulty separating effects that may develop from being exposed to cigarette smoke in utero from the effects of postnatal exposure due to living with adult smokers. Ideally, to research the effect of postnatal smoking, one would study children whose mothers did not smoke during pregnancy but who began smoking soon after giving birth. Few mothers meet these conditions, leaving most researchers to study children whose mothers smoked during pregnancy and who have continued to do so after the baby was born. Although some studies have not found lasting effects from birth,[139] many other studies have. The difference may be due to the extent to which other influences on growth are controlled or to the size of the sample. Using a sample of a few hundred children, Hardy and Mellits[140] found a 1 cm difference in length at 1 year of age that, though small, was statistically significant, but no differences at 4 and 7 years of age. In studies using larger samples, differences in height of about 1.5 cm at 3 years[141] and 5 years of age[142] were found. Analysis of the National Child Development Study (NCDS), which is a very large national sample from Britain, detected a deficit of approximately 1 cm in children's heights at 7 and 11 years associated with maternal smoking.[143] In one study of 3500 adolescents, the heights of 14-year-old girls were reduced by an average of nearly 1 cm, which was statistically significant, but the boys' heights did not differ significantly.[144] However, the NCDS sample found a small but significant reduction at age 16 years in male heights (about 0.9 cm), but not in females.[145] It seems that the difference of 1 cm that is present at birth becomes a smaller fraction of the variation in height that increases as individual differences in the tempo of height growth are expressed and reach their greatest magnitude at puberty.

The effect of passive smoking is small but significant in large samples. Rona and colleagues[146] examined the heights of children in relation to the number of smokers in the household (none, one or two) and corrected for birth weight to remove the effects of maternal smoking during pregnancy. Height declined with more smokers in the home, suggesting that passive smoking may affect postnatal growth.

Adipose tissue growth may also be reduced in relation to maternal smoking during pregnancy and postnatal exposure to cigarette smoke.[147] This finding is consonant with observations among adults that smokers are leaner[148] and their fat distribution tends to be more centripedal (located on the torso).[149,150] When adult smokers quit smoking, they add fat and attain a more peripheral or gynoid distribution (on the thighs, hips and arms). This, in turn, is consistent with the observation that cigarette smoking contains anti-estrogenic compounds such that female smokers tend to have fat patterns that resemble those more typical among males. The difference seen in 6–11-year-old children may be a late expression of an effect of prenatal exposure, or may be a response to postnatal exposure to cigarette smoke. In either case, adipose differences have not been found at birth.[129,151]

There is no doubt today that cigarette smoking is a powerful cause of reduced prenatal growth. Deficits are greatest in weight at birth but these appear to be made up during childhood, whereas the small deficit in length is not. Postnatal exposure to second-hand cigarette smoke seems to reduce height growth slightly, although more replication studies are needed. All studies of growth should consider the effects of smoking carefully, especially if the subject is the growth of the fetus.

10.6.2 Air Pollution

Air pollution is a ubiquitous form of pollution and a very heterogeneous category of materials. Most studies compare two or more settlements that differ in the severity of air pollution and many control well for differences in socioeconomic status. Most, but not all, of these studies report that height and weight growth are more favorable in less polluted areas.[152–155] Slower skeletal maturation has been observed in several studies.[155–157] It is possible that air pollution exerts an effect like high-altitude hypoxia, limiting the oxygen available for growth. Mikusek[158] found that girls from an air-polluted town were delayed in all growth dimensions except for chest development, a selective effect similar to the sparing of chest circumference growth seen in some studies of high-altitude Andean children.

The effect of air pollution begins prenatally. An early study of birth weight in Los Angeles, California, found that weights decreased in relation to the severity of the air pollution, and the effect was evident after controlling for some of the other large influences on birth weight (mother's cigarette smoking and socioeconomic status).[159] This finding has been well replicated,[160–162] but there also are a few instances where no effect has been found.[163] Some variation in results could be due to variation in the characteristics of the air pollution itself. The most recent work has tried to determine which components of air pollution may be responsible, the suspended particulate matter or the gases (sulfur oxides, nitrogen oxides and ozone), but the answer is not yet clear.

10.6.3 Organic Compounds

Organic pollutants include many insecticides and herbicides that have been used in agriculture and pest control. Dichlorodiphenyltrichloroethane (DDT) is a pesticide, highly effective in controlling mosquitos, which was banned in the USA in 1972, but is persistent and its metabolite (DDE) is found in the blood of many populations. Other pollutants were manufactured for use in various industries [e.g. polychlorinated biphenyls (PCBs), phthalates], and others, such as dioxin, are unintended by-products of manufacturing. Phthalates are plasticizers used in bottles, toys and personal care products. PCBs are a large group of similarly structured compounds with variation in toxicity and persistence in the environment and in the body. Some forms are very similar to dioxin. Polybromated diphenyl esters (PBDEs) are fire retardants added to a large variety of consumer items that leach into surrounding materials and now can be detected in many populations. Organic pollutants such as PCBs, dioxin and DDT are lipophilic; they are stored in fat cells and can be retained for years. They cross the placenta, and lactation is a significant source of exposure. They are also found in dietary items such as fish, meat and dairy products.

PCBs may affect endocrine function, physical growth, maturation and/or cognitive or behavioral development of children and youth. Evidence of the effects of PCBs in humans comes from two types of study: studies of *acute* poisoning, either food poisoning or an occupational accident, and studies of *chronic* low-level exposures, usually from ingestion of foods with slight but measurable contamination.

Ingestion of rice oil contaminated with a mixture of PCBs, dioxin and dibenzofurans poisoned thousands of adults and children in Japan in 1968 and in Taiwan in 1978–79, producing diseases called Yusho and Yucheng, respectively. Yusho/Yucheng infants have had higher rates of mortality and lower body weights at birth.[164] Even children born long after their mothers were exposed to the contaminated oil were more often born prematurely and small at birth,[165] probably because during gestation they were exposed to the toxicant mixture that had been stored in their mothers' fat tissue. Reduced postnatal growth also characterizes Yusho/Yucheng children.[164,166,167]

Studies of children born to women exposed to smaller amounts of PCBs over a long period have also found growth deficits. A common but not universal finding is that birth weight is reduced in response to PCB exposure.[168–173] The effect seems not to be due to prematurity but to less growth during gestation. Head circumference may also be reduced.[172] Three studies have related reduced birth weight to fetal exposure to dioxin and dioxin-like compounds.[174–176] PCB levels and BMI are related early in life[177] and at puberty.[178] DDE, a metabolite of the insecticide DDT, has been related to reduced birth weight and height in several studies,[179] but with increased height and weight for height in others.[177,178,180–182]

Not all studies agree (see, for example, Boas et al.[183]). It is important when interpreting conflicting results to account for differences in exposure among the studies. Certainly one would expect a smaller effect or none at all when the exposure is very low, and it is difficult to ascertain how exposure levels compare across studies because measurement techniques have changed substantially and only the most recent studies measure the contaminant compounds in similar ways. In addition, different effects may stem from differences in the timing of exposure: prenatal versus postnatal. Children exposed to PCBs from maternal consumption of fish from the Great Lakes were significantly lighter at 4 years of age, though not at 11 years of age.[184] The reduction at age 4 was related to the PCB level at birth, reflecting prenatal exposure, but not to their current PCB level. Another longitudinal study found greater height in girls at 5 years of age in a cohort in which there had been differences in size at birth.[168]

Clearly, not all studies of humans agree as to the size and direction of effects that these toxicants have. This creates questions about the influence of differences in levels of exposure, differences in the timing of exposure (prenatal or postnatal) and differences in the mixtures of compounds to which the sample was exposed. Excepting occupational exposures, mundane exposures tend to be a mixture of several compounds. While we tend to group all these persistent organic pollutants as "toxic", we have learned not to expect them to have similar effects on growth.

We do know that the alterations in growth are best explained as due to interference with hormonal signaling. Further, we know that different compounds affect signaling differently depending on the timing of exposure and their structure as some are agonistic and some antagonistic. Many studies of non-human animals have shown that hormone activity can be altered following exposure to PCBs and related compounds. Thyroid hormone signaling is especially important for normal physical and mental growth and development. Neurological effects are the most consistently reported effects of chronic PCB exposure.[185–187] A study of Mohawk adolescents, 10–17 years of age, found the combination of reduced thyroxine levels and increased thyroid-stimulating hormone levels (this usually signals low thyroid activity) in relation to levels of persistent PCBs that reflect past exposure, but not in relation to PCBs more reflective of current exposure. This suggests that prenatal or neonatal exposure may be influential.[188] In this same group, there were alterations in performance on some cognitive tests and particular effects on memory.[173]

Studies of the effects of phthalates in children have intensified recently as the material is found in more products and more populations. Effects of phthalates on the thyroid have been reported.[183] Growth and sexual development can also be affected.[183,189–194]

Many types of organic pollutant structurally resemble sex steroids as well as thyroid hormones. The possibility of deranged hormonal signaling of sexual development and reproduction has been a controversial area of research, but there is now sufficient evidence that it occurs in humans. Several studies conducted by different research groups and with different populations have reported that exposure to some pollutants accelerates the development of puberty, although the finding is not universal.[195–197]

Whether the effect is one induced by prenatal or postnatal exposure, or both, is not clear. Newborns were found to have significantly lower testosterone and estradiol levels in relation to their mothers' levels of dioxin, dibenzofuran (similar to dioxin) and dioxin-like PCBs.[185] Follow-up of boys in the Yucheng cohort who had substantial exposure to dioxin-like compounds showed reduced testosterone levels and increased FSH levels but no difference in Tanner stage.[186] Preliminary results from the same cohort found significantly reduced penile lengths, a possible effect of their exposure prenatally when critical sexual differentiation and development is occurring.[187]

A related compound, PBDE, is a new pollutant of concern. In adult men it has been strongly and inversely associated with a measure of androgen, with LH and FSH, and positively with inhibin B and sex steroid binding globulin.[198] This result and other similar findings strongly suggest that chemicals in our environment can affect levels of hormones directly involved in reproduction and development. In true experimental studies of non-human animals endocrine disruption is clearly evident and this establishes the biological plausibility of the associations seen in observational studies of human populations.[199] Research on children and PBDE is just beginning. If the effects seen in adults are present in children, they certainly could affect growth and sexual maturation.[200,201]

There is sufficient evidence to be concerned about PCB exposure in children and the fetus. Especially convincing are the controlled laboratory studies of higher primates and rodents that show reductions in growth and alterations in sexual development that are dependent on normally functioning hormonal systems.[202,203] Although it is not known whether a low-level exposure will produce effects on child development, we do see that high exposures, such as those from food heavily contaminated with PCBs, can produce predictable effects, and the possibility that the fetus is especially sensitive to PCBs, even at low levels, remains a very viable hypothesis.

10.6.4 Lead

Lead has been a common pollutant since it was first added to paint and gasoline. In the USA, lead burdens are higher among urban, disadvantaged, minority children because of their residence in older areas characterized by dilapidated housing having flaking leaded paint and with older roadways where cars burning leaded gasoline deposited lead in exhaust fumes. People also are exposed to lead from their mothers through transplacental passage and lactation. Lead is a legacy pollutant that has been transmitted to each generation through biological and social pathways.

Studies often indicate that size at birth is reduced and gestations are shorter with increased lead exposure,[204,205] although some studies do not detect such differences, perhaps owing to uncontrolled confounding variables such as maternal nutrition.[206–208] Studies in Cincinnati, Ohio, and Albany, New York, found a reduction in birth weight of nearly 200 g in relation to the log of maternal blood lead level.[209,210] Other studies have found reductions in head circumference.[211] A study of 43,000 recent births in New York state found reductions in birth weights of 61 to 87 g depending on the level of maternal lead, 0 versus 5 mg/dl and 6 versus 10 mg/dl, respectively.[212] Results from this extremely large study remind us that the inconsistencies in findings of pollutants and growth often can be explained simply by differences in the exposures of the populations studied.

Studies of birth weight are facilitated by routine collection of birth weight as part of public health surveillance. Studies of postnatal growth are less common. The largest studies have used national survey data from the USA. Data from the second National Health and Nutrition Examination Survey (NHANES) data (1976–1980) involving about 7000 children less than 7 years of age showed that lead level was negatively related to stature, weight and chest circumference after controlling for other important influences on growth.[213] Compared to children with a blood lead level of zero, children with the mean lead level were 1.5% shorter at the mean age of 59 months. The second large study used a data set of 7–12-year-old children from the Hispanic Health and Nutrition Examination Survey (1982–1984). Children whose blood lead levels were above the median for their age and sex were 1.2 cm shorter than those with lead below the median.[214] The third study used anthropometric data from the Third NHANES

(1988—1994) for non-Hispanic children 1—7 years of age, and found statistically significant reductions of 1.57 cm in stature and 0.52 cm in head circumference for each 10 μg of lead in the blood.[215] An analysis of 8—18-year-old girls in NHANES III found that those with moderately high lead levels were significantly shorter.

Studies with smaller samples have also found growth decrements in height, weight and/or head circumference of similar magnitude, indicating that the associations between growth and lead reported for national samples of US children may be present generally.[216—218] Other anthropometric dimensions may also be decreased.[219] When lead levels are reduced in a neighborhood through public health efforts, child growth may respond positively,[220] which suggests that early growth decrements may be reduced if the exposure is reduced early in life.

These studies are cross-sectional, that is, lead and stature were measured simultaneously, and consequently one could argue (and some have) that short children are simply exposed to more lead. However, experimental studies of non-human animals show very clearly that growth is reduced following lead exposure, which supports the latter explanation.

There have been few longitudinal studies of lead and growth. In the Cincinnati study, higher maternal blood lead levels coupled with higher infant lead levels were associated with poorer growth.[221—223] Similarly, when the children reached 33 months of age, two groups of children had decreased stature: those with low lead levels prenatally but high lead levels from 3 to 15 months, and the children with high lead levels in the prenatal *and* postnatal periods.

A study of infants in Albany, New York, also found that when lead levels increased between the pregnancy (maternal) level and the infant's own level at 12 months of age, there was poorer infant growth in weight and head circumference.[224] Reduced infant weight gain in relation to lead has been found in other studies as well.[225] Not all studies agree and this may be attributed to differences in methods and/or levels of exposure or control for normal sources of variation such as diet, cigarette smoking, etc.[226]

Several large studies with good measurement methods and control of relevant confounders have established quite clearly that lead exposure is related to delayed menarche and delay in attainment of Tanner stages.[227—231] Among African-American girls in the NHANES III study, lead-associated delays in reaching Tanner stages ranged from 2 to 6 months depending on the Tanner stage, and menarche was delayed by 3 months.[228] A study of 10—16.9-year-old girls of the Akwesasne Mohawk Nation in northern New York found delays in reaching menarche related to lead level. Girls were delayed on average by 10 months if their lead level was below the sample median compared to the average for those above the median.[224,231] Pubertal onset in Russian boys was delayed by some 6—10 months.[230] One study of girls in Poland found contrary results: earlier menarche was associated with higher lead levels.[232] However, in this study lead was not measured in the individual, but in the environment, thus raising concerns about the accuracy of exposure classification. Still, contrary results can reveal other influences and interaction effects, and if well constructed, should not be ignored.

A mechanism for the effect of lead on sexual maturation is not clear as yet, but one has been suggested through a study of girls in the NHANES III sample. Those with higher lead levels had lower levels of inhibin B, a marker of follicular development. Further studies into toxicants and sexual maturation are warranted to confirm this finding and elucidate the mechanism that produces the effect.

There is a better understanding of the mechanism for lead's effect on growth. Growth velocity can increase when children receive chelation therapy to remove lead from the body.[233] When children's lead levels have been reduced, their stimulated peak human GH levels are significantly higher compared to when lead levels are at a toxic level. In addition, among children with high levels of lead, insulin-like growth factor-1 is reduced with increasing lead level. These results help to make the statistical associations between lead and reduced height growth more understandable as true biological effects.

Over the past half-dozen years, more and more studies have shown relationships between lead levels and growth and maturation. Moreover, the effects are seen at quite low levels that would have been of little concern 20 years ago. This fact reinforces the idea that growth is sensitive to environmental inputs, including or especially anthropogenic ones, and at relatively low levels and through day-to-day, chronic exposures.

10.6.5 Radiation

High doses of radiation, as are used for some cancer treatments, do affect height growth.[234] In these studies radiation dose is both high and carefully measured, and the effect on growth is well established.

However, mundane exposure to radiation is far harder to measure. Individuals have difficulty recalling all their exposures to mundane sources (e.g. medical X-rays, airplane flights), thus introducing error in measuring exposure. One study found that women who had been exposed prenatally to medical X-rays were about 1.5 times more likely to experience menarche before the age of 10 years,[235] and others have detected postnatal growth retardation.[236]

In studies of people exposed through atomic bomb blasts, dose can be estimated by determining location in relation to the epicenter of the blast. Several studies have found that in utero exposure to an atomic bomb blast is associated with reduced head circumference, height and weight during childhood and adolescence and that the reduction is related to estimated dose.[236,237] One study of accidental exposure to a bomb test found that early postnatal exposure is detrimental as well.[238] In these early studies, maturation rate, if examined at all, was not affected significantly. The most recent work on growth and radiation examined growth at adolescence among survivors of the Hiroshima and Nagasaki atomic bombs.[239] From 10 to 18 years of age, total in utero exposure was related to a reduction of several centimeters in stature, but it was not possible to see whether exposure in a particular trimester was especially damaging.

Microwave radiation is quite different from radiation from bomb blasts or cancer treatment, but is commonly experienced through mobile phone use. The growth of children and adolescents using mobile phones has not been studied as yet, but reports of increased sleep disturbance related to greater phone use point to a possible indirect effect on growth (as described in Section 10.5).[240]

10.6.6 Noise Stress

Noise is a classic physiological stressor used in countless laboratory studies to induce stress responses in experimental animals. Studies of humans show that noise, defined as unwanted sound, stimulates the classic stress response as well. Thus, studies of noise are studies of stress.

Several studies have examined prenatal growth in relation to maternal exposure to noise from the workplace and have found small effects on birth weight or, in one case, none at all.[241−243] Studies of airports, where noise stress may be more severe or better measured, find fairly consistently that birth weight is depressed in relation to exposure.[244−250] Two studies have found similarly sized reductions in birth weight related to aircraft noise and both studies found these effects among female births but not males.[246,250] Other studies have not examined effects by gender, but further research may determine whether the effects of noise stress are modified by this factor.

Evidence for an environmental effect on growth depends on different lines of evidence (see Box 10.1) and these are present among the studies of noise stress. Large sample studies that have compared groups differing in exposure have found that high maternal noise exposure is negatively related to birth weight in a fairly consistent dose−response manner (Figure 10.2).[249] Examination of the rate of LBW and the rate of jet airplane flyovers at Kobe airport when jets were first introduced demonstrates the temporal relationship between exposure and effect. Before the introduction of jets, the rate of LBW near the airport was lower than the rest of Japan, but as soon as jet flights began, the rate of LBW increased markedly and the increase continued to parallel the increased number of jet take-offs. The temporal association strongly suggests that the jet take-offs are causally related to the change in the frequency of LBW.

Studies of postnatal growth are very few, perhaps owing to the difficulty in estimating noise exposure for postnatal life. The first such study found reduced heights and weights among children exposed to high noise from an airport in Japan. More recent studies have found a reduction in height at 3 years of age (Figure 10.3) and a reduction in soft-tissue dimensions in children at 5−12 years of age.[251]

An effect of noise seems plausible based on what we know about the relationship of high noise to the stress response and the relationship between stress and growth. Noise

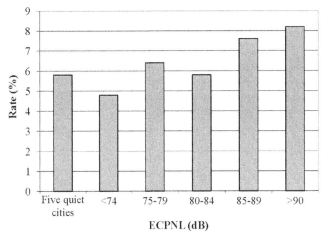

Figure 10.2 Percentage of low-birth-weight infants (< 2500 g) in 1969 according to mothers' exposure to aircraft noise measured as equivalent continuous perceived noise levels (ECPNL) (dB). (Source: *Adapted from Ando and Hattori.*[249])

activates the hypothalamic—pituitary—adrenal axis in the same way as other stressors do (Figure 10.4). Noise stress stimulates the autonomic nervous system and the pituitary gland which, in turn, affects the adrenal cortex, the thyroid and the gonads. Cortisol, thyroid hormones and sex steroids all affect growth and development. Thus, an effect of noise on growth is biologically plausible. Since noise is a form of stress, studying noise exposure is a way of learning about the effects of other kinds of stress as well.

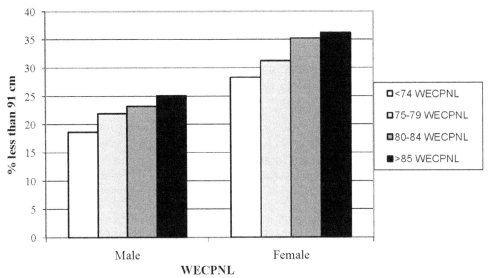

Figure 10.3 Percentage of 3-year-old children <91 cm tall by noise exposure measurement weighted equivalent continuous perceived noise levels (WECPNL). (Source: *Adapted from Schell and Ando.*[252])

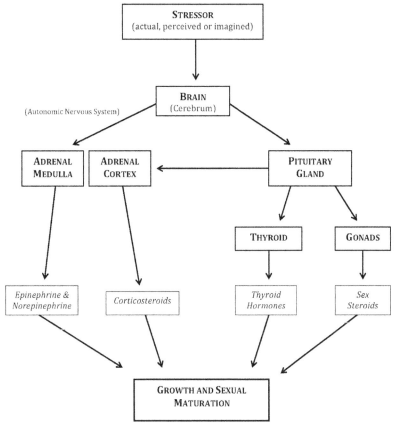

Figure 10.4 The biological stress response.

One general observation is that the effects of any pollutant depend on the dose. Effects of noise are not present in any study unless the exposures are quite high, perhaps over 100 dBA. Statements summarizing the relationship between noise and growth have to be careful about specifying the range of exposure observed. In general, statements about the relationship between any environmental factor, whether it is altitude or noise, should refer only to the ranges of exposures observed and should not extrapolate results to exposures above or below that range, otherwise results among studies will appear more inconsistent than they really are.

10.7 HOW DO WE INTERPRET DIFFERENCES IN GROWTH RELATED TO ENVIRONMENTAL FACTORS?

Recalling the two interpretations of growth reductions reviewed at the chapter's outset, it seems that the interpretation used depends on the environmental factor considered: slow and/or reduced growth is a disadvantage created by adverse health conditions (this is the growth monitoring or "biomedical" model), or slow and/or reduced growth is an adaptive response (i.e. beneficial) to features of the environment (the adaptation model). While there is no covering law to dictate which interpretation is appropriate in which circumstances, in general, it seems that growth reductions related to anthropogenic factors (lack of material resources for the child, including poor medical care and poor nutrition) tend to be interpreted with the growth monitoring model, while growth reductions related to features of the physical environment tend to be interpreted in an adaptive framework. This distinction is not foolproof. Air pollution, an anthropogenic factor, is also a product of volcanoes and other natural processes. Ultimately, the view that growth reductions do have adaptive benefits for the individual (affecting reproduction, or functioning such as cognition) will be determined by studies that seek to measure the adaptive benefits. In general, growth alterations may be seen as the result of trade-offs between resources for growth, reproduction or survival.

10.8 CONCLUSION

In addition to effects of nutrition and socioeconomic factors, the immediate physical environment can affect human physical growth and development. This conclusion is supported by many of the studies reviewed here on altitude, temperature and climate. Studies of pollutants also show effects on growth, although many of these studies have flaws that come from valued and important limitations on experiments with people. However, the results from numerous, carefully executed studies of non-human animals support the studies on humans. When compared to the effect of malnutrition, the effect of pollutants can seem small, but the size of the effect depends on the extent of exposure to the pollutant. If we clean up the environment, child growth will be little affected by air pollution, but if children grow up in an environment with many types of pollution, the effect of all the pollutants together may be large. Indeed, studies show that in industrialized countries the poor children have more exposure to pollutants, and the result can be impaired growth. It is wise to remember that exposure to many of the pollutants that affect growth are mediated by social factors, as is nutritional deprivation. Thus, growth can be considered as a monitor of the general quality of children's environments.

REFERENCES

1. Schell LM, Gallo MV, Ravenscroft J. Environmental influences on human growth and development: historical review and case study of contemporary influences. *Ann Hum Biol* 2009;**36**:459–77.
2. Lasker GW. Human biological adaptability. *Science* 1969;**166**:1480–6.
3. Schell LM. Human biological adaptability with special emphasis on plasticity: history, development and problems for future research. In: Mascie-Taylor CG, Bogin B, editors. *Human variability and plasticity*. Cambridge: Cambridge University Press; 1995. p. 213–37.
4. Huss-Ashmore R. Theory in human biology: evolution, ecology, adaptability, and variation. In: Stinson S, Bogin B, Huss-Ashmore R, O'Rourke D, editors. *Human biology: an evolutionary and biocultural perspective*. New York: John Wiley & Sons; 2000. p. 1–25.

5. Ulijaszek SJ, Huss-Ashmore R. *Human adaptability: past, present, and future*. Oxford: Oxford University Press; 1997. p. 1−336.

6. Tanner JM. Growth as a mirror of the condition of society: secular trends and class distinctions. In: Dubuc MB, Demirjian A, editors. *Human growth: a multidisciplinary review*. London: Taylor and Francis; 1986.

7. Schell LM, Magnus PD. Is there an elephant in the room? Addressing rival approaches to the interpretation of growth perturbations and small size. *Am J Hum Biol* 2007;**19**:606−14.

8. Roberts DF. *Climate and human variability. Module in anthropology No. 34*. Reading, MA: Addison-Wesley; 1973. p. 1−38.

9. Newman MT. The application of ecological rules to the racial anthropology of the aboriginal New World. *Am Anthropol* 1953;**55**:311−27.

10. Bogin B. *Patterns of human growth*. Cambridge: Cambridge University Press; 1988. p. 1−280.

11. Katzmarzyk PT, Leonard WR. Climatic influences on human body size and proportions: ecological adaptations and secular trends. *Am J Phys Anthropol* 1998;**106**:483−503.

12. Malina RM, Bouchard C. *Growth, maturation, and physical activity*. Champaign, IL: Human Kinetics; 1991. p. 1−520.

13. Eveleth PB. The effects of climate on growth. *Ann NY Acad Sci* 1966;**134**:750−9.

14. Palmer CE. Seasonal variation of average growth in weight of elementary school children. *Public Health Rep* 1933;**48**:211−33.

15. Tanner JM. *Growth at adolescence, with a general consideration of the effects of hereditary and environmental factors upon growth and maturation from birth to maturity*. Oxford: Blackwell; 1962. p. 1−325.

16. Bogin B. Monthly changes in the gain and loss of growth in weight of children living in Guatemala. *Am J Phys Anthropol* 1979;**51**:287−92.

17. Bogin B. Seasonal pattern in the rate of growth in height of children living in Guatemala. *Am J Phys Anthropol* 1978;**49**:205−10.

18. Marshall WA, Swan AV. Seasonal variation in growth rates of normal and blind children. *Hum Biol* 1971;**43**:502−16.

19. Marshall WA. The relationship of variations in children's growth rates to seasonal climatic variations. *Ann Hum Biol* 1975;**2**:243−50.

20. Vincent M, Dierickx J. Etude sur la croissance saisonnière des écoliers de Léopoldville. *Ann Soc Belg Med Trop* 1960;**40**. 837−44.

21. Griffin JE, Ojeda SR. *Textbook of endocrine physiology*. Oxford: Oxford University Press; 1996. 1−408.

22. Haddad JG, Hahn TJ. Natural and synthetic sources of circulating 25-hydroxyvitamin D in man. *Nature* 1973;**244**:515−7.

23. Stamp TCB, Round JM. Seasonal changes in human plasma levels of 25-hydroxyvitamin D. *Nature* 1974;**247**:563−5.

24. Johnston FE. Control of age at menarche. *Hum Biol* 1974;**46**:159−71.

25. Frisancho AR. Prenatal and postnatal growth and development at high altitude. *Human adaptation and accommodation*. Ann Arbor, MI: University of Michigan Press; 1993. p. 281−307.

26. Hall JE, Guyton AC. *Guyton and Hall textbook of medical physiology*. Philadelphia, PA: Saunders/Elsevier; 2011. p. 1−1120.

27. Beall CM. Optimal birthweights in Peruvian populations at high and low altitudes. *Am J Phys Anthropol* 1981;**56**:209−16.

28. West JB. The physiologic basis of high-altitude diseases. *Ann Intern Med* 2004;**141**:789−800.

29. Frisancho AR, Baker PT. Altitude and growth: a study of the patterns of physical growth of a high altitude Peruvian Quechua population. *Am J Phys Anthropol* 1970;**32**:279−92.

30. Aldenderfer MS. Modeling the neolithic on the Tibetan plateau. In: Madsen DB, editor. *Developments in quaternary science. Late quaternary climate change and human adaptation in arid China*. Oxford: Elsevier; 2007. p. 151−65.

31. McClung J. *Effects of high altitude on human birth: observations on mothers, placentas, and the newborn in two Peruvian populations*. Cambridge, MA: Harvard University Press; 1969. p. 1−168.

32. Haas JD. Maternal adaptation and fetal growth at high altitude in Bolivia. In: Greene LS, Johnston FE, editors. *Social and biological predictors of nutritional status, physical growth and neurological development*. New York: Academic Press; 1980. p. 257−90.

33. Kruger H, Arias-Stella J. The placenta and the newborn infant at high altitudes. *Am J Obstet Gynecol* 1970;**106**:586—91.
34. Wilson MJ, Lopez M, Vargas M, Julian C, Tellez W, Rodriguez A, et al. Greater uterine artery blood flow during pregnancy in multigenerational (Andean) than shorter-term (European) high-altitude residents. *Am J Physiol Regul Integr Comp Physiol* 2007;**293**:R1313—24.
35. Julian CG, Vargas E, Armaza JF, Wilson MJ, Niermeyer S, Moore LG. High-altitude ancestry protects against hypoxia-associated reductions in fetal growth. *Arch Dis Child Fetal Neonatal Ed* 2007;**92**:F372—7.
36. Julian CG, Galan HL, Wilson MJ, Desilva W, Cioffi-Ragan D, Schwartz J, et al. Lower uterine artery blood flow and higher endothelin relative to nitric oxide metabolite levels are associated with reductions in birth weight at high altitude. *Am J Physiol Regul Integr Comp Physiol* 2008;**295**:R90615.
37. Zamudio S, Postigo L, Illsley NP, Rodriguez C, Heredia G, Brimacombe M, et al. Maternal oxygen delivery is not related to altitude- and ancestry-associated differences in human fetal growth. *J Physiol* 2007;**582**:883—95.
38. Erzurum SC, Ghosh S, Janocha AJ, Xu W, Bauer S, Bryan NS, et al. Higher blood flow and circulating NO products offset high-altitude hypoxia among Tibetans. *Proc Natl Acad Sci USA* 2007;**104**:17593—8.
39. Schipani E. Hypoxia and HIF-1alpha in chondrogenesis. *Ann NY Acad Sci* 2006;**1068**:66—73.
40. Lampl M, Kuzawa CW, Jeanty P. Prenatal smoke exposure alters growth in limb proportions and head shape in the midgestation human fetus. *Am J Hum Biol* 2003;**15**:533—46.
41. Lampl M, Jeanty P. Exposure to maternal diabetes is associated with altered fetal growth patterns: a hypothesis regarding metabolic allocation to growth under hyperglycemic—hypoxemic conditions. *Am J Hum Biol* 2004;**16**:237—63.
42. Moore LG. Maternal O$_2$ transport and fetal growth in Colorado, Peru, and Tibet high-altitude residents. *Am J Hum Biol* 1990;**2**:627—37.
43. Davila RD, Julian CG, Wilson MJ, Browne VA, Rodriguez C, Bigham AW, et al. Do anti-angiogenic or angiogenic factors contribute to the protection of birth weight at high altitude afforded by Andean ancestry? *Reprod Sci* 2010;**17**:861—70.
44. Illsley NP, Caniggia I, Zamudio S. Placental metabolic reprogramming: do changes in the mix of energy-generating substrates modulate fetal growth? *Int J Dev Biol* 2010;**54**:409—19.
45. Lampl M. Cellular life histories and bow tie biology. *Am J Hum Biol* 2005;**17**:66—80.
46. Zamudio S, Torricos T, Fik E, Oyala M, Echalar L, Pullockaran J, et al. Hypoglycemia and the origin of hypoxia-induction reduction in human fetal growth. *PLoS ONE* 2010;**5**. e8551.
47. Hartinger S, Tapia V, Carrillo C, Bejarano L, Gonzales GF. Birth weight at high altitudes in Peru. *Int J Gynaecol Obstet* 2006;**93**:275—81.
48. Lichty JA, Ting RY, Bruns PD, Dyar E. Studies of babies born at high altitude. Part I: Relation of altitude to birth weight. *Am J Dis Child* 1957;**93**:666—9.
49. Haas JD, Baker PT, Hunt Jr EE. The effects of high altitude on body size and composition of the newborn infant in southern Peru. *Hum Biol* 1977;**49**:611—28.
50. Bennett A, Sain SR, Vargas E, Moore LG. Evidence that parent-of-origin affects birth-weight reductions at high altitude. *Am J Hum Biol* 2008;**20**:592—7.
51. Beall CM, Baker PT, Baker TS, Haas JD. The effects of high altitude on adolescent growth in southern Peruvian Amerindians. *Hum Biol* 1977;**49**:109—24.
52. Pawson IG. Growth characteristics of populations of Tibetan origin in Nepal. *Am J Phys Anthropol* 1977;**47**:473—82.
53. Clegg EJ, Pawson IG, Ashton EH, Flinn RM. The growth of children at different altitudes in Ethiopia. *Philos Trans R Soc Lond* 1972;**264**:403—37.
54. Pawson IG. Growth and development in high altitude populations: a review of Ethiopian, Peruvian, and Nepalese studies. *Proc R Soc Lond* 1976;**194**:83—98.
55. Beall CM, Decker MJ, Brittenham GM, Kushner I, Gebremedhin A, Strohl KP. An Ethiopian pattern of human adaptation to high-altitude hypoxia. *Proc Natl Acad Sci USA* 2002;**99**:17215—8.
56. Greksa LP. Age of menarche in Bolivian girls of European and Aymara ancestry. *Ann Hum Biol* 1990;**17**:49—53.

57. Dittmar M. Secular growth changes in the stature and weight of Amerindian schoolchildren and adults in the Chilean Andes, 1972—1987. *Am J Hum Biol* 1998;**10**:607—17.

58. Greksa LP, Spielvogel H, Caceres E. Effect of altitude on the physical growth of upper-class children of European ancestry. *Ann Hum Biol* 1985;**12**:225—32.

59. Leonard WR, Leatherman TL, Carey JW, Thomas RB. Contributions of nutrition versus hypoxia to growth in Nunoa, Peru. *Am J Hum Biol* 1990;**2**:613—26.

60. Leatherman TL, Carey JW, Thomas RB. Socioeconomic change and patterns of growth in the Andes. *Am J Phys Anthropol* 1995;**97**:307—21.

61. Mueller WH, Schull VN, Schull WJ, Soto P, Rothhammer F. A multinational Andean genetic and health program: growth and development in an hypoxic environment. *Ann Hum Biol* 1978;**5**:329—52.

62. Stinson S. The effect of high altitude on the growth of children of high socioeconomic status in Bolivia. *Am J Phys Anthropol* 1982;**59**:61—71.

63. Weitz CA, Garruto RM, Chin CT, Liu JC, Liu RL, He X. Growth of Qinghai Tibetans living at three different high altitudes. *Am J Phys Anthropol* 2000;**111**:69—88.

64. Weitz CA, Garruto RM, Chin CT, Liu JC. Morphological growth and thorax dimensions among Tibetan compared to Han children, adolescents and young adults born and raised at high altitude. *Ann Hum Biol* 2004;**31**:292—310.

65. Dang S, Yan H, Yamamoto S. High altitude and early childhood growth retardation: new evidence from Tibet. *Eur J Clin Nutr* 2008;**62**:342—8.

66. Weitz CA, Garruto RM. Growth of Han migrants at high altitude in central Asia. *Am J Hum Biol* 2004;**16**:405—19.

67. Bailey SM, Xu J, Feng JH, Hu X, Qui S, Zhang C. Tibetan children's upper and lower skeletal proportions are consistent with models of hypoxia inducible factor mediation of chondrocyte growth. *Am J Hum Biol* 2010.

68. Bailey SM, HU XM. High altitude growth differences among Chinese and Tibetan Children. In: Gilli G, Schell LM, Benso L, editors. *Human growth from conception to maturity*. London Smith-Gordon: 2002. p. 327—247.

69. Bailey SM, Xu J, Feng JH, Hu X, Zhang C, Qui S. Tradeoffs between oxygen and energy in tibial growth at high altitude. *Am J Hum Biol* 2007;**19**:662—8.

70. Pawson IG, Huicho L. Persistence of growth stunting in a Peruvian high altitude community, 1964—1999. *Am J Hum Biol* 2010;**22**:367—74.

71. Al-Shehri MA, Mostafa OA, Al-Gelban K, Hamdi A, Almbarki M, Altrabolsi H, et al. Standards of growth and obesity for Saudi children (aged 3—18 years) living at high altitudes. *West Afr J Med* 2006;**25**:42—51.

72. Tripathy V, Gupta R. Growth among Tibetans at high and low altitudes in India. *Am J Hum Biol* 2007;**19**:789—800.

73. Wan C, Shao J, Gilbert SR, Riddle RC, Long F, Johnson RS, et al. Role of HIF-1alpha in skeletal development. *Ann NY Acad Sci* 2010;**1192**:322—6.

74. Riddle RC, Khatri R, Schipani E, Clemens TL. Role of hypoxia-inducible factor-1alpha in angiogenic—osteogenic coupling. *J Mol Med* 2009;**87**:583—90.

75. Stinson S. Nutritional, developmental, and genetic influences on relative sitting height at high altitude. *Am J Hum Biol* 2009;**21**:606—13.

76. Greksa LP. Growth and development of Andean high altitude residents. *High Alt Med Biol* 2006;**7**:116—24.

77. Beall CM. Tibetan and Andean contrasts in adaptation to high-altitude hypoxia. *Adv Exp Med Biol* 2000;**475**:63—74.

78. Adesida AB, Grady LM, Khan WS, Millward-Sadler SJ, Salter DM, Hardingham TE. Human meniscus cells express hypoxia inducible factor-1alpha and increased SOX9 in response to low oxygen tension in cell aggregate culture. *Arthritis Res Ther* 2007;**9**:R69.

79. Beall CM, Song K, Elston RC, Goldstein MC. Higher offspring survival among Tibetan women with high oxygen saturation genotypes residing at 4,000 m. *Proc Natl Acad Sci USA* 2004;**101**:14300—4.

80. Greksa LP, Beall CM. Development of chest size and lung function at high altitude. In: Little MA, Haas JD, editors. *Human population biology*. New York: Oxford University Press; 1989. p. 222–38.

81. Semenza GL. Regulation of tissue perfusion in mammals by hypoxia-inducible factor 1. *Exp Physiol* 2007;**92**:988–91.

82. Hu CJ, Wang LY, Chodosh LA, Keith B, Simon MC. Differential roles of hypoxia-inducible factor 1alpha (HIF-1alpha) and HIF-2alpha in hypoxic gene regulation. *Mol Cell Biol* 2003;**23**:9361–74.

83. Murphy CL, Polak JM. Control of human articular chondrocyte differentiation by reduced oxygen tension. *J Cell Physiol* 2004;**199**:451–9.

84. Chen C, Pore N, Behrooz A, Ismail-Beigi F, Maity A. Regulation of glut1 mRNA by hypoxia-inducible factor-1. Interaction between H-ras and hypoxia. *J Biol Chem* 2001;**276**:9519–25.

85. Beall CM, Cavalleri GL, Deng L, Elston RC, Gao Y, Knight J, et al. Natural selection on EPAS1 (HIF2alpha) associated with low hemoglobin concentration in Tibetan highlanders. *Proc Natl Acad Sci USA* 2010;**107**:11459–64.

86. Iglowstein I, Jenni OG, Molinari L, Largo RH. Sleep duration from infancy to adolescence: reference values and generational trends. *Pediatrics* 2003;**111**:302–7.

87. Davis FC, Frank MG, Heller HC. Ontogeny of sleep and circadian rhythms. In: Turek FW, Zee PC, editors. *Regulation of sleep and circadian rhythms*. New York: Marcel Dekker; 1999. p. 19–80.

88. Culebras A. *The neurology of sleep. American Academy of Neurology Suppl 6(42)*. New York: Advanstar Communications; 1992. p. 6–8.

89. Van Cauter E, Spiegel K. Circadian and sleep control of hormonal secretions. In: Turek FW, Zee PC, editors. *Regulation of sleep and circadian rhythms*. New York: Marcel Dekker; 1999. p. 397–425.

90. Costin G, Kaufman FR, Brasel JA. Growth hormone secretory dynamics in subjects with normal stature. *J Pediatr* 1989;**115**:537–44.

91. Eastman CJ, Lazarus L. Growth hormone release during sleep in growth retarded children. *Arch Dis Child* 1973;**48**:502–7.

92. Mace JW, Gotlin RW, Beck P. Sleep related human growth hormone (GH) release: a test of physiologic growth hormone secretion in children. *J Clin Endocrinol Metab* 1972;**34**:339–41.

93. Finkelstein JW, Roffwarg HP, Boyar RM, Kream J, Hellman L. Age-related change in the twenty-four-hour spontaneous secretion of growth hormone. *J Clin Endocrinol Metab* 1972;**35**:665–70.

94. Vigneri R, D'Agata R. Growth hormone release during the first year of life in relation to sleep–wake periods. *J Clin Endocrinol Metab* 1971;**33**:561–3.

95. Beck W, Wuttke W. Diurnal variations of plasma luteinizing hormone, follicle-stimulating hormone, and prolactin in boys and girls from birth to puberty. *J Clin Endocrinol Metab* 1980;**50**:635–9.

96. Boyar RM, Rosenfeld RS, Kapen S, Finkelstein JW, Roffwarg HP, Weitzman ED, et al. Human puberty. Simultaneous augmented secretion of luteinizing hormone and testosterone during sleep. *J Clin Invest* 1974;**54**:609–18.

97. Kapen S, Boyar RM, Finkelstein JW, Hellman L, Weitzman ED. Effect of sleep–wake cycle reversal on luteinizing hormone secretory pattern in puberty. *J Clin Endocrinol Metab* 1974;**39**:293–9.

98. Parker DC, Judd HL, Rossman LG, Yen SSC. Pubertal sleep–wake patterns of episodic LH, FSH and testosterone release in twin boys. *J Clin Endocrinol Metab* 1975;**40**:1099–109.

99. Knutson KL. The association between pubertal status and sleep duration and quality among a nationally representative sample of US adolescents. *Am J Hum Biol* 2005;**17**:418–24.

100. Jenni OG, Molinari L, Caflisch JA, Largo RH. Sleep duration from ages 1 to 10 years: variability and stability in comparison with growth. *Pediatrics* 2007;**120**:e769–76.

101. Guilhaume A, Benoit O, Gourmelen M, Richardet JM. Relationship between sleep stage IV deficit and reversible HGH deficiency in psychosocial dwarfism. *Pediatr Res* 1982;**16**:299–303.

102. Carskadon MA, Acebo C. Regulation of sleepiness in adolescents: update, insights, and speculation. *Sleep* 2002;**25**:606–14.

103. Dahl RE, Lewin DS. Pathways to adolescent health sleep regulation and behavior. *J Adolesc Health* 2002;**31**:175–84.

104. Wolfson AR, Carskadon MA. Sleep schedules and daytime functioning in adolescents. *Child Dev* 1998;**69**:875–87.

105. Carskadon MA, Harvey K, Duke P, Anders TF, Litt IF, Dement WC. Pubertal changes in daytime sleepiness. *Sleep* 1980;**2**:453—60.

106. Sadeh A, Dahl RE, Shahar G, Rosenblat-Stein S. Sleep and the transition to adolescence: a longitudinal study. *Sleep* 2009;**32**:1602—9.

107. Marshall NS, Glozier N, Grunstein RR. Is sleep duration related to obesity? A critical review of the epidemiological evidence. *Sleep Med Rev* 2008;**12**:289—98.

108. Patel SR, Hu FB. Short sleep duration and weight gain: a systematic review. *Obesity (Silver Spring)* 2008;**16**:643—53.

109. Knutson KL, Van CE. Associations between sleep loss and increased risk of obesity and diabetes. *Ann NY Acad Sci* 2008;**1129**:287—304.

110. Asplund R, Aberg H. Body mass index and sleep in women aged 40 to 64 years. *Maturitas* 1995;**22**(1):1—8.

111. Jennings JR, Muldoon MF, Hall M, Buysse DJ, Manuck SB. Self-reported sleep quality is associated with the metabolic syndrome. *Sleep* 2007;**30**:219—23.

112. Cappuccio FP, Taggart FM, Kandala NB, Currie A, Peile E, Stranges S, et al. Meta-analysis of short sleep duration and obesity in children and adults. *Sleep* 2008;**31**:619—26.

113. Chen X, Beydoun MA, Wang Y. Is sleep duration associated with childhood obesity? A systematic review and meta-analysis. *Obesity (Silver Spring)* 2008;**16**:265—74.

114. Sugimori H, Yoshida K, Izuno T, Miyakawa M, Suka M, Sekine M, et al. Analysis of factors that influence body mass index from ages 3 to 6 years: a study based on the Toyama cohort study. *Pediatr Int* 2004;**46**:302—10.

115. Reilly JJ, Armstrong J, Dorosty AR, Emmett PM, Ness A, Rogers I, et al. Early life risk factors for obesity in childhood: cohort study. *BMJ* 2005;**330**:1357.

116. Lumeng JC, Somashekar D, Appugliese D, Kaciroti N, Corwyn RF, Bradley RH. Shorter sleep duration is associated with increased risk for being overweight at ages 9 to 12 years. *Pediatrics* 2007;**120**:1020—9.

117. Al MA, Lawlor DA, Cramb S, O'Callaghan M, Williams G, Najman J. Do childhood sleeping problems predict obesity in young adulthood? Evidence from a prospective birth cohort study. *Am J Epidemiol* 2007;**166**:1368—73.

118. Snell EK, Adam EK, Duncan GJ. Sleep and the body mass index and overweight status of children and adolescents. *Child Dev* 2007;**78**:309—23.

119. Taveras EM, Rifas-Shiman SL, Oken E, Gunderson EP, Gillman MW. Short sleep duration in infancy and risk of childhood overweight. *Arch Pediatr Adolesc Med* 2008;**162**:305—11.

120. Touchette E, Petit D, Tremblay RE, Boivin M, Falissard B, Genolini C, et al. Associations between sleep duration patterns and overweight/obesity at age 6. *Sleep* 2008;**31**:1507—14.

121. Spiegel K, Leproult R, L'hermite-Baleriaux M, Copinschi G, Penev PD, Van CE. Leptin levels are dependent on sleep duration: relationships with sympathovagal balance, carbohydrate regulation, cortisol, and thyrotropin. *J Clin Endocrinol Metab* 2004;**89**:5762—71.

122. Spiegel K, Tasali E, Penev P, Van CE. Brief communication: Sleep curtailment in healthy young men is associated with decreased leptin levels, elevated ghrelin levels, and increased hunger and appetite. *Ann Intern Med* 2004;**141**:846—50.

123. Silkworth JB, Brown Jr JF. Evaluating the impact of exposure to environmental contaminants on human health [published erratum appears in Clin Chem 1997;**43**:410]. *Clin Chem* 1996;**42**:1345—9.

124. Misra DP, Nguyen RHN. Environmental tobacco smoke and low birth weight: a hazard in the workplace. *Environ Health Perspect* 1999;**107**:897—904.

125. Kramer MS. Intrauterine growth and gestational duration determinants. *Pediatrics* 1987;**80**:502—11.

126. Mathai M, Vijayasri R, Babu S, Jeyaseelan L. Passive maternal smoking and birthweight in a south Indian population. *Br J Obstet Gynaecol* 1992;**99**:342—3.

127. Borlee I, Bouckaert A, Lechat MF, Misson CB. Smoking patterns during and before pregnancy. *Eur J Obstet Gynaecol Reprod Biol* 1978;**8**:171—7.

128. Schell LM, Hodges DC. Variation in size at birth and cigarette smoking during pregnancy. *Am J Phys Anthropol* 1985;**68**:549—54.

129. Harrison GG, Branson RS, Vaucher YE. Association of maternal smoking with body composition of the newborn. *Am J Clin Nutr* 1983;**38**:757–62.

130. Olsen J. Cigarette smoking in pregnancy and fetal growth. Does the type of tobacco play a role? *Int J Epidemiol* 1992;**21**:279–84.

131. Haste FM, Anderson HR, Brooke OG, Bland JM, Peacock JL. The effects of smoking and drinking on the anthropometric measurements of neonates. *Paediatr Perinat Epidemiol* 1991;**5**:83–92.

132. Schell LM, Relethford JH, Madan M, Naamon PBN, Hook EB. Unequal adaptive value of changing cigarette use during pregnancy for heavy, moderate, and light smokers. *Am J Hum Biol* 1994;**6**:25–32.

133. Christianson RE. Gross differences observed in the placentas of smokers and nonsmokers. *Am J Epidemiol* 1979;**110**:178–87.

134. Naeye RL. Effects of maternal cigarette smoking on the fetus and placenta. *Br J Obstet Gynaecol* 1978;**85**:732–7.

135. Mulcahy R, Murphy J, Martin F. Placental changes and maternal weight in smoking and nonsmoking mothers. *Am J Obstet Gynecol* 1970;**106**:703–4.

136. Andrews J. Thiocyanate and smoking in pregnancy. *J Obstet Gynaecol Br Commonw* 1973;**80**:810–4.

137. Field AE, Colditz GA, Willett WC, Longcope C, McKinlay JB. The relation of smoking, age, relative weight, and dietary intake to serum adrenal steroids, sex hormones, and sex hormone-binding globulin in middle-aged men. *J Clin Endocrinol Metab* 1994;**79**:1310–6.

138. Bremme K, Lagerström M, Andersson O, Johansson S, Eneroth P. Influences of maternal smoking and fetal sex on maternal serum oestriol, prolactin, hCG, and hPl levels. *Arch Gynecol Obstet* 1990;**247**:95–103.

139. Conter V, Cortinovis I, Rogari P, Riva L. Weight growth in infants born to mothers who smoked during pregnancy. *BMJ* 1995;**310**:768–76.

140. Hardy JB, Mellits ED. Does maternal smoking during pregnancy have a long-term effect on the child? *Lancet* 1972;**2**:1332–6.

141. Fox NL, Sexton M, Hebel JR. Prenatal exposure to tobacco: I. Effects on physical growth at age three. *Int J Epidemiol* 1990;**19**:66–71.

142. Wingerd J, Schoen EJ. Factors influencing length at birth and height at five years. *Pediatrics* 1974;**53**:737–41.

143. Butler NR, Goldstein H. Smoking in pregnancy and subsequent child development. *BMJ* 1973;**4**:573–5.

144. Rantakallio P. A follow up to the age of 14 of children whose mothers smoked during pregnancy. *Acta Paediatr Scand* 1983;**72**:747–53.

145. Fogelman K. Smoking in pregnancy and subsequent development of the child. *Child Care Health Dev* 1980;**6**:233–49.

146. Rona RJ, CdV Florey, Clarke JC, Chinn S. Parental smoking at home and height of children. *BMJ* 1981;**283**:1363.

147. Schell LM, Relethford JH, Hodges DC. Cigarette use during pregnancy and anthropometry of offspring 6–11 years of age. *Hum Biol* 1986;**58**:407–20.

148. Goldbourt U, Medalie JH. Characteristics of smokers, non-smokers and ex-smokers among 10,000 adult males in Israel. *Am J Epidemiol* 1977;**105**:75–86.

149. Shimokata H, Muller DC, Andres R. Studies in the distribution of body fat. III. Effects of cigarette smoking. *JAMA* 1989;**261**:1169–73.

150. Troisi RJ, Heinhold JW, Vokonas PS, Weiss ST. Cigarette smoking, dietary intake, and physical activity; effects on body composition – the Normative Aging Study. *Am J Clin Nutr* 1991;**53**:1104–11.

151. D'Souza SW, Black P, Richards B. Smoking in pregnancy: associations with skinfold thickness, maternal weight gain, and fetal size at birth. *BMJ* 1981;**282**:1661–3.

152. Danker-Hopfe H, Drobna M, Cermakova Z. Air pollution and growth of 3 to 7 year old children from Bratislava. *Conference proceeding, Soshowiec, Poland*; 1996.

153. Jedrychowski W, Flak E, Mroz E. The adverse effect of low levels of ambient air pollutants on lung function growth in preadolescent children. *Environ Health Perspect* 1999;**107**:669–74.

154. Antal A, Timaru J, Muncaci E, Ardevan E, Ionescu A, Sandulache L. Les variations de la reactivite de l'organisme et de l'etat de sante des enfants en rapport avec la pollution de l'air communal. *Atmos Environ* 1968;**2**:383–92.

155. Thielebeule U, Pelech L, Grosser P-J, Horn K. Body height and bone age of school children in areas of different air pollution concentration. *Z Gesamte Hyg* 1980;**26**:771—4.

156. Schlipkoter HW, Rosicky B, Dolgner R, Peluch L. Growth and bone maturation in children from two regions of the FRG differing in the degree of air pollution: results of the 1974 and 1984 surveys. *J Hyg Epidemiol Microbiol Immunol* 1986;**30**:353—8.

157. Schmidt P, Dolgner R. Interpretation of some results of studies in school-children living in areas with different levels of air pollution. *Zentralbl Bakteriol* 1977;**165**:539—47.

158. Mikusek J. Developmental age and growth of girls from regions with high atmospheric air pollution in Silesia. *Rocz Panstw Zakl Hig* 1976;**27**:473—81.

159. Williams L, Spence A, Tideman SC. Implications of the observed effects of air pollution on birth weight. *Soc Biol* 1977;**24**:1—9.

160. Nordstrom S, Beckman L, Nordenson I. Occupational and environmental risks in and around a smelter in northern Sweden. I. Variations in birth weight. *Hereditas* 1978;**88**:43—6.

161. Bobak M. Outdoor air pollution, low birth weight, and prematurity. *Environ Health Perspect* 2000;**108**:173—6.

162. Wilhelm M, Ghosh JK, Su J, Cockburn M, Jerrett M, Ritz B. Traffic-related air toxics and term low birth weight in Los Angeles County, California. *Environ Health Perspect* 2012;**120**:132—8.

163. Dolk H, Pattenden S, Vrijheid M, Thakrar B, Armstrong BG. Perinatal and infant mortality and low birth weight among residents near cokeworks in Great Britain. *Arch Environ Health* 2000;**55**:26—30.

164. Rogan WJ, Gladen BC, Hung K-L, Koong S-L, Shih L-Y, Taylor JS, et al. Congenital poisoning by polychlorinated biphenyls and their contaminants in Taiwan. *Science* 1988;**241**:334—6.

165. Yen YY, Lan SJ, Yang CY, Wang HH, Chen CN, Hsieh CC. Follow-up study of intrauterine growth of transplacental Yu-Cheng babies in Taiwan. *Bull Environ Contam Toxicol* 1994;**53**:633—41.

166. Guo YL, Lambert GH, Hsu C-C, Hsu MM. Yucheng: health effects of prenatal exposure to poly-chlorinated biphenyls and dibenzofurans. *Int Arch Occup Environ Health* 2004;**77**:153—8.

167. Guo YL, Lin CJ, Yao WJ, Ryan JJ, Hsu CC. Musculoskeletal changes in children prenatally exposed to polychlorinated biphenyls and related compounds (Yu-Cheng children). *J Toxicol Environ Health* 1994;**41**:83—93.

168. Hertz-Picciotto I, Charles MJ, James RA, Keller JA, Willman E, Teplin S. In utero polychlorinated biphenyl exposures in relation to fetal and early childhood growth. *Epidemiology* 2005;**16**:648—56.

169. Schell LM. Effects of pollutants on human prenatal and postnatal growth: noise, lead, polychlorinated compounds and toxic wastes. *Yearb Phys Anthropol* 1991;**34**:157—88.

170. Taylor PR, Stelma JM, Lawrence CE. The relation of polychlorinated biphenyls to birth weight and gestational age in the offspring of occupationally exposed mothers. *Am J Epdemiol* 1989;**129**:395—406.

171. Taylor PR, Lawrence CE, Hwang H-L, Paulson AS. Polychlorinated biphenyls: influence on birthweight and gestation. *Am J Public Health* 1984;**74**:1153—4.

172. Axmon A, Rylander L, Stromberg U, Dyremark E, Hagmar L. Polychlorinated biphenyls in blood plasma among Swedish female fish consumers in relation to time to pregnancy. *J Toxicol Environ Health A* 2001;**64**:485—98.

173. Newman J, Aucompaugh A, Schell LM, Denham M, DeCaprio AP, Gallo MV, et al. Akwesasne Task Force on the Environment. PCBs and cognitive functioning of Mohawk adolescents. *Neurotoxicol Teratol* 2006;**28**:439—45.

174. Patandin S, Koopman-Esseboom C, Weisglas-Kuperus N, Sauer PJJ. Birth weight and growth in Dutch newborns exposed to background levels of PCBs and dioxins. *Organohalogen Compounds* 1997;**34**:447—50.

175. Patandin S, Koopman-Esseboom C, De Ridder MAJ, Weisglas-Kuperus N, Sauer PJJ. Effects of environmental exposure to polychlorinated biphenyls and dioxins on birth size and growth in Dutch children. *Pediatr Res* 1998;**44**:538—45.

176. Vartiainen T, Jaakkola JJK, Saarikoski S, Tuomisto J. Birth weight and sex of children and the correlation to the body burden of PCDDs/PCDFs and PCBs of the mother. *Environ Health Perspect* 1998;**106**:61—6.

177. Verhulst SL, Nelen V, Hond ED, Koppen G, Beunckens C, Vael C, et al. Intrauterine exposure to environmental pollutants and body mass index during the first 3 years of life. *Environ Health Perspect* 2009;**117**:122—6.

178. Gallo MV, Ravenscroft J, Schell LM, DiCaprio A. Akwesasne Task Force on the Environment Environmental contaminants and growth of Mohawk adolescents at Akwesasne (abstract). *Acta Med Auxol* 2000;**32**:72.

179. Karmaus W, Asakevich S, Indurkhya A, Witten J, Kruse H. Childhood growth and exposure to dichlorodiphenyl dichloroethene and polychlorinated biphenyls. *J Pediatr* 2002;**140**:33—9.

180. Gladen BC, Ragan NB, Rogan WJ. Pubertal growth and development and prenatal and lactational exposure to polychlorinated biphenyls and dichlorodiphenyl dichloroethene. *J Pediatr* 2000;**136**. 490—6.

181. Su PH, Chen JY, Chen JW, Wang SL. Growth and thyroid function in children with in utero exposure to dioxin: a 5-year follow-up study. *Pediatr Res* 2010;**67**:205—10.

182. Wolff MS, Engel S, Berkowitz G, Teitelbaum S, Siskind J, Barr DB, et al. Prenatal pesticide and PCB exposures and birth outcomes. *Pediatr Res* 2007;**61**:243—50.

183. Boas M, Frederiksen H, Feldt-Rasmussen U, Skakkebaek NE, Hegedus L, Hilsted L, et al. Childhood exposure to phthalates: associations with thyroid function, insulin-like growth factor I, and growth. *Environ Health Perspect* 2010;**118**:1458—64.

184. Jacobson JL, Jacobson SW, Humphrey HEB. Effects of exposure to PCBs and related compounds on growth and activity in children. *Neurotoxicol Teratol* 1990;**12**:319—26.

185. Cao Y, Winneke G, Wilhelm M, Wittsiepe J, Lemm F, Furst P, et al. Environmental exposure to dioxins and polychlorinated biphenyls reduce levels of gonadal hormones in newborns: results from the Duisburg cohort study. *Int J Hyg Environ Health* 2008;**211**:30—9.

186. Hsu P-C, Lai T-J, Guo N-W, Lambert GH, Leon GY. Serum hormones in boys prenatally exposed to polychlorinated biphenyls and dibenzofurans. *J Toxicol Environ Health A* 2005;**68**: 1447 56.

187. Guo YL, Lambert GH, Hsu C-C. Growth abnormalities in the population exposed *in utero* and early postnatally to polychlorinated biphenyls and dibenzofurans. *Environ Health Perspect* 1995;**103**: 117—22.

188. Schell LM, Gallo MV, Denham M, Ravenscroft J, DeCaprio AP, Carpenter DO. Relationship of thyroid hormone levels to levels of polychlorinated biphenyls, lead, p, p'-DDE, and other toxicants in Akwesasne Mohawk youth. *Environ Health Perspect* 2008;**116**:806—13.

189. Swan SH, Main KM, Liu F, Stewart SL, Kruse RL, Calafat AM, et al. Decrease in anogenital distance among male infants with prenatal phthalate exposure. *Environ Health Perspect* 2005;**113**:1056—61.

190. Main KM, Mortensen GK, Kaleva MM, Boisen KA, Damgaard IN, Chellakooty M, et al. Human breast milk contamination with phthalates and alterations of endogenous reproductive hormones in infants three months of age. *Environ Health Perspect* 2006;**114**:270—6.

191. Zhang Y, Lin L, Cao Y, Chen B, Zheng L, Ge RS. Phthalate levels and low birth weight: a nested case—control study of Chinese newborns. *J Pediatr* 2009;**155**:500—4.

192. Huang AT, Batterman S. Formation of trihalomethanes in foods and beverages. *Food Addit Contam Part A Chem Anal Control Expo Risk Assess* 2009;**26**:947—57.

193. Wolff MS, Engel SM, Berkowitz GS, Ye X, Silva MJ, Zhu C, et al. Prenatal phenol and phthalate exposures and birth outcomes. *Environ Health Perspect* 2008;**116**:1092—7.

194. Durmaz E, Ozmert EN, Erkekoglu P, Giray B, Derman O, Hincal F, et al. Plasma phthalate levels in pubertal gynecomastia. *Pediatrics* 2010;**125**:e122—9.

195. Den Hond E, Roels HA, Hoppenbrouwers K, Nawrot T, Thijs L, Vandermeulen C, et al. Sexual maturation in relation to polychlorinated aromatic hydrocarbons: Sharpe and Skakkebaek's hypothesis revisited. *Environ Health Perspect* 2002;**110**:771—6.

196. Vasiliu O, Muttineni J, Karmaus W. In utero exposure to organochlorines and age at menarche. *Hum Reprod* 2004;**19**:1506—12.

197. Blanck HM, Marcus M, Rubin C, Tolbert PE, Hertzberg VS, Henderson AK, et al. Growth in girls exposed in utero and postnatally to polybrominated biphenyls and polychlorinated biphenyls. *Epidemiology* 2002;**13**:205—10.

198. Meeker JD, Johnson PI, Camann D, Hauser R. Polybrominated diphenyl ether (PBDE) concentrations in house dust are related to hormone levels in men. *Sci Total Environ* 2009;**407**:3425–9.

199. Faroon OM, Keith S, Jones D, de Rosa C. Effects of polychlorinated biphenyls on development and reproduction. *Toxicol Ind Health* 2001;**17**:63–93.

200. Chao HR, Wang SL, Lee WJ, Wang YF, Papke O. Levels of polybrominated diphenyl ethers (PBDEs) in breast milk from central Taiwan and their relation to infant birth outcome and maternal menstruation effects. *Environ Int* 2007;**33**:239–45.

201. Main KM, Skakkebaek NE, Virtanen HE, Toppari J. Genital anomalies in boys and the environment. *Best Pract Res Clin Endocrinol Metab* 2010;**24**:279–89.

202. Hany J, Lilienthal H, Sarasin A, Roth-Harer A, Fastabend A, Dunemann L, et al. Developmental exposure of rats to a reconstituted PCB mixture or aroclor 1254: effects on organ weights, aromatase activity, sex hormone levels, and sweet preference behavior. *Toxicol Appl Pharmacol* 1999;**158**:231–43.

203. Ahmad SU, Tariq S, Jalali S, Ahmad MM. Environmental pollutant Aroclor 1242 (PCB) disrupts reproduction in adult male rhesus monkeys (*Macaca mulatta*). *Environ Res* 2003;**93**:272–8.

204. Schell LM. Pollution and human growth: lead, noise, polychlorobiphenyl compounds and toxic wastes. In: Mascie-Taylor CG, Lasker GW, editors. *Applications of biological anthropology to human affairs*. Cambridge: Cambridge University Press; 1991. p. 83–116.

205. Pietrzyk JJ, Nowak A, Mitkowska Z, Zachwieja Z, Chlopicka J, Krosniak M, et al. Prenatal lead exposure and the pregnancy outcome. A case–control study in southern Poland. *Przegl Lek* 1996;**53**: 342–7.

206. Falcon M, Vinas P, Luna A. Placental lead and outcome of pregnancy. *Toxicology* 2003;**185**:59–66.

207. Awasthi S, Awasthi R, Srivastav RC. Maternal blood lead level and outcomes of pregnancy in Lucknow, north India. *Indian Pediatr* 2002;**39**:855–60.

208. Andrews KW, Savitz DA, Hertz-Picciotto I. Prenatal lead exposure in relation to gestational age and birth weight: A review of epidemiologic studies. *Am J Ind Med* 1994;**26**.13–32.

209. Bornschein RL, Grote J, Mitchell T, Succop PA, Dietrich KN, Krafft KM, et al. Effects of prenatal lead exposure on infant size at birth. In: Smith MA, Grant L, Sors AI, editors. *Lead exposure and child development*. Boston, MA: Kluwer; 1989. p. 307–19.

210. Schell LM, Stark AD. Pollution and child health. In: Schell LM, Ulijaszek SJ, editors. *Urbanism, health and human biology in industrialised countries*. Cambridge: Cambridge University Press; 1999. p. 136–57.

211. Rothenberg SJ, Schnaas-Arrieta L, Perez-Guerrero IA, Perroni-Hernandez E, Mercado-Torres L, Gomez-Ruiz C, et al. Prenatal and postnatal blood lead level and head circumference in children to three years: preliminary results from the Mexico City Prospective Lead Study. *J Expo Anal Environ Epidemiol* 1993;**3**(Suppl. 1):165–72.

212. Zhu M, Fitzgerald EF, Gelberg KH, Lin S, Druschel CM. Maternal low-level lead exposure and fetal growth. *Environ Health Perspect* 2010;**118**:1471–5.

213. Schwartz J, Angle CR, Pitcher H. Relationship between childhood blood lead levels and stature. *Pediatrics* 1986;**77**:281–8.

214. Frisancho AR, Ryan AS. Decreased stature associated with moderate blood lead concentrations in Mexican-American children. *Am J Clin Nutr* 1991;**54**:516–9.

215. Ballew C, Khan LK, Kaufmann R, Mokdad A, Miller DT, Gunter EW. Blood lead concentration and children's anthropometric dimensions in the Third National Health and Nutrition Examination Survey (NHANES III) 1988–1994. *J Pediatr* 1999;**134**:623–30.

216. Ignasiak Z, Slawinska T, Rozek K, Little BB, Malina RM. Lead and growth status of school children living in the copper basin of south-western Poland: differential effects on bone growth. *Ann Hum Biol* 2006;**33**:401–14.

217. Little BB, Snell LM, Johnston WL, Knoll KA, Buschang PH. Blood lead levels and growth status of children. *Am J Hum Biol* 1990;**2**:265–9.

218. Kafourou A, Touloumi G, Makropoulos V, Loutradi A, Papanagioutou A, Hatzakis A. Effects of lead on the somatic growth of children. *Arch Environ Health* 1997;**52**:377–83.

219. Lauwers M-C, Hauspie RC, Susanne C, Verheyden J. Comparison of biometric data of children with high and low levels of lead in the blood. *Am J Phys Anthropol* 1986;**69**:107–16.

220. Little BB, Spalding S, Walsh B, Keyes DC, Wainer J, Pickens S, et al. Blood lead levels and growth status among African-American and Hispanic children in Dallas, Texas G, 1980 and 2002: Dallas Lead Project II. *Ann Hum Biol* 2009;**36**:331–41.

221. Shukla R, Bornschein RL, Dietrich KN, Mitchell T, Grote J, Berger OG, et al. Effects of fetal and early postnatal lead exposure on child's growth in stature — the Cincinnati Lead Study. In: Lindberg S, Hutchinson T, editors. *Heavy metals in the environment.* Edinburgh: CEP Consultants; 1987. p. 210–2.

222. Shukla R, Bornschein RL, Dietrich KN, Buncher CR, Berger OG, Hammond PB, et al. Fetal and infant lead exposure: effects on growth in stature. *Pediatrics* 1989;**84**:604–12.

223. Shukla R, Dietrich KN, Bornschein RL, Berger O, Hammond PB. Lead exposure and growth in the early preschool child: a follow-up report from the Cincinnati Lead Study. *Pediatrics* 1991;**88**:886–92.

224. Schell LM, Denham M, Stark AD, Parsons PJ, Schulte EE. Growth of infants' length, weight, head and arm circumferences in relation to low levels of blood lead measured serially. *Am J Hum Biol* 2009;**21**:180–7.

225. Sanin LH, Gonzalez-Cossio T, Romieu I, Peterson KE, Ruiz S, Palazuelos E, et al. Effect of maternal lead burden on infant weight and weight gain at one month of age among breastfed infants. *Pediatrics* 2001;**107**:1016–23.

226. Kim R, Hu H, Rotnitzky A, Bellinger DC, Needleman HL. A longitudinal study of chronic lead exposure and physical growth in Boston children. *Environ Health Perspect* 1995;**103**:952–7.

227. Wu T, Buck GM, Mendola P. Blood lead levels and sexual maturation in US girls: the third national health and nutrition examination survey, 1988–1994. *Environ Health Perspect* 2003;**111**:737–41.

228. Selevan SG, Rice DC, Hogan KA, Euling SY, Pfahles-Hutchens A, Bethel J. Blood lead concentration and delayed puberty in girls. *N Engl J Med* 2003;**348**:1527–36.

229. Naicker N, Norris SA, Mathee A, Becker P, Richter L. Lead exposure is associated with a delay in the onset of puberty in South African adolescent females: findings from the Birth to Twenty cohort. *Sci Total Environ* 2010;**408**:4949–54.

230. Williams PL, Sergeyev O, Lee MM, Korrick SA, Burns JS, Humblet O, et al. Blood lead levels and delayed onset of puberty in a longitudinal study of Russian boys. *Pediatrics* 2010;**125**:e1088–96.

231. Denham M, Schell LM, Deane G, Gallo MV, Ravenscroft J, DeCaprio A. Akwesasne Task Force on the Environment. Relationship of lead, mercury, mirex, dichlorodiphenyldichloroethylene, hexachlorobenzene, and polychlorinated biphenyls to timing of menarche among Akwesasne Mohawk girls. *Pediatrics* 2005;**115**:e127–34.

232. Danker-Hopfe H, Hulanicka B. Maturation of girls in lead polluted areas. In: Hauspie R, Lindgren G, Falkner F, editors. *Essays on auxology. Welwyn Garden City: Castlemead*; 1995. p. 334–42.

233. Huseman CA, Varma MM, Angle CR. Neuroendocrine effects of toxic and low blood lead levels in children. *Pediatrics* 1992;**90**:186–9.

234. Ben Arush MW, Elhasid R. Effects of radiotherapy on the growth of children with leukemia. *Pediatr Endocrinol Rev PER* 2008;**5**:785–8.

235. Meyer MB, Tonascia JA. Long-term effects of prenatal X-ray of human females. I. Reproductive experience. *Am J Epidemiol* 1981;**114**:304–16.

236. Brent RL. Effects of ionizing radiation on growth and development. In: Klingberg MA, Weatherall JAC, Papier C, editors. *Epidemiologic methods for detection of teratogens.* New York: S. Karger; 1979. p. 147–83.

237. Burrow GN, Hamilton HB, Hrubec Z. *Study of adolescents exposed in utero to the atomic bomb*, Nagasaki: Japan. JAMA; 1965;**192**:97–104.

238. Sutow WW, Conard RA, Griffith KM. Growth status of children exposed to fallout radiation on Marshall Islands. *Pediatrics* 1965;**36**:721–31.

239. Nakashima E, Carter RL, Neriishi K, Tanaka S, Funamoto S. Height reduction among prenatally exposed atomic-bomb survivors: a longitudinal study of growth. *Health Phys* 1995;**68**:766–72.

240. Soderqvist F, Carlberg M, Hardell L. Use of wireless telephones and self-reported health symptoms: a population-based study among Swedish adolescents aged 15–19 years. *Environmental Health* 2008;**7**:18.

241. Hartikainen A-L, Sorri M, Anttonen H, Tuimala R, Laara E. Effect of occupational noise on the course and outcome of pregnancy. *Scand J Environ Health* 1994;**20**:444–50.

242. Hartikainen-Sorri A-L, Kirkinen P, Sorri M, Anttonen H, Tuimala R. No effect of experimental noise on human pregnancy. *Obstet Gynecol* 1991;**77**:611−5.

243. Wu T-N, Chen L-J, Lai J-S, Ko G-N, Shen C-Y, Chang P- Y. Prospective study of noise exposure during pregnancy on birth weight. *Am J Epidemiol* 1996;**143**:792−6.

244. Ando Y. Effects of daily noise on fetuses and cerebral hemisphere specialization in children. *J Sound Vib* 1988;**127**:411−7.

245. Rehm S, Jansen G. Aircraft noise and premature birth. *J Sound Vib* 1978;**59**:133−5.

246. Schell LM. The effects of chronic noise exposure on human prenatal growth. In: Borms J, Hauspie R, Sand A, Susanne C, Hebbelinck M, editors. *Human growth*. New York: Plenum Press; 1982. p. 125−9.

247. Schell LM. Environmental noise and human prenatal growth. *Am J Phys Anthropol* 1981;**56**:63−70.

248. Coblentz A, Martel A. Effects of fetal exposition to noise on the birth weight of children (abstract). *Am J Phys Anthropol* 1986;**69**:188.

249. Ando Y, Hattori H. Statistical studies on the effects of intense noise during human fetal life. *J Sound Vib* 1973;**27**:101−10.

250. Knipschild P, Meijer H, Sallé H. Aircraft noise and birth weight. *Int Arch Occup Environ Health* 1981;**48**:131−6.

251. Schell LM, Norelli RJ. Airport noise exposure and the postnatal growth of children. *Am J Phys Anthropol* 1983;**61**:473−82.

252. Schell LM, Ando Y. Postnatal growth of children in relation to noise from Osaka airport. *J Sound Vib* 1991;**151**:371−82.

SUGGESTED READING

High Altitude

Frisancho (1993)[25] is a classic introduction to high altitude adaptation, and Hall and Guyton (2011)[26] gives biomedical background on the physiology of respiration. Lampl (2005)[45] outlines theoretical debates over prenatal growth. Beall (2000)[77] is an excellent comparison of Andean and Tibetan adaptation. The quarterly journal High Altitude Medicine and Biology, available online, offers periodic reviews of problems confronting high-altitude populations or visitors.

Sleep

These articles are excellent examples of recent research:

Iglowstein I, et al. Sleep duration from infancy to adolescence: reference values and generational trends. *Pediatrics* 2003;**111**:302−7.

Van Cauter E, Spiegel K. Circadian and sleep control of hormonal secretions. In: Turek FW, Zee PC, editors. *Regulation of sleep and circadian rhythms*. New York: Marcel Dekker; 1999. p. 397−425.

Pollution

The journal Environmental Health Perspectives is a good source regarding pollution and health. For a review of polychlorinated biphenyls and health see:

Carpenter DO. Polychlorinated biphenyls (PCBs): routes of exposure and effects on human health. *Rev Environ Health* 2006;**21**:1−23.

For a recent review of pollution and human biology see any of these:

Schell LM. Industrial pollutants and human evolution. In: Muehlenbein MP, editor. *Human evolutionary biology*. Cambridge: Cambridge University Press; 2011. p. 566−80.

Schell LM, Burnitz KK, Lathrop PW. Pollution and human biology. *Ann Hum Biol* 2011;**3**:347−66.

Schell LM. In: Mascie-Taylor N, Yasukouchi A, Ulijaszek S, editors. *Human variation: from the laboratory to the field, Series. Impact of pollution on physiological systems: taking science from the laboratory to the field*, vol. 48. London: Taylor and Francis; 2010. p. 131−41.

INTERNET RESOURCES

Toxic substances and health: http://www.atsdr.cdc.gov/
Specific toxicants: http://www.atsdr.cdc.gov/toxprofiles/index.asp
Toxicants and the environment, including some information on human health effects: http://www.epa.gov/
Noise pollution: http://www.nonoise.org/
Sleep: http://www.sleepfoundation.org/
High altitude: http://www.altitude.org

CHAPTER 11

The Evolution of Human Growth

Barry Bogin
School of Sport, Exercise & Health Sciences, Loughborough University, Leicestershire LE11 3TU, UK

Contents

11.1 INTRODUCTION

If there is a "secret" to life, it is hidden in the process that converts a single cell, with its complement of deoxyribonucleic acid (DNA) into a multicellular organism composed of hundreds of different tissues, organs, behavioral capabilities and emotions. That process is no less wondrous when it occurs in an earthworm, a whale or a human being. This chapter will focus on the process of human growth and development; however, the reader must be made aware that much of what we know about human growth is derived

N.Cameron & B. Bogin (eds): Human Growth and Development, Second edition.
ISBN 978-0-12-383882-7, Doi: 10.1016/B978-0-12-383882-7.00011-8

from research on non-human animals. One reason for this is ethical limits on the kind of experimental research that may be performed on human beings. Another reason is the evolutionary history that connects all living organisms, meaning that growth processes occurring in non-human animals are often, but not always, identical or similar to those occurring in people.

Powerful genomic evidence for the common evolutionary origin of animal development came in 1984 with the discovery of the genomic homeobox.[1] The most common definition of a homeobox is a highly conserved sequence of about 180 DNA base pairs that code for a 60 amino acid segment of protein which regulates patterns of development. Genes containing homeoboxes are found in all eukaryotic genomes and are associated with cell differentiation and bodily segmentation during embryological development. Additional evidence for common evolutionary origins was published in 1995 with the identification of PAX6 as a master control gene for eye development in virtually all organisms that possess one or more eyes.[2–4] This discovery led to a new hypothesis about the monophyletic origin of the eye in evolution. The PAX6 eye gene is common to species as diverse as marine worms, squid, fruit flies, mice and humans. Other aspects of animal growth and development are not shared among all species. The unique features of the life cycle of different species, such as metamorphosis in insects and amphibians, or childhood and adolescence in human beings, attest to the ongoing evolution of life on earth.

Biological evolution is the continuous process of genetic and genomic adaptation of organisms to their environments. (As used here, the word "genetic" refers to structural genes coding for polypeptide chins of amino acids. The word "genomic" refers to the structural and regulatory DNA, epigenetic modification of the DNA and chromosomes, as well as all cellular factors which regulate gene expression.) Natural selection determines the direction of evolutionary change and operates by differential mortality between individual organisms prior to reproductive maturation and by differential fertility of mature organisms. Genetic, genomic and phenotypic adaptations that enhance the survival of individuals to reproductive age (defined as *fitness*), and that increase the production of similarly successful offspring, will increase in frequency in a population. The unique stages and events of human growth and development evolved because they conferred reproductive advantages to our species.

11.2 LIFE HISTORY AND STAGES OF THE LIFE CYCLE

Life history theory is a branch of biology that studies the selective forces that have guided the evolution of the schedule and duration of key events in an organism's lifetime, as these relate to investments in growth, reproduction and survivorship. Variation between species in these key events constitutes evolutionarily derived strategies to promote increased reproductive fitness. Each species has its own life history, that is, a pattern of allocation of energy towards growth, maintenance, reproduction, raising offspring to independence and avoiding death. It is important to emphasize that this pattern, or life history strategy, is not conscious or planned by the individual, groups of individuals or the species. Rather, life history constitutes the set of biological and behavioral traits that characterize a species and which evolved via a series of *trade-offs*, that is, costs to an organism's fitness in one trait due to investment in another trait.[5–9] An example of a trade-off between fertility rate and female adult body size in placental mammals is given in Figure 11.1. Fertility tends to decline as adult female body size increases, a negative correlation. This is the expected relationship because investment of energy and time to grow a larger body comes at the expense of investment in the next generation. Human beings are the exception and show a positive correlation between women's body size and fertility. Why this is so is not completely known, but is likely to relate to at least two novel characteristics: (1) the evolution of the human childhood and adolescence; and (2) the evolution of human biocultural systems of food exchange and reproduction. As will be explained in this chapter, the human life history stages of childhood and adolescence, stages not found in other species of mammals, enhance the reproductive success of

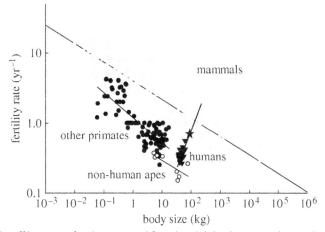

Figure 11.1 Trade-off between fertility rate and female adult body size in placental mammals. Fertility tends to decline as adult female body size increases, a negative correlation. Human beings are the exception and show a positive correlation between body size and fertility. The data are for 610 species of "mammals", excluding primates and bats; 101 species of "other primates", including lemurs, lorises, tarsiers, New World and Old World monkeys; 13 species of "non-human apes", including gibbons (nine species), orang-utans, chimpanzees, bonobos and gorillas (one species each); and 17 human societies including hunter–gatherers, horticulturalists, one pastoral group, and the Hutterites (star symbol), a Christian sect of North America. All the human societies are natural fertility populations, meaning that they use no effective contraception. Note that the non-human primates and apes are down-shifted, meaning that they have a slower life history including a longer period of growth and a later age at first reproduction than other mammals of the same body size. (Source: *Modified from Walker et al.*[10])

women. Similarly, human biocultural systems of cooperative food production and food exchange increase the energy available to support both body growth and greater fertility of the mother.

Living things on earth have greatly different life history strategies, and understanding what shapes these is one of the most active areas of research in whole-organism biology. Anthropologists, human biologists, physicians and others have become increasingly interested in explaining the significance of human life history. This interest is due to the discovery that several aspects of the human life cycle stand in sharp contrast to other species of social mammals, even other primates (Figure 11.1 is only one example of a contrast). Evolutionary theory needs to explain how humans successfully combined a vastly extended period of offspring dependency and delayed reproduction with helpless newborns, a short duration of breast feeding, an adolescent growth spurt and menopause. A central question is, did these characteristics evolve as a package or a mosaic? The present evidence suggests that human life history evolved as a mosaic and may have taken form over the past two million years.

Understanding the human condition requires a comparative approach, and here consideration is restricted to the mammals. We may use the same criteria to describe and define the stages of the life cycle for non-human mammals and the human species. Research on mammalian life history and its evolution spans the entire life cycle of an individual from conception to adulthood and extends across generations.[11,12] In this chapter the discussion is focused on the postnatal stages of the life cycle. Research with humans is especially concerned with changes in the rate of growth after birth and the timing of the onset of reproductive maturation (sometimes called "adulthood"). This is due to the fact that the majority of mammals progress from infancy to adulthood seamlessly, without any intervening stages. Furthermore, most mammals experience puberty, which is defined below as a set of neuroendocrine changes in the brain, after the peak velocity of postnatal growth. In contrast, human growth includes several stages between infancy and adulthood, and puberty occurs before the peak velocity of postnatal growth.

This typical mammalian pattern of postnatal growth is illustrated in Figure 11.2 using data for the mouse. Mammalian species living in highly social groups, such as wolves, wild dogs, lions, elephants and primates (e.g. the baboon, Figure 11.3) postpone adulthood by inserting a period of juvenile growth and behavior between infancy and adulthood.

Most mammalian biologists define *infancy* as the stage of life when the offspring are fed by nursing. *Adulthood* is the stage of life when the individual is capable of successful reproduction. The use of the word "successful" means that pregnancy, birthing and infant care are all possible and do not pose exceptional risk to the health of the mother or her offspring. *Juveniles* may be defined as "prepubertal individuals that are no longer dependent on their mothers (parents) for survival".[14] This definition is derived from

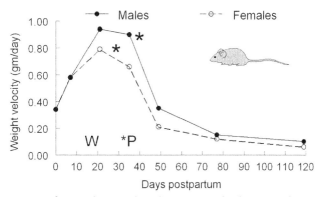

Figure 11.2 Velocity curves for weight growth in the mouse. In both sexes puberty (*; vaginal opening for females or spermatocytes in testes of males) occurs just after weaning (W) and maximal growth rate. Weaning takes place between days 15 and 20. Sexual maturity follows weaning by a matter of days.

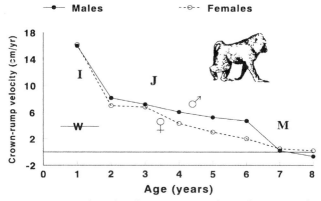

Figure 11.3 Baboon crown—rump length velocity. Letters indicate the stages of growth, I: infancy; J: juvenile; M: mature adult. Weaning (W) may take place any time between 6 and 18 months. (Source: *Adapted from Coelho.*[13])

ethological research with social mammals, especially non-human primates, but applies to the human species as well. The phrase "no longer dependent on their mothers" means that the young are weaned from feeding by lactation. However, some juvenile weanlings are, to a greater or lesser extent, dependent on their parents for food. Examples are lions, wild dogs, hyenas and a few species of primates, such as marmosets and tamarins. A common feature of these mammals is that they hunt prey and it takes time for the weaned juveniles to acquire independence in hunting skills. Even so, the major difference between all infant and juvenile mammals is that it is possible for the juveniles to survive the death of their adult caretakers. Ethnographic research shows that juvenile humans have the physical and cognitive abilities to provide much of their own food and to protect themselves from predation and disease. Younger infants and children cannot survive without the assistance of adolescent or adult caretakers. Humans have evolved a *childhood* stage of development, inserted between the infant and juvenile stages, and an *adolescence* stage between the juvenile and adult stages. More detailed definitions of human life history stages are given below.

11.3 HUMAN LIFE STAGES

It is here proposed that the stages of human life history between birth and adulthood are: infancy, childhood, juvenile and adolescent. Reproduction is best suited to take place during adulthood. Human women who live long enough universally experience menopause, an event that ends the reproductive capacity of the women but may mark the passage to a new stage of adult life that includes assisting the reproduction of younger women.

To visualize the amount and rate of growth that takes place during each of these stages, the growth in height (or length) for normal boys and girls is depicted in Figure 11.4; growth in weight follows very similar curves. The stages of growth are also outlined in Table 11.1. In Figure 11.4 the distance curve of growth, that is, the amount of

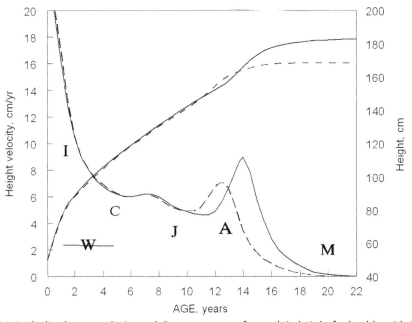

Figure 11.4 Idealized mean velocity and distance curves of growth in height for healthy girls (dashed lines) and boys (solid lines) showing the postnatal stages of the pattern of human growth. Note the spurts in growth rate at mid-childhood and adolescence for both girls and boys. Postnatal stages: I: infancy; C: childhood; J: juvenile; A: adolescence; M: mature adult; —W— : range of age for human weaning.[15,16]

Table 11.1 Stages in the human life cycle

Stage		Growth events/duration (approximate or average)
Prenatal life		
Fertilization		
	First trimester	Fertilization to 12th week: embryogenesis
	Second trimester	Fourth to sixth lunar month: rapid growth in length
	Third trimester	Seventh lunar month to birth: rapid growth in weight and organ maturation
Birth		
Postnatal life		
	Neonatal period	Birth to 28 days: extrauterine adaptation, most rapid rate of postnatal growth and maturation
	Infancy	Second month to end of lactation, usually by 36 months: rapid growth velocity, but with steep deceleration in growth rate, feeding by lactation, deciduous tooth eruption, many developmental milestones in physiology, behavior and cognition
	Childhood	Years 3—7: moderate growth rate, dependency on older people for care and feeding, mid-growth spurt, eruption of first permanent molar and incisor, cessation of brain growth by end of stage
	Juvenile	Years 7—10 for girls, 7—12 for boys: slower growth rate, capable of self-feeding, cognitive transition leading to learning of economic and social skills
	Puberty	Occurs at end of juvenile stage and is an event of short duration (days or a few weeks): reactivation of central nervous system of sexual development, dramatic increase in secretion of sex hormones
	Adolescence	The stage of development that lasts for 5—10 years after the onset of puberty: growth spurt in height and weight; permanent tooth eruption almost complete; development of secondary sexual characteristics; sociosexual maturation; intensification of interest in and practice of adult social, economic and sexual activities
Adulthood		
	Prime and transition	From 20 years old to end of child-bearing years: homeostasis in physiology, behavior and cognition; menopause for women by age 50
	Old age and senescence	From end of child-bearing years to death: decline in the function of many body tissues or systems

growth achieved from year to year, is labeled on the right y-axis. The velocity curve, which represents the rate of growth during any one year, is labeled on the left y-axis. Below the velocity curve are symbols that indicate the average duration of each stage of development. Clearly, changes in growth rate are associated with each stage of development. Each stage also may be defined by characteristics of the dentition, changes related to methods of feeding, physical and mental competencies, or maturation of the reproductive system and sexual behavior.

11.3.1 Infancy

Infancy is characterized by the most rapid velocity of growth of any of the postnatal stages. The infant's rate of growth is also characterized by a steep deceleration in velocity. As for all mammals, human infancy is the period when the mother provides all or some nourishment to her offspring via lactation or some culturally derived imitation of lactation. During infancy, the deciduous dentition (the so-called milk teeth) erupts through the gums. Human infancy ends when the child is weaned from the breast, which in preindustrialized societies occurs between 24 and 36 months of age.[17] By this age, all the deciduous teeth have erupted, even for very late-maturing infants. Complementary foods, which supplement mother's milk, are introduced to the infant's diet sometime before the age of 1 year after birth, but nursing provides important nutrients, immune substances and emotional support to the infant.[17]

Motor skills (i.e. what a baby can do physically) develop rapidly during infancy. Developmental milestones for sitting, standing and walking are shown in Figure 11.5. These motor development milestones are averages for human infants. Some infants may skip a stage, such as "hands-and-knees crawling", all of the stages overlap, and each stage is quite variable in duration. On average, by 3 years of age, the end of infancy, the youngster can run short distances, pour water from a pitcher, and manipulate small objects, such as blocks, well enough to control them. There is a similar progression of changes in the problem-solving, or cognitive, abilities of the infant. The development of the skeleton, musculature and nervous system accounts for all of these motor and cognitive advances. The rapid growth of the brain, in particular, is important. The human brain grows more rapidly during infancy than any of the other tissues or organs of the body depicted in Figure 11.6.

11.3.2 Childhood

The childhood stage follows infancy, encompassing the ages of about 3−6.9 years. Childhood may be defined by its own pattern of growth, feeding behavior, motor skills and cognitive development. The growth deceleration of infancy ends at the beginning of childhood, and the growth rate levels off at around 5−6 cm per year. This leveling-off in growth rate is unusual for mammals, because almost all other species continue a pattern of deceleration after infancy (Figures 11.2 and 11.3).

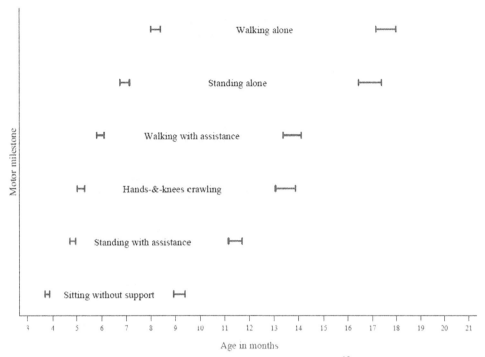

Figure 11.5 Windows of achievement for six gross motor milestones.[18] This figure is reproduced in the color plate section.

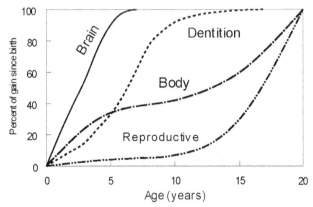

Figure 11.6 Growth curves for different body tissues. The "Brain" curve is for total weight of the brain.[19] The "Dentition" curve is the median maturity score for girls based on the seven left mandibular teeth (I1, I2, C, PM1, PM2, M1, M2).[20] The "Body" curve represents growth in stature or total body weight, and the "Reproductive" curve represents the weight of the gonads and primary reproductive organs.[21]

This slow and steady rate of human growth maintains a relatively small-sized body during the childhood years. In terms of feeding, children are weaned from the breast or bottle but still depend on older people for specially prepared food and protection. Most mammalian species move from infancy and its association with dependence on nursing to a stage of independent feeding. Postweaning dependency is, by itself, not a sufficient criterion to define human childhood, as several species of social mammals, especially carnivores (such as lions, wild dogs and hyenas) and some species of primates also have this. Human childhood is defined by a suite of features, not all of which are found for the social carnivores and non-human primates. A list of many of these features is given in Box 11.1. Human children require specially prepared foods because of the immaturity of their dentition, the small size of their stomachs and intestines, and the rapid growth of their brain (Figure 11.6). The human brain is especially important. The human newborn uses 87% of its resting metabolic rate (RMR) for brain growth and function.[22] By the age of 5 years, the percentage RMR usage is still high at 44%, whereas in the adult human, the figure is between 20 and 25% of RMR. At comparable stages of development, the RMR values for the chimpanzee are about 45, 20 and 9%, respectively.

The human constraints of immature dentition and small digestive system necessitate a childhood diet that is easy to chew and swallow and low in total volume. The child's relatively large and active brain, which is about four times the size of an adult chimpanzee's brain (Figure 11.7), requires that the low-volume diet be dense in carbohydrates, lipids and proteins, all of which may provide energy (glucose) for the brain. Children do not yet have the motor and cognitive skills to prepare such a diet for themselves. Children are also especially vulnerable to predation and disease and thus require protection. Children will not survive in *any* society if deprived of the care provided by older individuals. So-called wolf children and even street children, who are sometimes alleged to have lived on their own, are either myths or not children at all. A search of the literature finds no case of a child (i.e. a youngster under the age of 6.5 years) living alone, either in the wild or on urban streets.

Box 11.1 Human childhood: a list of defining traits — no other mammalian species has this entire suite of features

- Slow and steady rate of growth and relatively small body size
- A large, fast-growing brain
- Higher RMR than any other mammalian species
- Immature dentition
- Motor immaturity
- Cognitive immaturity
- Both adrenarche and the mid-growth spurt

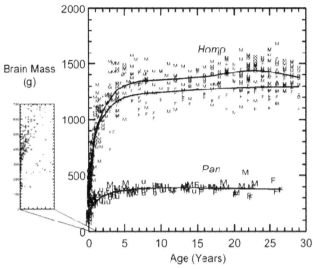

Figure 11.7 Brain-mass growth data for humans (*Homo sapiens*) and chimpanzees (*Pan troglodytes*). Brain mass increases during the postnatal period in both species. Lines represent best-fit Lowess regressions through the data points. M: males; F: females; U: sex unidentified. The human regressions separate into male (upper) and female (lower) curves. The inset shows brain-mass growth for each species during the first postnatal year. *(Source: Leigh.[23] Reproduced with kind permission of the author.)*

Two of the important physical developmental milestones of childhood are the replacement of the deciduous teeth with the eruption of the first permanent teeth, and completion of brain growth (in weight). First molar eruption takes place, on average, between the ages of 5.5 and 6.5 years in most human populations. Eruption of the central incisor quickly follows, or sometimes precedes, the eruption of the first molar. By the transition to the juvenile stage, usually at the age of 7 years, most juveniles have their first molars in occlusion and permanent incisors have begun to replace "milk" incisors. Along with growth in size and strength of the jaws and the muscles for chewing, these new teeth provide sufficient capabilities to eat a diet similar to that of adults. At this stage of development, not only is the new juvenile capable dentally of processing an adult–type diet but also the nutrient requirements for brain growth diminish as the rate of brain growth slows (Figure 11.7). Moreover, cognitive and emotional capacities mature to new levels of self-sufficiency. Language and symbolic thinking skills mature rapidly, social interaction in play and learning becomes common, and the 7-year-old individual can perform many basic tasks, including food preparation, with little or no supervision.

Another feature of the childhood phase of growth associated with these physical and mental changes is the modest acceleration in growth velocity at about 6–8 years, called the mid-growth spurt (shown in Figure 11.4). The mid-growth spurt has been linked with an endocrine event called adrenarche, the progressive increase in the secretion of adrenal androgen hormones.[24] It is unclear whether adrenal androgens produce the mid-growth spurt in height, but these androgens seem to cause the appearance of a small amount of axillary and pubic hair. It is hypothesized that the increase in adrenal hormones during the juvenile stage may help to maintain human brain development.[25] The mechanism controlling adrenarche is not understood because no known hormone appears to cause it. Whatever its cause, a mid-growth spurt has not been detected in chimpanzees or other primates, and it may be unique to humans.

Adrenal androgens also seem to regulate the development of body fatness and fat distribution, and in humans adrenarche has been related to the "adiposity rebound" at the transition between the childhood and juvenile stages of the life cycle. The adiposity rebound describes the increase in body fatness that takes place between the ages of 5 and 7 years. The adiposity rebound is usually measured as the body mass index (BMI), which is calculated as (weight in kilograms/height in meters^2). In many human populations the BMI is a reasonable proxy measure of body fatness. From late infancy though childhood, there is usually a decline in BMI, but then in late childhood fatness begins to increase.[26,27] There is some evidence that early adiposity rebound leads to a greater risk for later life overweight and obesity.

The physical changes induced by adrenarche are accompanied by a change in cognitive function, called the "5- to 7-year-old shift" by some psychologists, or the shift from the preoperational to concrete operational stage, using the terminology of Piaget. This shift leads to new learning and work capabilities in the juvenile. Adrenarche and the human mid-growth spurt may function as a life history event, marking a transition from the childhood to the juvenile growth stage which is visible to other members of human social groups.

11.3.3 Juvenile

The juvenile stage is characterized by a deceleration in rate of growth in height and the slowest rate of growth since birth. The human juvenile stage begins at about 7 years old. In girls, the juvenile period ends, on average, at about the age of 10, 2 years before it usually ends in boys, the difference reflecting the earlier onset of adolescence in girls.

The evolution of the juvenile stage of primates is associated with both social complexity, especially larger social groups, and dietary complexity, including foraging for fruits and seeds.[28] Studies of juvenile primates and human juveniles in many cultures indicate that much social learning takes place during this stage, and a "learning hypothesis" has often been proposed to account for the evolution of the juvenile stage. It has also been argued that the primary reason for the evolution of juvenility appears to be a strategy to avoid death from competition with older individuals while living in a social group.[29] Another possible explanation for a juvenile stage for social mammals may be called the "dominance hypothesis". Research in wild and captive primates shows that high-ranking individuals in the social hierarchy can suppress and inhibit the reproductive maturation of low-ranking individuals. The inhibition may be due to the stress of social intimidation acting directly on the endocrine system, or the suppression may be secondary to inadequate nutrition due to feeding competition. Juveniles are almost always low-ranking members of primate social systems. In the past, individuals with slow growth and delayed reproductive maturation after infancy may have survived to adulthood more often than individuals with rapid growth and maturation, and thus the juvenile stage may have evolved. Whatever the cause, Alexander[30] points out that in broad perspective, "juvenile life has two main functions: to get to the adult stage without dying and to become the best possible adult".

Human juvenile boys and girls in traditional societies (that is, foragers, horticulturalists, pastoralists and non-mechanized agriculturalists) learn a great deal about important adult activities, including the production of food and methods of infant and child care. Juveniles of industrialized societies may learn some of these skills and also spend a great deal of time in schools and other systems of formal education (e.g. sport and music instruction). The completion of growth in weight of the brain and the onset of new cognitive competencies allow for this increased intensity of juvenile learning. In addition, because juveniles are prepubertal, they can attend to this kind of social learning without the distractions caused by sexual maturation.

11.3.4 Adolescence

Human adolescence is the stage of life when social and sexual maturation takes place. Adolescence begins with puberty, or more technically with gonadarche, which is the final "on" of the on—off—on pattern of the gonadotrophin-releasing hormone (GnRH) pulse generator of the hypothalamus (Figure 11.8a). The transition from the juvenile to the adolescent stage requires not only the renewed production of GnRH but also its secretion from the hypothalamus in a specific frequency and amplitude of pulses (Figure 11.8b).

None of these hormonal changes can be seen without sophisticated technology, but the effects of gonadarche can be noted easily as visible and audible signs of sexual maturation. One such sign is a sudden increase in the density of pubic hair (indeed, the term "puberty" is derived from the Latin *pubescere*, "to grow hairy"). In boys, the increased density and darkening of facial hair and the deepening of the voice (voice "breaking" or "cracking") are other signs of puberty. In girls, visible signs are pubic hair and the development of the breast bud, the first stage of breast development. The pubescent boy or girl, his or her parents, and relatives, friends, and sometimes everyone else in the social group, can observe one or more of these signs of early adolescence.

The adolescent stage also includes development of secondary sexual characteristics, such as development of the external genitalia, sexual dimorphism in body size and composition (Figure 11.9), and the onset of greater interest and practice of adult patterns of sociosexual and economic behavior. These physical and behavioral changes occur with puberty in many species of social mammal. What makes human adolescence unusual among the primates are two important differences. The first is the length of time between age at puberty and age at first birth. Humans take, on average, at least 10 years for this transition.[33–35] On a worldwide basis, and throughout history, the average age

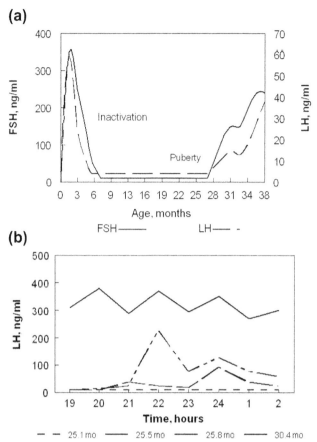

Figure 11.8 (a) Pattern of secretion of follicle-stimulating hormone (FSH) and luteinizing hormone (LH) in a male rhesus monkey (genus *Macaca*). The testes of the monkey were removed surgically at birth. The curves for FSH and LH indicate the production and release of gonadotrophin-releasing hormone (GnRH) from the hypothalamus. After the age of 3 months (i.e. during infancy), the hypothalamus is inactivated. Puberty takes place at about 27 months, and the hypothalamus is reactivated. (b) Development of hypothalamic release of GnRH during puberty in a male rhesus monkey with testes surgically removed. At 25.1 months (mo) of age (bottom line), the hypothalamus remains inactivated. At 25.5 and 25.8 mo (next two lines up from the bottom), modest hypothalamic activity is observed, indicating the onset of puberty. By 30.4 mo (top line), the adult pattern of LH release is nearly achieved. This pattern shows increases in both the number of pulses of release and the amplitude of release. In human beings, a very similar pattern of infant inactivation and late juvenile reactivation of the hypothalamus takes place. (Source: *adapted, with some simplification, from Plant.*[31])

for the first external manifestations of puberty for healthy girls is 9 years and first birth is at 19 years. On the same basis, boys show external signs of puberty, on average, at 11 years and fatherhood at 21–25 years. The evolutionary reasons for delay between puberty and first birth or fatherhood are discussed below. The point to make here is that monkeys and apes typically take less than 3 years to make the transition from puberty to parenthood.

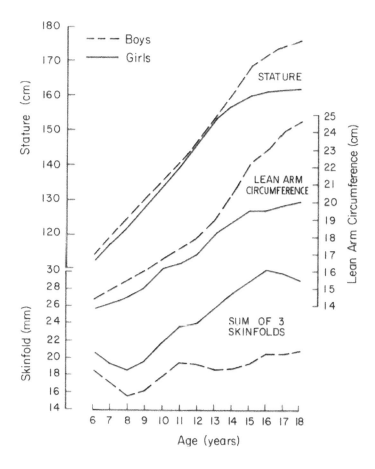

Figure 11.9 Mean stature, mean lean arm circumference, and median of the sum of three skinfolds for Montreal boys and girls. Notice that sexual dimorphism increases markedly after puberty (~12–13 years old).[32]

The second human difference is that during this life stage, both boys and girls experience a rapid acceleration in the growth velocity of almost all skeletal tissue: the adolescent growth spurt (Figure 11.4, velocity curve). The magnitude of this acceleration in growth was calculated for a sample of healthy Swiss boys and girls measured annually between the ages of 4 and 18 years. At the peak of their adolescent growth spurt, the average velocity of growth in height was +9.0 cm/year (3.5 inches/year) for boys and +7.1 cm/year (2.8 inches/year) for girls.[36] Similar average values are found for adolescents in all human populations. No other primate species, not even chimpanzees, exhibit this pattern of skeletal growth (Figure 11.10). It is important to point out that these are average human values and that there is considerable variation in the size of the spurt at peak height velocity

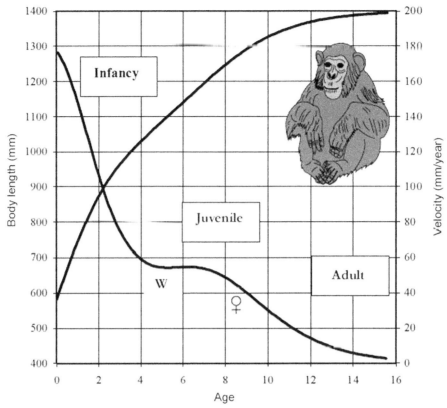

Figure 11.10 A model of distance and velocity curves for chimpanzee growth in body length: infancy, juvenile and mature adult. This is based on the longitudinal study of captive chimpanzee growth conducted by Hamada and Udono.[37] In the wild, weaning (W) usually takes place between 48 and 60 months of age.[38]

(the maximum rate of growth during adolescence) and in the duration of adolescence. In some studies of adolescent growth, up to 10% of seemingly healthy girls showed no discernible growth spurt. These were usually late-maturing girls, and late maturation is associated with a smaller growth spurt. The absence of an adolescent growth spurt in healthy, well-nourished boys is rare, even in late maturers.

Most primate species have rapid growth in length and body weight during infancy and then a declining rate of growth from weaning to adulthood. Some primate species may show a rapid acceleration in soft tissue growth at puberty, especially of muscle mass in male monkeys and apes. Sexually maturing non-human primates may have skeletal spurts in the face, for example, due to the eruption of large canine teeth in male baboons.[33,39] However, unlike humans, other primate species have either no adolescent acceleration in total skeletal growth or a very small increase in growth rate.[40,41] The human skeletal growth spurt is unequaled by other species, and when viewed graphically, the duration and intensity of the growth spurt defines human adolescence (Figure 11.4): it is a species-specific characteristic. Human adolescence, however, is more than skeletal growth. It is also a stage of the life cycle defined by several changes in behavior and cognition that are found only in our species. These changes are discussed later in this chapter.

11.3.5 Adulthood

Adolescence ends and early adulthood begins with the completion of the growth spurt, the attainment of adult stature and the achievement of full reproductive maturity, meaning both physical and psychosocial maturity. Height growth stops when the long bones of the skeleton (femur, tibia, etc.) and the vertebral bodies of the backbone lose their ability to increase in length. Usually this occurs when the epiphysis, the growing end of the bone, fuses with the diaphysis, the shaft of the bone. The fusion of epiphysis and diaphysis is stimulated by the gonadal hormones, the androgens and estrogens. These are also the hormones that promote reproductive maturation of the body and the brain.

The production of viable spermatozoa in boys and viable oocytes in girls is achieved during adolescence, but these events mark only the early stages, not the completion, of reproductive maturation. Socioeconomic and psychobehavioral maturation must accompany physiological development. All of these developments coincide, on average, by the age of about 19 years in women and 21–25 years in men. Possible evolutionary reasons for the human sequence of biological, social and psychological development towards adulthood are discussed below.

The course of growth and development during the prime reproductive years of adulthood are relatively uneventful. Most tissues of the body lose the ability to grow by hyperplasia, but many may grow by hypertrophy. Exercise training can increase the size of skeletal muscles, and caloric oversufficiency certainly will increase the size of adipose tissue. However, the most striking feature of the prime adult stage of life is its stability, or homeostasis, and its resistance to pathological influences, such as disease-promoting organisms and psychological stress. Adulthood contrasts with the preceding stages of life, which were characterized by change and greater susceptibility to pathology.

11.3.6 Late Life Stage

Old age and senescence follow the prime years of adulthood. The aging period is one of gradual or sometimes rapid decline in the ability to adapt to environmental stress. The pattern of decline varies greatly between individuals. Although specific molecular, cellular and organismic changes can be measured and described, not all changes occur in all people. Unlike the biological regulation of growth and development prior to adulthood, the aging process appears to follow no species-specific uniform plan. Menopause may be the only event of the later adult years that is experienced universally by women who live past 50 years of age; men have no similar event. The biology and possible value of menopause are discussed later in this chapter.

There are many theories about the aging process and about why we must age at all. Crews and Ice[42] discuss some of them and show that aging is a multicausal process. We may state here simply that the inability of all cell types — renewing, expanding and static — to use nutrients and repair damage leads to aging, senescence and death.

11.4 EVOLUTION OF THE HUMAN LIFE CYCLE

11.4.1 Why do New Life Stages Evolve?

In *Size and Cycle*, Bonner[43] develops the idea that the stages of the life cycle of an individual organism, a colony or a society are "the basic unit of natural selection". Bonner's focus on life-cycle stages follows in the tradition of many of the nineteenth century embryologists who proposed that speciation is often achieved by altering rates of growth of existing life stages and by adding or deleting stages. Bonner shows that the presence of a stage and its duration in the life cycle relate to such basic adaptations as locomotion, reproductive rates and food acquisition. From this theoretical perspective, it is profitable to view the evolution of human childhood, adolescence and perhaps menopause as adaptations for both feeding and reproduction.

11.4.2 Why Childhood?

Consider the data shown in Figure 11.11, which depicts several hominoid developmental landmarks. Compared with living apes, human beings experience developmental delays in eruption of the first permanent molar, age at menarche and age at first birth. However, humans have a shorter infancy and a shorter birth interval. In apes and traditional human societies the infancy stage, that is, the time of feeding by lactation, is virtually equal to the interval between births.

Many mammals, including most primate species, wean infants and terminate the infancy stage about the time the first permanent molar (M1) erupts. This timing makes sense, because the mother must nurse her current infant until it can process and consume an adult diet, which requires at least some of the permanent dentition.

The data for the social carnivores and some non-human primates show that the correlation between M1 eruption and weaning can be modified by other factors, especially continued dependence of the young on help from the mother. Such modification is most extreme for chimpanzees and orang-utans. The interval between successful births for chimpanzee females averages nearly 5 years and for orang-utan mothers, who may be under nutritional stress, it averages 6–8 years in the wild.[45] But, first permanent molar eruption in these apes takes place at about age 3.1 years in captivity. Even in the wild chimpanzee M1 eruption occurs at about the age of 4.[46] Birth spacing in these two ape species, and some other social and predatory mammals, may be delayed until well after the age at which the first permanent teeth erupt in the current infant.

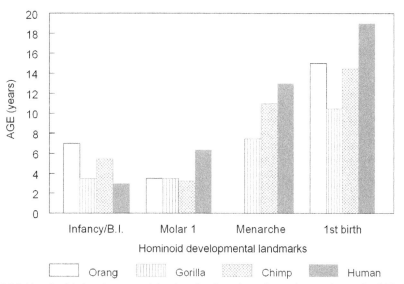

Figure 11.11 Hominoid developmental landmarks. Data based on observations of wild-living individuals or, for humans, healthy individuals from various cultures. Infancy/B.I.: period of dependency on mother for survival, usually coincident with mean age at weaning and/or a new birth (where B.I. is birth interval); Molar 1: mean age at eruption of first permanent molar; Menarche: mean age at first estrus or menstrual bleeding; 1st birth: mean age of females at first offspring delivery. Orang, *Pongo pygmaeus*; gorilla, *Gorilla gorilla*; chimp, *Pan troglodytes*; human, *Homo sapiens*.[44] This figure is reproduced in the color plate section.

The human species is a striking exception to this relationship between permanent tooth eruption and birth interval. Women in traditional societies wait, on average, 3 years between births, not the 6 years expected on the basis of the average age at M1 eruption. The short birth interval gives humans a distinct advantage over other apes: human women can produce and rear two offspring through infancy in the time it takes chimpanzees or orang-utans to produce and rear one offspring. By reducing the length of the infancy stage of life (i.e. lactation) and by developing the special features of the human childhood stage, humans have the potential for greater lifetime fertility than any other ape.

The answer to the question "Why childhood?" is that the evolution of childhood gave human beings a reproductive advantage over other apes. Human childhood decreases the interbirth interval for women. However, childhood comes with a significant trade-off: human children are still dependent on older individuals for feeding, needing foods that are specially chosen and prepared, and protection. Without sufficient care children will suffer and die. Humans deal with the special needs of children by spreading the needs for feeding and protection across a number of people, including juveniles, adolescents and adults. The mother of the child does not have to provide 100% of offspring nutrition and care directly.

Cooperative childcare, also called cooperative breeding, seems to be a human universal.[33,47] Key features of human cooperation are food sharing and division of labor by age and sex. The contribution of food and labor may be quantified in terms of energy (kilocalories). The relatively slow rate of body growth and small body size of children reduces competition with adults for food resources. Slow-growing small children require less food per day than larger juveniles, adolescents and adults.[48] Thus, although provisioning children is time consuming, it is not as onerous a task of investment as it would be, for instance, if both brain and body growth were rapid simultaneously.

Another way is to quantify the food and labor sharing in terms of reproductive outcomes, such as birth and survival of offspring to adulthood. Reiches et al.[49] do just this by proposing a "pooled energy budget" hypothesis. They define the pooled energy budget as, "… the combined energetic allocations of all members of a reproductive community that might result in direct or indirect reproductive effort" (p. 424). According to Reiches and colleagues, the pooled energy budget allows human women to sustain a higher fertility (see Figure 11.1) and greater survival of their offspring than would be possible for energetically isolated individuals. Energetic isolation, relative to the human condition, is the case for adult female chimpanzees, bonobos, gorillas and orang-utans. Females in these ape species are, generally, competitors. In terms of survival, about 50—60% of infants and children in human forager societies reach adulthood. In the great apes the survival averages less than 40%.[40]

One human example of pooled energy budgeting comes from the Hadza society, African hunters and gatherers, where grandmothers and great aunts supply a significant amount of food and care to children.[51,52] Another example is Agta society (Philippine hunter—gatherers), where women hunt large game animals but still retain primary responsibility for child care. They accomplish this dual task by living in extended family groups — two or three brothers and sisters, their spouses, children and parents — and sharing the child care. Among the Maya of Guatemala (horticulturists and agriculturists), many people live together in extended family compounds. Women of all ages work together in food preparation, clothing manufacture and child care (Bogin, field observations, Figure 11.12). In some societies, fathers provide significant child care, including the Agta and the Aka pygmies, hunter—gatherers of central Africa. Summarizing the data from many human societies, Lancaster and Lancaster[53] called this kind of cooperative child care and feeding "the hominin adaptation", because no other primate or mammal does all of this. [The term "hominin" refers to living humans and those fossil species capable of facultative (part-time) or obligate (full-time) bipedal locomotion. See Wood and Lonergan[54] for further definition.] Bogin[55,56] calls it biocultural reproduction, because the human style of cooperative child care enhances the social, economic, political, religious and ideological well-being of the society as much as it contributes to biological fitness.

Phenotypes and Plasticity

Childhood may also be viewed as a mechanism that allows for more precise "tracking" of ecological conditions by allowing more time for developmental plasticity. The fitness of

Figure 11.12 Cooperative care of children by women and juvenile girls. The example is from the Kaqchikel-speaking Maya region of Guatemala. The women perform household and food preparation duties while the juvenile girls play with and care for the children. *(Photograph by Barry Bogin)*. This figure is reproduced in the color plate section.

a given phenotype (i.e. the physical features and behavior of an individual) varies across the range of variation of an environment. When phenotypes are fixed early in development, such as in mammals that mature sexually soon after weaning (e.g. rodents), environmental change and high mortality are positively correlated. Mammals with more plasticity in biological and social characteristics (carnivores, elephants, primates) prolong the developmental period by adding a juvenile stage between infancy and adulthood. Adult phenotypes develop more slowly in these mammals because the juvenile stage lasts for years. These social mammals experience a wider range of environmental variation, such as seasonal variation in temperature and rainfall. They also experience years of food abundance and food shortage as well as changes in the number of predators and in types of diseases. The result on the phenotype may be a better conformation between the individual and the environment. For human beings, the additional 4 years of developmental time added by childhood has the potential to increase the quality of the young and the chances of surviving to adulthood. This potential is only realized when the biological, social and emotional needs of children are provided by the older members of their society.

11.4.3 Why Adolescence?

An adolescent stage of human growth may have evolved by the processes of natural selection and sexual selection. Both types of selection were identified by Charles Darwin. Where natural selection operates to increase the frequency of genotypes and phenotypes that confer reproductive advantages on the individuals possessing them, sexual selection,

"... depends on the advantage which certain individuals have over other individuals of the same sex and species, in exclusive relation to reproduction".[57] (Today, we would replace the word "reproduction" with "mating". In many primate species a good deal of mating is more about social relations rather than a reproductive event, such as fertilization.) Darwin also wrote that, "There are many other structures and instincts which must have been developed through sexual selection — such as the weapons of offence and the means of defence possessed by the males for fighting with and driving away their rivals — their courage and pugnacity — their ornaments of many kinds — their organs for producing vocal or instrumental music — and their glands for emitting odours; most of these latter structures serving only to allure or excite the female".[58] It is known today that sexual selection also works for females, meaning that female-specific physical and behavioral traits may evolve via competition between the females for mating opportunities with males.

Both childhood and adolescent stages must ultimately be shaped by natural selection. Childhood could be selected because it allows hominin females to give birth at shorter intervals, but producing offspring is only a small part of reproductive fitness. Rearing the young to their own reproductive maturity is a surer indicator of success. Adolescence may be key in helping the next generation rear its own young successfully. The adolescence stage may provide boys and girls with a life history strategy to survive to adulthood, the immediate benefit, and as a longer-term benefit to practice complex social skills required for effective mating and parenting. Sexual selection may have acted on the "mating", and natural selection on the "parenting". There are, of course, trade-offs associated with the adolescence stage, such as a delay in the onset of reproduction and risks for damage and death. These risks, it seems, have been outweighed by the reproductive advantages of adding adolescence to human life history.

Studies of yellow baboons, toque macaques and chimpanzees show that between 50 and 60% of first-born offspring die in infancy, and more die before reaching adulthood (see reference 22 for citations for these statistics and those that follow in this section). By contrast, in hunter—gatherer human societies, between 39% (the Hadza of eastern Africa) and 44% (the !Kung of southern Africa) of offspring die in infancy. Studies of wild baboons show that whereas the infant mortality rate for the first born is 50%, mortality for the second born drops to 38%, and for the third and fourth born reaches only 25%.[59] The difference in infant survival is, in part, likely due to experience and knowledge gained by the mother with each subsequent birth.

Such maternal information is internalized by human females during their juvenile and adolescent stages, giving the adult women a reproductive edge. The initial human advantage may seem small, but it means that up to 21 more people than baboons or chimpanzees survive out of every 100 first-born infants — more than enough over the vast course of evolutionary time to make the evolution of human adolescence an overwhelmingly beneficial adaptation.

During adolescence, human girls refine some skills learned as juveniles and also add new skills that will be of benefit during adulthood. In human societies, juvenile girls often are expected to provide significant amounts of child care for their younger siblings, whereas in most other social mammal groups, the juveniles are often segregated from adults and infants. Thus, human girls enter adolescence with considerable knowledge of the needs of young children, and can add to that knowledge. New skills acquired by adolescent girls relate to the economic, sexuality, social and political roles of women. Adolescent girls are better able to learn and practice these new skills because they look physically mature, and are treated as such, several years before they actually become fertile.

The adolescent growth spurt serves as a signal of maturation. Early in the spurt, before peak height velocity is reached, girls develop pubic hair and fat deposits on breasts, buttocks and thighs. They appear to be maturing sexually. About a year after peak height velocity, girls experience menarche, an unambiguous external signal of internal reproductive system development. Menarche, however, does not signal fertility as most girls experience 1—3 years with partial anovulatory menstrual cycles. Following menarche, the frequency of these anovulatory cycles, which do not release an egg cell from the ovary, is initially high and then declines over time. One study of the fertility development of 200 normal girls in Finland, aged 7—17 years, examined the participants on two occasions between 1 and 5 years apart. Ovulation was determined by a hormonal assessment.[60] It was found that, "In postmenarchal girls, about 80% of the cycles were anovulatory in the first year after menarche, 50% in the third and 10% in the sixth year" (p. 107). These findings have been replicated in other studies.[61] Pregnancy is possible in the immediate postmenarche years, but its likelihood is low compared with later ages.[62] Nevertheless, the dramatic changes of early adolescence stimulate the girls to participate in adult social, sexual and economic behavior, and stimulate the adults around these girls to encourage, or at least worry about, this participation. All human societies instruct, regulate, restrict and attempt to normalize the behavior of adolescent girls (and boys) to conform towards desirable and acceptable social standards.

Practicing to become an adult has a future benefit, but does adolescence also have an immediate benefit for the adolescent? Adolescence adds a reproductive delay of 6 years or more compared with the chimpanzee and would not be likely to evolve unless there were some immediate benefits. Most evolutionary benefits relate to feeding or reproduction and, as adolescents do relatively little of the latter, the benefits seem more likely to relate to feeding. Human juveniles may hunt, gather or produce some of their own food intake, but overall they require provisioning to achieve energy balance. This is the case in many traditional societies, such as the Ache, Hiwi, !Kung foragers,[50] the Maya farmers of Mexico and Guatemala,[63] as well as many historical and contemporary urban—industrial societies.[33,34] In contrast, human adolescents are capable of producing sufficient quantities of food to exceed their own energy requirements (Figure 11.13).[64] Some of the

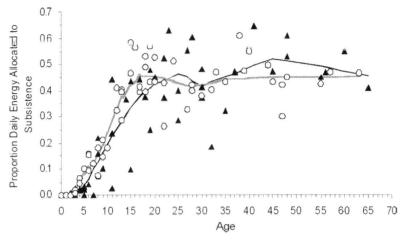

Figure 11.13 Proportion of an individual's total daily energetic expenditure allocated to subsistence activity (food production and other basic activities for survival), calculated for Maya males (triangles, thinner black line) and females (open circles, thicker gray line). By the age of 15 years the level of food production reaches adult values.[64]

food that adolescents produce may be used to fuel their own growth and development, creating larger, stronger and healthier bodies. The surplus production is shared with other members of the social group, including younger siblings, parents and other immediate family members (defining families in the broad ethnographic sense). One immediate benefit of adolescence is that the adolescents are economically valuable for their services in food production and other contributions to the biocultural reproduction of their family and society. For this they receive care and protection to safeguard their health and survival. This is an important additional immediate benefit because adolescents are immature in terms of sociocultural knowledge and experience and still require some care and protection.[50,65,66]

Two Pathways Through Adolescence

The growth and development of girls and boys during adolescence follows two distinctly different paths. Girls look reproductively mature years before they are in fact fertile. In contrast, boys are fertile years before they look reproductively mature. There seem to be biocultural benefits associated with each of these pathways.

Adolescent Growth and Development of Girls

Full reproductive maturation in human women is not achieved until about 5 years after menarche. For example, the average age at menarche for girls living in the USA is 12.4 years, which means that the average age at full sexual maturation occurs between the ages of 17 and 18 years. Although adolescents younger than these ages can have babies, both the teenage mothers and the infants are at added risk because of the reproductive immaturity of the mother, especially for low-birth-weight infants, premature births and high blood pressure in the mother. The likelihood of these poor outcomes declines, and the chance of successful pregnancy and birth increases markedly, after the age of 15 years, dropping to a minimum after the age of 18 years.[67]

One reason for this is related to another feature of human growth not found in the African apes: female fertility tracks the growth of the pelvis. Human menarche may be predicted by biiliac width, the distance between the left and right iliac crests of the pelvis.[68,69] A median width of 24 cm seems to be needed for menarche in American girls living in Berkeley, California, USA, Kikuyu girls of East Africa, and Bundi girls of highland New Guinea, even though the pelvic width constant is attained at different ages in these three cultures, about 13, 16 and 17 years old, respectively. Differences in nutrition and disease account for the substantial spread in growth rates and age of sexual maturity across these human groups. Moerman[70] also reported a special human relationship between growth in pelvic size and reproductive maturation. She found that the crucial variable for successful first birth is size of the pelvic inlet, the bony opening of the birth canal. Moerman measured pelvic X-rays from a sample of healthy, well-nourished American girls who achieved menarche between 12 and 13 years, although they did not attain adult pelvic inlet size until 17—18 years of age. Quite unexpectedly, she found that the adolescent growth spurt, which occurs before menarche, does not influence the size of the pelvis in the same way as the rest of the skeleton. Rather, the female pelvis has its own slow pattern of growth, which continues for several years after adult stature is achieved. Cross-cultural studies of reproductive behavior suggest that human societies acknowledge (consciously or not) this special pattern of pelvic growth. The age at marriage and first childbirth clusters around 18—19 years for women from such diverse cultures as the Kikuyu of Kenya, Mayans of Guatemala, Copper Eskimos of Canada, and both the colonial and contemporary USA.[33,34]

Why the human pelvis follows this unusual pattern of growth is not clearly understood. It probably relates to the trade-off between bipedalism and reproduction. Unlike human beings, females of many non-human primate species give birth successfully before the mother's own body growth terminates (http://pin.primate.wisc.edu/factsheets). Bipedalism is known to have changed the shape of the human pelvis from the basic ape-like shape. Apes have a cylindrical-shaped pelvis, but humans have a bowl-shaped pelvis. The human shape is more efficient for bipedal locomotion but less efficient for reproduction because it restricts the size of the birth canal.[71] The required pelvic architecture in terms of size and shape for successful reproduction may take longer to develop in bipedal humans than in the non-human primates.

Why do Boys have Adolescence?

Boys become fertile well before they assume the size and the physical characteristics of men. Analysis of urine samples from boys 11—16 years old shows that they begin producing sperm at a median age of 13.4 years.[72] Yet cross-cultural evidence indicates that few boys successfully father children until they are into their third decade of life. The National Center for Health Statistics of the USA, for example, reports that only 3.09% of liveborn infants in 1990 were fathered by men less than 20 years of age. In Portugal, for the years 1990, 1994 and 1999, the percentage of fathers under 20 years of age was always below 3%. In 2001, Portugal stopped presenting results concerning the percentage of fathers below 20 because there were too few of them.[73] Among the traditional Kikuyu of East Africa, men do not marry and become fathers until about the age of 25 years, although they become sexually active after their circumcision rite at around age 18. Among the Ache, traditional foragers of the forests of Paraguay, adolescent boys do not become net food producers until the age of 17 years, and they do not marry until about age 20 years.[74] In the central Canadian Arctic, Inuit people living as traditional hunters did not even consider an adolescent boy ready for marriage until he was 17–18 years old.[75] Even then, the adolescent had to provide bride service to his prospective in-laws for several years before he became a father. These delays in fatherhood occurred despite the fact that there was considerable pressure to reproduce because of, "the slim margin of survival in the pre-contact period ...".[75, p. 270]

The explanation for the lag between sperm production and fatherhood is likely to be due to the fact that the average boy of 13.4 years is only beginning his adolescent growth spurt (Figure 11.4). Growth researchers have documented that in terms of physical appearance, physiological status, psychosocial development and economic productivity, the 13-year-old boy is still quite immature. Anthropologists working in many diverse cultural settings report that few women (and more importantly from a cross-cultural perspective, few prospective in-laws) view the teenage boy as a biologically, economically and socially viable husband and father.

The adolescence of boys is not wasted time in either a biological or social sense. The obvious and the subtle psychophysiological effects of testosterone and other androgen hormones that are released after gonadal maturation may "prime" boys to be receptive to their future roles as men. Alternatively, it is possible that physical changes provoked by the endocrines provide a social stimulus to older members of the group to encourage the boys towards adulthood. Whatever the case, early in adolescence, sociosexual feelings including guilt, anxiety, pleasure and pride intensify. At the same time, adolescent boys become more interested in adult activities, adjust their attitude to parental figures, and think and act more independently. In short, they begin to behave like men. However — and this is where the survival advantage may lie — they still look like boys. One might say that a healthy, well-nourished 13.5-year-old human male, at a median height of 160 cm (62 inches) "pretends" to be more childlike than he really is.[76]

During the adolescent years, boys are shorter than girls of roughly the same chronological age, furthering an immature image (Figure 11.4).[77] Even more to the point is that the spurt in muscle mass of adolescent males does not occur until an average age of 17 years (Figure 11.9).[78] At peak height velocity during the skeletal growth spurt the typical adolescent boy has achieved 91% of his adult height, but only 72% of his adult lean body mass. Since most of the lean body mass is voluntary muscle tissue, adolescent boys cannot do the work of men. This is one important reason why adults of the Kikuyu, the Inuit and many other cultures do not even think of younger adolescent boys as man-like.

As Schlegel and Barry[66] found in their cross-cultural survey, adolescent boys are usually encouraged to associate and "play" with their age mates rather than associate with adult men. During these episodes of "play" these juvenile-looking adolescent males are protected from many dangers of adult male life and they can practice behaving like adult men before they are actually perceived as adults. The activities that take place in these adolescent male peer groups include the type of productive, economic, aggressive/militaristic and sexual behaviors that older men perform. The sociosexual antics of adolescent boys are often considered to be more humorous than serious, yet they provide the experience to fine-tune their sexual and social roles before either their lives, or those of their offspring, depend on them. For example, competition between men for women favors the older, more experienced man. As such competition may be fatal, the juvenile-like appearance of the immature, but hormonally primed, adolescent male may be life-saving, as well as educational.

11.4.4 When did Childhood and Adolescence Evolve?

The stages of the life cycle may be studied directly only for living species. However, we can postulate on the life cycle of extinct species. Such inferences for the hominins are, of course, hypotheses based on comparative anatomy, comparative physiology, comparative ethology and archaeology. The hominin fossil record now appears to stretch back to more than 6 million years ago (mya) with the appearance of *Sahelanthropus*, *Orrorin* and *Ardipithecus* in Africa.[54] There is an ever-expanding list of fossils and scientific names for these fossils. The following discussion provides a simplified summary of current research. A more detailed review may be found in Bogin and Smith.[56]

Figure 11.14 is a summary of the evolution of the human pattern of growth and development. This figure must be considered as "a work in progress", because only the data for *Pan* (chimpanzees) and *Homo sapiens* are known with certainty. Known ages for eruption of the M1 are given for *Pan* and *H. sapiens*. Estimated ages for M1 eruption in other species were calculated by Smith and Tompkins.[79] Brain size is another crucial influence on life history evolution, and known or estimated adult brain sizes are given at the top of each bar. These values are averages based on skulls from the fossil record.

These earliest known forms appear to have had brains no larger than the living bonobo and chimpanzee, a mere 300–350 cm^3 in *Ardipithecus ramidus* (ca. 4.4 mya) and less than 400 cm^3 in *Sahelanthropus*. We begin to know something more directly of infant and juvenile life by the appearance of *Australopithecus afarensis* about 3.9 mya. The Dikika infant fossil, about 3 years of age at death by its teeth, already has a brain size of 330 cm^3;

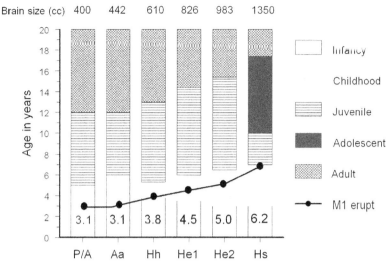

Figure 11.14 The evolution of hominin life history during the first 20 years of life. Abbreviations of the pongid and hominin taxa are: P/A: *Pan, Australopithecus afarensis*; Aa: *Australopithecus africanus*; Hh: *Homo habilis*; He1: early *Homo erectus*; He2: late *Homo erectus*; Hs: *Homo sapiens*; M1 erupt: eruption of first molar.[33] This figure is reproduced in the color plate section.

adult *A. afarensis* reached an adult brain size of about 400 cm^3 with a pattern of dental development little different from extant apes. Therefore, the chimpanzee and *A. afarensis* are depicted as sharing the typical tripartite stages of postnatal growth of social mammals: infant, juvenile, adult. In Figure 11.14, the duration of each stage and the age at which each stage ends are based on empirical data for the chimpanzee.

A probable descendant of *A. afarensis* is the fossil species *A. africanus*, dating from about 3 mya. To achieve the larger adult brain size of *A. africanus* (average of 442 cm^3) may have required an addition to the length of the fetal and/or infancy periods. Figure 11.14 indicates an extension to infancy of 1 year. With a body and brain size near that of the chimpanzee, it is likely that *A. africanus* and other closely related species followed a pattern of growth and development very similar to that of chimpanzees. Moreover, detailed study of incremental growth in teeth confirms that in *A. africanus* the M1 erupted at about 3.1 years of age.[80]

At about 2.2 mya, we find fossils with several more human-like traits, larger cranial capacities and greater manual dexterity. Also dated to about this time are stone tools of the Oldowan tradition. To date, manufacture of stone tools can be traced approximately 2.5 mya, although the use of (possibly natural) stone to cut carcasses has been pushed back as far as 3 mya.[81] Given the biological and cultural developments associated with these fossils they are considered by most paleontologists to be members of the genus *Homo*. The rapid expansion of adult brain size during the time of *H. habilis* (650–800 cm^3) might have been achieved with further expansion of both the fetal and infancy periods. However, the insertion of a brief childhood stage into hominin life history may have occurred.[82,83]

As indicated in Figure 11.14, *H. habilis* childhood began after the eruption of M1 and lasted for about 1 year. *Homo habilis* children would need to be supplied with special weaning foods and there is archaeological evidence for just such a scenario. *Homo habilis* seems to have intensified its dependence on stone tools. There is considerable evidence that some of these tools were used to scavenge animal carcasses, especially to break open long bones and extract bone marrow. This behavior may be interpreted as a strategy to feed children the essential amino acids, some of the minerals and, especially, the fat (dense source of energy) that are required for growth of the brain and body.

Further brain size increase occurred during the time of *H. erectus*, which began about 1.8 mya. The earliest adult specimens have mean brain sizes of 826 cm^3, but many individual adults had brain sizes between 850 and 900 cm^3. From other fossils of *H. erectus*, we even know that the timing of M1 eruption has evolved, changing from about 3–3.5 years in the earliest hominins to about 4.5 years.[84] In all, the fossil record indicates that *H. erectus* evolved slowed general maturation, more helpless infants, larger adult brains, larger body size and increasing sophisticated tools — all of which adds up to an adaptive network of higher quality offspring, long and intense learning, and reliance on complex behavior.[85] Taken together, this evidence points to an enormous energy demand for reproduction in an *H. erectus* female, a demand that necessitated a change towards a human-like strategy to care for newly weaned offspring. A human-like childhood period would at first diminish the reproductive cost of this high-energy strategy, and eventually allow the strategy to become even more extreme. As shown in Figure 11.14, the *H. erectus* infancy period shrinks to below that of chimpanzees and the childhood period expands, which would have given *H. erectus* a greater reproductive advantage than any previous hominin. The fact that *H. erectus* populations certainly did increase in size and began to spread throughout Africa and into other regions of the world suggests that fundamental changes in life history had already begun.

Later *H. erectus*, with average adult brain sizes of 983 cm^3, are depicted in Figure 11.14 with further expansion of the childhood stage. In addition to bigger brains (some individuals had brains as large as 1100 cm^3), the archaeological record for later *H. erectus* shows increased complexity of technology (tools, fire and shelter) and social organization. These technosocial advances, and the increased reliance on learning that occurred with these advances, may well be correlates of changes in biology and behavior associated with further development of the childhood stage of life.[44]

The evolutionary transition to archaic and, finally, modern *H. sapiens* expands the childhood stage to its current dimension. Note that M1 eruption becomes one of the events that coincides with the end of childhood. This is the roughly the point at which many mammals become independent juveniles, and as discussed earlier, in humans is the period that introduces significant biological, cognitive, behavioral and social changes.

With the appearance of *H. sapiens* comes evidence for the full gamut of human cultural capacities and behaviors. The *H. sapiens* grade of evolution also sees the addition of an adolescent stage to postnatal development. The single most important feature defining human adolescence is the skeletal growth spurt that is experienced by virtually all boys and girls. There is no compelling evidence for a human-like adolescent growth spurt in any living ape or any hominin prior to *H. sapiens*.[76,84] There are several fossils of a species called *Homo antecessor*, found in Spain and dated to between 800,000 and 960,000 years before present (BP).[86] Based on an analysis of tooth formation, this species seems to have a pattern of dental maturation much like modern humans, but there is as yet no juvenile with both teeth and skeleton for an in-depth analysis of growth patterns.

For later hominins like the Neanderthals, we have one fossil in which the associated dental and skeletal remains needed to assess adolescent growth are preserved. Le Moustier 1, found in 1908 in western France, is dated at between 42,000 and 37,000 years BP. Thompson and Nelson[87] estimate that this individual was male, with a dental age of 15.5 ± 1.25 years. His M2 was erupted, which for a living human would correspond to an adolescent. However, Le Moustier 1 had achieved about 87% of adult femur length with only the stature of an 11-year-old living juvenile. The dental age of 15.5 years and the stature age of 11 years are in very poor agreement, and indicate that Le Moustier 1 seems short for his dental age and may not have followed a human pattern of adolescent growth. Indeed, recent work finds evidence for more rapid growth and development in Neanderthals than expected for modern humans.[88,89]

To date, the earliest fossil evidence of modern human growth and development comes from an archaic *H. sapiens* fossil from Jebel Irhoud, Morocco, dated at about 160,000 BP. The estimated age at death is 7.8 years based on the microstructure of tooth formation, and this estimate is consistent with the "… degree of eruption, developmental stage, and crown formation time …" of the fossil compared with modern *H. sapiens*. This is the earliest *H. sapiens* fossil for which the estimated age of death matches the assigned human dental age.[90] While this is strong evidence of a well-established human childhood and possibly a fully human pattern of growth, it does not confirm the existence of adolescence.

11.4.5 The Valuable Grandmother, or Could Menopause Evolve?

In addition to childhood and adolescence, human life history has another unusual aspect: menopause combined with the potential of many years of healthy postreproductive life. One generally accepted definition of menopause is "the sudden or gradual cessation of the menstrual cycle subsequent to the loss of ovarian function".[91] The process of menopause is closely associated with but distinct from the adult female postreproductive stage of life.

Wild-living non-human primate females do not share the universality of human menopause, and human males have no comparable life history event. Some elderly wild-living primates do experience a degradation of fertility and may even cease reproductive cycling prior to death, but their postreproductive lifespan is usually less than 1 year.[92,93] In contrast to the non-human female primates, the human female reproductive system is "shut down" well before other systems of the body. Human women may be healthy at menopause and may have 20 or more years of relatively vigorous and active life following menopause. Why are human women different from the females of other primate species?

Human reproduction usually ends before menopause. In traditional societies, such as the !Kung (foragers of the Kalahari Desert), the Dogon of Mali and the rural-living Maya of Guatemala, women rarely give birth after the age of 40 and almost never give birth after 44. Even in the USA from 1960 to today, with modern health care, good nutrition and low levels of hard physical labor, women rarely give birth after age 45: in the year 2007 the number of births to US women aged 45–49 years was 7349 out of a total of 4,317,119 total births, or 0.17%. Births to women over 45 years old are rare even among social groups attempting to maximize lifetime fertility, such as the Old Order Amish.

In industrialized nations, as among the !Kung, Dogon and Maya, menopause occurs well after this fertility decline. A study by Reynolds and Obermeyer[94] found that the median age of natural menopause in Spain is estimated at 51.7 years, and in the USA it is 52.6 years.

There are several hypotheses to explain both menopause and long postreproductive life. Many of these propose that both menopause and postreproductive lifespan must have coevolved. Some versions of the "grandmother hypothesis" depend on the evolution of human menopause.[42,52] However, in a review of the data for mammals, Cohen[95] finds that reproductive cessation prior to death is, in fact, widespread across different species of female mammals. To be sure, the human condition of menopause followed by one or more decades of postreproductive lifespan is an unusual and extreme example. There is no need, however, to propose any special hypothesis for the evolution of human menopause. As noted by Bogin and Smith,[44] and developed more fully by Cohen, it appears that a 50-year age barrier exists to female primate fertility because by that age, the reserve of oocytes is sufficiently depleted to end reproductive cycles.

Several researchers find that across mammalian species, somatic (body) and reproductive senescence tend to follow separate paths and occur on different developmental schedules.[96,97] Cohen[95] proposes that the timing of somatic versus reproductive senescence is probably determined by the maximization of reproductive performance and survival early in adulthood. This proposal is formalized as the "reserve capacity hypothesis" by Crews and Bogin.[98,99] According to this hypothesis, the building of reserve capacity before the onset of reproduction in early adulthood is why humans live long past the point of reproductive decline. Reserve capacity is defined as the amount of physical, cognitive, social and emotional resources that exceed the minimum for immediate survival. Greater reserve capacity leads to a healthier body and brain, greater social well-being, enhanced biocultural reproduction and a longer lifespan.

The new human life history stages of childhood and adolescence can result in more reserve capacity. This is because there is more time to grow, develop and mature as an individual person and as an integrated member of the social group. The conditions for healthy growth must, of course, exist if there is to be a surplus of reserve capacity. It should not be necessary here to document the consequences of poor nutrition, disease, neglect, abuse, civil unrest and other adverse circumstances on reduced human health, reduced expectations for survival, lower successful reproduction and shorter lifespan.

Reproductive cessation and a significant period of postreproductive life may be a shared feature of many mammals, but the unusually long postreproductive life of human women has some unique effects on human life history and human biocultural reproduction. Given the biological age limit to fertility, one reproductive strategy open to postmenopausal human women is to provide increasing amounts of aid to their existing offspring and, if available, their grand-offspring. The ethnographic evidence shows that postreproductive women often do this and are beneficial to the survival of children and grandchildren.[100,101] Grandmothers provide food, child care, and a repertoire of knowledge and life experiences that assist in the education of their grandchildren. Older men may also provide support to their existing children and grandchildren, but men do not experience anything like menopause and may continue siring offspring well past the age of 50 years. Even so, research in the wealthy nations does show that children with little or no contact with grandparents (male or female) suffer more abuse and neglect than children with regular multigenerational interaction — another testament to the value of grandparents.

In sum, the inevitabilities of primate biology, combined with the creativity of human culture, allow women of our species to develop biocultural strategies to take greatest advantage of a postreproductive life stage. Viewed in this biocultural context, the period of human female postreproductive lifespan may be added to human childhood and adolescence as a distinctive stage of the human life history.

11.5 LIFE-CYCLE TRADE-OFFS AND RISKS FOR CHILDREN, ADOLESCENTS AND POSTMENOPAUSAL WOMEN

The evolution of new human structures, functions and stages of life history may have brought about many biocultural benefits; however, these also incur trade-offs and risks. A detailed review of the hazards may be found in Bogin and Smith.[56] A summary of the trade-offs and risks is given here.

The benefits of childhood need to be tempered against the hazards of dependency on older individuals for food and protection. In the past 20 years, the global epidemic of the human immunodeficiency virus (HIV)/acquired immunodeficiency syndrome (AIDS) has claimed the life of millions of young adults, often leaving their children with no biological parent. Africa has been burdened most severely, with at least 50 million orphans as the legacy of AIDS and other diseases, as well as war and high rates of death in pregnancy and childbirth.[102] The "charms" of children and childhood do not provide for total security.

Human adolescence comes with a new set of risks. Among the most common and serious of these are psychiatric and behavioral disorders. The onset of such problems tends to peak during the adolescent life history stage.[103] Most mammalian species terminate all brain growth well before sexual maturation, but human adolescents show enlargement and pruning of some brain regions leading to structural changes in the cerebral cortex. One hypothesis for the increase in brain-related disorders posits that these cortical changes leave the adolescent brain, "… more or less sensitive to reward".[103, p. 947]

The evolution of long postreproductive lifespan is associated with several risks for older women (Crews and Ice[42] review these in greater detail). The hormonal changes that occur with the cessation of ovarian function may bring about several degenerative diseases, such as osteoporosis. Postmenopausal women may suffer emotional diseases due to hormonal changes and also due to changes in social and economic roles. Adequate training and social support can help during the transition to postreproductive life, but not all societies provide these. Older women and men also may be forced to retire from productive activities and this can be a source of biocultural stress that can exacerbate the normal degenerative process of aging.

11.6 SUMMARY

This chapter takes a life history approach to the study of human growth and development. The postnatal life cycle of the social mammals, including the non-human primates, has three basic stages of postnatal development: infant, juvenile and adult. The human life cycle, however, is best described by six stages: infant, child, juvenile, adolescent, adult and postreproductive woman. It is hypothesized that the new life stages of the human life cycle represent feeding and reproductive specializations of the genus *Homo*. Hominins prior to the genus *Homo* probably had life histories more similar to living apes than humans. These hominins seem to have lived at a pace nearly twice as fast as modern humans. Analyses of bones and teeth of early hominins who died as subadults suggest that the new life stage of childhood evolved after appearance of the genus *Homo*, possible with the evolution of *H. habilis* or *H. erectus*. The adolescence stage may have evolved nearly 1 million years ago with *H. antecessor*, or as late as around 160,000 years ago with the appearance of archaic *H. sapiens*. It is clear that the fossil evidence suggests that the elements of human life history evolved as a mosaic over more than 2 million years. Moreover, the complete "package" of modern human life history, with our distinctive biocultural reproduction and long postreproductive lifespan for women, took shape with the evolution of modern *Homo sapiens* around 125,000 BP, and not before. Finally, some of the risks of each stage of human life history were discussed. Our biocultural nature means that we can make choices to ignore these risks or take action to ameliorate them.

REFERENCES

1. McGinnis W, Levine MS, Hafen E, Kuroiwa A, Gehring WJ. A conserved DNA sequence in homoeotic genes of the *Drosophila* antennapedia and bithorax complexes. *Nature* 1984;**308**:128–33.
2. Halder G, Callaerts P, Gehring WJ. New perspectives on eye evolution. *Curr Opin Genet Dev* 1995;**5**:602–9.
3. Callaerts P, Halder G, Gehring WJ. PAX-6 in development and evolution. *Annu Rev Neurosci* 1997;**20**:483–532.
4. Gehring WJ. *Master control genes in development and evolution: the homeobox story*. New Haven, CT: Yale University Press; 1998.
5. Kozlowski J, Wiegert RG. Optimal allocation to growth and reproduction. *Theor Popul Biol* 1986;**29**:16–37.
6. Stearns SC. Trade-offs in life-history evolution. *Funct Ecol* 1989;**3**:259–68.
7. Stearns SC. *The evolution of life histories*. Oxford: Oxford University Press; 1992.
8. Charnov EL, Berrigan D. Why do primates have such long life spans and so few babies? *Evol Anthropol* 1993;**1**:191–4.
9. Roff D. *The evolution of life histories: theory and analysis*. New York: Chapman & Hall; 1992.
10. Walker RS, Gurven M, Burger O, Hamilton MJ. The trade-off between number and size of offspring in humans and other primates. *Proc Biol Sci* 2008;**275**:827–33.
11. Kuzawa CW, Sweet E. Epigenetics and the embodiment of race: developmental origins of US racial disparities in cardiovascular health. *Am J Hum Biol* 2009;**21**:2–15.
12. Varela-Silva MI, Azcorra H, Dickinson F, Bogin B, Frisancho AR. *Influence of maternal stature, pregnancy age, and infant birth weight on growth during childhood in Yucatan*. Mexico: a test of the intergenerational effects hypothesis; *Am J Hum Biol* 2009;**21**:657–663.
13. Coelho AM. Baboon dimorphism: growth in weight, length, and adiposity from birth to 8 years of age. In: Watts ES, editor. *Nonhuman primate models for human growth*. New York: Alan R. Liss; 1985. p. 125–59.
14. Pereira ME, Altmann J. Development of social behavior in free-living nonhuman primates. In: Watts ES, editor. *Nonhuman primate models for human growth and development*. New York: Alan R. Liss; 1985. p. 217–309.
15. Prader A. Biomedical and endocrinological aspects of normal growth and development. In: Borms J, Hauspie RR, Sand A, Susanne C, Hebbelinck M, editors. *Human growth and development*. New York: Plenum; 1984. p. 1–22.
16. Bock RD, Thissen D. Statistical problems of fitting individual growth curves. In: Johnston FE, Roche AF, Susanne C, editors. *Human physical growth and maturation, methodologies and factors*. New York: Plenum; 1980. p. 265–90.
17. Sellen DW. Lactation, complementary feeding and human life history. In: Hawkes Paine RR, editor. *The evolution of human life history*. Santa Fe, NM: School of American Research Press; 2006. p. 155–96.
18. World Health Organization Multicentre Growth Reference Study Group. WHO motor development study: windows of achievement for six gross motor development milestones. *Acta Paediatr Suppl* 2006;**450**:86–95.
19. Cabana T, Jolicoeur P, Michaud J. Prenatal and postnatal growth and allometry of stature, head circumference, and brain weight in Québec children. *Am J Hum Biol* 1993;**5**:93–9.
20. Dentition Demirjian A. In: Falkner F, Tanner JM, editors. *Human growth. Postnatal growth*, Vol. 2. New York: Plenum; 1986. p. 269–98.
21. Scammon RE. The measurement of the body in childhood. In: Harris JA, Jackson CM, Paterson DG, Scammon RE, editors. *The measurement of man*. Minneapolis, MN: University of Minnesota Press; 1930. p. 173–215.
22. Leonard WR, Robertson ML. Evolutionary perspectives on human nutrition: the influence of brain and body size on diet and metabolism. *Am J Hum Biol* 1994;**6**:77–88.
23. Leigh SR. Brain growth, life history, and cognition in primate and human evolution. *Am J Primatol* 2004;**62**:139–64.

24. Bogin B, Campbell B. Adrenarche. http://carta.anthropogeny.org/moca/topics/adrenarche
25. Campbell BC. Adrenarche and the evolution of human life history. *Am J Hum Biol* 2006;**18**:569—89.
26. Cole TJ. Children grow and horses race: is the adiposity rebound a critical period for later obesity? *BMC Pediatrics* 2004;**4**:6.
27. Hochberg Z. Evo-devo of child growth II: Human life history and transition between its phases. *Eur J Endocrinol* 2009;**160**:135—41.
28. Walker R, Burger O, Wagner J, Von Rueden CR. Evolution of brain size and juvenile periods in primates. *J Hum Evol* 2006;**51**:480—9.
29. Janson CH, Van Schaik CP. Ecological risk aversion in juvenile primates: slow and steady wins the race. In: Perieira ME, Fairbanks LA, editors. *Juvenile primates: life history, development, and behavior.* New York: Oxford University Press; 1993. p. 57—74.
30. Alexander RD. *How did humans evolve? Reflections on the uniquely unique species. Special Publication No. 1.* Ann Arbor, MI: University of Michigan Museum of Zoology; 1990.
31. Plant TM. Hypothalamic control of the pituitary—gonadal axis in higher primates: key advances over the last two decades. *J Neuroendocrinol* 2008;**20**:719—26.
32. Baughn B, Brault-Dubuc M, Demirjian A, Gagnon G. Sexual dimorphism in body composition changes during the pubertal period: as shown by French—Canadian children. *Am J Phys Anthropol* 1980;**52**:85—94.
33. Bogin B. *Patterns of human growth.* 2nd ed. Cambridge: Cambridge University Press; 1999.
34. Bogin B. *The growth of humanity.* New York: Wiley-Liss; 2001.
35. Walker R, Gurven M, Hill K, Migliano A, Chagnon N, De Souza R, et al. Growth rates and life histories in twenty-two small-scale societies. *Am J Hum Biol* 2006;**18**:295—311.
36. Largo RH, Gasser Th, Prader A, Stutzle, W, Huber PJ. Analysis of the adolescent growth spurt using smoothing spline functions. *Ann Hum Biol* 1978;**5**:421—34.
37. Hamada Y, Udono T. Longitudinal analysis of length growth in the chimpanzee (*Pan troglodytes*). *Am J Phys Anthropol* 2002;**118**:268—84.
38. Pusey AE. Mother—offspring relationships in chimpanzees after weaning. *Anim Behav* 1983;**31**:363—77.
39. Leigh SR. The evolution of human growth. *Evol. Anthropol* 2001;**10**:223—6.
40. Watts ES, Gavan JA. Postnatal growth of nonhuman primates: the problem of the adolescent spurt. *Hum Biol* 1982;**54**:53—70.
41. Hamada Y, Udono T. Longitudinal analysis of length growth in the chimpanzee (*Pan troglodytes*). *Am J Phys Anthropol* 2002;**118**:268—84.
42. Crews DE, Ice GH. Aging, senescence, and human variation. In: Stinson S, Bogin B, O'Rourke D, editors. *Human biology: an evolutionary and biocultural approach.* 2nd ed. New York: Wiley-Blackwell; 2012. p. 639—94.
43. Bonner JT. *Size and cycle.* Princeton, NJ: Princeton University Press; 1965.
44. Bogin B, Smith BH. Evolution of the human life cycle. *Am J Hum Biol* 1996;**8**:703—16.
45. Van Schaik C. *Among orangutans: red apes and the rise of human culture.* Cambridge, MA: Belknap Press; 2004.
46. Zihlman A, Bolter D, Boesch C. Wild chimpanzee dentition and its implications for assessing life history in immature hominin fossils. *Proc Natl Acad Sci USA* 2004;**101**:10541—3.
47. Hrdy SB. *Mother nature: a history of mothers, infants, and natural selection.* New York: Pantheon; 1999.
48. Gurven M, Walker R. Energetic demand of multiple dependents and the evolution of slow human growth. *Proc Biol Sci* 2006;**273**:835—41.
49. Reiches MW, Ellison PT, Lipson SF, Sharrock KC, Gardiner E, Duncan LG. Pooled energy budget and human life history. *Am J Hum Biol* 2009;**21**:421—9.
50. Kaplan H, Hill K, Lancaster J, Hurtado AM. A theory of human life history evolution: diet, intelligence, and longevity. *Evol Anthropol* 2000;**9**:156—85.
51. Blurton Jones NG, Smith LC, O'Connel JF, Handler JS. Demography of the Hadza, an increasing and high density population of savanna foragers. *Am J Phys Anthropol* 1992;**89**:159—81.
52. Hawkes K, O'Connell JF. Blurton Jones NG, Alvarez H, Charnov EL. Grandmothering, menopause, and the evolution of human life histories. *Proc Natl Acad Sci USA* 1998;**95**:1336—9.

53. Lancaster JB, Lancaster CS. Parental investment: the hominid adaptation. In: Ortner DJ, editor. *How humans adapt*. Washington, DC: Smithsonian Institution Press; 1983. p. 33−65.

54. Wood B, Lonergan NL The hominin fossil record: taxa, grades and clades. *J Anat* 2008;**212**:354−76.

55. Bogin B. Evolution of human growth. In: Muehlenbein M, editor. *Human evolutionary biology* Cambridge: Cambridge University Press; 2010. p. 379−95.

56. Bogin B, Smith BH. Evolution of the human life cycle. In: Stinson S, Bogin B, O'Rourke D, editors. *Human biology: an evolutionary and biocultural approach*. 2nd ed. New York: Wiley-Blackwell; 2012. p. 517−87.

57. Darwin C. *The expression of emotions in man and other animals*. London: John Murray; 1872. p. 276.

58. Darwin C. *The expression of emotions in man and other animals*. London: John Murray; 1872. p. 257−258.

59. Altmann J. *Baboon mothers and infants*. Cambridge, MA: Harvard University Press; 1980.

60. Apter D. Serum steroids and pituitary hormones in female puberty: a partly longitudinal study. *Clin Endocrinol* 1980;**12**:107−20.

61. Ibáñez L, de Zegher F, Potau N. Anovulation after precocious pubarche: early markers and time course in adolescence. *J Clin Endocrinol Metab* 1999;**84**:2691−5.

62. Ellison PT, Bogin B, O'Rourke MT. Demography, Part 2: Population growth and fertility regulation. In: Stinson S, Bogin B, O'Rourke D, editors. *Human biology: an evolutionary and biocultural approach*. 2nd ed. New York: Wiley-Blackwell; 2012. p. 759−805.

63. Kramer KL. Variation in juvenile dependence: helping behaviour among Maya children. *Hum Nat* 2002;**13**:299−325.

64. Kramer KL, Ellison PT. Pooled energy budgets: resituating human energy allocation trade-offs. *Evol Anthropol* 2010;**19**:136−47.

65. Bogin B. Why must I be a teenager at all? *New Sci* 1993;**137**:34−8. 6 March.

66. Schlegel A, Barry H. *Adolescence: an anthropological inquiry*. New York: Free Press; 1991.

67. Kramer KL, Lancaster JB. Teen motherhood in cross-cultural perspective. *Ann Hum Biol* 2010;**37**:613−28.

68. Ellison PT. Skeletal growth, fatness, and menarcheal age: a comparison of two hypotheses. *Hum Biol* 1982;**54**:269−81.

69. Worthman CM. Biocultural interactions in human development. In: Perieira ME, Fairbanks LA, editors. *Juvenile primates: life history, development, and behavior*. New York: Oxford University Press; 1993. p. 339−57.

70. Moerman ML. Growth of the birth canal in adolescent girls. *Am J Obstet Gynecol* 1982;**143**:528−32.

71. Rosenberg KR, Trevathan WR. The evolution of human birth. *Sci Am* 2001;**285**(5):72−7.

72. Muller J, Nielsen CT, Skakkebaek NE. Testicular maturation and pubertal growth and development in normal boys. In: Tanner JM, Preece MA, editors. *The physiology of human growth*. Cambridge: Cambridge University Press; 1989. p. 201−7.

73. Instituto Nacional de Estatística. *Resultados definitivos: a natalidade em Portugal, 2001 − Informação à Comunicação Social − Destaque de 4 de Julho de 2001*. Lisbon: Instituto Nacional de Estatística; 1999, 2001.

74. Hill K, Kaplan H. Trade offs in male and female reproductive strategies among the Ache. Parts 1 and 2. In: Betzig L, Borgerhoff-Mulder M, Turke P, editors. *Human reproductive behavior: a Darwinian perspective*. Cambridge: Cambridge University Press; 1988. p. 277−89, 291−305.

75. Condon RG. The rise of adolescence: social change and life stage dilemmas in the central Canadian Arctic. *Hum Org* 1990;**49**:266−79.

76. Smith BH. Physiological age of KMN-WT 15000 and its significance for growth and development of early *Homo*. In: Walker AC, Leakey RF, editors. *The Nariokotome Homo erectus skeleton*. Cambridge, MA: Belknap Press; 1993. p. 195−220.

77. Tobias PV. Puberty, growth, malnutrition and the weaker sex − and two new measures of environmental betterment. *Leech* 1970;**40**:101−7.

78. Malina RM. Growth of muscle tissue and muscle mass. Human growth. In: Falkner F, Tanner JM, editors. *Postnatal growth*., vol. 2. New York: Plenum; 1986. p. 77−99.

79. Smith BH, Tompkins RL. Toward a life history of the Hominidae. *Annu Rev Anthropol* 1995;**25**:257−79.

80. Dean MC, Lucas VS. Dental and skeletal growth in early fossil hominins. *Ann Hum Biol* 2010;**36**:545—61.

81. McPherron SP, Alemseged Z, Marean CW, Wynn JG, Reed D, Geraads D, et al. Evidence for stone-tool-assisted consumption of animal tissues before 3.39 million years ago at Dikika, Ethiopia. *Nature* 2010;**466**:857—60.

82. Tardieu C. Short adolescence in early hominids: infantile and adolescent growth of the human femur. *Am J Phys Anthropol* 1998;**197**:163—78.

83. Tardieu C. Development of the human hind limb and its importance for the evolution of bipedalism. *Evol Anthropol* 2010;**19**:174—86.

84. Dean MC, Smith BH. Growth and development of the Nariokotome youth, KNM-WT 15000. In: Grine FE, Fleagle JG, Leakey RE, editors. *The first humans: origin and evolution of the genus* Homo. Berlin: Springer; 2009. p. 101—20.

85. Aiello L, Key C. Energetic consequences of being a *Homo erectus* female. *J Hum Biol* 2002;**14**:551—65.

86. Bermúdez de Castro JM, Martinón-Torres M, Prado L, Gómez-Robles A, Rosell J, López-Polín L, et al. New immature hominin fossil from European Lower Pleistocene shows the earliest evidence of a modern human dental development pattern. *Proc Natl Acad Sci USA* 2010;**107**: 11739—44.

87. Thompson JL, Nelson AJ. Relative postcranial development of Neanderthals. *J Hum Evol* 1997;**32**:a23—4.

88. Smith TM, Toussaint M, Reid DJ, Olejniczak AJ, Hublin J-J. Rapid dental development in a Middle Paleolithic Belgian Neanderthal. *Proc Natl Acad Sci USA* 2007;**104**:20220—5.

89. Smith TM, Tafforeauc P, Reidd DJ, Pouech J, Lazzarib V, Zermenoa JP, et al. *Dental evidence for ontogenetic differences between modern humans and Neanderthals* 2010;**107**:20923—8.

90. Smith TM, Tafforeau P, Reid DJ, Grün R, Eggins S, Boutakiout M, et al. Earliest evidence of modern human life history in North African *Homo sapiens*. *Proc Natl Acad Sci USA* 2007;**104**: 6128—33.

91. Timiras PS. *Developmental physiology and aging*. New York: MacMillan; 1972.

92. Margulis SW, Atsalis S, Bellem A, Wielebnowski N. Assessment of reproductive behavior and hormonal cycles in geriatric western lowland gorillas. *Zoo Biol* 2007;**26**:117—39.

93. Atsalis S, Margulis S. Primate reproductive aging: from lemurs to humans. *Interdiscip Top Gerontol* 2008;**36**:186—94.

94. Reynolds RF, Obermeyer CM. Age at natural menopause in Spain and the United States: results from the DAMES project. *Am J Hum Biol* 2005;**17**:331—40.

95. Cohen AA. Female post-reproductive lifespan: a general mammalian trait. *Biol Rev Camb Philos Soc* 2008;**79**:733—50.

96. Finch CE. Evolution and the plasticity of aging in the reproductive schedules in long lived animals: the importance of genetic variation in neuroendocrine mechanisms. *Horm Brain Behav* 2002;**4**:799—820.

97. Shanley DP, Kirkwood TBL. Evolution of the human menopause. *BioEssays* 2001;**23**:282—7.

98. Crews. *Human senescence: evolutionary and biocultural perspectives*. Cambridge: Cambridge University Press; 2003.

99. Bogin B. Childhood, adolescence, and longevity: a multilevel model of the evolution of reserve capacity in human life history. *Am J Hum Biol* 2009;**21**:567—77.

100. Sear R, Mace R, McGregor IA. Maternal grandmothers improve nutritional status and survival of children in rural Gambia. *Proc R Soc Biol Sci B* 2000;**267**:1641—7.

101. Hawkes K, Paine RR. *The evolution of human life history*. Santa Fe, NM: School of American Research Press; 2006.

102. Grady D. Death in birth. *New York Times* June 2009;**25**.

103. Paus T, Keshavan M, Giedd JN. Why do many psychiatric disorders emerge during adolescence? *Nat Rev Neurosci* 2008;**9**:947—57.

SUGGESTED READING

Bogin B. *Patterns of human growth*. 2nd ed; Cambridge: Cambridge University Press; 1999.

Hawkes K, Paine RR. *The evolution of human life history*. Santa Fe, NM: School of American Research Press; 2006.

Smith BH. Life history and the evolution of human maturation. *Evol Anthropol* 1992;**1**:134–42.

INTERNET RESOURCES

Center for Academic Research and Training in Anthropogeny (CARTA): http://carta.anthropogeny.org/
Anthropogeny is the investigation of the origin of man (humans) Oxford English Dictionary; 2006 (1839 Hooper Med Dict, Anthropogeny, the study of the generation of man). This website provides up-to-date information about research on the phenomenon of the human species.

Museum of Comparative Anthropogeny (MOCA): http://carta.anthropogeny.org/moca/about
This collection of comparative information regarding humans and our closest evolutionary cousins (chimpanzees, bonobos, gorillas and orang-utans, i.e. "great apes"), with an emphasis on uniquely human features, is of special value.

Human Life-History Project, University of Sheffield, UK: http://www.huli.group.shef.ac.uk/study.html
This site reports the work of researchers focusing on variation in reproductive success, longevity and the strategies applied by individuals to maximize their evolutionary success in all kinds of human populations.

CHAPTER 12

Early Environments, Developmental Plasticity and Chronic Degenerative Disease

Christopher W. Kuzawa

Department of Anthropology and Cells 2 Society, The Center on Social Disparities and Health, Northwestern University, Evanston, IL 60208, USA

Contents

12.1 INTRODUCTION

In recent generations, populations around the globe have experienced dramatic shifts in the burden of disease, with infections increasingly replaced by chronic degenerative diseases as the major causes of morbidity and mortality. Explaining these trends has been a central problem for demographers, epidemiologists and biological anthropologists for some time.[1] Four decades ago, Omran[2] proposed a demographic explanation for these trends in the concept of the "epidemiologic transition". He noted that when a population succeeds in controlling infant mortality related to early life infections and malnutrition, life expectancy increases, and as a result a greater percentage of the population survives long enough to be affected by chronic degenerative diseases that only emerge at older ages.

N. Cameron & B. Bogin (eds): Human Growth and Development, Second edition.
ISBN 978-0-12-383882-7, Doi: 10.1016/B978-0-12-383882-7.00012-X

In addition to this role of population aging, modifiable lifestyle or environmental factors such as diet, physical activity, excess weight gain, smoking and stress also influence the development of many chronic degenerative diseases. Although some of these factors may be viewed as lifestyle "choices", in many societies, one's exposure to both good and bad environmental and lifestyle influences is often powerfully shaped and constrained by societal factors, such as poverty, inequality, class and various forms of discrimination.[3] Taken together, the aging of the global population, combined with these lifestyle and social—structural influences, have traditionally been viewed as explaining the rising global burden of chronic diseases and their differential impacts among different groups within societies.

This chapter will survey a new literature that is bringing a fresh perspective to our understanding of chronic disease epidemiology. Research in recent decades has shown that prenatal nutrition, stress and other early life factors can influence risk for developing conditions such as hypertension, diabetes, heart attack and stroke in adulthood. These relationships reflect the sensitivity of developmental biology to environmental experiences, which can have lingering effects that influence biology and health later in the life cycle. This research suggests that an individual's risk of developing many adult chronic conditions may be established, in part, by experiences much earlier in the life cycle, often beginning before birth. By extension, some of the burden of disease in the current generation of adults may be traced to the social and environmental experiences of their mothers and other recent ancestors.

This chapter will first review evidence from human populations that developmental responses to early life environments can influence adult risk for many common adult chronic degenerative conditions, with a primary focus on the cardiovascular diseases of hypertension, diabetes, heart attack and stroke. It will continue by briefly reviewing some of the developmental and epigenetic mechanisms known to contribute to these relationships, and then briefly explore the hypothesis that these sensitivities in developmental biology may have evolved to allow individuals to cope with changing environmental conditions. The chapter concludes by considering the insights that this literature sheds on two central problems in public health: the rise of chronic disease in populations experiencing rapid nutritional or lifestyle transition, and the patterns of health disparity that map onto social gradients of inequality related to class, ethnicity or socially defined race.

12.2 DEVELOPMENTAL ORIGINS OF HEALTH AND DISEASE: EVIDENCE AND MECHANISMS

12.2.1 Early Environments, Developmental Biology and Adult Health

Several decades ago, researchers in the UK observed that the risk of dying from cardiovascular disease (CVD), or of suffering from conditions that precede CVD, such as hypertension or diabetes, is highest among individuals who were light as newborns.[4-6] Now, hundreds of human studies have replicated similar findings relating lower birth weight to later CVD in populations across the globe, many using longitudinal designs that follow cohorts of individuals over decades as they age.[7-13] It is now well supported that individuals who were born small are more likely to have hypertension,[14] insulin resistance and diabetes,[15,16] abnormal cholesterol profiles,[17] a high-risk visceral pattern of fat deposition[18] and elevated risk of CVD mortality.[19,20]

Infancy and childhood nutrition and growth also predict adult biological and health outcomes. Not unlike birth size, small size in infancy is associated with higher CVD risk in adulthood, while breast-fed infants have lower rates of hypertension, obesity and diabetes as adults.[21,22] There is also evidence that prenatal and postnatal exposures have interactive effects on adult health. For instance, being born small but gaining weight rapidly during childhood predicts the same cluster of adult chronic diseases.[23,24] Thus, it appears that the combination of small birth size and rapid weight gain during postnatal life may be an especially high-risk scenario with respect to developing adult CVD (Figure 12.1).

Although much of this research has focused on nutritional stress, psychological stressors experienced by the mother during or even before pregnancy can lead to similar changes in disease risk in her adult offspring, and these effects can occur even in the

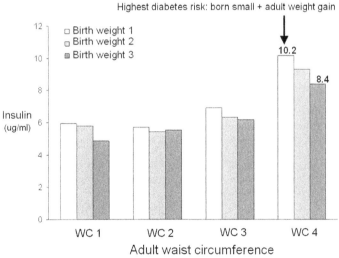

Figure 12.1 How adult fasting insulin relates to birth weight and adult waist circumference among young men living in Cebu City, the Philippines (unpublished data). Men with more abdominal body fat as adults have higher fasting insulin, indicating a higher risk of developing diabetes in the future. Note that the inverse relationship between birth weight and fasting insulin is strongest among the men who are heaviest as adults, and that the highest diabetes risk is found in men who were light at birth but then gained the most weight by adulthood.

absence of changes in birth weight. The fetus is normally shielded from exposure to the glucocorticoid hormone cortisol (a key stress hormone) produced by the mother's body by placental enzymes that inactivate the hormone. The placental capacity to buffer the fetus can be exceeded when the mother is severely stressed, leading to fetal exposure to maternal stress hormones. This, in turn, can contribute to reduced birth size by either directly reducing fetal growth rate or leading to early pregnancy termination.[25] When the fetus is exposed to high levels of cortisol, this can lead to similar changes in CVD risk as observed after fetal nutrient restriction, including high blood pressure, changes in stress reactivity, a tendency to deposit fat abdominally and resistance to the effects of insulin.[26] Collectively, this research is making clear that nutritional or psychosocial stress experienced by the fetus before birth can influence the risk of developing CVD and other chronic degenerative diseases in adulthood.

12.2.2 Mechanisms of Developmental Programming

What biological mechanisms might account for these relationships? Many of these studies relate adult health to birth weight, which is inherently challenging to interpret as a biological measure. Because birth weight partly reflects genetic factors, a relationship between birth weight and adult biology or disease risk could simply reflect the effects of any genes that influence both fetal growth rate and metabolic or physiological processes that contribute to chronic disease risk in adulthood. For instance, insulin not only is related to glucose metabolism and risk of diabetes, but also helps to regulate fetal growth rate. Thus, if individuals within a population vary in which insulin-influencing genes they carry, this could result in a correlation between fetal growth and adult risk of diabetes simply as a result of genetic correlations.[27,28] Although birth weight is a complex multifactorial phenotype,[29] there is now extensive evidence that the relationship between birth outcomes such as birth weight and adult chronic disease are not simply due to genetic influences of this sort.

First, heritabilities for birth weight tend to be quite low. Based upon twin registries, heritabilities for birth weight are typically reported in the range 0.2—0.4 (e.g. Refs 30—32), with national birth weight registry studies finding similar estimates (0.31 for birth weight and 0.27 for birth length in all Norwegian births from 1967 to 2004[33]). This implies that most of the variance in birth weight found within these populations traces to factors other than shared genetic ancestry. Other studies show that maternal influences such as nutritional status, exposure to stress or other factors influencing blood flow to the endometrial lining or placenta are important determinants of a baby's birth size.[34] Importantly, among monozygotic twins, who share identical genomes, the twin born lighter has elevated risk for adverse changes in body composition, risk for diabetes and hypertension later in life,[35,36] showing that differences in birth size predict adult CVD risk among genetically identical siblings.

Perhaps the most important evidence that gestational experiences shape future adult health comes from animal model research, which has used experimentally induced stressors to replicate many of the disease outcomes found in relation to lower birth weight in human populations.[37] For instance, restricting the nutritional intake of pregnant rats, mice or sheep, or directly restricting blood flow to the fetus, increases postnatal blood pressure, cholesterol, abdominal fat deposition and diabetes risk in offspring.[38,39]

Several types of biological adjustment are made by the developing fetus in response to prenatal stressors that contribute to these long-term changes in disease risk. All are examples of *developmental plasticity*, which may be defined as the capacity of the developing body to modify its structure and function in response to environmental or behavioral experiences. The most straightforward mechanism of plasticity involves changes in growth of a tissue or an organ as reflected in size or cell number. For instance, the kidneys of prenatally undernourished individuals tend to be smaller and have fewer nephrons, which increases the risk of hypertension and renal failure in adulthood.[40,41] Similarly, changing the number or type of muscle cells can modify the body's ability to clear glucose from the bloodstream, leading to changes in insulin sensitivity and diabetes risk.[42]

One increasingly well-studied set of mechanisms linking early environments with adult health involves *epigenetic* changes, which are defined as chemical modifications that change the pattern of gene expression in a specific tissue or organ without changing the nucleotide sequences of the DNA.[43,44] Several epigenetic mechanisms have received considerable attention for their likely role as links between early environments and adult health. Chemical modification of histone proteins that the DNA strands are wound around in the cell nucleus can lead to tighter or looser DNA packing in the region of specific genes, reducing or enhancing gene expression, respectively. Methyl groups can also be attached in regions adjacent to specific gene promoters ("methylation"), which can impede binding by transcription factors and thereby silence gene expression in that cell.[45]

Experimental studies using animal models show that modification of nutritional or other characteristics of prenatal or early postnatal rearing environments can lead to durable epigenetic changes that persist into later life to influence biology and underlying processes that contribute to disease risk.[44,46,47] For instance (Figure 12.2A), restricting the protein intake of pregnant rats reduces methylation of the promoter region of the gene that encodes an important stress hormone receptor [glucocorticoid receptor (GR)] in the liver of adult offspring.[48] By reducing methylation, which silences gene expression, this intervention *increases* expression of this receptor, enhancing the liver's metabolic response to stress.[49] In a similar rat model, maternal protein restriction was found to reduce methylation of the angiotensinogen receptor gene in the adrenal gland. The resulting *enhanced* capacity for expression of this gene could contribute to the high blood pressure observed in these animals.[50]

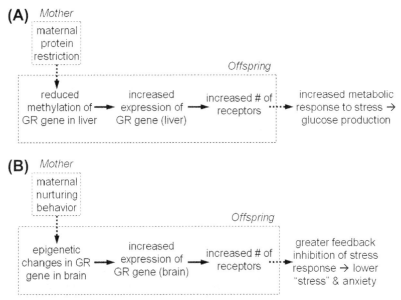

Figure 12.2 Two examples of how early maternal experience or behavior can shape offspring biology via epigenetic changes in gene regulation. (A) When pregnant rats are fed a protein-restricted diet, this can reduce methylation of the receptor that binds to and senses stress hormones (glucocorticoids) in the liver of offspring.[48] Because methylation generally suppresses gene expression, reducing methylation increases expression at the GR gene and thus increases the number of glucocorticoid receptors expressed in the liver. When glucocorticoids increase as a result of stress, these animals have an accentuated capacity to produce glucose for use as energy. This can help them cope with the stressor, but can also heighten risk of diabetes. (B) When rats are raised by nurturing females, they exhibit complex epigenetic changes that increase expression of glucocorticoid receptors, this time in hippocampal (brain) neurons.[51] When these animals experience stress later in life, the increased number of receptors allows the brain to quickly sense rising hormone levels and to shut down further stress hormone production. This contributes to a blunted stress response that reduces anxiety.

The postnatal environment also has important influences on the epigenome. Well-described rat studies have shown that a nurturing maternal rearing style can lead to epigenetic changes in the brain of offspring (Figure 12.2B), lowering their reactivity to stress and reducing anxiety as adults.[51,52] In humans, untreated maternal depression or famine exposure during pregnancy has been shown to predict similar epigenetic changes in offspring, suggesting that comparable epigenetic processes may link early experiences with adult health in humans.[53,54]

In summary, we now have good confidence that the widely documented relationships between early life measures such as birth weight and later CVD partially reflect the effects of the gestational and infancy environments on the development of biological systems, including effects on how the body manages glucose and lipids, deposits fat, regulates blood pressure and responds to stress (for further review see Gluckman et al.[55]) (Figure 12.3). These effects typically reflect changes in the growth and development of

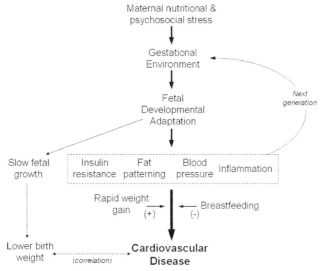

Figure 12.3 Developmental origins of adult chronic disease. Maternal nutritional or psychosocial stressors influence the nutrient and hormonal characteristics of the gestational environment experienced by the developing fetus, which can durably modify multiple biological functions and elevate adult risk for cardiovascular disease (CVD). The arrow relating lower birth weight with CVD risk is dashed to indicate that the relationship is only correlational rather than causal. CVD risk is also elevated by rapid postnatal weight gain, and may be reduced by being breast fed. Some of the adverse adult biological effects of compromised early environments may in turn influence the maternal/gestational environment experienced by the next generation before birth, potentially perpetuating patterns of adverse health across generations.

specific organs and tissues or modifications in the regulation of hormones, metabolism or physiology. They are increasingly being traced to durable, environmentally induced epigenetic changes in the chromosomes that modify gene expression in specific tissues or organs without modifying the DNA itself.

12.3 DEVELOPMENTAL PLASTICITY AS A MEANS OF ADAPTATION

12.3.1 Early Life Developmental Plasticity and Adaptation to Ecological Change

Why might the body modify its developmental biology in response to early life stressors? Some of the lingering effects of early experience on adult health simply reflect unintended side-effects of adaptations made by the fetus to improve its chances of surviving a nutritionally stressful prenatal environment. For instance, a smaller fetus has lower nutritional needs, and the reduced ability of their smaller muscle mass to clear glucose from the blood stream could spare energy for the glucose-hungry brain, which is fragile and large relative to the body during early life.[56,57] Another possibility, which in some ways may be the most straightforward, is incomplete buffering of the fetus against maternal stressors.[58] A failure of the mother's body, or the placenta, to fully buffer the developing body from nutritional or other forms of stress can lead to impairments with long-term unintended side-effects on biological function and health.

Although these non-functional explanations for plasticity clearly are important, it has also been speculated that developmental plasticity could, in some instances, allow the fetus to prepare for conditions likely to be experienced *after* birth[59–62] (Figure 12.4). Some of the adjustments made by the nutritionally stressed fetus in utero, such as a tendency to deposit more abdominal body fat, and the reduced response of muscle to insulin that spares glucose for use elsewhere in the body, could provide advantages after birth if the postnatal environment is also nutritionally stressful.[61,64] In addition, other systems that change biological settings in response to early environments, such as stress physiology,[51] immunity[65] and reproductive biology,[66] might also be "fine-tuned" in response to early experience.

One challenge to this idea comes from the fact that humans have a long lifespan. Because we typically live for many decades, any conditions that we experience during a few months of early development, such as gestation or early infancy, may not be reliable cues of environments likely to be experienced decades in the future.[60,67] One intriguing possibility is that it is precisely the brief and early timing of many of the body's periods of heightened developmental sensitivity that paradoxically *helps* the developing organism overcome the challenge of reliably predicting conditions well into the future.[58,68] Here, the idea is that the mother's physiology could buffer the fetus against the day-to-day, month-to-month or seasonal fluctuations in the environment, while passing along information about local conditions that is more stable and reliable. Because the mother's biology and behavior have been modified by her lifetime of experiences, the nutrients, hormones and other resources that she transfers to the fetus in utero, or to her infant via breast milk, could correlate with her average experiences more than what she is experiencing during any week or month of gestation itself.[60,67]

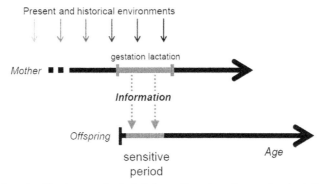

Figure 12.4 Maternal–offspring ecological information transfer and adaptation. The mother's biology and behavior embody a record of her cumulative environmental and social experiences, which can be conveyed to offspring as developmental information via nutrients, hormones, rearing behavior and other cues. *(Source: From Kuzawa and Quinn,[63] with permission.)*

Perhaps the best evidence for such a capacity to convey average, rather than transient, ecological information, comes from studies of the effects of a mother's nutrition on the birth weight of her baby.[60] Studies generally find that birth weights tend to be lighter in populations in which nutrition has been marginal for multiple generations. Despite this evidence for environmental influence on fetal growth rate and birth size, supplementing pregnant women generally has minimal effects on the birth weight of offspring. Thus, it appears that long-term history in an environment may be an important influence on the resources transferred in support of offspring growth, but that fluctuations in intake during pregnancy itself, reflected for instance in dietary supplementation, have comparably modest effects.

This *phenotypic inertia* — reflecting the lingering biological but non-genetic effects of the mother's average experiences in the past — could allow the fetus to track those dynamic features of environments that are relatively stable on the timescale of decades or several generations (see Refs 60 and 68). This could allow adjustment to environmental changes that are too rapid to result in modifications in gene frequencies via natural selection, which requires many generations, but that are too chronic to be buffered efficiently by reversible homeostatic processes. In this way, the mother's body could pass biological "memories", reflecting her own lifetime of experiences in the local environment, to her developing offspring, allowing developmental adjustments to be made in anticipation of conditions likely to be experienced locally.

12.4 IMPLICATIONS OF DEVELOPMENTAL PROGRAMMING FOR HUMAN HEALTH DISPARITIES

The preceding sections have surveyed some of the evidence that early life stressors can have health effects that linger into adulthood and in some instances may even transcend the present generation to be passed on to offspring. These findings hold promise to help explain why health and disease tend to relate strongly to environmental, social and economic conditions both within and between populations. There is growing evidence that early environment-triggered developmental plasticity can help to clarify two broad problems in public health: disease transitions in populations experiencing rapid cultural, nutritional or lifestyle change, and health disparities within populations marked by chronic inequality and social stratification related to class, ethnicity or social "race". The discussion ends with a brief review of the evidence that developmental processes contribute to each of these public health issues.

12.4.1 High-Risk Scenario 1: When Early Life Undernutrition is Followed by Adult Weight Gain

The above text discussed the hypothesis that the fetus has a capacity to "anchor" its nutritional expectations to a gestational signal (e.g. hormones, nutrients) of average recent nutrition as experienced by the mother. This might allow the developing organism to modify its own nutritional expenditure, as reflected in its growth rate, body size and other traits, as locally experienced nutritional conditions change. It is easy to see how metabolic changes that could be favorable to a nutritionally stressed fetus or infant — such as sparing glucose or depositing more fat in the abdomen — might also plant the seeds for heightened risk of developing metabolic diseases if that individual ends up gaining weight rapidly during childhood or as an adult. Thus, any context in which individuals routinely face nutritional stress before birth or during infancy but then gain excess weight during later childhood or as an adult should be associated with a high susceptibility of developing metabolic disease.

The now common finding that CVD risk is highest among individuals who were born small but later put on weight is consistent with this idea.[18,69] Under what societal conditions might this pattern of early dearth followed by later excess be especially prominent or influential within a population? One way is as a result of rapid cultural, political or economic transition.[70] In many societies, industrialization of farming is increasing affordability of cheap calories,[71] while populations are also increasingly relying upon automobiles and other forms of transportation to move from place to place.[72] As individuals take in more calories while expending fewer during the day, weight gain is inevitable. When the transition to relative caloric excess takes place within a single generation, individuals raised under austere nutritional conditions during early life may go on to gain excess weight as older children or adults and have heightened CVD risk as a result. Consistent with this model, stunting — a measure of early life undernutrition — has been shown to be a risk factor for metabolic syndrome and obesity in populations experiencing rapid nutritional transition.[73–75]

Poor early life nutrition may also coexist with adult overnutrition simply because nutritional stressors are often concentrated during periods of heightened nutritional vulnerability early in the life cycle. In contrast to trends towards positive adult energy balance and weight gain in many global populations, the nutritional experiences of infants and young children are often more strongly influenced by common communicable diseases and their underlying social determinants, such as sanitation, crowding and the availability of clean water. That nutritional stress around the age of weaning is often severe is revealed by the mammalian strategy of depositing extra body fat after birth in preparation for weaning. Among mammals, humans give birth to the fattest babies on record, which may help us to prepare for this weaning stress, which is accentuated in our species owing to the need to provide a constant supply of energy for our unusually large and energetically fragile brains.[76]

It is an unfortunate fact that in many developing economies today, nutritional stressors at this early age tend to be common — tracing to factors such as diarrhea and respiratory tract infections — despite the fact that those same individuals may later experience excess weight gain as adults. Because infancy nutrition remains tightly linked to social conditions related to poverty, while the availability of cheap calories is increasingly common and driving adult weight gain, many individuals may now experience early life nutrition stress followed by adult caloric excess even in the absence of rapid societal transition. This is reflected, for instance, in the common co-occurrence of obese and malnourished individuals in the same household within some low-income populations.[77] The body's developmental response to early nutritional stressors can help to explain why these populations often have high rates of cardiovascular and other metabolic diseases.[70]

12.4.2 High-Risk Scenario 2: The Social Origins of Health Disparities Related to Ethnicity, Class and Race

In addition to helping to explain heightened CVD risk in scenarios of early life undernutrition followed by later nutritional excess, the developmental origins framework can help to explain why CVD and related metabolic diseases tend to map onto social categories such as ethnicity, class and race. These "health disparities" are among the most pressing of contemporary public health issues.[3]

As one well-studied example, African-Americans on average have a higher burden of many CVDs, including hypertension and diabetes, when compared to other demographic subgroups in the USA. When studies find that self-identified race is still a significant predictor of these conditions after statistically adjusting for various lifestyle and socio-economic characteristics, some researchers have been tempted to conclude that genetic factors might explain the black—white difference in health. The problem is that there is in fact very little evidence for a genetic contribution to these health differences,[78,79] which instead relate powerfully to social and environmental factors. Importantly, these factors often reflect influences beyond the control of individuals, such as unequal patterns of opportunity or stress. For instance, chronic stressors such as discrimination, or living in segregated low-income neighborhoods with high crime rates or few safe opportunities for exercise, are important influences on stress levels, the rate of weight gain and the prevalence of high blood pressure.[80]

These effects of unhealthy environments on adult health are not surprising, and in fact are well established. Where these traditional effects of stressful environments converge with the present story is in the realization that they also contribute to poor birth outcomes and compromised gestational environments, which have health effects that can linger into adulthood, and even transcend the present generation of adults. Indeed, African-Americans not only have higher rates of adult CVD, but are also disproportionately affected by the early life antecedents to these conditions, such as a lower mean birth weight, intrauterine growth retardation and premature delivery. Importantly, these early life health disparities have also been linked to experiences of stress and discrimination rather than to genes.[81,82]

Bringing these threads of evidence together suggests that the developmental and intergenerational processes discussed above are likely to be an important part of the story of US health disparities.[03] Imagine the following sequence of effects. First, a pregnant mother experiences chronic stress that elevates her production of stress hormones (e.g. cortisol) during pregnancy. As the level of this hormone rises, the ability of the placenta to shield the fetus from it is exceeded, and the fetus is exposed to high levels of this maternal hormone. This modifies various aspects of developmental biology, for instance by changing how the offspring's body regulates stress hormones, glucose homeostasis, blood pressure or fat deposition. Some of these changes involve epigenetic or developmental modifications in the regulation of organs, tissues or metabolism, which are relatively durable. Later in life, the offspring — now an adult — has higher glucose, insulin, blood pressure and stress hormones as a result of these early life effects. In this way, stressors experienced unequally by the adults of one generation (the mother during pregnancy) might contribute to adult health disparities in the next generation of offspring.

Unfortunately, the story does not stop here, however, because among these adult offspring are females who become pregnant and have offspring of their own. How might the

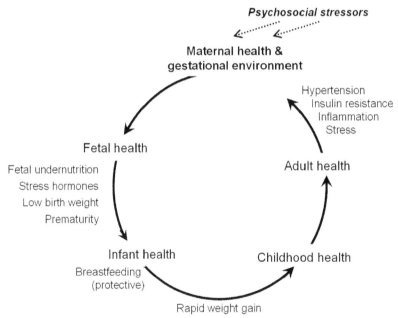

Figure 12.5 A life-course, intergenerational model of health disparities. A mother's experience of stressors influences biological settings and health of her offspring. In female offspring, some of these changes persist into adulthood to influence the gestational environment experienced by the grand-offspring. Thus, developmental responses to early environments can perpetuate patterns of health disparity not only across life cycles but potentially also across generations. *(Source: Modified after Kuzawa.[68])*

original stress, experienced by the mother, affect the health of her grand-offspring? The simple answer is that we do not know because studies have yet to investigate this definitively in human populations. But, there are good reasons to suspect that the effects of the original stressor could be passed on, albeit more weakly, across several generations. This is because some of the long-term effects of an adverse gestational environment on adult health in offspring, such as insulin resistance, high blood pressure or inflammation, can negatively affect the gestational environments experienced by the next generation, and can also lead to low-birth-weight deliveries (Figure 12.5). In this way, life-course influences of early life stressors can be transformed into intergenerational pathways for the perpetuation of health disparities across generations.[84] This type of intergenerational transmission is believed to help explain why conditions such as gestational diabetes can influence health in multiple generations of offspring, potentially amplifying obesity or diabetes rates across generations.[83]

12.5 CONCLUSION

This chapter has discussed how fetal and infancy stressors can influence developmental biology to modify one's risk of developing many common degenerative diseases as one ages, including but not limited to hypertension, diabetes, heart attacks and stroke. This research shows how adult health in one generation may be linked with the environmental experiences of recent ancestors, especially the mother during and before pregnancy. This chapter also surveyed the biological mechanisms underlying these effects, and considered the insights that these findings bring to our understanding of two common contexts for socially driven disease in today's human populations. The first involves situations in which an individual experiences nutritional or infectious disease stressors early in life but subsequently gains weight rapidly during childhood or adulthood owing to caloric excess and positive energy balance. The second example is the tendency for health inequality to map onto social gradients of privilege, opportunity, discrimination and stress within societies, as exemplified by the stark differences in health that typically relate to class, ethnicity and socially defined race. The developmental origins framework shows one set of mechanisms by which social inequalities can become embodied physically as health inequalities in the next generation, operating through the effects of maternal biology on offspring development. Collectively, these findings point to the long-term benefits to society of ensuring adequate nutrition, health care and buffering of stress among pregnant women and their young offspring.

REFERENCES

1. Barrett R, Kuzawa C, McDade T, Armelagos G. Emerging and re-emerging infectious diseases: the third epidemiologic transition. *Annu Rev Anthropol* 1998;**27**:247–71.
2. Omran AR. The epidemiologic transition. A theory of the epidemiology of population change. *Milbank Mem Fund Q* 1971;**49**:509–38.

3. Marmot M, Ryff CD, Bumpass LL, Shipley M, Marks NF. Social inequalities in health: next questions and converging evidence. *Soc Sci Med* 1997;**1982**(44):901–10.

4. Barker DJ, Osmond C, Golding J, Kuh D, Wadsworth ME. Growth in utero, blood pressure in childhood and adult life, and mortality from cardiovascular disease. *BMJ* 1989;**298**:564–7.

5. Barker DJ, Osmond C. Infant mortality, childhood nutrition, and ischaemic heart disease in England and Wales. *Lancet* 1986;**i**:1077–81.

6. Barker D. *Mothers, babies, and disease in later life*. London: BMJ Publishing; 1994.

7. Adair LS, Kuzawa CW, Borja J. Maternal energy stores and diet composition during pregnancy program adolescent blood pressure. *Circulation* 2001;**104**:1034–9.

8. Dalziel SR, Parag V, Rodgers A, Harding JE. Cardiovascular risk factors at age 30 following pre-term birth. *Int J Epidemiol* 2007;**36**:907–15.

9. Gupta M, Gupta R, Pareek A, Bhatia R, Kaul V. Low birth weight and insulin resistance in mid and late childhood. *Indian Pediatr* 2007;**44**:177–84.

10. Law CM, Egger P, Dada O, Delgado H, Kylberg E, Lavin P, et al. Body size at birth and blood pressure among children in developing countries. *Int J Epidemiol* 2001;**30**:52–7.

11. Levitt NS, Lambert EV, Woods D, Hales CN, Andrew R, Seckl JR, et al. Impaired glucose tolerance and elevated blood pressure in low birth weight, nonobese, young South African adults: early programming of cortisol axis. *J Clin Endocrinol Metab* 2000;**85**:4611–8.

12. Miura K, Nakagawa H, Tabata M, Morikawa Y, Nishijo M, Kagamimori S. Birth weight, childhood growth, and cardiovascular disease risk factors in Japanese aged 20 years. *Am J Epidemiol* 2001;**153**:783–9.

13. Tian JY, Cheng Q, Song XM, Li G, Jiang GX, Gu YY, et al. Birth weight and risk of type 2 diabetes, abdominal obesity and hypertension among Chinese adults. *Eur J Endocrinol* 2006;**155**:601–7.

14. Adair L, Dahly D. Developmental determinants of blood pressure in adults. *Annu Rev Nutr* 2005;**25**:407–34.

15. Eriksson JG, Forsén T, Tuomilehto J, Jaddoe VW, Osmond C, Barker DJ. Effects of size at birth and childhood growth on the insulin resistance syndrome in elderly individuals. *Diabetologia* 2002;**45**:342–8.

16. Yajnik CS. Early life origins of insulin resistance and type 2 diabetes in India and other Asian countries. *J Nutr* 2004;**134**:205–10.

17. Kuzawa CW, Adair LS. Lipid profiles in adolescent Filipinos: relation to birth weight and maternal energy status during pregnancy. *Am J Clin Nutr* 2003;**77**:960–6.

18. Oken E, Gillman MW. Fetal origins of obesity. *Obes Res* 2003;**11**:496–506.

19. Leon DA, Lithell HO, Vâgerö D, Koupilová I, Mohsen R, Berglund L, et al. Reduced fetal growth rate and increased risk of death from ischaemic heart disease: cohort study of 15 000 Swedish men and women born 1915–29. *BMJ* 1998;**317**:241–5.

20. Huxley R, Owen CG, Whincup PH, Cook DG, Rich-Edwards J, Smith GD, et al. Is birth weight a risk factor for ischemic heart disease in later life? *Am J Clin Nutr* 2007;**85**:1244–50.

21. Arenz S, Ruckerl R, Koletzko B, von Kries R. Breast-feeding and childhood obesity – a systematic review. *Int J Obes Relat Metab Disord* 2004;**28**:1247–56.

22. Lawlor DA, Riddoch CJ, Page AS, Andersen LB, Wedderkopp N, Harro M, et al. Infant feeding and components of the metabolic syndrome: findings from the European Youth Heart Study. *Arch Dis Child* 2005;**90**:582–8.

23. Adair LS, Cole TJ. Rapid child growth raises blood pressure in adolescent boys who were thin at birth. *Hypertension* 2003;**41**:451–6.

24. Ong KK. Size at birth, postnatal growth and risk of obesity. *Horm Res* 2006;**65**(Suppl. 3):65–9.

25. Challis JR, Bloomfield FH, Bocking AD, Casciani V, Chisaka H, Connor K, et al. Fetal signals and parturition. *J Obstet Gynaecol Res* 2005;**31**:492–9.

26. Seckl JR, Meaney MJ. Glucocorticoid programming. *Ann NY Acad Sci* 2004;**1032**:63–84.

27. Hattersley AT, Tooke JE. The fetal insulin hypothesis: an alternative explanation of the association of low birthweight with diabetes and vascular disease. *Lancet* 1999;**353**:1789–92.

28. Freathy RM, Weedon MN, Bennett A, Hypponen E, Relton CL, Knight B, et al. Type 2 diabetes TCF7L2 risk genotypes alter birth weight: a study of 24,053 individuals. *Am J Hum Genet* 2007;**80**:1150–61.

29. Kuzawa CW. Modeling fetal adaptation to nutrient restriction: testing the fetal origins hypothesis with a supply—demand model. *J Nutr* 2004;**134**:194—200.

30. Whitfield JB, Treloar SA, Zhu G, Martin NG. Genetic and non-genetic factors affecting birth-weight and adult body mass index. *Twin Res* 2001;**4**:365—70.

31. Vlietinck R, Derom R, Neale MC, Maes H, van Loon H, Derom C, et al. Genetic and environmental variation in the birth weight of twins. *Behav Genet* 1989;**19**:151—61.

32. Baird J, Osmond C, MacGregor A, Snieder H, Hales CN, Phillips DI. Testing the fetal origins hypothesis in twins: the Birmingham twin study. *Diabetologia* 2001;**44**:33—9.

33. Lunde A, Melve KK, Gjessing HK, Skjaerven R, Irgens LM. Genetic and environmental influences on birth weight, birth length, head circumference, and gestational age by use of population-based parent—offspring data. *Am J Epidemiol* 2007;**165**:734—41.

34. Gluckman PD, Hanson MA. Maternal constraint of fetal growth and its consequences. *Semin Fetal Neonatal Med* 2004;**9**:419—25.

35. Bo S, Cavallo-Perin P, Scaglione L, Ciccone G, Pagano G. Low birthweight and metabolic abnormalities in twins with increased susceptibility to type 2 diabetes mellitus. *Diabet Med* 2000;**17**:365—70.

36. Iliadou A, Cnattingius S, Lichtenstein P. Low birthweight and type 2 diabetes: a study on 11 162 Swedish twins. *Int J Epidemiol* 2004;**33**:948—53.

37. Symonds ME, Mostyn A, Pearce S, Budge H, Stephenson T. Endocrine and nutritional regulation of fetal adipose tissue development. *J Endocrinol* 2003;**179**:293—9.

38. Langley-Evans SC, Langley-Evans AJ, Marchand MC. Nutritional programming of blood pressure and renal morphology. *Arch Physiol Biochem* 2003;**111**:8—16.

39. McMillen IC, Robinson JS. Developmental origins of the metabolic syndrome: prediction, plasticity, and programming. *Physiol Rev* 2005;**85**:571—633.

40. Lampl M, Kuzawa CW, Jeanty P. Infants thinner at birth exhibit smaller kidneys for their size in late gestation in a sample of fetuses with appropriate growth. *Am J Hum Biol* 2002;**14**:398—406.

41. Luyckx VA, Brenner BM. Low birth weight, nephron number, and kidney disease. *Kidney Int* 2005;(Suppl. 97):S68—77.

42. Jensen CB, Storgaard H, Madsbad S, Richter EA, Vaag AA. Altered skeletal muscle fiber composition and size precede whole-body insulin resistance in young men with low birth weight. *J Clin Endocrinol Metab* 2007;**92**:1530—4.

43. Jenuwein T, Allis CD. Translating the histone code. *Science* 2001;**293**:1074—80.

44. Waterland RA, Michels KB. Epigenetic epidemiology of the developmental origins hypothesis. *Annu Rev Nutr* 2007;**27**:363—88.

45. Berger SL. The complex language of chromatin regulation during transcription. *Nature* 2007;**447**:407—12.

46. Gluckman PD, Hanson MA, Beedle AS. Non-genomic transgenerational inheritance of disease risk. *Bioessays* 2007;**29**:145—54.

47. Thayer ZM, Kuzawa CW. Biological memories of past environments: epigenetic pathways to health disparities. *Epigenetics* 2011;**6**:798—803.

48. Lillycrop KA, Phillips ES, Jackson AA, Hanson MA, Burdge GC. Dietary protein restriction of pregnant rats induces and folic acid supplementation prevents epigenetic modification of hepatic gene expression in the offspring. *J Nutr* 2005;**135**:1382—6.

49. Lillycrop KA, Slater-Jefferies JL, Hanson MA, Godfrey KM, Jackson AA, Burdge GC. Induction of altered epigenetic regulation of the hepatic glucocorticoid receptor in the offspring of rats fed a protein-restricted diet during pregnancy suggests that reduced DNA methyltransferase-1 expression is involved in impaired DNA methylation and changes in histone modifications. *Br J Nutr* 2007;**97**:1064—73.

50. Bogdarina I, Welham S, King PJ, Burns SP, Clark AJ. Epigenetic modification of the renin—angiotensin system in the fetal programming of hypertension. *Circ Res* 2007;**100**:520—6.

51. Weaver IC, Cervoni N, Champagne FA, D'Alessio AC, Sharma S, Seckl JR, et al. Epigenetic programming by maternal behavior. *Nat Neurosci* 2004;**7**:847—54.

52. Diorio J, Meaney MJ. Maternal programming of defensive responses through sustained effects on gene expression. *J Psychiatry Neurosci* 2007;**32**:275—84.

53. Oberlander TF, Weinberg J, Papsdorf M, Grunau R, Misri S, Devlin AM. Prenatal exposure to maternal depression, neonatal methylation of human glucocorticoid receptor gene (NR3C1) and infant cortisol stress responses. *Epigenetics* 2008;**3**:97—106.

54. Heijmans BT, Tobi EW, Stein AD, Putter H, Blauw GJ, Susser ES, et al. Persistent epigenetic differences associated with prenatal exposure to famine in humans. *Proc Natl Acad Sci USA* 2008;**105**:17046—9.

55. Gluckman PD, Hanson MA, Cooper C, Thornburg KL. Effect of in utero and early-life conditions on adult health and disease. *N Engl J Med* 2008;**359**:61—73.

56. Hales C, Barker D. Type 2 (non-insulin dependent) diabetes mellitus: the thrifty phenotype hypothesis. *Diabetologia* 1992;**35**:595—601.

57. Kuzawa CW. Beyond feast—famine: brain evolution, human life history, and the metabolic syndrome. In: Muehlenbein M, editor. *Human evolutionary biology.* Cambridge: Cambridge University Press; 2010. p. 518—27.

58. Kuzawa C, Thayer Z. Timescales of human adaptation: the role of epigenetic processes. *Epigenomics* 2011;**3**:221—34.

59. Bateson P. Fetal experience and good adult design. *Int J Epidemiol* 2001;**30**:928—34.

60. Kuzawa CW. Fetal origins of developmental plasticity: are fetal cues reliable predictors of future nutritional environments? *Am J Hum Biol* 2005;**17**:5—21.

61. Gluckman PD, Hanson M. *The fetal matrix: evolution, development, and disease.* New York: Cambridge University Press; 2005.

62. Wells J. Environmental quality, developmental plasticity and the thrifty phenotype: a review of evolutionary models. *Evol Bioinform* 2007;**3**:109—20.

63. Kuzawa C, Quinn E. Developmental origins of adult function and health: evolutionary hypotheses. *Annu Rev Anthropol* 2009;**38**:131—47.

64. Kuzawa C, Gluckman P, Hanson M, Beedle A. Evolution, developmental plasticity, and metabolic disease. In: Stearns S, Koella K, editors. *Evolution in health and disease.* 2nd ed. Oxford: Oxford University Press; 2008. p. 253—64.

65. McDade TW, Rutherford J, Adair L, Kuzawa CW. Early origins of inflammation: microbial exposures in infancy predict lower levels of C-reactive protein in adulthood. *Proc Biol Sci* 2010;**277**: 1129—37.

66. Kuzawa CW, McDade TW, Adair LS, Lee N. Rapid weight gain after birth predicts life history and reproductive strategy in Filipino males. *Proc Natl Acad Sci USA* 2010;**107**:16800—5.

67. Wells JC. The thrifty phenotype hypothesis: thrifty offspring or thrifty mother? *J Theor Biol* 2003;**221**:143—61.

68. Kuzawa C. The developmental origins of adult health: intergenerational inertia in adaptation and disease. In: Trevathan W, Smith E, McKenna J, editors. *Evolutionary medicine and health: new perspectives.* New York: Oxford University Press; 2008. p. 325—49.

69. Fagerberg B, Bondjers L, Nilsson P. Low birth weight in combination with catch-up growth predicts the occurrence of the metabolic syndrome in men at late middle age: the Atherosclerosis and Insulin Resistance study. *J Intern Med* 2004;**256**:254—9.

70. Victora CG, Adair L, Fall C, Hallal PC, Martorell R, Richter L, et al. Maternal and child undernutrition: consequences for adult health and human capital. *Lancet* 2008;**371**:340—57.

71. Drewnowski A, Popkin B. The nutrition transition: new trends in the global diet. *Nutr Rev* 1997;**55**:31—43.

72. Popkin BM, Gordon-Larsen P. The nutrition transition: worldwide obesity dynamics and their determinants. *Int J Obes Relat Metab Disord* 2004;**28**(Suppl. 3):S2—9.

73. Popkin BM, Richards MK, Montiero CA. Stunting is associated with overweight in children of four nations that are undergoing the nutrition transition. *J Nutr* 1996;**126**:3009—16.

74. Steyn K, Bourne L, Jooste P, Fourie JM, Rossouw K, Lombard C. Anthropometric profile of a black population of the Cape Peninsula in South Africa. *East Afr Med J* 1998;**75**:35—40.

75. Florencio TT, Ferreira HS, Cavalcante JC, Luciano SM, Sawaya AL. Food consumed does not account for the higher prevalence of obesity among stunted adults in a very-low-income population in the Northeast of Brazil (Maceio, Alagoas). *Eur J Clin Nutr* 2003;**57**:1437—46.

76. Kuzawa CW. Adipose tissue in human infancy and childhood: an evolutionary perspective. *Am J Phys Anthropol* 1998;(Suppl. 27):177—209.

77. Garrett JL, Ruel MT. Stunted child—overweight mother pairs: prevalence and association with economic development and urbanization. *Food Nutr Bull* 2005;**26**:209—21.

78. Gravlee CC, Non AL, Mulligan CJ. Genetic ancestry, social classification, and racial inequalities in blood pressure in Southeastern Puerto Rico. *PLoS ONE* 2009;**4**(9):e6821.

79. Cooper RS, Kaufman JS. Race and hypertension: science and nescience. *Hypertension* 1998;**32**:813—6.

80. Williams DR, Collins C. US socioeconomic and racial differences in health: patterns and explanations. *Annu Rev Sociol* 1995;**21**:349—86.

81. Alexander GR, Kogan MD, Himes JH, Mor JM, Goldenberg R. Racial differences in birthweight for gestational age and infant mortality in extremely-low-risk US populations. *Paediatr Perinat Epidemiol* 1999;**13**:205—17.

82. Collins Jr JW, Wu SY, David RJ. Differing intergenerational birth weights among the descendants of US-born and foreign-born whites and African Americans in Illinois. *Am J Epidemiol* 2002;**155**:210—6.

83. Kuzawa CW, Sweet E. Epigenetics and the embodiment of race: developmental origins of US racial disparities in cardiovascular health. *Am J Hum Biol* 2009;**21**:2—15.

84. Drake A, Walker B. The intergenerational effects of fetal programming: non-genomic mechanisms for the inheritance of low birth weight and cardiovascular risk. *J Endocrinol* 2004;**180**:1—16.

SUGGESTED READING

Barker DJ, Osmond C, Golding J, Kuh D, Wadsworth ME. Growth in utero, blood pressure in childhood and adult life, and mortality from cardiovascular disease. *BMJ* 1989;**298**:564—7.
A classic early study demonstrating a link between size at birth and adult cardiovascular disease mortality.

Drake A, Walker B. The intergenerational effects of fetal programming: non-genomic mechanisms for the inheritance of low birth weight and cardiovascular risk. *J Endocrinol* 2004;**180**:1—16.
Presents a model for the intergenerational perpetuation of the adverse effects of maternal stress across multiple generations.

Gluckman PD, Hanson MA, Cooper C, Thornburg KL. Effect of in utero and early-life conditions on adult health and disease. *N Engl J Med* 2008;**359**:61—73.
Reviews evidence for developmental influences on adult health and disease.

Kuzawa C, Quinn E. Developmental origins of adult function and health: evolutionary hypotheses. *Annu Rev Anthropol* 2009;**38**:131—47.
Reviews evidence that early developmental plasticity allows organisms to adapt to the environment.

Victora CG, Adair L, Fall C, Hallal PC, Martorell R, Richter L, et al. Maternal and child undernutrition: consequences for adult health and human capital. *Lancet* 2008;**371**:340—57.
Reviews evidence for long-term effects of early environments on adult health in five cohort studies from lower and middle income nations.

Waterland RA, Michels KB. Epigenetic epidemiology of the developmental origins hypothesis. *Annu Rev Nutr* 2007;**27**:363—88.
Reviews evidence for epigenetic contributions to the developmental origins of adult disease.

INTERNET RESOURCES

Official webpage of the International Society for Developmental Origins of Health and Disease: http://www.mrc.soton.ac.uk/dohad/

CHAPTER *13*

Leg Length, Body Proportion, Health and Beauty[1]

Barry Bogin

School of Sport, Exercise and Health Sciences, Loughborough University, Leicestershire LE11 3TU, UK

Contents

13.1 INTRODUCTION

This chapter begins by placing the study of human body size and shape in a historical context. Human beings display a variety of sizes, shapes, colors, temperaments and other phenotypic characteristics. Professional anthropologists, physicians and others have debated the cause and significance of human phenotypic variation for centuries. Much of the historical discourse focused on concepts of "race" and some of the dispute centered on the human status of various living groups of people.[1,2] Serious proposals about the

---

This chapter is an updated version of an article which appeared in the International Journal of Environmental Research and Public Health, www.mdpi.com/journal/ijerph, an open access journal. The original citation is: Bogin B, Varela-Silva MI. Leg length, body proportion, and health: a review with a note on beauty. Int J Environ Res Public Health 2010;7:1047–5; doi:10.3390/ijerph7031047

N.Cameron & B. Bogin (eds): Human Growth and Development, Second edition.
ISBN 978-0-12-383882-7, Doi: 10.1016/B978-0-12-383882-7.00013-1
343

hierarchy of humanness appeared as recently as 1962 with the publication of *The Origin of Races* by Carleton Coon,[3] Professor of Anthropology at the University of Pennsylvania. Coon divided living peoples of the world into five "races" based, in part, on body size and proportions. According to Coon, the Australian Aborigines (designated "Australoids" by Coon) have exceptionally long legs in proportion to stature, and African pygmies ("Congoids" in Coon's taxonomy) have exceptionally short stature, long arms relative to leg length, and especially short lower legs. In Coon's words, "Their manner of dwarfing verges in the achondroplastic …".[3, p. 653] Moreover, Coon asserted that both "races" crossed the species threshold between *Homo erectus* and *Homo sapiens* only in the last 10,000—50,000 years. In contrast, Coon proposed that ancient Europeans (called "Caucasoids" by Coon) had crossed the *H. sapiens* threshold about 200,000 years ago. According to Coon, the ancient Europeans were "normal" in size and shape, able to "… sit in any western European restaurant without arousing particular comment except for their table manners".[3, p. 582]

These claims of race-based human taxonomy, including Coon's time thresholds for homo sapienation, have been discredited by paleontological and genomic research showing the antiquity of modern human origins within Africa, as well as the essential genomic African nature of all living human beings.[4–6] Coon's claim that African pygmies have "achondroplastic proportions" is also wrong. Shea and Bailey[7] show that African pygmies are reduced in overall size and have a body shape that is allometrically proportional to the size reduction.

Discarding the racist history of the study of human morphology allowed research to focus on more meaningful biological, medical, social and aesthetic implications of human body size and shape. This chapter reviews the evidence that human body shape, especially the length of the legs relative to total stature, is an important indicator for epidemiology and environmental public health. It is found that across the human species, as well as within geographic, social and ethnic groups of people, relative leg length reflects nutritional status and health during the years of physical growth and also has biologically and statistically significant associations with risks for morbidity and mortality in adulthood.

13.2 LEG LENGTH DEFINED

A strict anatomical definition of leg length is the length of the femur plus the tibia. Owing to the bipedal nature of the human species, "leg length" often is measured as: (femur + tibia + the height of the foot, from the tibia—talus articulation to the ground). Alternatively, the phrase "lower limb length" may be used to denote this linear dimension. In this chapter, "leg length" is used to denote any of the measurements described in next section. This is done because in a living human being it is difficult to measure anatomical leg length. The maximum length of the femur is measured from its head, at the proximal end, to its medial condyle, at the distal end. In life, the femur and pelvic bones overlap and the head of the femur is difficult to assess owing to its articulation within the acetabulum. A high degree of body fatness may make these bony landmarks difficult, or impossible, to access. Consequently, leg length is often defined by an easier to measure dimension such as iliac height (IH) and subischial leg length (SLL). It is also possible to measure leg length via the combination of thigh length (TL) and knee height (KH). Some studies employ only one of these measures as the indicator of leg length.

Each of these measurements can be transformed in ratios, generally in relation to total stature and sitting height (SH) to give indications of body proportions. This chapter will discuss the sitting height ratio (SHR), relative subischial leg length (RSLL) and knee height ratio (KHR).

13.3 PRACTICAL METHODS AND TECHNIQUES

This section presents a brief description of the anthropometric methods required to obtain various measures of leg length. More detail of the methods may be found in Lohman et al.[8] and the NHANES anthropometric manual (see Internet Resources). The purpose in providing these descriptions is to show the variety of methods employed to estimate leg length, biases that may be associated with each method, the limits of comparability between methods, and the variety of anatomical growth centers and different biological growth processes that underlie the concept of "leg length".

- **Iliac height (IH)**: The distance between the summit of the iliac crest and the floor (Figure 13.1).
- **Subischial leg length (SLL)**: The difference between stature and sitting height. It assumes that in a seated position the proximal landmark corresponds to the hip joint, which is very difficult to locate (Figure 13.1).
- **Thigh length (TL)**: The distance between the proximal end of the greater trochanter and the distal lateral femoral condyle. Because in living humans it is difficult to locate these joints, TL is measured from the mid-point of the inguinal ligament to the proximal edge of the patella (Figure 13.2). In overweight or obese people with excessive abdominal subcutaneous fat it may be difficult to find the inguinal ligament. Moreover, social and ethical prohibitions may prevent access to the site of the inguinal ligament.
- **Knee height (KH)**: The distance between the superior surface of the patella and the floor (Figure 13.3) or from the heel to, "...the anterior surface of the right thigh, above the condyles of the femur, about two inches above the patella" (NHANES Anthropometric Manual, see Internet resources). These two methods for measuring knee height provide different results. The second method is preferred for measuring people over 60 years of age.
- **Sitting height ratio (SHR)**: SHR is calculated as (Sitting Height/Height) $\times$ 100. It defines the percentage of total stature that is comprised by head and trunk

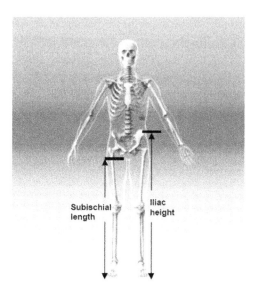

Figure 13.1 Iliac height and subischial length. *(Source: Roger Harris/Science Photo Library, royalty-free image, labeling added by the authors.)* This figure is reproduced in the color plate section.

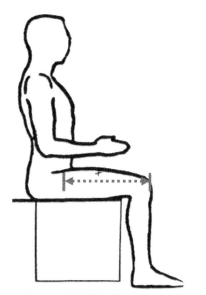

Figure 13.2 Thigh length. *(Source: from NHANES Anthropometric Manual.)*

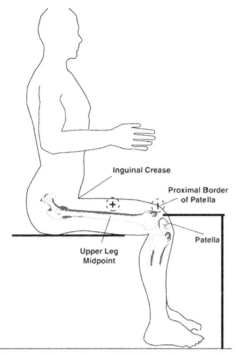

Figure 13.3 Knee height. *(Source: Adapted from NHANES anthropometric manual.)*

[see Figure 13.4 for details on sitting height (SH) measurement]. The remaining portion of the body will be the length of the legs. The lower the SHR the relatively longer the legs are. SHR allows individuals with different heights to be compared in terms of the percentage of the body that is composed by the relative length of legs. Because it is SH dependent, this measure can be overestimated in individuals with high levels of gluteofemoral fat, therefore underestimating the relative contribution of the lower limb to total stature.[9] There are international references[10] that allow the comparison of any values and the conversion of SHR raw data into percentiles and Z-scores.

- **Relative subischial leg length (RSLL)**: RSLL is calculated as H − SH/H × 100, where H is overall height. It defines the percentage of total stature that is comprised by the legs. The lower the RSLL the shorter the legs. There are no international reference values and it requires a harder computation of values of stature and sitting height.

- **Knee height ratio (KHR)**: KHR is calculated as KH/H × 100. It defines the percentage of total stature that is comprised by the lower segment of the leg (tibia + foot height). The higher the KHR the longer the leg segment. There are no international reference values.

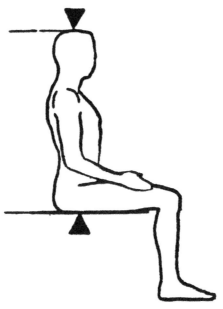

Figure 13.4 Sitting height is measured from the vertex of the head to the seated buttocks. *(Source: From NHANES anthropometric manual.)*

13.4 EVOLUTIONARY BACKGROUND OF HUMAN BODY SHAPE

The human species is distinguished from the non-human primates by several anatomical features. Among these are proportions of the arms and legs relative to total body length. The human difference is illustrated in Figure 13.5. In proportion to total body length, measured as stature, modern human adults have relatively long legs and short arms. Quantitative differences between adult humans, chimpanzees (*Pan troglodytes*) and bonobos (*Pan paniscus*) are given in Table 13.1. The combined values for the inter-membral index and the humerofemoral index show that humans have leg bones aver-aging 34% longer then the non-human apes, relative to the length of arm bones. The primary reason for this is human bipedal locomotion, a behavior that evolved at least by 4.4 million years ago (mya), as shown in the fossil hominin species *Ardipithecus ramidus*. Leg length must approximate 50% of total stature to achieve the biomechanical efficiency of the human striding bipedal gait. In modern humans this happens at the end of the childhood life history stage, which occurs at about 7 years of age.[13] By adulthood, human species-specific body proportions allow for not only the bipedal striding gait, but also — as has been observed, experimentally tested or speculatively proposed — for technological manipulation,[14] more efficient thermoregulation in a tropical savannah environment,[15–18] the freeing of the hands for carrying objects and infants,[19] for

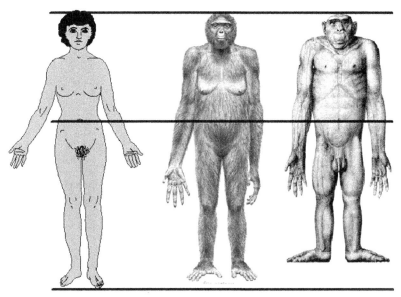

Figure 13.5 Approximate body proportions of *Homo sapiens*, *Ardipithecus ramidus* (4.4 mya hominin, probable life appearance) and *Pan troglodytes* (chimpanzee). The figures are aligned at the crown of the head and the umbilicus to approximate a constant trunk length. Relative to trunk length, humans have the longest legs and shortest arms. *(Source: Homo sapiens, SlideWrite Plus, 4.1, with authorization; Ardipithecus ramidus, Science 2 October 2009, © J.H. Matternes, http://www.jay-matternes.com/; Pan troglodytes, Schultz, 1933,[11] with permission of the publisher, http://www.schweizerbart.de)*

long-distance running,[20] and for gesticulation, communication, language and social—emotional contact.[21]

Human adult body proportions are brought about by differential growth of the body segments.[22] At birth, head length is approximately one-quarter of total body length, while at 25 years of age the head is only approximately one-eighth of the total length. There are also proportional changes in the length of the limbs, which become longer

Table 13.1 Long bone indices of humans and chimpanzees[12]

Species	Intermembral index	Humerofemoral index
Human (male)	69.7	71.4
Human (female)	68.5	69.8
Chimpanzee (male)	108.0	101.1
Chimpanzee (female)	109.4	102
Bonobo (male and female)	102.2	98.0

All indices are based on measurements of the maximum length of the long bones. Intermembral index = [(humerus + radius) × 100]/(femur + tibia); Humerofemoral index = (humerus × 100)/femur.

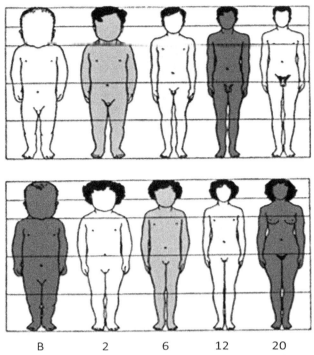

Figure 13.6 Changes in body proportion during human growth after birth. Ages for each profile are, from left to right, newborn, 2 years, 6 years, 12 years, 25 years. The hair style and shading of the cartoon silhouettes are for artistic purposes and are not meant to imply any ethnic, eco-geographical, or "racial" phenotypic characteristics of the human species. (Source: *Provided courtesy of Dr J.V. Basmajian.*)

relative to total body length during the years of growth.[23] The cartoons of Figure 13.6 show the typical changes that take place in people from birth to age 25 years. Human beings follow a cephalocaudal gradient of growth and development, the pattern common to most mammals. There are, however, some species-specific features of human body plan development. In a classic 1926 article, Schultz[24] published his sketches of the body proportions of hominoid fetuses, reproduced here as Figure 13.7. The human fetus "of the 4th month" has shorter legs than the chimpanzee, orang-utan or gibbon. This assumes that Schultz's estimates of development for the non-human apes are correct (see Figure 13.7 legend). Another difference in proportion, not noted by Schultz, is the size of the cranium relative to the face, which is larger in the human fetus than in the chimpanzee, orang-utan or gibbon.

This human pattern of change in body proportions during gestation to birth and then to adulthood may be explained, in part, by the evolution of bipedalism inter-acting with the evolution of a large and complex brain. Apes have a pattern of brain growth that is rapid before birth and relatively slow after birth. Humans have rapid

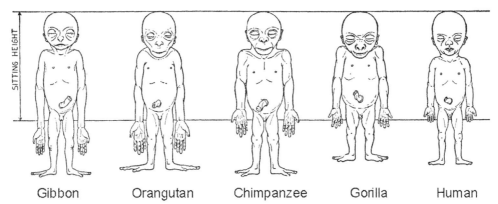

Figure 13.7 Schultz's sketches of the body proportions of hominoid fetuses. The original legend for this figure states, "All the figures have the same sitting height. The human fetus is the 4th month, the gorilla and the gibbon fetus correspond in development to the human fetus, but the chimpanzee and the orang fetus are slightly more advanced in their growth". (Source: *Schultz, 1926,[24] p. 465−6, accessed from http://www.jstor.org/stable/2808286*)

brain growth both before and after birth.[25,26] Human newborns are bigger brained than any of the apes, although not so much bigger in terms of brain−body mass ratio (Table 13.2). The human−ape differences in brain mass and brain−body mass ratio are much larger at adulthood, and many of these differences are achieved by 6.9 years of age.[13,26] More than brain mass, it is brain metabolic activity that is, perhaps, the crucial difference. The human newborn uses 87% of its resting metabolic rate (RMR) for brain growth and function. By the age of 5 years, the percentage RMR usage is still high at 44%, whereas in the adult human, the figure is between 20 and 25% of RMR. At comparable stages of development, the RMR values for the chimpanzee are about 45, 20 and 9%, respectively.[28] With such high metabolic demands from its brain, the human infant and child may well have been naturally selected to make trade-offs in the allocation of limited nutrients, oxygen and other resources required to grow the brain versus other body parts. Trade-offs between the growth,

Table 13.2 Neonatal and adult brain weight and total body weight for the great apes and humans

| Species | Neonatal mass (g) | | | Adult mass (g) | | |
	Brain	Body	Brain/body ratio	Brain	Body	Brain/body ratio
Pongo (orang-utan)	170.3	1728.0	0.10	413.3	53,000.0	0.008
Pan (chimpanzee)	128.0	1756.0	0.07	410.3	36,350.0	0.011
Gorilla	227.0	2110.0	0.11	505.9	126,500.0	0.004
Homo sapiens	384.0	3300.0	0.12	1250.0	44,000.0	0.284

Adult body weight is the average of male and female weight. Data from Harvey et al.[27]

development and maturation of body parts are common across the diversity of life histories of animal and plant species,[29–31] including the human species.[32,33] From this perspective, the ultimate level reason that human leg growth is delayed during fetal and infant development is that it allows for rapid growth of the brain.

The proximate level controls of the trade-offs in the growth of body segments and organs are not well known. Genetic, hormonal and nutrient supply factors are likely to be involved. In a review of bone growth biology, Rauch[34, p. 194] states, "Bone growth in length is primarily achieved through the action of chondrocytes in the proliferative and hypertrophic zones of the growth plate. Longitudinal growth is controlled by systemic, local paracrine and local mechanical factors. With regard to the latter, a feedback mechanism must exist which ensures that bone growth proceeds in the direction of the predominant mechanical forces. How this works is unknown at present". It is known that the length of the proliferative columns in the growth plate correlates with the length of limbs, "… a species with long legs and short arms has longer columns at the knees and shorter at the elbows than an oppositely proportioned species".[35, p. 21]

Quantitative trait locus (QTL) mapping of laboratory mice has identified genomic regions associated with phenotypic differences in length of the femur, tibia, humerus and ulna.[36] Changes in genomic growth regulation, such as Hox expression patterns, are known to be associated with the growth of primate forearm segments.[37] Changes in the sensitivity of bone growth plates to growth-promoting and -inhibiting factors at different times during development, and at different sites of the skeleton, are also known to be responsible for differential growth of body segments.[38,39] A further speculation is that blood circulation of the fetus may contribute to the brain versus leg growth trade-off. Blood in the fetal ascending aorta has higher oxygen saturation than does the blood descending to the common iliac artery (Figure 13.8). In addition, the umbilical arteries carry some of the blood descending towards the leg back to the placenta. This pattern of fetal circulation is common to most mammals and is likely to be evolutionarily ancient. Combined with the more recently evolved metabolic demands of the human fetal brain, the ancient circulatory pattern may leave the human fetal legs with reduced supply of oxygen and nutrients, further slowing leg growth and development compared with more cephalic regions of the body. This would explain the human—ape differences seen in Figure 13.7, but the author of this chapter can find no experimental support for this proposal. There is human clinical case study evidence that increased blood flow to the limbs is associated with greater amount of growth.[41] In addition, a study of children aged 8—11 years living at high altitude in Tibet found that both, "… absolute and relative tibia length was significantly reduced in children with lower blood oxygen saturation".[32, p. 662] This study provides very clear evidence of a trade-off between oxygen availability and body proportions, with low oxygen availability favoring growth of the head and trunk.

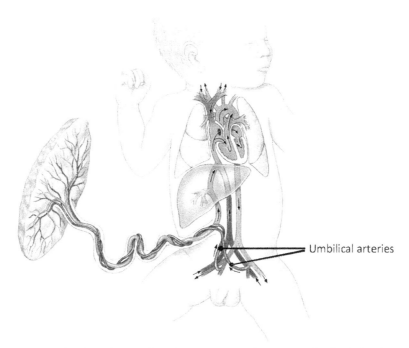

Umbilical arteries

Figure 13.8 Human fetal circulation. The relative amount of oxygen in the fetal blood is greatest in the upper thorax, neck and head, indicated by the red color of the vessels ascending from the heart. Blood flowing to the abdomen and legs is less well oxygenated; indicated by the violet color of the vessels descending from the heart. *(Source: Adapted from Martini and Bartholomew.[40])* This figure is reproduced in the color plate section.

13.5 SIZE AND SHAPE IN LIVING HUMANS

The general pattern of human body shape development is a species-specific characteristic. Historical artwork, sculpture and anatomical drawings from Renaissance Europe[42,43] and pre-Columbian Mexico[44] show fundamental commonalities in the depiction of body shape of late-term fetuses, newborns and infants. Discrete populations of living humans, however, present a diversity of body sizes and shapes. Mean stature for populations of adults varies from minimum values for the Efe Pygmies of Africa at 144.9 cm for men and 136.1 cm for women[45] to the maximum values for the Dutch of Europe at 184.0 cm for men and 170.6 cm for women.[46] There are also biologically and statistically significant variations between human populations in body shape. Eveleth and Tanner[47,48] published data for body proportions and leg length, estimated via the sitting height ratio, from dozens of human populations, distributed across most geographical regions of the world (Figure 13.9). The sitting height ratio (SHR) is a commonly used measure of body proportion. Measured stature minus sitting height may also be used to estimate leg length, but this measure does not standardize for total height, making it

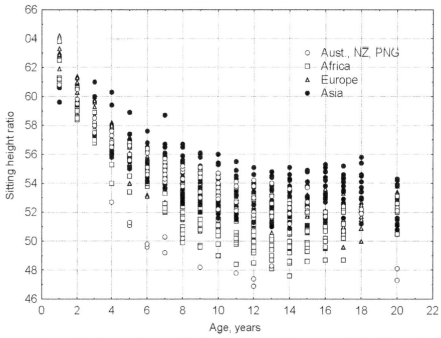

Figure 13.9 Sitting height ratio by age for the four geographic groups defined by Eveleth and Tanner.[47,48] Age 20 includes data for adults over the age of 18 years. A larger sitting height ratio (SHR) indicates relatively short legs for total stature. *(Source: Authors' original figure.)*

difficult to compare individuals with different statures. Mean SHR for populations of adults varies from minimum values, i.e. relatively long legs, for Australian Aborigines (SHR = 47.3 for men and 48.1 for women) to the maximum SHR values, i.e. relatively short legs, for Guatemala Maya men and Peruvian women (SHR = 54.6 and 55.8).

Making sense of these worldwide comparisons is difficult because of the differences in lifestyle, environment and genomics. Two well-known ecogeographic principles, Bergmann's and Allen's rules, are often cited as primary causes for the global patterns of human body shape variation. Bergmann,[49] in 1847, observed that closely related mammalian species, such as bears, have greater body mass in colder climates. Allen[50] added in 1877 that the limbs and tails of such species tend to be shorter in cold climates and longer in warmer environments. Large body mass and relatively short extremities increase the ratio of volume-to-surface area and provide for a body shape that maximizes metabolic heat retention in a mammal. Conversely, in warmer temperatures, relatively long extremities increase surface areas relative to volume and allow for greater heat loss. It has been shown experimentally that mice and other non-human mammals reared in warmer temperature experience greater bone tissue growth and longer limb bones.[51] The usual explanation for this is greater vascularization, allowing for greater oxygen and nutrient perfusion. Recent experimental research shows, however, that even in the absence of vasculature, in vitro culture of chondrocytes from mouse metatarsal bone show a positive correlation between higher environmental temperature and, "… greater proliferation and extracellular matrix volume …".[51, p. 19348]

Bergmann's and Allen's rules apply, to some extent, to the human species. In 1953, Roberts[52] published an analysis showing a significant relationship between body mass and latitude for human beings, with groups of people living at higher latitudes having greater body mass than those living closer to the equator. Twenty-five years later, Roberts[53] updated and reaffirmed these findings. Other research shows that people living in colder regions also tend to have shorter limbs relative to total stature, compared with groups of people living in warmer regions.[17,47,48]

These climate relationships, however, are not perfect. A reanalysis of Roberts' data by Katzmarzyk and Leonard[54] modifies the importance of climate as the primary molder of human body shape. Katzmarzyk and Leonard analyze the sitting height ratio of 165 human groups studied between 1960 and 1996. All of the human data analyzed by Roberts were collected prior to 1953. Katzmarzyk and Leonard show that the more recently studied groups still follow the ecological principles of body shape, but that the association with climate has been attenuated since Robert's study. The slopes of the best fitting linear regression lines for the relation of mean annual temperature to sitting height ratio arc half those reported by Roberts. Katzmarzyk and Leonard[54, p. 483] state that, "… although climatic factors continue to be significant correlates of worldwide variation in human body size and morphology, differential changes in nutrition among tropical, developing world populations have moderated their influence". The authors define the nutritional changes as modifications in diet and lifestyle, especially the introduction of western foods and behaviors. They point out that, "… climate may shape morphology through its influence on food availability and nutrition [meaning that] linear builds of tropical populations are the consequence of nutritional [factors] rather than thermal stress …".[54, p. 491–2] In this case, during the years of growth and development, more or less total food intake, more or less of any essential nutrient, more or less physical activity (and the type of activity) could influence body shape. Guatemala Maya, for example, consume only approximately 80% of the total energy needed for healthy growth, and 20.4% of the population are also iodine deficient.[55] Iodine deficiency during infancy and childhood results in reduced leg length, especially the distal femur, the tibia and the foot.[56] Maya children and adults spend considerable time and energy at heavy labor,[57] which diverts available energy in the diet away from growth. This nutrition and lifestyle combination is known to reduce total stature and leg length.[58]

The body shape of people may have a genetic basis, especially for human groups who have resided in the same environment for many generations. A comparison of stature and body proportion between blacks (African-Americans) and whites (European-Americans) in the USA provides an example of genome—environment interactions and their effect on growth.[59] The first National Health and Nutrition Examination Survey

(NHANES I) of the USA gathered anthropometric data on a nationally representative sample of blacks and whites aged 18–74 years. When the data are adjusted for differences between the two ethnic groups in income, education, urban or rural residence and age, there is no significant difference in average height between black and white men. Nor is there a significant difference in average height between black and white women.

Although white and black adults in the USA have the same average stature, when education, income and other variables are controlled, the body proportions of the two groups are different. Krogman[60] found that for the same height, blacks living in Philadelphia, USA, had shorter trunks and longer extremities than whites, especially the lower leg and forearm. Hamill et al.[61] found that this was also true for a national sample of black and white youths 12–17 years old, and it is the case for adults 20–49 years old measured for the NHANES III survey, 1988–1994.[9] A genomic contribution to the body proportion differences between blacks and whites seems likely, as the blacks tend to have more sub-Sahara African genomic origins than the whites.

Few if any specific genes for human body proportions are known. In a statistical pedigree analysis of two human samples, Livshits et al.[62] estimate that between 40% and 75% of interindividual variation in the body proportions they studied (adjusted for age and sex) are attributable to "genetic effects". These may be better described as familial effects because the authors analyzed families and also because they found significant common environmental effects for siblings as well as significant sex by age interactions. The range of the sources of variation in the analysis makes it difficult to compute simple genetic variance.

Even if specific genotypes are discovered, their direct contribution to normal ethnic (so-called "racial") variation in human body shape may be relatively small. At 40 weeks' gestation, fetuses identified as African-Americans have, on average, longer legs than fetuses identified as European-Americans.[24] But the difference, as measured by (total length/crown–rump length), is less than 1%. In an analysis of the data shown in Figure 13.9, Bogin et al.[63] estimated the contribution of geographical origin to the variance in the SHR to be 0.04, which accords well with genomic estimates for variation in total stature of 0.04–0.06.[64] Forensic anthropologists and physicians in the USA have often used "race-specific" body proportions to ascribe an African-American or European/Asian-American ethnicity to a skeleton.[65,66] Feldesman and Fountain[67] tested the utility of the femur length/stature ratio to correctly identify 798 femur/stature pairs of skeletons of known ethnicity. They found that the mean, "… 'Black' femur/stature ratio is statistically significantly different from those of 'Whites' and 'Asians'".[67, p. 207] However, discriminant function and cluster analysis shows that the variation within groups is so large that coherence to so-called "races" defined by geographical origin is poor, with results barely better than chance. Feldesman and Fountain conclude by stating that using "race-specific" body proportions to identify unknown skeletons would result in a high number of incorrect attributions of ethnicity.

A more promising approach to understanding the control of human body proportions comes from pan-human and pan-mammalian genomic research. Hox genes and homeobox sequences, and a growing number of growth and signaling factors, are known to regulate the growth of body segments[68, 147] and these genes are shared across all taxa. There is observational and experimental evidence that Hoxd expression is linked with forearm, hand and digit length differences in the apes.[37] The short stature homeobox-containing gene (SHOX) is another genomic region that may be relevant to human body proportions. SHOX, located on the distal ends of the X and Y chromosomes, encodes a homeodomain transcription factor responsible for a significant proportion of long-bone growth.[69] Turner's syndrome (45, XO karyotype) results in approximately 20 cm deficit in stature. Some studies find that legs are disproportionately affected,[70,71] but other studies find no disproportion.[72] More specific candidate genes for body shape are known from some non-human mammals[73,74] and in insects.[29]

Another very active area of research is epigenetic regulation of body growth.[75] Epigenetic effects may act through a number of actions on gene expression, from the level of the genome (e.g. DNA methylation and histone modification), proteome (e.g. micro-RNA regulation) and environment (e.g. climate, diet and physical activity) interactions. These epigenetic effects may well play the major role in determination of human size and shape (see Chapter 7).

13.6 DEVELOPMENTAL PLASTICITY

Plasticity refers to the concept that the development of the phenotype of an organism is responsive to variations in the quality and quantity of environmental factors required for life.[76] This concept is used here to mean that during the years of growth and development humans can grow more or less of various tissues and come to be adults of various sizes and shapes. As adults these sizes and shapes are largely fixed, especially for total stature and the length of body segments. Human growth is highly plastic during the years of growth and development, responding to the overall quality of living conditions.[13] From the perspective of developmental plasticity, leg length, both in terms of absolute size and relative to total stature, is an indicator of the quality of the environment for growth during infancy, childhood and the juvenile years of development.

The reason for this is the general principle that those body parts growing the fastest will be most affected by a shortage of nutrients, infection, parasites, physical or emotional trauma, and other adverse conditions. The cephalocaudal principle of growth as applied to the human species means that the legs, especially the tibia, are growing quickly relative to other body segments from birth to 7 years of age. Relatively short leg length in groups of adolescents and adults, therefore, is likely to be due to adversity during infancy and childhood leading to competition between body segments, such as trunk versus limbs and between organs and limbs. In the simplest case, such competition may be for the limited nutrients available during growth.[33,58,63] More complex explanations for competition relate to aspects of the thrifty phenotype hypothesis,[77,78] the intergenerational influences hypothesis,[79,80] the fetal programming hypothesis[81] and the predictive adaptive response hypothesis.[82,83] Discussion of these hypotheses is beyond the scope of this review (see Chapter 12 as well as reference 33 and other articles in the same issue of the journal for such discussion), but in essence each of these hypotheses predicts that the vital organs of the head, thorax and abdomen of the body will be protected from adversity at the expense of the less vital tissues of the limbs.

13.7 USE OF LEG LENGTH IN HUMAN BIOLOGY AND ENVIRONMENTAL EPIDEMIOLOGY

Dr Isabella Leitch[84] was the first medical researcher to propose that a ratio of leg length to total stature could be a good indicator of the early life nutritional history and general health of an individual. Leitch[84, p. 145] wrote, "… it would be expected on general principles that children continuously underfed would grow into underdeveloped adults … with normal or nearly normal size head, moderately retarded trunk and relatively short legs". Reviewing the literature available at the time (pre-1950), Leitch found that improved nutrition during infancy and childhood did result in a greater increase in leg length than in total height or weight. One of the critical studies in her review is the Carnegie UK Dietary and Clinical Survey, which recorded height, weight and iliac height (IH). When the participants were grouped by age and family expenditure on food, it was found that IH, "… was consistently better than total height for indicating [food] expenditure group".[84, p. 213] Leitch also reported that longer-legged children were less susceptible to bronchitis, which was then a scourge of poorly fed children.

Leitch was careful to state that leg length per se is not a direct cause of better or worse health and that children and adults with relatively short legs may be quite healthy. She viewed greater leg length as a correlate of an improved constitution. This view anticipates current biomedical research on the development of somatic and cognitive reserve capacity[85–87] in relation to health and rate of senescence. The reserve capacity hypothesis posits that during human growth and development the somatic and cognitive systems usually "overshoot" their minimally necessary capacity for sustaining life of the individual. By overshooting this necessary capacity an individual has reserve capacity which may be channeled into greater growth, better health, more successful reproduction, social and economic success, and slower rates of senescence. Leg length relative to total stature may be one indicator of overall reserve capacity of a person or a group of people.

13.7.1 Leg Length and Human Environmental Health

Many studies support Leitch's findings and hypothesis.[88–101] In the past 10 years the number of publications on the relationship of leg length to human health has increased at a rapid rate. A systematic review of these studies is not provided here; instead, some of the literature is sampled to provide an overview of research.

Table 13.3 summarizes several recent studies that show how leg length and body proportion ratios are powerful indicators of the quality of the environment and of the plasticity of the human body. The table provides only a few studies, of which there are dozens. What is important to note is that regardless of the specific leg measure taken, longer leg length is associated with better environments, better nutrition, higher socioeconomic status and better general health, overall.

Poor childhood health, insufficient diet, adverse family circumstances and maternal smoking during pregnancy are each known to reduce leg length.[106,113–116] Frisancho et al.[109] emphasize the environmental effects in a study that finds that the leg length of Mexican-Americans aged 2–17 years is significantly associated with the socioeconomic status of their families. In that study, individuals from better-off families have significantly longer legs, but equal trunk length, when compared with boys and girls from poorer families. Dangour[117] reports similar findings for two tribes of Amerindian children living in Guyana. The tribes are both of low socioeconomic status, but differ markedly in the quality of their living conditions. Children in the tribe with better living conditions are taller than their age-mates in the other tribe. The difference in stature is due almost entirely to differences in leg length, as there are no significant differences in sitting height between the tribes.

The present authors' studies are of Maya families from Guatemala who migrated to the USA from the late 1970s to the early 1990s.[33,58,118–121] In Guatemala the Maya are subjected to chronic adversity in the form of poor nutrition, heavy workloads, contaminated drinking water, infectious disease, limited education opportunities and state-supported violence. In the USA the Maya tend to have low socioeconomic status and work at physically demanding jobs, but benefit from safe drinking water, copious food availability, public education, health care and relative safety. Births to Maya immigrant women created a sizeable number of Maya-American children. The height and sitting height of 5–12-year-old children ($n = 431$) were measured in 1999 and 2000, and from these measurements leg length was estimated. These data were compared with a sample of Maya children of the same ages living in Guatemala measured in 1998 ($n = 1347$). Maya-American children are currently 11.54 cm taller and 6.83 cm longer-legged, on average, than Maya children living in Guatemala. The values indicate that about 60% of the increase in stature is due to longer legs.

13.8 LEG LENGTH AND RISK FOR MORBIDITY AND MORTALITY

Decomposing stature into its major components is proving to be a useful strategy to assess the antecedents of disease, morbidity and death in adulthood.[122–125] Human leg length, however it is measured, trunk length and their proportions [e.g. relative leg length or the

Table 13.3 Summary of a few studies published since 2000 employing measures of leg length in relation to early life living conditions and health

Measure of "leg length"	Sample size	Sample	Results	Source
IH	Total 2209 (M 1062, F 1147)	2–14 years. Extracted from The Boyd Orr Survey. Children from 1343 working-class families in England and Scotland, measured between 1937 and 1939	M&F: positive association with length of breastfeeding, decreasing numbers of children in the household and increasing household income. Overall, the individual components of stature mostly associated with childhood environment were leg length (measured as IH) and foot length (not in the scope of this entry)	Whitley et al., 2008[112]
	Total 916 (M 376, F 540)	65+ years inhabitants of Kwangju, South Korea, assessed in 2003	Shorter limb length is associated with markers of lower early-life socioeconomic status and is associated with dementia later in life, especially in women	Kim et al., 2008[103]
SLL	Total 2338 (M 1040, F 1298)	30–59 years (UK)	M&F: inverse association with systolic BP, diastolic BP, total cholesterol and fibrinogen. Direct association with FEV, FVC, BW and BMI	Gunnell et al., 2003[104]

Total 10,308 (M 6895, F 3413)	35–55 years (London, UK)	M&F: strong inverse association with pulse pressure and systolic BP. Strong positive association with lower total/HDL cholesterol ratio, triglycerides and 2-hour glucose. M: strong inverse association with total cholesterol. F: strong inverse association with diastolic BP	Ferrie et al., 2006[105]
Total 3262	Longitudinal study, births from 3 to 9 March 1946 (21 assessment occasions between birth and 53 years). MRC National Survey of Health & Development (UK)	M&F: positive association with mother's and father's height, BW. SLL greater among individuals from non-manual social class and among individuals who were breast fed	Wadsworth et al., 2002[106]
Total 5900	The 1958 British Birth Cohort. Participants assessed at birth and at ages 7, 11, 16, 23, 32, 42 and 45 years	Adult SLL associated with parental height, birth weight. Taller prepubertal stature associated with higher SLL. Maternal smoking during pregnancy resulted in lower adult SLL. Overall, adult SLL is related to a greater extent than trunk length to early life factors and prepubertal height	Li et al., 2007[107]

(Continued)

Table 13.3 Summary of a few studies published since 2000 employing measures of leg length in relation to early life living conditions and health—cont'd

Measure of "leg length"	Sample size	Sample	Results	Source
KH	Total 50 (M 27, F 23)	Infants grouped by gestation time at birth: <28 weeks, 28—31 weeks, 32—36 weeks, >36 weeks. Births occurred in 2004—2005, in the neonatal intensive care, Christchurch, New Zealand	Changes in KH (using a kneemometer) correlate very well with changes in weight. If gain in weight is achieved, normal linear growth may be assumed. Because of this, kneemometry is not a useful addition to routine measurements of growth in the neonatal unit	Dixon et al., 2008[108]
SHR	Total 2985 (M 1465, F 1520)	2—17 years. Mexican Americans (NHANES III, USA)	M&F: individuals with relatively short legs in proportion to total stature are poorer than longer-legged individuals (poverty income ratio)	Frisancho et al., 2001[109]
	Total 1472 (M 747, F 707)	6—13 years, Oaxaca, Southern Mexico. Urban in 1972: total 409 (M 218, F 173); Rural in 1978: total 363 (M 179, F 184); Urban in 2000: total 339 (M 173, F 166); Rural in 2000: total 361 (M 177, F 184)	Positive time trend in leg length from 1972 to 2000 in both rural and urban settings	Malina et al., 2004[110]

	Sample	Description	Findings	Reference
	Total 2003 (M 2003, F 0)	7–16 years. Two cross-sectional surveys among school-aged boys from Kolkata, India. 1982–1983 ($n = 816$); 1999–2002 ($n = 1187$)	Positive time trend in relative leg length. Boys measured in 1999–2002 had longer legs in proportion to total stature than their counterparts in 1983–1983	Dasgupta et al., 2008[111]
	Total 1995 (M 977, F 1018)	5–12 years. Maya migrants to the USA in 1992 ($n = 211$), Maya migrants to the USA in 2000 ($n = 431$), Maya in Guatemala in 1998 ($n = 1353$)	Leg length is a sensitive indicator of the quality of the environment. Maya children in the USA show longer legs in proportion to stature than their counterparts in Guatemala. By 2000, Maya migrants to the USA were 11.54 cm taller and 6.83 cm longer-legged than Maya children in Guatemala	Bogin et al., 2002[58]
RSLL	Total 273	Intergenerational sample. Parents' generation: total 165 (M 80, F 85). Offspring generation: total 108 (M 49, F 59). From Auckland and Taipei	Is an effective marker of intergenerational changes	Floyd 2008[112]
KHR	Total 273	Intergenerational sample. Parents' generation: total 165 (M 80, F 85). Offspring generation: total 108 (M 49, F 59). From Auckland and Taipei	Is an effective marker of intergenerational changes. Lower leg growth, as represented by KHR, is similar to changes in overall leg length in sensitivity to environmental change	Floyd 2008[112]

IH: iliac height; SLL: subischial leg length; KH: knee height; SHR: sitting height ratio; RSLL: relative subischial leg length; KHR: knee height ratio; M: male; F: female; MRC: Medical Research Council; NHANES: National Health and Nutrition Examination Survey; BP: blood pressure; FEV: forced expiratory volume; FVC: forced vital capacity; BW: birth weight; BMI: body mass index; HDL: high-density lipoprotein.

sitting height ratio (sitting height/stature)] are associated with epidemiological risk for several diseases and syndromes. Relatively short legs and short stature due to relatively short legs may increase the risk for overweight (fatness), coronary heart disease and diabetes.[105,114,124–127] These same proportions are associated with liver dysfunction (increased levels of the liver enzymes alanine aminotransferase, gamma-glutamyl-transferase, aspartate transaminase and alkaline phosphatase).[128] In a systematic review of the literature prior to 2001, Gunnell et al.[129] find that some cancers, such as prostate and testicular cancer, premenopausal breast cancer, endometrial cancer and colorectal cancer, are statistically more likely in adults with greater stature and relatively long legs. These authors report that the positive relationship between leg length and risk for these cancers may be due to the effects of insulin-like growth factor-1 (IGF-1). Gunnell and colleagues write that, "… raised levels of IGF-I are associated with increased risks of prostate, breast, and colorectal cancers. The most potent cell survival factor controlling apoptosis is insulin-like growth factor I (IGF-I). Raised levels of IGF-I and reduced levels of its main binding protein, insulin-like growth factor (IGF)-binding protein 3, may diminish this defense against a range of cancers".[129, p. 313, citations in the original omitted] Since 2001 several more reports of a relationship between IGF-1, IGF-1 receptors and cancer risk have been published,[130,131] as well as associations between IGF-2 and IGF-2 receptors and cancer risk.[132] A search of PubMed.gov using the terms "cancer, IGF" leads to more than 4100 articles published in the past 10 years. This is an active area of research, often reporting contradictory findings, but not reviewed further in the present article.

There are complications in the relationship between leg length, health, socioeconomic status and better environments for growth. One such complication is noted by Schooling et al.[133,134] in an analysis of a cross-sectional sample of 9998 Chinese people aged at least 50 years and measured in 2005–2006. Sitting height (SH) and height (H) were measured and leg length estimated as H − SH. The growth environment for the 50+-year-old adults was estimated via a questionnaire asking about own education, father's occupation, parental literacy and parental possessions. The authors find that leg length and height, but not sitting height, vary with some childhood conditions. Participants with two literate parents who owned more possessions have longer legs. Unexpectedly, the participants' education level and their father's occupation have no effect on height or leg length. Higher scores for these variables do associate with an earlier age at menarche for female participants. The authors explain that earlier menarche for girls, and earlier puberty for boys, will terminate growth at an earlier age. Other research shows that earlier menarche results in relatively short leg length.[145–147] This may explain why higher socioeconomic status of the participants and their parents, as measured by education and father's occupation, did not associate with longer leg length. That parental literacy and possessions did associate with leg length indicates that researchers must focus on factors that are socially and historically relevant to the population under study, rather than a generic measure of socioeconomic status.

Another complication is noted by Padez et al.,[138] who analyzed the growth status of Mozambique adolescents. The sample comprised 690 boys and 727 girls, aged between 9 and 17 years, from Maputo, the capital city. The sample is divided between those living in the center of Maputo (higher socioeconomic status) and those living in the slums on the periphery of the city. Height, weight and sitting height were measured and the sitting height ratio was calculated. The hypothesis that relative leg length is more sensitive than total stature as an indicator of environmental quality is not uniformly confirmed. Overall, mean stature is greater for the center group than the slum group, but relative leg length as measured by the sitting height ratio does not differ. Compared with African-American references (NHANES II), all center girls, 9—14-year-old slum girls, all slum boys and the oldest center boys show relatively short legs. These findings show that within the Mozambique sample, relative leg length is not sensitive enough to distinguish the quality of the living environment. A reason for this is that Mozambique was a colony of Portugal until 1975. Civil unrest and warfare characterized the late colonial period and the post-independence period until a peace settlement was concluded in 1992. It is possible that all socioeconomic status groups exposed to the civil war within the country suffered sufficiently to reduce relative leg length compared with the better-off African-American reference sample.

13.9 LEG LENGTH AND BEAUTY

"The legs, besides being a very important functional unit, are also an important sexual attraction in themselves, and in all cultures they have a preponderant place in the concept of beauty".[139, p. 505] A concern with body proportion has deep roots in European history. Building on the work of Vitruvius, a first century BC Roman architect and writer, Leonardo da Vinci (b1452—d1519) developed canons, or rules, for drawing human proportions. According to these canons, human body height is to be the length of eight heads, with an additional one-quarter head for neck length. Leg length is to be four head lengths. Leonardo's "Vitruvian Man" (c. 1487) is the iconic illustration of the canons. Albrecht Dürer (b1471—d1528), a German artist, devised technology to draw both the canonical forms and many variations as observed in nature. With his geometric methods, Dürer could draw any manner of human variation in size or proportion. He applied his method to drawings of men, women, children and infants. Including women and children in this type of methodological work was an innovation, as most artists followed the teachings of Cennino Cennini (c. b1400) who wrote that women do "… not have any set proportion".[42, p. 202] Children, it seems, were too inconsequential for Cennini to even mention!

After the year 1600, the post-Renaissance painters begin to depict children with normal proportions and also with growth pathologies. The Flemish artist Van Dyck depicts three normal children in the painting *The Children of Charles I* (1635). The painting *The Maids of Honor (Las Meninas)* by Diego Velazquez (1656) depicts a normal child, a woman with achondroplastic dwarfism (normal sized head and trunk with short arms and legs) and a man with growth hormone deficiency dwarfism (proportionate reduction in size of all body parts). At the time of these paintings, of course, the biological control of normal and pathological growth in size and proportion was not known.

Edmund Burke, the British statesman and philosopher, published in 1756 the essay, "The philosophical inquiry into the origin of our ideas on the sublime and beautiful". One part of this essay is subtitled, "Proportion not the cause of beauty in the human species". Burke argued that people with body proportions outside the canon of Leonardo might still be considered beautiful. He held the human leg to be especially handsome, "I believe nobody will think the form of a man's leg so well adapted to running, as those of a horse, a dog, a deer, and several other creatures; at least they have not that appearance: yet, I believe, a well-fashioned human leg will be allowed to far exceed all these in beauty". One is left to wonder which human legs are "well-fashioned". Perhaps Burke meant those that are relatively straight and long: contra-indicating rickets, suggesting good health and nutrition in childhood, and predicting fecundity in adult women.

The intersection of biomedical and aesthetic concern with the beauty of human leg is still strong today. Leitch[84] mentions that the adult survivors of early-life undernutrition and disease may be very "tough" individuals, but they are short-legged and not beautiful. The quote from Cuenca-Guerra and colleagues[139] that opened this section is from an article on the surgical use of calf implants to enhance leg attractiveness. There is a burgeoning literature on the scientific analysis of beauty and the medical means to enhance it, much of which focuses on body proportion and leg length.[140–142]

13.10 CONCLUSION

Cosmetic surgery, elevated heels on shoes and other clever styles of clothing can make legs more attractive, but these techniques do not overcome the fundamental linkages between leg length and human health. A broad review of the literature indicates that there is good evidence that adults with skeletal disproportions, especially high SHR (short legs), are at greater risk for coronary heart disease via hypercholesterolemia, impaired glucose and insulin regulation, increased pulse pressure and systolic blood pressure, and higher fibrinogen levels.[105] Some cancers are associated with relatively long legs.

Early-life undernutrition and disease, especially between birth and the age of 7–8 years, account for relatively short legs in adults, but still do not explain why they are at greater risk for disease and mortality at earlier ages than the longer-legged adults. An association between childhood stunting and adult overweight is becoming well known.

A prospective 3-year study of stunted Brazilian boys and girls, 11–15 years old, finds that they gain more fat mass and less lean body mass compared with non-stunted peers.[143] Brazilian adult women with short stature and disproportionately short legs have high risk for obesity.[144] The reason for these associations with fatness seems to be tied to impaired fat oxidation in stunted children.[145] Fasting respiratory quotient (RQ = the ratio of the volume of carbon dioxide produced by an organism to the volume of oxygen consumed) is significantly higher, and hence fat oxidation is lower, leading to greater body fat stores in the stunted group. Other contributors may be impairment of appetite control associated with early malnutrition and lower resting and postprandial energy expenditure.[146]

13.11 SUMMARY

Early-life undernutrition and disease not only reduces leg length relative to total stature, but may also alter human physiology towards a phenotype with a deranged metabolism. Understanding the nature of metabolic impairments may provide help to explain the relationship between measures of leg length with risks for overweight/ obesity, diabetes, hypertension, low bone density, coronary heart disease, other human pathologies, and premature mortality. Edmund Burke may have found relatively short legs to be capable of beauty, but the epidemiological evidence finds them to be a risk for health.

REFERENCES

1. Gould SJ. *The mismeasure of man.* New York: Norton; 1981.
2. Marks J. *Human biodiversity: genes, race, and history.* New York: Aldine De Gruyter; 1995.
3. Coon C. *The origin of races.* New York: Knopf; 1962.
4. Tishkoff SA, Kidd KK. Implications of biogeography of human populations for "race" and medicine. *Nat Genet* 2004;**36**:S21–7.
5. Ramachandran S, Deshpande O, Roseman CC, Rosenberg NA, Feldman MW, Cavalli-Sforza LL. Support from the relationship of genetic and geographic distance in human populations for a serial founder effect originating in Africa. *Proc Natl Acad Sci USA* 2005;**102**:15942–7.
6. Tattersall I. Out of Africa: modern human origins special feature: human origins: out of Africa. *Proc Natl Acad Sci USA* 2009;**106**:16018–21.
7. Shea BT, Bailey RC. Allometry and adaptation of body proportions and stature in African pygmies. *Am J Phys Anthropol* 1996;**100**:311–40.
8. Lohman TG, Roche AF, Martorell R. *Anthropometric standardization reference manual.* Champaign, IL: Human Kinetics; 1988.
9. Bogin B, Varela-Silva MI. Fatness biases the use of estimated leg length as an epidemiological marker for adults in the NHANES III sample. *Int J Epidemiol* 2008;**8**:201–9.
10. Frisancho AR. *Anthropometric standards. an interactive nutritional reference of body size and body composition for children and adults.* Ann Arbor, MI: University of Michigan Press; 2008.
11. Schultz AH. Die Körporproportionen der erwachsenen catarrhinen Primaten, mit spezieller Berüchsichtigung der Menschenaffen. *Anthropologischer Anzeiger* 1933;**10**:154–85.
12. Aiello L, Dean MC. *Human evolutionary anatomy.* London: Academic Press; 1990.

13. Bogin B. *Patterns of human growth*. 2nd ed. Cambridge: Cambridge University Press; 1999.
14. Darwin C. *The descent of man, and selection in relation to sex*. London: John Murray; 1981.
15. Underwood CR, Ward EJ. The solar radiation area of man. *Ergonomics* 1966;**9**:155−68.
16. Newman RW. Why man is such a sweaty and thirsty naked animal: a speculative review. *Hum Biol* 1970;**42**:12−27.
17. Ruff C. Variation in human body size and shape. *Annu Rev Anthropol* 2002;**31**:211−32.
18. Frisancho AR. *Human adaptation and accommodation*. Ann Arbor, MI: University of Michigan Press; 1993.
19. Zihlman A. Woman the gatherer: the role of women in early hominid evolution. In: Sandra M, editor. *Gender and anthropology: critical reviews for teaching and research*. Washington, DC: American Anthropological Association; 1989. p. 23−43.
20. Bramble DM, Lieberman DE. Endurance running and the evolution of Homo. *Nature* 2004;**18**:345−52.
21. Corballis MC. *From hand to mouth: the origins of language*. Princeton, NJ: Princeton University Press; 2002.
22. Scammon RE, Calkins LA. *The development and growth of the external dimensions of the human body in the fetal period*. Minneapolis, MN: University of Minnesota Press; 1929.
23. Scammon RE. The measurement of the body in childhood. In: Harris JA, Jackson CM, Paterson DG, Scammon RE, editors. *The measurement of man*. Minneapolis, MN: University of Minnesota Press; 1930. p. 173−215.
24. Schultz AH. Fetal growth of man and other primates. *Q Rev Biol* 1926;**1**:465−521.
25. Martin RD. *Human brain evolution in an ecological context. 52nd James Arthur Lecture*. New York: American Museum of Natural History; 1983.
26. Leigh SR. Brain growth, life history, and cognition in primate and human evolution. *Am J Primatol* 2004;**62**:139−64.
27. Harvey P, Martin RD, Cluton-Brock TH. Life histories in comparative perspective. In: Smuts B, Cheney DL, Seyfarth RM, Wrangham RW, Struhsaker TT, editors. *Primate societies*. Chicago, IL: University of Chicago Press; 1983. p. 181−96.
28. Leonard WR, Robertson ML. Evolutionary perspectives on human nutrition: the influence of brain and body size on diet and metabolism. *Am J Hum Biol* 1994;**6**:77−88.
29. Klingenberg CP, Nijhout HF. Competition among growing organs and developmental control of morphological asymmetry. *Proc R Soc Lond* 1998;**265**:1135−9.
30. Charnov EL. *Life history invariants*. Oxford: Oxford University Press; 1993.
31. Stearns SC. *The evolution of life histories*. Oxford: Oxford University Press; 1992.
32. Bailey SM, Xu J, Feng JH, Hu X, Zhang C, Qui S. Tradeoffs between oxygen and energy in tibial growth at high altitude. *Am J Hum Biol* 2007;**19**:662−8.
33. Bogin B, Varela Silva MI, Rios L. Life history trade-offs in human growth: adaptation or pathology? *Am J Hum Biol* 2007;**19**:631−42.
34. Rauch F. Bone growth in length and width: the yin and yang of bone stability. *J Musculoskelet Neuronal Interact* 2005;**5**:194−201.
35. Tanner JM. A historical perspective on human auxology. *Humanbiol Budapest* 1994;**25**:9−22.
36. Norgard EA, Jarvis JP, Roseman CC, Maxwell TJ, Kenney-Hunt JP, Samocha KE, et al. Replication of long bone length QTL in the F9−F10 LG, SM advanced intercross. *Mamm Genome* 2009;**20**:224−35.
37. Reno PL, McCollum MA, Cohn MJ, Meindl RS, Hamrick M, Lovejoy CO. Patterns of correlation and covariation of anthropoid distal forelimb segments correspond to Hoxd expression territories. *J Exp Zool Mol Dev Evol* 2008;**310B**:240−58.
38. Kajantie E. Insulin-like growth factor (IGF)-I, IGF binding protein (IGFBP)-3, phosphoisoforms of IGFBP-1 and postnatal growth in very-low-birth-weight infants. *Horm Res* 2003;**60**:124−30.
39. Serrat MA, Lovejoy CO, King D. Age- and site-specific decline in insulin-like growth factor-1 receptor expression is correlated with differential growth plate activity in the mouse hindlimb. *Anat Rec* 2007;**290**:375−81.
40. Martini FH, Bartholomew EF. *Essentials of anatomy and physiology*. San Francisco, CA: Pearson Education; 2007.

41. Boros SJ, Nystrom J, Thompson T, Reynolds J, Williams H. Leg growth following umbilical artery catheter-associated thrombus formation: a 4-year follow-up. *J Pediatr* 1975;**87**:973—6.

42. Boyd E. Origins of the study of human growth. In: Savara BS, Schilke JF, editors. Eugene, OR: University of Oregon Press; 1980.

43. Tanner JM. *A history of the study of human growth*. Cambridge: University of Cambridge Press; 1981.

44. Tate C, Bendersky G. Olmec sculptures of the human fetus. *Perspect Biol Med* 1999;**42**:303—32.

45. Dietz WH, Marino B, Peacock NR, Bailey RC. Nutritional status of Efe pygmies and Lese horti-culturalists. *Am J Phys Anthropol* 1989;**78**:509—18.

46. Fredriks AM, van Buuren S, Burgmeijer RJ, Meulmeester JF, Beuker RJ, Brugman E, et al. Continuing positive secular growth change in The Netherlands 1955—1997. *Pediatr Res* 2000;**47**:316—23.

47. Eveleth PB, Tanner JM. *Worldwide variation in human growth*. Cambridge: Cambridge University Press; 1976.

48. Eveleth PB, Tanner JM. *Worldwide variation in human growth*. 2nd ed. Cambridge: Cambridge University Press; 1990.

49. Bergmann K. Über die Verhältnisse der Wärmeökonomie der Thiere zu ihrer Grösse. *Göttinger Studien* 1847;**3**:95—108.

50. Allen JA. The influence of physical conditions in the genesis of species. *Rad Rev* 1877;**1**:108—40.

51. Serrat MA, King D, Lovejoy CO. Temperature regulates limb length in homeotherms by directly modulating cartilage growth. *Proc Natl Acad Sci USA* 2008;**105**:19348—53.

52. Roberts DF. Bodyweight, race, and climate. *Am J Phys Anthropol* 1953;**11**:533—58.

53. Roberts DF. *Climate and human variability*. 2nd ed. Menlo Park, CA: Cummings; 1978.

54. Katzmarzyk PT, Leonard WR. Climatic influences on human body size and proportions: ecological adaptations and secular trends. *Am J Phys Anthropol* 1998;**106**:483—503.

55. Bogin B, Keep R. Eight thousand years of economic and political history in Latin America revealed by anthropometry. *Ann Hum Biol* 1999;**26**:333—51.

56. Andersen H. The influence of hormones on human development. In: Falkner F, editor. *Human development*. Philadelphia, PA: WB Saunders; 1966. p. 184—221.

57. Kramer K. Variation in juvenile dependence: helping behavior among Maya children. *Hum Nat* 2002;**13**:299—325.

58. Bogin B, Smith PK, Orden AB, Varela Silva MI, Loucky J. Rapid change in height and body proportions of Maya American children. *Am J Hum Biol* 2002;**14**:753—61.

59. Fulwood R, Abraham S, Johnson C. *Height and weight of adults ages 18—74 years by socioeconomic and geographic variables. Vital and health statistics, Series 11, No. 224. DHEW Pub. No. (PHS) 81-1674*. Washington, DC: US Government Printing Office; 1981.

60. Krogman WM. Growth of the head, face, trunk, and limbs in Philadelphia white and Negro children of elementary and high school age. *Monogr Soc Res Child Dev* 1970;**20**:1—91.

61. Hamill PVV, Johnston FE, Lemshow S. *Body weight, stature, and sitting height: white and Negro youths. 12—17 years, United States. DHEW Publication No. (HRA) 74-1608*. Washington, DC: US Government Printing Office; 1973.

62. Livshits G, Roset A, Yakovenko K, Trofimov S, Kobyliansky E. Genetics of human body size and shape: body proportions and indices. *Ann Hum Biol* 2002;**29**:271—89.

63. Bogin B, Kapell M, Varela Silva MI, Orden AB, Smith PK, Loucky J. How genetic are human body proportions?. In: Dasgupta P, Hauspie R, editors. *Perspectives in human growth, development and maturation*. Dordrecht: Kluwer; 2001. p. 205—21.

64. Aulchenko YS, Struchalin MV, Belonogova NM, Axenovich TI, Weedon MN, Hoffman A, et al. Predicting human height by Victorian and genomic methods. *Eur J Hum Genet* 2009;**17**:1070—5.

65. Holliday TW, Falsetti AB. A new method for discriminating African-American from European-American skeletons using postcranial osteometrics reflective of body shape. *J Forensic Sci* 1999;**44**:926—30.

66. Martorell R, Malina RM, Castillo RO, Mendoza FS. Body proportions in three ethnic groups: children and youths 2—17 years in NHANES and HHANES. *Hum Biol* 1988;**60**:205—22.

67. Feldesman MR, Fountain RL. Race specificity and the femur/stature ratio. *Am J Phys Anthropol* 1996;**100**:207—24.

68. Mark M, Rijli FM, Chambon P. Homeobox genes in embryogenesis and pathogenesis. *Pediatr Res* 1997;**42**:421—9.

69. Blum WF, Crowe BJ, Quigley CA, Jung H, Cao D, Ross JL, et al. SHOX Study Group. Growth hormone is effective in treatment of short stature associated with short stature homeobox-containing gene deficiency: two-year results of a randomized, controlled, multicenter trial. *J Clin Endocrinol Metab* 2007;**92**:219—28.

70. Neufeld ND, Lippe BM, Kaplan SA. Disproportionate growth of the lower extremities. A major determinant of short stature in Turner's syndrome. *Am J Dis Child* 1978;**132**:296—8.

71. Ogata T, Inokuchi M, Ogawa M. Growth pattern and body proportion in a female with short stature homeobox-containing gene overdosage and gonadal estrogen deficiency. *Eur J Endocrinol* 2002;**147**:249—54.

72. Hughes PC, Ribeiro J, Hughes IA. Body proportions in Turner's syndrome. Arch Dis Child **986**; 61:506-507

73. Anderssen L, Haley CS, Ellegren H, Knott SA, Johansson M, Andersson K, et al. Genetic mapping of quantitative trait loci for growth and fatness in pigs. *Science* 1994;**262**:1771—4.

74. Quignon P, Schoenebeck JJ, Chase K, Parker HG, Mosher DS, Johnson GS, et al. Fine mapping a locus controlling leg morphology in the domestic dog. *Quant Biol* 2009;**74**:327—33.

75. Tost J. DNA methylation: an introduction to the biology and the disease-associated changes of a promising biomarker. *Methods Mol Biol* 2009;**507**:3—20.

76. Lasker GW. Human biological adaptability. *Science* 1969;**166**:1480—6.

77. Hales CN, Barker DJ. Type 2 (non-insulin-dependent) diabetes mellitus: the thrifty phenotype hypothesis. *Diabetologia* 1992;**35**:595—601.

78. Wells JCK. The thrifty phenotype as an adaptive maternal effect. *Biol Rev* 2007;**82**:143—72.

79. Emanuel I. Maternal health during childhood and later reproductive performance. *Ann NY Acad Sci* 1986;**477**:27—39.

80. Varela-Silva MI, Frisancho AR, Bogin B, Chatkoff D, Smith P, Dickinson F, et al. Behavioral, environmental, metabolic and intergenerational components of early life undernutrition leading to later obesity in developing nations and in minority groups in the USA. *Coll Antropol* 2007;**31**:315—9.

81. Barker DJP, Eriksson JG, Forsén T, Osmond C. Fetal origins of adult disease: strength of effects and biological basis. *Int J Epidemiol* 2002;**31**:1235—9.

82. Gluckman PD, Hanson MA. *The fetal matrix*. Cambridge: Cambridge University Press; 2005.

83. Gluckman PD, Hanson MA, Beedle AS. Early life events and their consequences for later disease: a life history and evolutionary perspective. *Am J Hum Biol* 2007;**19**:1—19.

84. Leitch I. Growth and health. *Br J Nutr* 1951;**5**:142—151. (Reprinted with commentaries in Int J Epidemiol 2001;**30**:212—25.)

85. Crews DE. *Human senescence: evolutionary and biocultural perspectives*. New York: Cambridge University Press; 2003.

86. Larke A, Crews DE. Parental investment, late reproduction, and increased reserve capacity are associated with longevity in humans. *J Phys Anthropol* 2006;**25**:119—31.

87. Bogin B. Childhood, adolescence, and longevity: a multilevel model of the evolution of reserve capacity in human life history. *Am J Hum Biol* 2009;**21**:567—77.

88. Thomson AM, Duncan DL. The diagnosis of malnutrition in man. *Nutr Abstr Rev* 1954;**24**:1—18.

89. Wolanski N. Parent—offspring similarity in body size and proportions. *Stud Hum Ecol* 1979;**3**:7—26.

90. Ramos Rodríguez RM. El significado del miembro superior una hipótesis a considerar. *Bol Med Hosp Infant Mex* 1981;**38**:373—7.

91. Ramos Rodríguez RM. Algunos aspectos de proporcionalidad lineal de una población del estado de Oaxaca. *An Antropol* 1990;**27**:85—96.

92. Tanner JM, Hayashi T, Preece MA, Cameron N. Increase in length of leg relative to trunk in Japanese children and adults from 1957 to 1977: comparison with British and with Japanese Americans. *Ann Hum Biol* 1982;**9**:411—23.

93. Buschang PH, Malina RM, Little BB. Linear growth in Zapotec schoolchildren: growth status and early velocity for leg length and sitting height. *Ann Hum Biol* 1986;**13**:225—34.

94. Dickinson F, Cervera M, Murguía R, Uc L. Growth, nutritional status and environmental change in Yucatan, Mexico. *Stud Hum Ecol* 1990;**9**:135—49.

95. Gurri FD, Dickinson F. Effects of socioeconomic, ecological, and demographic conditions on the development of the extremities and the trunk: a case study with adult females from Chiapas. *J Hum Ecol* 1990;**1**:125—38.

96. Murguía R, Dickinson F, Cervera M, Uc L. Socio-economic activities, ecology and somatic differences in Yucatan, Mexico. *Stud Hum Ecol* 1990;**9**:111—34.

97. Bolzán AG, Guimarey LM, Pucciarelli HM. Crecimiento y dimorfismo sexual de escolares según la ocupación laboral paterna. *Arch Latinoam Nutr* 1993;**43**:132—8.

98. Wolanski N, Dickinson F, Siniarska A. Biological traits and living conditions of Maya Indian and non-Maya girls from Merida, Mexico. *Int J Anthropol* 1993;**8**:233—46.

99. Siniarska A. Family environment and body build in adults of Yucatan Mexico. *Am J Phys Anthropol* 1995;(Suppl. 20):196.

100. Wolanski N. Household and family as environment for child growth. Cross cultural studies in Poland, Japan, South Korea and Mexico. In: Wright SD, Meeker DE, Griffore R, editors. *Human ecology: progress through integrative perspectives*. Bar Harbor, ME: Society for Human Ecology; 1995. p. 140—52.

101. Jantz LM, Jantz RL. Secular change in long bone length and proportion in the United States, 1800—1970. *Am J Phys Anthropol* 1999;**110**:57—67.

102. Whitley E, Gunnell G, Davey-Smith G, Holly JMP, Martin RM. Childhood circumstances and anthropometry: the Boyd Orr cohort. *Ann Hum Biol* 2008;**35**:518—34.

103. Kim J-M, Stewart R, Shin I-S, Kim SW, Yang S-J, Yoon J-S. Associations between head circumference, leg length and dementia in a Korean population. *Int J Geriatr Psychiatry* 2008;**23**:41—8.

104. Gunnell D, Whitley E, Upton MN, McConnachie A, Davey-Smith G, Watt GC. Associations of height, leg length, and lung function with cardiovascular risk factors in the Midspan Family Study. *J Epidemiol Community Health* 2003;**57**:141—6.

105. Ferrie JE, Langenberg C, Shipley MJ, Marmot MG. Birth weight, components of height and coronary heart disease: evidence from the Whitehall II study. *Int J Epidemiol* 2006;**35**:1532—42.

106. Wadsworth ME, Hardy RJ, Paul AA, Marshall SF, Cole TJ. Leg and trunk length at 43 years in relation to childhood health, diet and family circumstances: evidence from the 1946 national birth cohort. *Int J Epidemiol* 2002;**31**:383—90.

107. Li L, Dangour AL, Power C. Early life influences on adult leg and trunk length in the 1958 British Birth Cohort. *Am J Hum Biol* 2007;**19**:836—43.

108. Dixon B, Darlow B, Prickett T. How useful is measuring neonatal growth? *J Paediatr Child Health* 2008;**44**:444—8.

109. Frisancho AR, Guilding N, Tanner S. Growth of leg length is reflected in socio-economic differences. *Acta Med Auxol* 2001;**33**:47—50.

110. Malina RM, Pena Reyes ME, Tan SK, Buschang PH, Little BB, Koziel S. Secular change in height, sitting height and leg length in rural Oaxaca, southern Mexico: 1968—2000. *Ann Hum Biol* 2004;**31**:615—33.

111. Dasgupta P, Saha R, Nubé M. Changes in body size, shape and nutritional status of middle class Bengali boys of Kolkata, India, 1982—2002. *Econ Hum Biol* 2008;**6**:75—94.

112. Floyd B. Intergenerational gains in relative knee height as compared to gains in relative leg length within Taiwanese families. *Am J Hum Biol* 2008;**20**:462—4.

113. Gunnell DJ, Smith GD, Frankel SJ, Kemp M, Peters TJ. Socio-economic and dietary influences on leg length and trunk length in childhood: a reanalysis of the Carnegie (Boyd Orr) survey of diet and health in pre-war Britain (1937—39). *Paediatr Perinat Epidemiol* 1998;**12**:96—113.

114. Lawlor DA, Davey-Smith G, Ebrahim S. Association between leg length and offspring birthweight: partial explanation for the trans-generational association between birthweight and cardiovascular disease: findings from the British Women's Heart and Health Study. *Paediatr Perinat Epidemiol* 2003;**17**:148—55.

115. Martin RM, Davey-Smith G, Frankel S, Gunnell D. Parents' growth in childhood and the birth weight of their offspring. *Epidemiology* 2004;**15**:308—16.

116. Leary S, Davey Smith G, Ness A. ALSPAC Study Team. Smoking during pregnancy and components of stature in offspring 2006. *Am J Hum Biol* 2006;**18**:502—12.

117. Dangour AD. Growth of upper- and lower-body segments in Patamona and Wapishana Amerindian children (cross-sectional data). *Ann Hum Biol* 2001;**28**:649—63.

118. Bogin B, Rios L. Rapid morphological change in living humans: implications for modern human origins. *Comp Biochem Physiol A* 2003;**136**:71—84.

119. Bogin B, Varela-Silva MI. Anthropometric variation and health: a biocultural model of human growth. *J Child Health* 2003;**1**:149—72.

120. Smith PK, Bogin B, Varela-Silva MI, Orden AB, Loucky J. Does immigration help or harm children's health? The Mayan case. *Soc Sci Q* 2002;**83**:994—1002.

121. Smith PK, Bogin B, Varela-Silva MI. Economic and anthropological assessments of the health of children in Maya families in the United States. *Econ Hum Biol* 2003;**1-2**:145—60.

122. Han TS, Hooper JP, Morrison CE, Lean ME. Skeletal proportions and metabolic disorders in adults. *Eur J Clin Nutr* 1997;**51**:804—9.

123. Gunnell DJ, Davey-Smith G, Frankel S, Nanchahal K, Braddon FE, Pemberton J, et al. Childhood leg length and adult mortality: follow up of the Carnegie (Boyd Orr) Survey of Diet and Health in Pre-war Britain. *J Epidemiol Community Health* 1998;**52**:142—52.

124. Jarvelin MR. Fetal and infant markers of adult heart diseases. *Heart* 2000;**84**:219—26.

125. Smith GD, Greenwood R, Gunnell D, Sweetnam P, Yarnell J, Elwood P. Leg length, insulin resistance, and coronary heart disease risk: the Caerphilly Study. *J Epidemiol Community Health* 2001;**55**:867—72.

126. Langenberg C, Hardy R, Kuh D, Wadsworth ME. Influence of height, leg and trunk length on pulse pressure, systolic and diastolic blood pressure. *J Hypertens* 2003;**21**:537—43.

127. Lawlor DA, Taylor M, Davey-Smith G, Gunnell D, Ebrahim S. Associations of components of adult height with coronary heart disease in postmenopausal women: the British women's heart and health study. *Heart* 2004;**90**:745—9.

128. Fraser A, Ebrahim S, Smith GD, Lawlor DA. The associations between height components (leg and trunk length) and adult levels of liver enzymes. *J Epidemiol Community Health* 2008;**62**:48—53.

129. Gunnell D, Okasha M, Smith GD, Oliver SE, Sandhu J, Holly JM. Height, leg length, and cancer risk: a systematic review. *Epidemiol Rev* 2001;**23**:313—42.

130. Ogilvy-Stuart AL, Gleeson H. Cancer risk following growth hormone use in childhood: implications for current practice. *Drug Saf* 2004;**27**:369—82.

131. Lima GA, Corrêa LL, Gabrich R, Miranda LC, Gadelha MR. IGF-I, insulin and prostate cancer. *Arq Bras Endocrinol Metabol* 2009;**53**:969—75.

132. Weng CJ, Hsieh YH, Tsai CM, Chu YH, Ueng KC, Liu YF, et al. Relationship of insulin-like growth factors system gene polymorphisms with the susceptibility and pathological development of hepatocellular carcinoma. *Ann Surg Oncol* 2010;**17**:1808—15.

133. Schooling CM, Jiang CQ, Heys M, Zhang WS, Adab P, Cheng KK, et al. Are height and leg length universal markers of childhood conditions? The Guangzhou Biobank cohort study. *J Epidemiol Community Health* 2008;**62**:607—14.

134. Schooling CM, Jiang CQ, Heys M, Zhang WS, Lao XQ, Adab P, et al. Is leg length a biomarker of childhood conditions in older Chinese women? The Guangzhou Biobank Cohort Study. *J Epidemiol Community Health* 2008;**62**:160—6.

135. Osuch JR, Karmaus W, Hoekman P, Mudd L, Zhang J, Haan P, et al. Association of age at menarche with adult leg length and trunk height: speculations in relation to breast cancer risk. *Ann Hum Biol* 2010;**37**:76—85.

136. Lorentzon M, Norjavaara E, Kindblom JM. Pubertal timing predicts leg length and childhood body mass index predicts sitting height in young adult men. *J Pediatr* 2011;**3**:452—7.

137. McIntyre MH. Adult stature, body proportions and age at menarche in the United States National Health and Nutrition Survey (NHANES) III. *Ann Hum Biol* 2011;**38**:716—20.

138. Padez C, Varela-Silva MI, Bogin B. Height and relative leg length as indicators of the quality of the environment among Mozambican juveniles and adolescents. *Am J Hum Biol* 2009;**21**:200—9.
139. Cuenca-Guerra R, Daza-Flores JL, Saade-Saade AJ. Calf implants. *Aesthetic Plast Surg* 2009;**33**:505—13.
140. Fan J, Liu F, Wu J, Dai W. Visual perception of female physical attractiveness. *Proc Biol Sci* 2004;**271**:347—52.
141. Weeden J, Sabini J. Physical attractiveness and health in western societies: a review. *Psychol Bull* 2005;**131**:635—53.
142. Gründl M, Eisenmann-Klein M, Prantl L. Quantifying female bodily attractiveness by a statistical analysis of body measurements. *Plast Reconstr Surg* 2009;**123**:1064—71.
143. Martins PA, Hoffman DJ, Fernandes MT, Nascimento CR, Roberts SB, Sesso R, et al. Stunted children gain less lean body mass and more fat mass than their non-stunted counterparts: a prospective study. *Br J Nutr* 2004;**92**:819—25.
144. Velásquez-Meléndez G, Silveira EA, Allencastro-Souza P, Kac G. Relationship between sitting-height-to-stature ratio and adiposity in Brazilian women. *Am J Hum Biol* 2005;**17**:646—53.
145. Hoffman DJ, Sawaya AL, Verreschi I, Tucker KL, Roberts SB. Why are nutritionally stunted children at increased risk of obesity? Studies of metabolic rate and fat oxidation in shantytown children from São Paulo, Brazil. *Am J Clin Nutr* 2000;**72**:702—7.
146. Sawaya AL, Martins PA, Baccin Martins VJ, Florêncio TT, Hoffman D, Franco MdCP, et al. Malnutrition, long-term health and the effect of nutritional recovery. In: Kalhan SC, Prentice AM, Yajnik CS, editors. *Emerging societies — coexistence of childhood malnutrition and obesity. Nestlé Nutrition Institute Workshop Series, Pediatric Program. Nestec,* Vol. 63. Basel: Vevey/S. Karger; 2009; p. 95—108.
147. Bénazet JD, Zeller R. Vertebrate limb development: moving from classical morphogen gradients to an integrated 4-dimensional patterning system. *Cold Spring Harbor Perspect Biol* 2009;**1**:a001339.

SUGGESTED READING

Bayley N, Davis FC. Growth changes in bodily size and proportions during the first three years: a developmental study of 61 children by repeated measurements. *Biometrika* 1935;**27**:26—87.
  This is a classic paper on the topic.
Eveleth PB, Tanner JM. *World-wide variation in human growth.* Cambridge: Cambridge University Press; 1976.
Eveleth PB, Tanner JM. *World-wide variation in human growth.* 2nd ed. Cambridge: Cambridge University Press; 1990.
  These two references provide a large compendium of data on human growth, including body proportions.
Sabharwal S, Green S, McCarthy J, Hamdy RC. What's new in limb lengthening and deformity correction. *J Bone Joint Surg Am* 2011;**93**:213—21.

INTERNET RESOURCES

Body proportions: http://en.wikipedia.org/wiki/Body_proportions.
NHANES anthropometric manual: http://www.cdc.gov/nchs/data/nhanes/nhanes3/cdrom/NCHS/MANUALS/ANTH.
Physical attractiveness: http://en.wikipedia.org/wiki/Physical_attractiveness.
  Many other websites may be found via any search engine using the terms "human body proportions", leg length" and "leg length and health".

CHAPTER 14

Physical Activity as a Factor in Growth and Maturation

Robert M. Malina★,★★

★Department of Kinesiology and Health Education, University of Texas at Austin, TX 78712, USA
★★Department of Kinesiology, Tarleton State University, Stephenville, TX 76402, USA

Contents

N. Cameron & B. Bogin (eds): Human Growth and Development, Second edition.
ISBN 978-0-12-383882-7, Doi: 10.1016/B978-0-12-383882-7.00014-3

14.1 INTRODUCTION

Interactions of genes, hormones, nutrients and energy maintain the biological processes of growth and maturation. Genes, hormones, energy and nutrients interact among themselves, and also with the environments in which the individual is reared and lives. Many genes have been identified with growth and maturation, and many more will be identified. A key, of course, is the understanding of the mechanisms whereby genetic variants exert their effects on the processes of growth and maturation and interact with the environments in which children grow and mature.

A number of environmental factors, including the home, also affect growth and maturation. The shared cultural environment of the home includes present lifestyle characteristics and those transmitted from parents to children through modeling, education, socioeconomic status, and so on. These are important sources of variation in growth and maturation, especially transgenerational secular changes, both positive and negative. The home environment influences energy and nutrient availability and in turn food consumption. A key is the understanding of the mechanisms whereby cultural practices interact with genes and hormones to influence growth and maturation.

Physical activity is an environmental factor that has historically been viewed as important for healthy growth.[1] In one of the first comprehensive reviews of "exercise and growth", it was suggested that: "… certain minima of muscular activity are essential for supporting normal growth and for maintaining the protoplasmic integrity of the tissues. What these minima mean in terms of intensity and duration of activity has not been ascertained".[2, p. 459] Interest in the potential influence of physical activity on growth and maturation has a long tradition in auxology and sport sciences.[3–9] At present, physical activity and sedentary behavior are of considerable interest to public health, medicine and education, largely in the context of the obesity epidemic in the pediatric population and potentially negative health consequences of a physically inactive lifestyle.[10,11]

Physical activity as a factor that may affect growth and maturation is the focus of this chapter. It is essential, nevertheless, to recognize that physical activity is only one of many factors that may affect these processes. The influence of regular physical activity on children and adolescents is considered in the context of three general areas: first, commonly used indicators of growth and maturation; second, indicators of health status (chronic disease risk) and physical fitness; and third, the special case of intensive training for sport at young ages.

14.2 PHYSICAL ACTIVITY AND SEDENTARY BEHAVIOR

At present, the public health and biomedical community is focusing on physical activity in the context of health promotion and disease prevention and sedentary behavior as a major risk factor, among others, for cardiovascular and metabolic disease. The educational community highlights activity in the context of physical education as a component of the overall school experiences of youth.

Physical activity and sedentary behavior occur in many contexts. Both are important avenues for learning, enjoyment, social interactions and self-understanding. Currently, evidence and opinion suggest an imbalance in the direction of increased inactivity and reduced activity underlying the emergence of metabolic risk factors for cardiovascular disease and the current epidemic of obesity in youth.

Physical activity is a multidimensional behavior that is viewed most often in terms of energy expenditure [metabolic equivalents (METS), oxygen consumption]. It involves gross bodily movements associated with significant increases in energy expenditure above resting levels and associated stresses and strains of weight bearing and ground reaction forces. Physical fitness (performance and health related) and motor skill (proficiency in a variety of movements) are important dimensions of activity.[12,13] Context is a dimension of physical activity that is often overlooked, or perhaps undervalued. Context refers to types and settings of activity, and includes play, physical education, exercise, sport, work, and others. Contexts per se and meanings attached to them vary with age among youth and also among and within different cultural groups.[14]

Sedentary behavior or physical inactivity also has several dimensions. Public health and medicine view inactivity in terms of insufficient energy expenditure, force generation and health-related physical fitness. Sedentary behavior, however, has a major cultural component. Many forms of inactivity are highly valued by societies — school, study, reading, music, art, television viewing, video games, personal computers and the like — and opportunities for being physically inactive have increased over time. Motorized transport is another form of physical inactivity that is highly valued by major segments of society.

Discussions of physical activity generally focus on school-aged youth, about 6—18 years old,[12] although interest in relationships among motor development, movement proficiency and physical activity among preschool children is increasing.[15,16] Physical activity as used in many studies refers to habitual physical activity, i.e. the level of physical activity that characterizes the lifestyle of the individual. It is often quantified in terms of amount of time in an activity (hours/day or hours/week), an activity score and/or time or energy expended in moderate to vigorous activities. Estimates are ordinarily based on questionnaires, interviews, diaries, accelerometers, pedometers, or a combination of methods.[12,13]

There is a need to qualify and quantify physical activity programs for children and adolescents if the effects of physical activity upon growth and maturation are to be identified and partitioned from other factors known to affect these outcomes. This requires more specific details about number of sessions per week, duration of activity sessions or distance covered in a session, intensity of the activity or energy cost, and type of activity, e.g. sprint, speed, endurance or strength, or some combination thereof.

14.3 APPROACHES TO THE STUDY OF PHYSICAL ACTIVITY

Three general approaches have been used to evaluate the potential influence of physical activity on growth and maturation. The first compares characteristics of children and adolescents who are habitually physically active to those who are not active or, more appropriately, are less active. Criteria for habitual activity and inactivity vary among studies. Inactive young people generally participate in normal daily activities, often including physical education. They are not inactive; they are inactive relative to those who engage in physical activities on a more regular basis.

Comparisons of young athletes and non-athletes are occasionally used to make inferences about the influence of physical activity. It is assumed that the athletes have been active on a regular basis in sport-specific training, and differences relative to non-athletes arc attributed to the activity involved in the training program. This approach has several problems including, in particular, variation among sport or sport disciplines and issues related to subject selectivity and retention. Successful young athletes, especially more elite athletes, are different from non-athletes, quite often in size and maturation.[13,17,18] The selectivity of artistic gymnastics, figure skating and ballet for specific physical characteristics is well documented. Hence, it is not correct to generalize from athletes to the general population of children and adolescents.

The second approach is experimental. It involves comparison of individuals exposed to a specific physical activity program (experimental group) and those not exposed to the program (control group). Experimental studies vary in power. Many earlier studies did not carefully match subjects in the experimental and control groups and simply did before-and-after comparisons. The physical activity stimulus also varied among studies in type, intensity and duration. Selection of subjects, variation in growth and maturity status, control of outside activity, and other factors make comparison of experimental studies difficult. With the more recent focus on health benefits of physical activity, there is more rigorous emphasis on specifying the frequency, intensity, duration and type of physical activity needed to bring about specific health benefits.[10] The randomized control design is the ideal in this regard. The design attempts to control for pre-existing differences among subjects by randomly assigning them to experimental (intervention) and control (non-intervention) groups.

A special type of experimental study involves the use of extreme unilateral activity, for example, specialized use of the arm as in racket sports, to illustrate the effects of physical activity on limb muscle, skeletal and adipose tissues. The individual is his or her own control, as the dominant limb (experimental, active limb) is compared to the non-dominant limb (control, less active limb). In the context of growth, these studies are largely retrospective. Differences between limbs are related to training history, specifically the onset of training in racket sports during childhood.

The third approach is correlational. It considers the relationship between an estimate of habitual physical activity and an indicator of growth and maturation or disease risk, e.g. the correlation between level of physical activity and subcutaneous fatness. Correlational studies, of course, do not permit causal statements.

Studies of physical activity in children and adolescents are also limited by the fact that the activity stimulus is rarely monitored over several years. Most experimental studies are specific and relatively short term, e.g. 15 or 20 weeks of endurance training at a specific heart rate, or 8 or 12 weeks of resistance training. Rarely is a specific activity program monitored over several years. Training for sport — the systematic, specialized practice for a specific sport or sport discipline — often spans several years during childhood and adolescence. Except for swimming, which often records distances covered in training, programs for young athletes rarely specify the sport-specific training demands and practices. Hours per week are a vague quantification of sport training.

14.4 PHYSICAL ACTIVITY, HEIGHT, WEIGHT AND BODY COMPOSITION

14.4.1 Height

Regular physical activity has no apparent effect on attained height and rate of growth in height. Longitudinal data on active and inactive boys followed from childhood through adolescence, and girls followed during childhood, indicate, on average, either no differences, or only small differences in height between the active and less active youth.[19–21] Although some early data suggested an increase in stature with regular activity, the observed changes were usually quite small and were based on studies that did not control for subject selection and for maturity status at the time of training or at the time of making the comparisons.[3,8]

On the other hand, regular physical activity does not have a negative effect on growth in height. This is relevant because several studies of young athletes have attributed short stature and a slower growth rate to training for sport.[18,22] Some have accepted these observations at face value and concluded that training for athletic competition may slow down or even stunt growth in stature. However, several important factors are not considered, including selection criteria for some sports, differential dropout or exclusion, and interindividual variation in biological maturation. The growth of young athletes is considered later in the chapter.

14.4.2 Body Weight and Composition

Differences in body weights of active and less active boys and girls are generally small and not significant.[19–21] However, the composition of the body mass can be potentially influenced by regular activity. Early studies of body composition used primarily the two-compartment model, body weight = fat-free mass (FFM) + fat mass (FM), whereas more recent studies partition FFM into soft lean tissues (skeletal muscle in particular) and bone minerals.[13]

It is often suggested that regular physical activity is associated with a decrease in FM and an increase in FFM. Indicators of fatness are considered later in this section. Both sexes have a significant adolescent spurt in FFM, males more so than females,[13] so that it is difficult to partition effects of training on FFM from expected changes associated with growth and maturation, specifically during adolescence. Nevertheless, advances in analytical techniques provide insights.

Two early studies indicate several of the problems in partitioning physical activity effects from those associated with growth and maturation. In the first,[23,24] 40 boys were divided into three groups with different physical activity and/or training programs and were followed longitudinally from 11 to 18 years. The sample size at the start of the study was 143, so that over 8 years about 100 boys dropped out of the study. The active boys ($n = 8$, >6 hours/week) included six selected for basketball and two for athletics and participated in supervised training. The other two groups had less regular physical activity: moderate activity ($n = 18$, 4 hours/week in sport activity but not on a regular basis) and limited activity ($n = 13$, <2 hours/week in unsystematic sport activity, including physical education). Activity levels of the three groups were described slightly differently in a related report,[25] as active: 4 hours/week 11–15 years, 6 hours/week 15–18 years; moderate activity: 2 hours/week 11–15 years, 3 hours/week 15–18 years; limited activity: 1 hour/week 11–15 years, no regular activity 15–18 years.

The active boys were especially taller than boys in the other two groups at the start and throughout the study, and heavier from 13–18 years. The groups differed slightly in densitometric estimates of FFM and percentage body fat at the beginning of the study, but during the course of the study and at its end, the most active boys had significantly more FFM and less fat than the moderately and least active boys. The latter two groups differed to a small extent in FFM, but the group with limited physical activity had greater relative fatness. The active group was also advanced in skeletal maturity and attained peak height velocity at an earlier age, showing a growth pattern characteristic of early-maturing boys. Their greater heights and larger gains in FFM compared to the other groups were probably related to their advanced maturity status, which was not controlled in the comparisons.

The second study involved 5 months of endurance activities designed to increase the maximal aerobic power of nine boys, 11–13 years of age.[26,27] Significant gains were noted in potassium concentration, measured by whole-body counting of potassium-40. The boys gained, on average, 0.5 kg in weight and 12 g of potassium. A 12 g increase in potassium, an index of FFM and in particular muscle mass, would correspond to a gain of about 4 kg of muscle mass, which would indicate that the 0.5 kg gain in body weight was accompanied by a loss of about 3 kg of fat during the endurance training program. Relative to growth in height over 5 months, the gain in potassium after the aerobic training program was 6% greater than expected, whereas the gain in weight was 5% less than expected. The boys gained an average of 3.5 cm in height over 5 months, suggesting that the adolescent spurt probably occurred in some boys over this interval. FFM is, of course, highly correlated with height.[13] Variation in sexual maturation is an additional confounder. With one exception, the largest gains in height and potassium were observed in boys who were in mid- to late puberty based on measurement of testicular volume at the end of the study.[13] The observed gains in FFM probably reflected those that accompany the adolescent spurt and sexual maturation in boys, and not necessarily the effect of the aerobic activity program. This is of relevance because endurance programs are not often associated with gains in FFM.

These early studies illustrate the difficulties in partitioning changes in FFM associated with physical activity or short-term training programs from those that accompany normal growth and maturation during male adolescence. More recently, the Saskatchewan Pediatric Bone Mineral Accrual Study applied multilevel modeling, allowing for height and biological maturity (years from peak height velocity), to longitudinal observations for lean tissue mass based on dual X-ray absorptiometry (DXA) in boys and girls.[28] Peak velocity of growth in lean tissue mass in this sample of Canadian youth occurred, on average, shortly after peak height velocity.[29] Allowing for height and biological maturity, physical activity had an independent effect on growth in lean tissue mass in both sexes. At the same level of activity, the effect of physical activity on lean tissue mass was greater in boys than in girls, especially for estimated lean tissue accrual in the arms (boys 70 ± 27 g, girls 31 ± 15 g) and trunk (boys 249 ± 91 g, girls 120 ± 58 g) compared to the legs (boys 198 ± 60 g, girls 163 ± 40 g). Nevertheless, given the relatively large gains in FFM or lean tissue mass during adolescence,[13] the proportion attributed to physical activity is relatively small.

Results from these and other shorter-term studies are consistent with resistance training programs which are designed to improve muscular strength. Allowing for variable durations of training programs, changes in estimates of lean tissue mass (limb girths and areas, FFM) with resistance training in children and adolescents were variable and in most cases minimal.[30]

It is commonly assumed that regular physical activity has a favorable influence on indicators of fatness or adiposity — skinfold thicknesses, body mass index (BMI) and percentage body fat (%Fat) — in children and adolescents. Results of correlation and regression analyses in young people of mixed weight status (normal weight, overweight, obese) indicate a low to moderate relationship between habitual physical activity and adiposity. Statistics were reasonably consistent across studies considering the mix of methods used to measure/estimate physical activity; most of the variance in adiposity was not explained by physical activity. Nevertheless, young people who engaged in more physical activity on a regular basis, specifically vigorous activities, tended to have less adiposity than those who engaged in less activity.[10,31] Studies utilizing skinfold thicknesses need to be viewed in the context of three potential confounders: extremity but not trunk skinfolds decline in thickness, on average, during male but not female adolescence; changes in extremity skinfolds are related to the time of peak height velocity; and measurement variability.[13]

Enhanced and experimental physical activity programs in normal-weight youth appear to have a minimal effect on adiposity, but the issue of activity volume has not been systematically addressed. It is possible that normal-weight youth require a greater activity volume, as suggested by several studies of obese youth which used 80 minutes/day of moderate to vigorous physical activity.[10] On the other hand, physical activity interventions with overweight and obese youth resulted in reductions in overall adiposity and in visceral (abdominal) adiposity. The programs included a variety of activities (largely aerobic) of moderate and vigorous intensity, three to five times per week, for 30 to 60 minutes' duration. The most consistent favorable effects of physical activity on fatness were found in studies that used more direct estimates of body composition, specifically DXA estimates of %Fat and magnetic resonance imaging of visceral adiposity, in contrast to the BMI, skinfolds per se or %Fat predicted from skinfold thicknesses.[10,31]

In contrast to FFM and fatness, evidence from a variety of cross-sectional and longitudinal studies indicates a beneficial effect of regular physical activity on bone mineral content in young people.[10] Most data are derived from prepubertal children of both sexes and youth in the early stages of puberty, and from girls more often than boys. Among postpubertal youth or those nearing maturity, the influence of physical activity, although generally positive, is more variable. In the longitudinal series from the Saskatchewan Pediatric Bone Mineral Accrual Study followed through the adolescent growth spurt, peak velocity of growth in bone mineral occurred, on average, more than 6 months after peak height velocity.[29] Nevertheless, young people of both sexes who were active during the interval of maximal growth gained more bone mineral than their less active peers.[32] Active boys and girls had 9% and 17% greater total body bone mineral, respectively, than their less active peers after peak velocity of bone mineral accrual. The results thus indicated an enhanced effect of physical activity on bone mineral accrual during the period of rapid growth in both boys and girls.

Physical activity interventions aimed at augmenting bone mineral are consistent with observations based on comparisons of active and less active children and adolescents. The programs generally met two or three times per week for moderate- to high-intensity activities, weight-bearing activities of a longer duration (45—60 minutes) and/or high impact activities over a shorter duration (10 minutes).[10] More recent data suggest a positive role of physical activity in enhancing bone strength in youth. Changes in bone geometry observed in three-dimensional imaging indicated a substantial increase in bone strength. Moreover, bone strength was related to habitual physical activity, but short bouts of activity may have been as effective as sustained activity in youth.[33]

The benefits of regular physical activity on bone mineral content have implications for longer-term skeletal health. Bone mineral established during childhood and adolescence is a determinant of bone mineral status in adulthood.[13]

14.5 PHYSICAL ACTIVITY AND BIOLOGICAL MATURATION

Biological maturation is a highly individual characteristic that is variable in timing and tempo. Skeletal maturation is the only system of maturity assessment that spans the period of growth from childhood to late adolescence. The hand–wrist area is used most often. Age at peak height velocity is ordinarily derived from longitudinal data that span adolescence, although a protocol for estimating time before or after peak velocity (maturity offset) is available.[34] Secondary sex characteristics are useful only when they overtly manifest. Details of assessment are considered elsewhere.[13]

It is difficult to quantify the effects of regular physical activity on maturity indicators commonly used in growth studies, especially during adolescence. The same hormones regulate somatic, skeletal and sexual maturation during adolescence, so that regular activity, if it has an effect, should influence the different indicators of maturity in a generally similar manner.

14.5.1 Skeletal Maturation

Although regular physical activity enhances the accrual of bone mineral, it does not influence skeletal maturation of the hand and wrist. Active and non-active boys followed longitudinally from 13 to 18 years did not differ in skeletal age.[21]

14.5.2 Somatic Maturation

Age at peak height velocity (PHV) is not affected by level of habitual physical activity. Small samples of boys classified as physically active and less active for the years before and during the adolescent growth spurt do not differ in estimated age at PHV.[20,21,35,36] The magnitude of PHV in active and less active boys also does not differ and is within the range of variation in growth velocity during the adolescent spurt. By inference, peak velocity of growth in height during the adolescent growth spurt is not affected by regular activity. Corresponding data are not available for girls classified by level of habitual physical activity.

14.5.3 Sexual Maturation

Discussions of the potential influence of regular physical activity on sexual maturation focus most often on females, and specifically on age at menarche, a late event in the pubertal sequence. Menarche, of course, has cultural and social significance in the lives of adolescent girls.

Longitudinal data on the sexual maturation of habitually active and non-active boys and girls are not available. Some epidemiological data suggest an association between habitual physical activity and later menarche,[37,38] but other data do not.[39] The association is not strong and is confounded by many other factors known to influence age at menarche.[7,18] In contrast, much of the discussion on the association between physical activity and age at menarche focuses on later mean ages at menarche which are commonly reported in athletes in many, but not in all sports.[18,40] This is considered later in the chapter.

14.6 PHYSICAL ACTIVITY AND PHYSICAL FITNESS

Two components of physical fitness have received most attention in the context of the influence of regular physical activity: aerobic fitness and muscular strength and endurance.

14.6.1 Aerobic Fitness

Data from both cross-sectional and longitudinal studies indicate higher levels of aerobic fitness, measured as maximal aerobic power ($\dot{V}O_{2max}$) or endurance runs, in active youth than in in less active youth. In experimental studies of young people from 8 years through adolescence, continuous, vigorous physical activity has a favorable effect on maximal aerobic power.[10] Programs generally involved continuous, vigorous activity (e.g. 80% of maximal heart rate), 3 days per week and 30—45 minutes per session. The associated gain in $\dot{V}O_{2max}$ was about 10% (3—4 ml/kg/min). Limited longitudinal data suggest an enhanced effect of physical activity during the interval of maximal growth in height on $\dot{V}O_{2max}$.[20] Habitually active boys and boys with average activity gained more in absolute $\dot{V}O_{2max}$ than boys low in habitual activity, but the three groups were clearly separated in relative $\dot{V}O_{2max}$ during the adolescent spurt. Per unit mass, peak oxygen uptake (ml/kg/minute) showed the following gradient: active > average > less active.

14.6.2 Muscular Strength and Endurance

Cross-sectional and longitudinal data are equivocal regarding the association of physical activity with muscular strength and endurance among children and adolescents, with one exception. Longitudinal observations indicate better upper body muscular strength and endurance (flexed arm hang) in active compared to less active boys.[21] Experimental data show significant gains in muscular strength and endurance in children and adolescents with resistance training programs involving a variety of progressive activities that incorporate reciprocal and large muscle groups.[30] Most programs involved sessions of

30–45 minutes' duration, 2 or 3 days per week, with a rest day between sessions. Programs generally varied from 8 to 12 weeks in duration. Results of resistance programs show some degree of specificity. Larger gains in strength are associated with protocols of relatively high resistance and low repetitions, while greater gains in muscular endurance are associated with protocols of relatively low resistance and high repetitions. An essential ingredient in the safety of strength training protocols is adult supervision.[30]

14.7 PHYSICAL ACTIVITY AND CHRONIC DISEASE RISK

Current interest in health benefits of physical activity often focuses on selected risk factors for chronic cardiovascular and metabolic diseases. The influence of physical activity on several risk factors in children and adolescents is considered subsequently.

14.7.1 Lipids and Lipoproteins

Cross-sectional and longitudinal studies indicate relatively weak associations between level of physical activity and total cholesterol, high-density lipoprotein cholesterol (HDL-C), low-density lipoprotein cholesterol (LDL-C) and triglycerides. The relationships, though weak, are best for physical activity and HDL-C and triglycerides. The observational data are consistent with a variety of intervention studies which show a weak, beneficial influence of moderate to vigorous physical activity, 40 minutes per day, 5 days per week over 4 months on HDL-C and triglycerides; such programs have no influence on total cholesterol and LDL-C. It is possible that a more sustained volume of activity may be needed to beneficially influence lipids and lipoproteins.[10]

School-based intervention programs were generally not effective in improving lipid and lipoprotein profiles. Such programs may be confounded, in part, by children who had relatively normal values of lipids and lipoproteins at the start of the intervention.[10]

14.7.2 Blood Pressures

There is no clear association between physical activity and blood pressures in normotensive young people, i.e. those with normal blood pressures. However, aerobic training programs have a beneficial effect on the blood pressures of hypertensive youth. The programs ranged in duration from 12 to 32 weeks. Aerobic training programs may also reduce blood pressures in those with mild essential hypertension. Limited data for resistance (strength) training indicate no effect on the blood pressures of hypertensive youth.[10]

14.7.3 Other Indicators of Cardiovascular Health

Physical activity may be related to other, less commonly studied indicators of cardiovascular health among children and adolescents. Relationships between physical activity and fibrinogen level and C-reactive protein are weak in youth, while the relationships between physical activity and endothelial function are inconclusive. On the other hand, aerobic training increases resting vagal tone in obese youth.[41] Vagal tone is reflected in beat-to-beat variability in the RR interval of an electrocardiogram. It is called heart rate variability and is a marker of cardiac parasympathetic activity. Among adults, parasympathetic activity is relatively low in the obese, is high in the endurance trained and increases in response to regular training. Implications of this indicator for the development of cardiovascular disease are uncertain, but low parasympathetic activity is a strong predictor of mortality after myocardial infarct.

14.7.4 Metabolic Risk

The clustering of risk factors for cardiovascular disease — low HDL-C, high triglycerides, elevated blood pressures, impaired glucose metabolism, insulin resistance, obesity and abdominal obesity — is commonly labeled the metabolic syndrome. The syndrome places individuals at elevated risk for type 2 diabetes and cardiovascular morbidity.[42] Risk factors that define the metabolic syndrome are increasingly prevalent in young people, especially the obese.[43,44]

In a largely non-obese sample of youth, a favorable metabolic profile (lower blood pressures, total cholesterol, triglycerides and glycemia; higher HDL-C; lower skinfolds) was independently associated with high physical activity and low physical inactivity and with high aerobic fitness.[45] Evidence from the European Youth Heart Study, a multi-center, international, cross-sectional study, showed consistent associations between physical activity and cardiorespiratory fitness, on the one hand, and a better metabolic profile, on the other.[46–51] Data from the European Youth Heart Study indicated interactions between physical activity and cardiorespiratory fitness affecting the metabolic profile,[49] stronger relationships between cardiorespiratory fitness and reduced metabolic risk than between physical activity and risk,[50] and independent inverse associations between aerobic fitness and metabolic risk and between habitual physical activity and metabolic risk.[51] Overall, those who were regularly active and/or who had high aerobic fitness tended to present a favorable metabolic risk profile. Fatness was an independent risk factor for metabolic risk — those who were leaner and with less central adiposity (measured indirectly as waist circumference) tended to present a more favorable metabolic risk profile.

The preceding studies were cross-sectional and demonstrated important associations between physical activity and metabolic risk. What is the influence of physical activity programs on metabolic risk factors individually or clustered? Effects of physical activity on several individual risk factors were noted earlier, e.g. adiposity, lipids and lipoproteins and blood pressures. Experimental physical activity programs improved the metabolic risk profile of overweight and obese youth, with a reduction in adiposity, insulin, triglycerides and inflammation markers; improved insulin sensitivity, lipid profile and cardiorespiratory fitness; and increased heart rate variability.[41,52–55] The favorable responses to regular activity were lost, however, when obese youth were no longer involved in regular physical activity,[54] i.e. when they reverted to a lifestyle without or with reduced physical activity. The results highlight the need for physical activity on a regular basis for metabolic health.

14.8 TRAINING FOR SPORT AND THE GROWTH AND MATURATION OF YOUNG ATHLETES

Definition and classification of young sport participants as athletes are variable. Samples vary considerably and can range from general participants to youth classified as select, elite, or junior national. Sport by its nature is exclusive. Samples of athletes in late childhood and early adolescence are very different than those in late adolescence. Many youngsters have dropped out voluntarily or have been excluded by the sport system.

In order to evaluate the potential influence of training for a specific sport on the growth and maturation of young athletes, it is essential to have a grasp of their status relative to the general population of children and adolescents.

14.8.1 Size Attained

Artistic gymnasts of both sexes consistently present, on average, a profile of short stature. Figure skaters of both sexes also present shorter statures, on average, although data are not extensive. Female ballet dancers tend to have shorter statures during childhood and early adolescence, but catch up to non-dancers in late adolescence. Athletes of both sexes in other sports have, on average, statures that equal or exceed reference medians. Body mass presents a similar pattern. Gymnasts, figure skaters and ballet dancers of both sexes consistently show lighter body mass. Gymnasts and figure skaters have appropriate mass for height, while ballet dancers have low mass for height. A similar trend is evident in female distance runners. Young athletes in other sports tend to have body masses that, on average, equal or exceed the reference medians.[13,17,18]

14.8.2 Maturity Status

With few exceptions, male athletes in a variety of sports tend to be average (on time) or advanced in sexual and skeletal maturation. Data are more available for skeletal age, although there is some variation dependent on method of assessment. Other than gymnasts, who present later skeletal maturation, late- and early-maturing boys are about equally represented in samples of male athletes in several sports at 10–11 years of age. With increasing age during adolescence, the number of later maturing male athletes declines while the number of early-maturing and skeletally mature athletes increases.[56]

However, late-maturing boys are often successful in some sports in later adolescence (16—18 years), e.g. track and basketball, which emphasizes catch-up in skeletal maturation, reduced significance of maturity-associated variation in body size in the performances of boys in late adolescence, and/or different composition of the samples of athletes in early, mid- and late adolescence.

Skeletal maturation data for female athletes are available primarily for artistic gymnasts and swimmers, with limited information for girls in other sports. Late- and early-maturing girls are about equally represented among artistic gymnasts in late childhood. During adolescence, girls late and on time (average) in skeletal maturation dominate samples of elite gymnasts; early-maturing girls are a minority. The data suggest different samples at late childhood and late adolescent chronological ages. The majority of swimmers younger than 14 years of age are on time in skeletal maturation, with several more early- than late-maturing girls. Among swimmers aged 14—15 years, equal numbers are on time or mature, while most swimmers aged 16—17 years are skeletally mature.[56]

Limited longitudinal observations for male athletes are generally consistent with the data for skeletal age, i.e. age at PHV tends to be average or earlier in male athletes. PHVs for boys active in sports do not differ from those in less active boys, and are well within the range of means for boys in longitudinal studies of non-athletes.[13] Data for two samples of girls active in sport indicate ages at PHV and PHVs that approximate the means for the general population, while ages at PHV are later in artistic gymnasts and PHVs are only slightly less than estimates for European girls.[13,22,57]

The pubertal progress of boys and girls active in sport suggests no differences in tempo of sexual maturation compared to non-athletes. Mean intervals for progression from one pubertal stage to the next or across two stages are similar to those for non-active youth,[58] and are within the range of normal variation in longitudinal studies of non-athletes. The interval between age at PHV and menarche for girls active in sport and non-active girls does not differ, and is similar to those for several samples of non-athletic girls, with mean intervals of 1.2—1.5 years.[59]

Most discussions of maturation of female athletes focus on menarche, which is a late pubertal event. Later mean ages at menarche are reported in athletes in many, but not all, sports.[13,18,40] There is, however, confusion about later ages at menarche in athletes, which is related in part to the methods of estimating age at menarche. Only prospective and status quo data deal with young athletes in the process of maturing. Prospective data for athletes followed from prepuberty through puberty are ordinarily short term and limited to small, select samples; a potentially confounding issue is selective dropout or exclusion. Status quo data for young athletes provide sample or population estimates, but these samples generally include athletes of different skill levels at younger ages and more select athletes at older ages. The majority of mean/median ages at menarche of adolescent athletes (prospective, status quo samples) are within the range of normal variation and, with the exception of artistic gymnasts, figure skaters and ballet dancers, tend to approximate the mean for the general population. In contrast, most data for athletes are retrospective and are based on samples of mature (postmenarcheal) late-adolescent and adult athletes.

14.8.3 Body Composition

FFM follows a growth pattern similar to that for height and body mass so that variation in FFM in young athletes varies with body size.[13] It is perhaps for this reason that most studies of body composition of young athletes have focused on %Fat, although more recently attention has shifted to bone mineral.

Allowing for variation among samples and in methodology, individual differences in maturity status per se and the timing and tempo of maturation, several trends can be noted in young athletes.[60] Most of the data for young athletes are derived from individuals in contrast to team sports. Athletes tend to have a lower %Fat than non-athletes of the same chronological age, but there is considerable overlap between athletes and the reference, especially in males. Few samples of female athletes have a %Fat that approximates the reference and only two are higher (track and field throwing event athletes and handball players). Estimates of %Fat overlap considerably among samples of athletes in different sports, but there is variability in estimates within a sport, allowing for age-associated changes. Considering the sports represented, there appears to be more variation in %Fat among female than male athletes in later adolescence.

In contrast to FFM and %Fat, there is considerable current interest in the influence of regular training for sport on bone health in young athletes. Comparisons of athletes with non-athletes indicate higher bone mineral content in the former. Potential confounding factors are selection for sport, differential dropout and limited control of biological maturity status. In contrast to cross-sectional comparisons, several longitudinal studies indicate higher bone mineral content and bone mineral density in athletes of both sexes compared to the general population of young people.[60] The long-term influence of sport training on skeletal tissue is especially apparent in the dominant versus non-dominant arms of racket sport athletes.[61–63] Extreme unilateral activity is associated with localized increases in bone mineral accretion in young racket sport athletes.

14.8.4 Indicators of Chronic Disease Risk

Data relating cardiovascular and metabolic risk indicators to training in young athletes are not extensive. Young athletes aged around 9–18 years in several sports that included a significant endurance component showed, on average, a better cholesterol profile of blood lipids (lower triglycerides) and lipoproteins (higher HDL-C, lower LDL-C) than the general population. The profiles of athletes, however, showed considerable variability, including some trained athletes with dyslipidemia. Regular endurance training was not associated with a more favorable HDL-C profile in adolescent athletes.[64] On the other hand, a study of female athletes 9–15 years of age indicated higher levels of HDL-C and a higher HDL-C/total cholesterol ratio in gymnasts compared to runners and non-athletes.[65]

14.8.5 Does Training Influence Growth and Maturation of Young Athletes?

Given the available information on the growth and maturity status of young athletes in a variety of sports and the many factors that can influence growth and maturation, it is difficult to implicate systematic training as a significant factor affecting height and biological maturation either positively or negatively. However, it is likely that regular physical activity associated with training for sport exerts beneficial effects on body composition and physical fitness. Data on risk factors for cardiovascular and metabolic complications for young athletes are limited.

Later mean ages at menarche in athletes are often attributed to regular training for sport before menarche, i.e. training "delays" menarche. Use of the term "delay" in the context of physical activity or training is misleading because it implies that regular physical activity or training "causes" menarche to be later than normal. The data dealing with the inferred relationship between training and later menarche are based on association and are retrospective, and do not permit such a conclusion. Two comprehensive discussions of physical activity and female reproductive health have concluded with two observations: first, menarche occurs, on average, later in athletes in some sports compared to non-athletes, and second, the relationship between later menarche and training for sport is not causal.[66,67] According to one consensus statement, "… it has yet to be shown that exercise delays menarche in anyone".[66, p. S288] Many factors are known to influence the timing of menarche and rarely are they considered in studies of athletes.[7,18,40]

The short stature and later maturation of elite female artistic gymnasts is the example most often attributed to intensive training at young ages.[18,22] However, it is interesting that no mention is made of male gymnasts who also have short stature and later maturation. Given the available data, it is difficult to establish training as the causative factor for short stature and later maturation of female gymnasts. Allowing for variation in sampling and methodology, constitutional factors in the selection and retention processes of the respective sports (and perhaps others) need careful consideration. If early training can compromise the growth and maturation of gymnasts, it is essential to partition potential negative effects of training from the extremely selective physical and aesthetic criteria of the sport, requisite skills, differential dropout or exclusion, and the complexities of the tightly controlled elite sport environment before causality can be established.[13,18,22]

14.9 OVERVIEW

The "business of growing up" through childhood and adolescence places many demands on the individual and some of these demands are conflicting from the perspective of physical activity. Many demands placed upon young people are socially sanctioned forms of physical inactivity, e.g. school, homework, non-school reading, television and video games, extracurricular classes (tutoring, music, art), and others. Motorized transport is also a highly valued form of inactivity. There is a need for systematic study of physical inactivity or sedentary behaviors in their many forms and contexts. Both inactivity and activity have different meanings and contexts in childhood and adolescence, and they are generally independent of each other.

The growing and maturing individual adapts to the stresses imposed by physical activity. The stimulus of regular activity and responses to the activity are, to a large extent, not sufficient to significantly alter the processes of growth and maturation as they are ordinarily monitored. Thus, physical activity has no apparent effect on stature or on commonly used indicators of biological maturation. It is, however, an important factor in the regulation of body weight and specifically fatness and bone mineral, physical fitness and several indicators of chronic disease risk.

A confounding factor in evaluating the effects of physical activity on youth is individual differences in normal growth and maturation. Many of the variables of interest change with normal growth and maturation and are influenced by individual differences, especially timing and tempo of sexual maturation and the adolescent growth spurt. Several variables of interest (e.g. bone mineral content, muscular strength, HDL-C, adipose tissue) have their own growth patterns and some have adolescent spurts which vary in timing and tempo.

The beneficial effects of physical activity on indicators of disease risk appear to differentiate between "healthy" and "unhealthy" youth. Among "healthy" youth (i.e. normal weight and blood pressure), evidence suggests relatively small effects on adiposity, lipids and blood pressures. It is possible that a greater volume of activity may be needed to induce greater effects in healthy youth. On the other hand, health benefits of systematic physical activity are generally more apparent among "unhealthy" youth — the obese, hypertensive, and those with features of the metabolic syndrome. Among "unhealthy" youth, physical activity programs have a beneficial effect on adiposity in the obese, blood pressures in the hypertensive, and insulin, triglycerides and adiposity in obese youth with the metabolic syndrome.

Physical fitness and many cardiovascular and metabolic risk factors are affected by obesity in children and adolescents. A key issue, therefore, is the prevention of "unhealthy weight gain" in young people. Specification of "unhealthy weight gain" among children and adolescents is difficult since body weight is expected to increase with normal growth and maturation. There is a need, nevertheless, to address the issue in this age group.

Limited longitudinal data indicate smaller gains in the BMI in physically active youth,[68] while two longitudinal studies have implications for the role of physical activity in the prevention of excess weight gain at different phases of growth. More active children between 4 and 11 years have less fatness in early adolescence and may also have a later adiposity rebound,[69] and an increase in physical activity during adolescence may limit the accrual of fat mass in males, but not in females.[70] Maintenance of smaller gains in the BMI over time through physical activity may prevent unhealthy weight gain and, in turn, reduce the risk of overweight or obesity.

Physical activity is presumably important in normal growth and maturation, but it is not known how much activity is necessary. Apparently, the day-to-day activities of childhood and adolescence are adequate to maintain the integrity of growth and maturation processes, with the possible exception of adipose tissue.

Most of the experimental protocols in studies of fatness, aerobic fitness and risk factors used continuous activity programs, with the exception of studies of bone health and muscular strength and endurance. Activities of children, especially young children, are, however, primarily intermittent. There is a need for data that examine the effects of high-intensity, intermittent activity protocols.

Intervention and experimental studies of the influence of physical activity on indicators of fitness and disease risk were generally focused on the effects of the respective programs. The studies did not address the issue of the amount of activity that is needed to maintain the beneficial effects of activity programs. The beneficial effects associated with systematic physical activity are often lost or markedly reduced when the program stops.

Activity needs vary with age during childhood and adolescence. It is likely that the emphasis during childhood should be on general physical activity, and movement skills in particular. As children make the transition into puberty and adolescence, their capacity for continuous activities increases and activity can be more prescriptive, with an emphasis on fitness and health.[10,71]

14.10 SUMMARY

The influence of regular physical activity on children and adolescents is considered in the context of commonly used indicators of growth and maturation, health status and physical fitness, and training for sport.

Physical activity has no apparent effect on stature or on indicators of biological maturation, but it is an important factor in the regulation of body weight and specifically fatness and bone mineral, physical fitness and several indicators of chronic disease risk.

Systematic training for sport does not affect height and biological maturation. Menarche occurs later in athletes in some but not all sports, but the relationship between later menarche and training is not causal. Constitutional factors in the selection and retention of young athletes in a sport need careful consideration. Training for sport has beneficial effects on body composition and physical fitness. Data on chronic disease risk factors in young athletes are limited.

REFERENCES

1. Steinhaus AH. Chronic effects of exercise. *Physiol Rev* 1933;**13**:103—47.
2. Rarick GL. Exercise and growth. In: Johnson WR, editor. *Science and medicine of exercise and sports.* New York: Harper and Brothers; 1960. p. 440—65.
3. Malina RM. Exercise as an influence upon growth: review and critique of current concepts. *Clin Pediatr* 1969;**8**:16—26.
4. Malina RM. The effects of exercise on specific tissues, dimensions and functions during growth. *Stud Phys Anthropol* 1979;**5**:21—2.
5. Malina RM. Growth, exercise, fitness, and later outcomes. In: Bouchard C, Shephard RJ, Stephens T, Sutton JR, McPherson BD, editors. *Exercise, fitness, and health: a consensus of current knowledge.* Champaign, IL: Human Kinetics; 1990. p. 637—53.
6. Malina RM. Physical activity: relationship to growth, maturation, and physical fitness. In: Bouchard C, Shephard RJ, Stephens T, editors. *Physical activity, fitness, and health.* Champaign, IL: Human Kinetics; 1994. p. 918—30.
7. Malina RM. Growth and maturation: interactions and sources of variation. In: Mascie-Taylor CGN, Yasukouchi A, Ulijaszek S, editors. *Human variation from the laboratory to the field. Boca Raton FL.* CRC Press, Taylor and Francis Group; 2010. p. 199—218.
8. Bailey DA, Malina RM, Rasmussen RL. Postnatal growth. In: Falkner F, Tanner JM, editors. *The influence of exercise, physical activity, and athletic performance on the dynamics of human growth. Human growth,* vol. 2. New York: Plenum; 1978. p. 475—505.
9. Bailey DA, Malina RM, Mirwald RL. Physical activity and growth of the child. Human growth. In: Falkner F, Tanner JM, editors. *Postnatal growth neurobiology.* 2nd ed, vol. 2. New York: Plenum; 1986. p. 147—70.
10. Strong WB, Malina RM, Blimkie CJR, Daniels SR, Dishman RK, Gutin B, et al. Evidence based physical activity for school youth. *J Pediatr* 2005;**146**:732—7.
11. Physical Activity Guidelines Committee. *Physical Activity Guidelines Advisory Committee Report 2008.* Part G Section 9: Youth. Washington, DC: Department of Health and Human Services; 2008. p. G-1—9—33.
12. Malina RM, Katzmarzyk PT. Physical activity and fitness in an international standard for preadolescent and adolescent children. *Food Nutr Bull* 2006;**27**(Suppl. 4):S295—313.
13. Malina RM, Bouchard C, Bar-Or O. *Growth, maturation, and physical activity.* 2nd ed. Champaign, IL: Human Kinetics; 2004.
14. Malina RM. Biocultural factors in developing physical activity levels. In: Smith AL, Biddle SJH, editors. *Youth physical activity and inactivity: challenges and solutions.* Champaign, IL: Human Kinetics; 2008. p. 141—66.
15. Oliver M, Schofiled GM, Kolt GS. Physical activity in preschoolers: understanding prevalence and measurement issues. *Sports Med* 2007;**37**:1045—70.
16. Riethmuller AM, Jones RA, Okely AD. Efficacy of interventions to improve motor development in young children: a systematic review. *Pediatrics* 2009;**124**:e1—11.
17. Malina RM. Physical growth and biological maturation of young athletes. *Exerc Sport Sci Rev* 1994;**22**:389—433.
18. Malina RM. Growth and maturation of young athletes — is training for sport a factor?. In: Chan K, Micheli LJ, editors. *Sports and children.* Hong Kong: Williams and Wilkins Asia-Pacific; 1998. p. 133—61.
19. Saris WHM, Elvers JWH, van't Hof MA, Binkhorst RA. Changes in physical activity of children aged 6 to 12 years. In: Rutenfranz J, Mocellin R, Klimt F, editors. *Children and exercise XII.* Champaign, IL: Human Kinetics; 1986. p. 121—30.
20. Mirwald RL, Bailey DA. *Maximal aerobic power.* London, Ontario: Sports Dynamics; 1986.

21. Beunen GP, Malina RM, Renson R, Simons J, Ostyn M, Lefevre J. Physical activity and growth, maturation and performance: a longitudinal study. *Med Sci Sports Exerc* 1992;**24**:576—85.

22. Malina RM. Growth and maturation of elite female gymnasts: is training a factor. In: Johnston FE, Zemel B, Eveleth PB, editors. *Human growth in context*. London: Smith-Gordon; 1999. p. 291—301.

23. Pařizkova J. Longitudinal study of the relationship between body composition and anthropometric characteristics in boys during growth and development. *Glasnik Antropoloskog Drustva Jugoslavije* 1970;**7**:33—8.

24. Mirwald RL, Baxter-Jones ADG, Bailey DA, Beunen G. An assessment of maturity from anthropometric measurements. *Med Sci Sports Exerc* 2002;**34**:689—94.

25. Pařizkova J. *Body fat and physical fitness*. The Hague: Martinus Nijhoff; 1977.

26. Šprynarova S. Longitudinal study of the influence of different physical activity programs on functional capacity of the boys from 11 to 18 years. *Acta Paediatr Belg* 1974;(Suppl. 28): 204—13.

27. Von Dobeln W, Eriksson BO. Physical training, maximal oxygen uptake and dimensions of the oxygen transporting and metabolizing organs in boys 11—13 years of age. *Acta Paediatr Scand* 1972;**61**:653—60.

28. Eriksson BO. Physical training, oxygen supply and muscle metabolism in 11—13 year old boys. *Acta Physiol Scand Suppl* 1972;**384**:1—48.

29. Baxter-Jones ADG, Eisenmann JC, Mirwald RL, Faulkner RA, Bailey DA. The influence of physical activity on lean mass accrual during adolescence: a longitudinal analysis. *J Appl Physiol* 2008;**105**:734—41.

30. Iuliano-Burns S, Mirwald RL, Bailey DA. Timing and magnitude of peak height velocity and peak tissue velocities for early, average, and late maturing boys and girls. *Am J Hum Biol* 2001;**13**:1—8.

31. Malina RM. Weight training in youth-growth, maturation, and safety: an evidence-based review. *Clin J Sports Med* 2006;**16**:478—87.

32. Malina RM, Howley E, Gutin B. Body mass and composition. Report prepared for the Youth Health Subcommittee. *Physical Activity Guidelines Advisory Committee*; 2007.

33. Bailey DA, McKay HA, Mirwald RL, Crocker PRE, Faulkner RA. A six year longitudinal study of the relationship of physical activity to bone mineral accrual in growing children: the University of Saskatchewan Bone Mineral Accrual Study. *J Bone Miner Res* 1999;**14**:1672—9.

34. MacDonald H, Kontulainenm S, Petit M, Janssen P, McKay H. Bone strength and its determinants in pre- and early pubertal boys and girls. *Bone* 2006;**39**:598—608.

35. Kobayashi K, Kitamura K, Miura M, Sodeyama H, Murase Y, Miyashita M, et al. Aerobic power as related to body growth and training in Japanese boys: a longitudinal study. *J Appl Physiol* 1978;**44**:666—72.

36. Šprynarova S. The influence of training on physical and functional growth before, during and after puberty. *Eur J Appl Physiol* 1987;**56**:719—24.

37. Moisan J, Meyer F, Gingras S. Leisure physical activity and age at menarche. *Med Sci Sports Exerc* 1991;**23**:1170—5.

38. Merzenich H, Boeing H, Wahrendorf J. Dietary fat and sports activity as determinants for age at menarche. *Am J Epidemiol* 1993;**138**:217—24.

39. Moisan J, Meyer F, Gingras S. A nested case—control study of the correlates of early menarche. *Am J Epidemiol* 1990;**132**:953—61.

40. Malina RM. Menarche in athletes: a synthesis and hypothesis. *Ann Hum Biol* 1983;**10**:1—24.

41. Gutin B, Barbeau P, Litaker MS, Ferguson M, Owens S. Heart rate variability in obese children: relations to total body and visceral adiposity, and changes with physical training and detraining. *Obes Res* 2000;**8**:12—9.

42. Grundy SM. Metabolic syndrome: a multiplex cardiovascular risk factor. *J Clin Endocrinol Metab* 2007;**92**:399—404.

43. Cook S, Aunger P, Li C, Ford ES. Metabolic syndrome rates in United States adolescents, from the National Health and Nutrition Examination Survey 1999—2002. *J Pediatr* 2008;**152**:165—70.

44. Cook S, Weitzman M, Auinger P, Nguyen M, Dietz WH. Prevalence of a metabolic syndrome phenotype in adolescence: findings from the third National Health and Nutrition Examination Survey 1988—1994. *Arch Pediatr Adolesc Med* 2003;**157**:821—7.

45. Katzmarzyk PT, Malina RM, Bouchard C. Physical activity, physical fitness, and coronary heart disease risk factors in youth: the Quebec Family Study. *Prev Med* 1999;**29**:555–62.

46. Andersen LB, Harro M, Sardinha LB, Froberg K, Ekelund U, Brage S, et al. Physical activity and clustered cardiovascular risk in children: a cross-sectional study (The European Youth Heart Study). *Lancet* 2006;**368**:299–304.

47. Andersen LB, Sardinha LB, Froberg K, Riddoch CJ, Page AS, Anderssen SA. Fitness, fatness and clustering of cardiovascular risk factors in children from Denmark, Estonia and Portugal: the European Youth Heart Study. *Int J Pediatr Obes* 2008;**1**(Suppl. 3):58–66.

48. Anderssen SA, Cooper AR, Riddoch C, Sardinha LB, Harro M. Brage Sm Andersen LB. Low cardiorespiratory fitness is a strong predictor for clustering of cardiovascular disease risk factors in children independent of country, age and sex. *Eur J Cardiovasc Prev Rehab* 2007;**14**:526–31.

49. Brage S, Wedderkopp N, Ekelund U, Franks PA, Wareham NJ, Andersen LB, et al. Features of the metabolic syndrome are associated with objectively measured physical activity and fitness in Danish children. *Diabetes Care* 2004;**27**:2141–8.

50. Rizzo NS, Ruiz JR, Hurtig-Wennlof A, Ortega FB, Sjostrom M. Relationship of physical activity, fitness, and fatness with clustered metabolic risk in children and adolescents: the European Youth Heart Study. *J Pediatr* 2007;**150**:388–94.

51. Ekelund U, Anderssen SA, Frobert K, Sardinha LB, Andersen LB, Brage S, European Youth Heart Study Group. Independent associations of physical activity and cardiorespiratory fitness with metabolic risk factors in children: the European Youth Heart Study. *Diabetologia* 2007;**50**:1832–40.

52. Gutin B, Yin Z, Johnson M, Barbeau P. Preliminary findings of the effect of a 3-year after-school physical activity intervention on fitness and body fat: the Medical College of Georgia Fitkid Project. *Int J Pediatr Obes* 2008;**3**:3–9.

53. Bell LM, Watts K, Siafarikas A, Thompson A, Ratnam N, Bulsara M, et al. Exercise alone reduces insulin resistance in obese children independently of changes in body composition. *J Clin Endocrinol Metab* 2007;**92**:4230–5.

54. Carrel AL, Clark RR, Peterson SE, Nemeth BA, Sullivan J, Allen DB. Improvement of fitness, body composition, and insulin sensitivity in overweight children in a school-based exercise program. *Arch Pediatr Adolesc Med* 2005;**159**:963–8.

55. Nassis GP, Papantakou K, Skenderi K, Triandafillopoulou M, Kavouras SA, Yannakoulia M, et al. Aerobic exercise training improves insulin sensitivity without changes in body weight, body fat, adiponectin, and inflammatory markers in overweight and obese girls. *Metab Clin Exp* 2005;**54**:1472–9.

56. Malina RM. Skeletal age and age verification in youth sport. *Sports Med* 2011;**41**:925–47.

57. Thomis M, Claessens AL, Lefevre J, Philippaerts R, Beunen GP, Malina RM. Adolescent growth spurts in female gymnasts. *J Pediatr* 2005;**146**:239–44.

58. Malina RM, Woynarowska B, Bielicki T, Beunen G, Eweld D, Geithner CA, et al. Prospective and retrospective longitudinal studies of the growth, maturation, and fitness of Polish youth active in sport. *Int J Sports Med* 1997;**3**(Suppl. 18):S179–85.

59. Geithner CA, Woynarowska B, Malina RM. The adolescent spurt and sexual maturation in girls active and not active in sport. *Ann Hum Biol* 1998;**25**:415–23.

60. Malina RM, Geithner CA. Body composition of young athletes. *Am J Lifestyle Med* 2011;**5**:262–78.

61. Kannus P, Sievanen H, Vuori I. Physical loading, exercise, and bone. *Bone* 1996;**18**:1S–3S.

62. Bass SL, Saxon L, Daly RM, Turner CH, Robling AG, Seeman E, et al. The effect of mechanical loading on the size and shape of bone in pre-, peri-, and post-pubertal girls: a study in tennis players. *J Bone Miner Res* 2002;**17**:2274–80.

63. Daly RM, Saxon L, Turner CH, Robling AG, Bass SL. The relationship between muscle size and bone geometry during growth and in response to exercise. *Bone* 2004;**34**:281–7.

64. Eisenmann JC. Blood lipids and lipoproteins in child and adolescent athletes. *Sports Med* 2002;**32**:297–307.

65. Vasankari T, Lehtonen-Veromaa M, Möttönen T, Ahotupa M, Irjala K, Heinonen O, et al. Reduced mildly oxidized LDL in young female athletes. *Atherosclerosis* 2000;**151**:399–405.

66. Loucks AB, Vaitukaitis J, Cameron JL, Rogol AD, Skrinar G, Warren MP, et al. The reproductive system and exercise in women. *Med Sci Sports Exerc* 1992;**24**:S288—93.

67. Clapp JF, Little KD. The interaction between regular exercise and selected aspects of women's health. *Am J Obstet Gynecol* 1995;**173**:2—9.

68. Berkey CA, Rockett HRH, Gillman MW, Colditz GA. One-year changes in activity and in inactivity among 10- to 15-year old boys and girls: relationship to change in body mass index. *Pediatrics* 2003;**111**:836—43.

69. Moore LL, Gao D, Bradlee ML, Cupples LA, Sundarajan-Ramamurti A, Proctor MH, et al. Does early physical activity predict body fat change throughout childhood? *Prev Med* 2003;**37**:10—7.

70. Mundt CA, Baxter-Jones ADG, Whiting SJ, Bailey DA, Faulkner RA, Mirwald RL. Relationships of activity and sugar drink intake on fat mass development in youths. *Med Sci Sports Exerc* 2006;**38**:1245—54.

71. Malina RM. Fitness and performance: adult health and the culture of youth. In: Park RJ, Eckert HM, editors. *New possibilities, new paradigms? Academy of Physical Education Papers No. 24*. Champaign, IL: Human Kinetics; 1991. p. 30—8.

SUGGESTED READING

Malina RM. Physical growth and biological maturation of young athletes. *Exerc Sport Sci Rev* 1994;**22**:389—433.

Malina RM. Weight training in youth — growth maturation and safety: an evidence-based review. *Clin J Sports Med* 2006;**16**:478—87.

Malina RM. Skeletal age and age verification in youth sport. *Sports Med* 2011;**41**:925—47.

Malina RM, Geithner CA. Body composition of young athletes. *Am J Lifestyle Med* 2011;**5**:262—78.

Malina RM, Katzmarzyk PT. Physical activity and fitness in an international standard for preadolescent and adolescent children. *Food Nutr Bull* 2006;**27**(Suppl. 4):S295—313.

Steinhaus AH. Chronic effects of exercise. *Physiol Rev* 1933;**13**:103—47.

Strong WB, Malina RM, Blimkie CJR, Daniels SR, Dishman RK, Gutin B, et al. Evidence based physical activity for school youth. *J Pediatr* 2005;**146**:732—7.

CHAPTER 15

Comparative and Evolutionary Perspectives on Human Brain Growth

William R. Leonard*, J. Josh Snodgrass**, Marcia L. Robertson*

*Department of Anthropology, Northwestern University, Evanston, IL 60208, USA
**Department of Anthropology, University of Oregon, Eugene, OR 97403, USA

Contents

15.1 INTRODUCTION

The evolution of large human brain size has had important implications for many aspects of human biology, including our species' distinctive patterns of growth and development. What is extraordinary about these large brains is their high metabolic costs. As shown in Table 15.1, brain tissue has a very high energy demand per unit weight, roughly 16 times that of muscle tissue.[1,2] On average, adult humans spend some 400 kcal/day on brain metabolism.[1] Yet, despite the fact that humans have much larger brains per body weight than other primates or mammals, the resting energy demands for the human body are no more than for any other mammal of equivalent size.[3] Consequently, humans expend a much larger share of their resting metabolic rate (RMR) to "feed their brains" than other primates or mammals.[4]

To support the high metabolic demands of their large brains, humans have diets of much higher quality — more dense in calories and nutrients — than other primates.[3] On average, humans consume higher levels of dietary fat than other primates[5] and much higher levels of key fatty acids that are critical to brain development.[6,7] Moreover, humans are distinctive in their developmental changes in body composition, having

N. Cameron & B. Bogin (eds): Human Growth and Development, Second edition.
ISBN 978-0-12-383882-7, Doi: 10.1016/B978-0-12-383882-7.00015-5

Table 15.1 Weight, metabolic rate, total energy costs and energy costs as percentage of resting metabolic rate (RMR) for selected organs in the human body

Organ	Weight (kg)	Metabolic rate (kcal/kg/hour)	Total energy costs (kcal/day)	Proportion of RMR (%)
Brain	1.40	12.17	409	23
Liver	1.60	11.88	456	25
Kidney	0.30	15.94	115	6
Skeletal muscle	28.30	0.75	509	28

Source: Data from Holliday,[1] based on a 70 kg adult male.

higher levels of body fatness than other primate species, and these differences are particularly evident early in life.

Humans achieve their large brain sizes through a pattern of brain growth that is also distinctive from other primates. In all primate species (including humans), the brain grows most rapidly before birth. Yet, while brain growth in other primates slows dramatically after birth, in humans, that rapid rate of prenatal brain growth continues for about the first year after birth.[8,9] Consequently, compared to other primates, humans are born relatively "underdeveloped", with brain sizes that are a much smaller proportion of their final adult brain size.[8,10] In addition, the very high energy demands of continued rapid rates of "fetal-like" brain growth during the first year after birth place unique metabolic and nutritional constraints on human infants.

This chapter draws on comparative information from human and primate biology to explore the influence of brain size and metabolism on human growth and development. It begins by examining how variation in brain size influences metabolic demands and dietary/nutritional patterns among modern primates. Next, it briefly considers the ecological and nutritional factors that have promoted the evolution of human brain size. The chapter then examines patterns of early life brain growth in humans and how they differ from those of other primates. Finally, it explores how the high energy demands of brain growth in early childhood shape nutritional needs, patterns of growth and development, and body size and composition.

15.2 COMPARATIVE PERSPECTIVES ON BRAIN SIZE AND METABOLISM

The high energy costs of large human brains are evident in Figure 15.1, which shows the *allometric* (scaling) relationship between brain weight (g) and RMR (kcal/day) for humans, 36 other primate species and 22 non-primate mammalian species. The solid line is the best-fit regression for non-human primate species, and the dashed line denotes the best-fit regression for the non-primate mammals. The data point for humans is shown with a star.

The differences in the regressions imply that for a given metabolic rate, primates have systematically larger brains than other mammals. Humans, in turn, are outliers on the

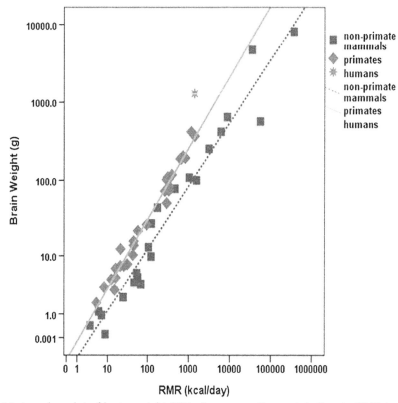

Figure 15.1 Log–log plot of brain weight (BW; g) versus resting metabolic rate (RMR; kcal/day) for humans, 36 other primate species and 22 non-primate mammalian species. The primate regression line is systematically and significantly elevated above the non-primate mammal regression. For a given RMR, primates have brain sizes that are three times those of other mammals, and humans have brains that are three times those of other primates. This figure is reproduced in the color plate section.

primate regression, having much larger brains than expected for their RMR. In caloric terms, this means that brain metabolism accounts for around 20–25% of RMR in adult humans, compared to about 8–10% in other primate species, and roughly 3–5% for non-primate mammals.

To accommodate the metabolic demands of their large brains, humans consume diets that are more dense in energy and nutrients than other primates of similar size.

Figure 15.2 shows the association between dietary quality and body weight in living primates, including modern human foragers. The diet quality (DQ) measure shown here was developed by Sailer et al.,[11] and reflects the relative proportions (by volume) of different types of food source, ranging from low-quality, energy-poor items like leaves, bark and stems, to foods that are higher in quality such as ripe fruit and animal material.

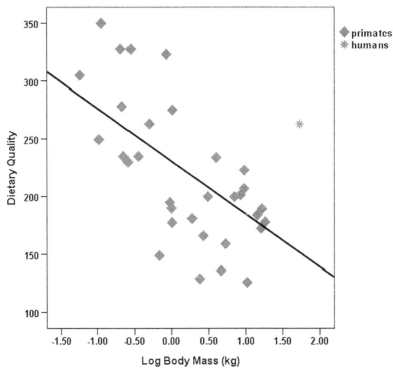

Figure 15.2 Plot of diet quality (DQ) versus log-body mass for 33 primate species. DQ is inversely related to body mass [$r = -0.59$ (total sample), -0.68 (non-human primates only); $p < 0.001$], indicating that smaller primates consume relatively high-quality diets. Humans have systematically higher quality diets than predicted for their size. (Source: *Adapted from Leonard et al.[4]*). This figure is reproduced in the color plate section.

The index ranges from a minimum of 100 (a diet of all leaves) to a maximum of 350 (a diet of all animal material).

There is a strong inverse relationship between DQ and body mass across primates, with smaller primates relying on energy-rich food such as insects, saps and gums, whereas large-bodied primates rely on low-quality plant foods, such as foliage. The diets of modern human foragers fall substantially above the regression line in Figure 15.2, implying that humans have systematically higher DQs than expected for a primate of this size. In fact, the staple foods for all human societies are much more nutritionally dense than those of other large-bodied primates. Although there is considerable variation in the diets of modern human foraging groups, recent analyses by Cordain and colleagues[12] have shown that modern human foragers derive fully 45–65% of their dietary energy intake from animal foods. In comparison, modern great apes obtain the bulk of their diet from low-quality plant foods. Gorillas derive over 80% of their diet from fibrous foods such as leaves and bark.[13] Even among common chimpanzees (*Pan troglodytes*), only

Table 15.2 Proportion of dietary energy intake derived from fat, protein and carbohydrates (CHO) in selected human populations, chimpanzees and gorillas

Species/group	Fat (%)	Protein (%)	CHO (%)	References
Humans (*Homo sapiens*)				
USA (2000)	33	14	53	Briefel and Johnson, 2004[15]
Modern foragers	28–58	19–35	22–40	Cordain et al., 2000[12]
Chimpanzee (*Pan troglodytes*)	6	21	73	Richard, 1985;[13] Tutin and Fernandez, 1992, 1993;[16,17] Popovich et al., 1997[5]
Gorilla (*Gorilla gorilla*)	3	24	73	Popovich et al., 1997[5]

about 5–10% of calories are derived from animal foods, including insects.[14] This higher quality diet means that we need to eat less volume of food to obtain the energy and nutrients we require.

Table 15.2 presents comparative data on *macronutrient* (i.e. fat, protein and carbohydrate) intakes of selected human groups, compared to those of chimpanzees and gorillas living in the wild. The dietary information for human populations was derived from US national data[15] and from a recent review of the diets of present-day hunter–gatherers (foragers) by Cordain et al.[12] Data for chimpanzees and gorillas were derived from foraging studies in the wild[5,13,16,17] and compositional analysis of commonly consumed food items.[5] Today's foraging societies derive between 28 and 58% of their daily energy intakes from dietary fat. Groups living in more northern climes (e.g. Inuits) derive a larger share of their diet from animal foods, and thus have higher daily fat intakes. Conversely, tropical foraging populations generally have lower fat intakes because they obtain more of their diet from plant foods. In comparison, Americans and other populations of the industrialized world fall within the range seen for hunter–gatherers, deriving about one-third of their daily energy intake from fat.[15]

In contrast to the levels seen in human populations, the great apes obtain only a small share of calories from dietary fat. Popovich and colleagues[5] estimate that western lowland gorillas derive approximately 3% of their energy from dietary fats. Chimpanzees appear to have higher fat intakes than gorillas (about 6% of dietary energy), but they are still well below the low end of the modern forager range. Thus, the higher consumption of meat and other animal foods among human hunter–gatherers is associated with diets that are higher in fat and more dense in energy.

Beyond the energetic benefits associated with greater animal consumption, the human diet also provides higher levels of key fatty acids that are critical to brain development. Mammalian brain growth is dependent upon sufficient amounts of two long-chain polyunsaturated fatty acids (LC-PUFAs): docosahexaenoic acid (DHA) and arachidonic acid (AA).[6,7] Species with relatively large brain sizes have greater requirements for DHA and AA.[7] Since mammals have a limited capacity to synthesize these fatty acids, dietary sources of DHA and AA appear to be limiting nutrients that constrained the evolution of larger brain size in many mammalian lineages.[7,18] Cordain and colleagues[6] have shown that the wild plant foods that make up the diets of most large-bodied primates contain little or no AA and DHA. In contrast, animal foods (e.g. fish, mammalian muscle tissue and organ meat) provide moderate to high levels of these fatty acids.

The link between brain size and dietary quality is seen in Figure 15.3, which shows relative brain size versus relative dietary quality for the 31 different primate species for which metabolic, brain size and dietary data are available. Both the brain size and diet quality measures have been standardized relative to body weight. There is a strong positive relationship ($r = 0.63$, $p < 0.001$) between the amount of energy allocated to the brain and the caloric density of the diet. Across all primates, larger brains require higher quality diets. Humans fall at the positive extremes for both parameters, having the largest relative brain size and the highest quality diet. Thus, the high costs of the large,

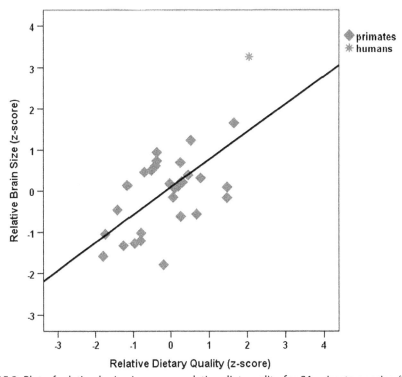

Figure 15.3 Plot of relative brain size versus relative diet quality for 31 primate species (including humans). Primates with higher quality diets for their size have relatively large brain size ($r = 0.63$, $p < 0.001$). Humans represent the positive extremes for both measures, having large brain:body size and a substantially higher quality diet than expected for their size. (Source: *Adapted from Leonard et al.[4]*). This figure is reproduced in the color plate section.

metabolically expensive human brain are partially offset by the consumption of a diet that is more dense in energy and fat than those of other primates of similar size.

This relationship implies that the evolution of large human brains would have necessitated the adoption of a sufficiently high-quality diet (including meat and energy-rich fruits) to support the increased metabolic demands of *encephalization*. Evidence from the human fossil record is consistent with this model: the first major burst of evolutionary change in *hominid* brain size occurred with the emergence and evolution of early members of the genus *Homo* between about 1.7 and 2.0 million years ago (mya) (Table 15.3).[4] Before this, our earlier hominid ancestors, the australopithecines, showed only modest brain size evolution from an average of 400 to 510 g over a 2-million-year span from 4 to 2 mya. With the evolution of the genus *Homo* there was rapid change, with brain sizes of, on average, around 600 g in *Homo habilis* (at 2.4–1.6 mya) and 800–900 g in early members of *Homo erectus* (at 1.8–1.5 mya). Although the relative brain size of *H. erectus* has not yet reached the size of modern humans, it is outside the range seen among other living primate species.

The evolution of *H. erectus* in Africa is widely viewed as a major adaptive shift in human evolution.[19,20] Indeed, what is remarkable about the emergence of *H. erectus* in East Africa at 1.8 million years is that there were (1) marked increases in both brain and body size, and (2) the evolution of human-like body proportions at the same time as (3) a reduction in face and tooth sizes.[21,22] These trends clearly suggest major energetic and dietary shifts: (1) the large body sizes necessitating greater daily energy needs; (2) bigger brains suggesting the need for a higher quality diet; (3) and the facial and dental changes suggesting that they were consuming a different mix of foods from their australopithecine ancestors.

The archeological record provides evidence that this occurred with *H. erectus*, as this species is associated with stone tools and the development of the first rudimentary

Table 15.3 Geological age, brain size, and estimated male and female body weights of selected fossil hominid species

Species	Geological age (mya)	Brain size (g)	Body weight (kg) Male	Female
Australopithecus afarensis	3.9–3.0	438	45	29
Australopithecus africanus	3.0–2.4	452	41	30
Australopithecus boisei	2.3–1.4	521	49	34
Australopithecus robustus	1.9–1.4	530	40	32
Homo habilis	1.9–1.6	612	37	32
Homo erectus (early)	1.8–1.5	863	66	54
Homo erectus (late)	0.5–0.3	980	60	55
Homo sapiens	0.4–0.0	1350	58	49

mya: million years ago.
Source: Data derived from Leonard et al.[4]

hunting and gathering economy. Meat appears to have been more common in the diet of *H. erectus* than it was in the australopithecines, with mammalian carcasses probably being acquired through both hunting and scavenging.[23,24] Increasingly sophisticated stone tools (i.e. the Acheulean industry) emerged around 1.6–1.4 mya, improving the ability of these hominids to process animal and plant materials.[25] These changes in diet and foraging behavior would not have turned our ancestors into carnivores; however, the addition of even modest amounts of meat to the diet (10–20% of dietary energy) combined with the sharing of resources that is typical of hunter–gatherer groups would have significantly increased the quality and stability of the diet of *H. erectus*.

15.3 HUMAN BRAIN GROWTH: PATTERNS AND METABOLIC CONSEQUENCES

Large adult human brain size is achieved primarily by extending the rapid, proliferative period of brain growth through the first year after birth. Thus, human infants are born *altrically* (relatively underdeveloped for their age) with a brain size that is a much smaller proportion of final adult size than in other primates. This pattern is evident in Figure 15.4, which shows the relationship between adult brain size and newborn (neonate) brain size in humans and 31 other primate species.[26] On average, adult brain sizes in non-human primates are about 2.3 times the size of those of newborns. Among humans, this ratio is significantly larger — about 3.5 — with brain size increasing from about 400 g at birth to 1400 g in adulthood.[8,9]

Figure 15.5 presents the growth velocities in brain size (g/year) and body size (height; cm/year) for boys and girls under the age of 10 years.[27,28] Growth in both the brain and body is most rapid under the age of 1 year; however, growth rates in the brain decline much more rapidly, such that by 5–6 years of age, children are close to reaching their adult brain sizes.

The high metabolic costs of our large developing brain create nutritional constraints for human infants. During infancy and early childhood, the energy demands of large brains are extreme because brain growth is most rapid and brain:body weight ratios are much larger than in adulthood. These points are highlighted in Table 15.4, which presents age changes in brain size (g), body size (kg), RMR (kcal/day) and percentage of RMR allocated to brain metabolism from birth to adulthood. Whereas brain metabolism accounts for 20–25% of resting needs in adults, in an infant of under 10 kg it is using upwards of 60%.[1]

Figure 15.6 shows the consequences of rapid early life growth in brain and body size for total daily energy requirements in children under 10 years, based on the World Health Organization's most recent recommendations on human energy needs.[29] Total energy needs are standardized per kilogram of body weight, and reflect the combined costs of RMR, growth and activity. Note that during the first few months after birth,

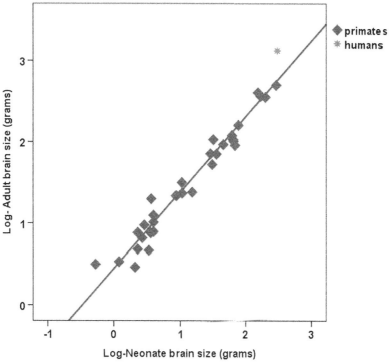

Figure 15.4 Log–log plot of adult brain size (g) versus neonate brain size (g) for humans and 31 other primate species. Brain growth of human infants is much faster than in other primate species. Consequently, adult human brain sizes average about 3.5 times those of newborns, compared to 2.3 times newborn size in other primates. (Source: *Data derived from Barton and Capellini.*[26]). This figure is reproduced in the color plate section.

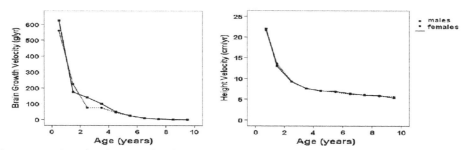

Figure 15.5 Growth velocities of (a) brain size (g/year) and (b) height/length (cm/year) for children 0–10 years of age. Growth rates for both brain and body size are most rapid early in life. Rates of brain growth, however, decline more rapidly than body growth such that by 5–6 years of age, children are approaching their adult brain sizes. (Source: *Brain growth data are from Leigh;*[27] *height growth data from Baumgartner et al.*[28])

Table 15.4 Body weight, brain weight, resting metabolic rate (RMR) and percentage of RMR allocated to brain metabolism (BrMet) for humans from birth to adulthood

Age	Body weight (kg)	Brain weight (g)	RMR (kcal/day)	BrMet (%)
Newborn	3.5	400	161	73
3 months	5.5	650	300	64
18 months	11.0	1045	590	53
5 years	19.0	1235	830	44
10 years	31.0	1350	1160	34
Adult male	70.0	1400	1800	23
Adult female	50.0	1360	1480	27

Source: All data are from Holliday.[1]

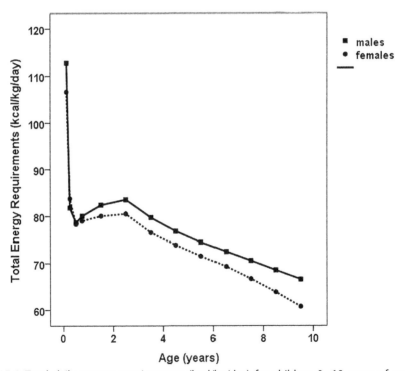

Figure 15.6 Total daily energy requirements (kcal/kg/day) for children 0–10 years of age. Rapid rates of growth in brain and body size contribute to very high energy requirements early in life (>100 kcal/kg/day). By age 10 years, weight-specific energy requirements decline by 40% (65–70 kcal/kg/day), and by adulthood, they are less than half those observed in early infancy. (Source: *Data from FAO/WHO/UNU.*[29])

the energy demands for the developing infant are extraordinary — averaging more than 100 kcal/kg/day in both boys and girls. By the age of 9 years, the energy requirements have declined by about 40% to 70 kcal/kg/day in boys and 65 kcal/kg/day in girls. Once adulthood is reached, daily energy requirements are about 35–40 kcal/kg/day in women, and between 40–50 kcal/kg/day for men.

The high nutrient density and digestibility of breast milk help to support the rapid growth of the brain during early postnatal life. Human breast milk provides about 70 kcal, 4.2 g of fat, 7.3 g of carbohydrate (mostly as the sugar lactose) and 1.3 g of protein per 100 g consumed.[30] The energy and macronutrient content of breast milk varies across populations, with heavier women tending to produce milk with higher fat content.[31] However, the degree of variation in macronutrient composition tends to be small relative to the degree of variation observed in maternal body weight and composition. Consequently, mothers who are marginally nourished may show physiological accommodations (such as utilizing their own fat and muscle stores or reducing their RMRs) during lactation that allow them to produce sufficient quality and quantity of breast milk to support the developing infant.[32]

Breast milk also contains significant amounts of the key essential fatty acids — DHA and AA — that are critical for promoting brain growth. Indeed, greater intakes of these LC-PUFAs are associated with improved brain growth, cognitive development and immune function.[33] In addition, since humans have a limited capacity to synthesize these essential fatty acids, variation in DHA and AA concentrations in breast milk are strongly associated with maternal dietary patterns.[34] Consequently, current dietary guidelines encourage pregnant and lactating women to consume foods containing these LC-PUFAs.[33]

15.4 BRAIN GROWTH AND BODY COMPOSITION

In addition to the nutritional resources provided by breast milk, distinctive aspects of infant body composition help to support the energy demands of the developing brain. Human infants are born with the highest body fat levels of any mammalian species, and continue to gain fat during their early postnatal life.[35,36] These high levels of adiposity in early life thus coincide with the periods of greatest metabolic demand of the brain.

Figure 15.7 shows changes in percentage body fatness during the first 48 months of life based on longitudinal data collected by Dewey and colleagues.[36] From birth to around 9 months, infant body fatness increases from 16% to about 26%. Thus, during early postnatal life, human infants continue to store additional energy as body fat to support their rapidly growing brains. Between 12 and 48 months of age, the period during which children transition from breast milk to solid foods, percentage body fat declines to about 16%.

For young children growing up in impoverished conditions in the developing world, obtaining sufficient energy and nutrients to sustain rapid rates of growth in both the brain

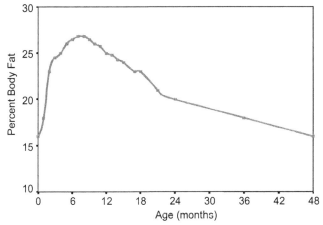

Figure 15.7 Changes in percentage body fatness of human infants from birth to 48 months. Body fatness increases from 16 to 26% during the first 9 months, and then declines to ~16% by 4 years of age. *(Source: Data from Dewey et al.[36]).* This figure is reproduced in the color plate section.

and body can be particularly challenging.[37,38] Research on children in the developing world suggests that chronic, mild to moderate undernutrition has a relatively small impact on a child's fatness. Instead of taking away the fat reserves, nutritional needs appear to be down-regulated by substantially reducing rates of growth in height/length, producing the common problem of infant/childhood "growth stunting" or growth failure that is ubiquitous among impoverished populations of the developing world.[39]

Figure 15.8 shows an example of this process based on growth data collected from young girls of the indigenous population of lowland Bolivia (the Tsimane').[40] Note that early in life the stature of Tsimane' girls closely approximates the US median, but by the

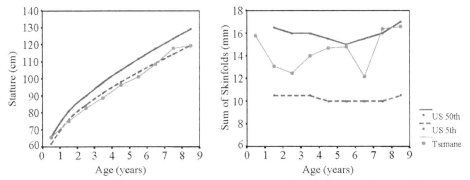

Figure 15.8 Patterns of physical growth in stature (cm) and body fatness (as sum of triceps and subscapular skinfolds, mm) in girls of the Tsimane' of lowland Bolivia. Growth of Tsimane' girls is characterized by marked linear growth stunting, whereas body fatness compares more favorably to US norms. *(Source: Data from Foster et al.[40]).* This figure is reproduced in the color plate section.

age of 3—4 years it has dropped below the 5th centile, where it will track for the rest of life. In contrast, body fatness (as measured by the sum of the triceps and subscapular skinfolds) compares more favorably to US norms, tracking between the 15th and 50th US centiles. The problem of early childhood growth failure is the product of both increased infectious disease loads and reduced dietary quality, which is particularly acute during the weaning period.

Recent work has suggested a mechanism to explain how body fatness is preserved under conditions of early-life growth stunting. Working among impoverished populations of Brazil, Hoffman and colleagues[41] found that children who were growth stunted had significantly lower RMRs and rates of fat metabolism than their "non-stunted" peers. Under resting conditions, the stunted children derived only 25% of the energy needs from fat, compared to 34% in the non-stunted group. These researchers hypothesize that the impaired fat metabolism of the stunted children is associated with a reduction in insulin-like growth factor-1 (IGF-1) that is commonly observed with poor childhood growth.[41,42] IGF-1 has been shown to increase cellular lipid metabolism;[43] hence, significant reductions in IGF-1 during growth can be expected to result in decreased fat consumption.

Overall, key aspects of human growth and development of body composition are shaped by the very high metabolic demands of brain metabolism early in life. Human infants are born altricially and, unlike other primates, continue rapid brain growth into early postnatal life.[9,10] To provide energy reserves for the high metabolic demands of large, rapidly growing brains, human infants are born with high body fat levels, and continue to gain fat during the first year of postnatal life. Furthermore, under conditions of chronic nutritional stress, human infants show the capacity to preserve brain metabolism by (1) down-regulating linear growth, (2) reducing fat oxidation, and (3) increasing fat storage. These adaptive responses are evidenced in the preservation of body fatness among "growth-stunted" children, and in the tendency of stunted children to gain weight and body fatness later in life.[44]

15.5 SUMMARY

The evolution of large human brain size has had important implications for the biology of our species. Humans expend a much larger share of their resting energy budget on brain metabolism than other primates or non-primate mammals. Comparative analyses of primate dietary patterns indicate that the high costs of large human brains are supported, in part, by diets that are relatively rich in energy and fat. Compared to other large-bodied apes, modern humans derive a much larger share of their dietary energy from fat. Among living primates, the relative proportion of metabolic energy allocated to the brain is positively correlated with dietary quality. Humans fall at the positive end of this relationship, having both a very high-quality diet and a large brain.

The human fossil record indicates that major changes in both brain size and diet occurred in association with the emergence of the genus *Homo* between 2.0 and 1.7 mya in Africa. With the evolution of early *H. erectus* 1.8 mya, there is evidence of an important adaptive shift: the evolution of the first hunting and gathering economy, characterized by greater consumption of animal foods and sharing of food within social groups. *Homo erectus* was human-like in body size and proportions, and had a brain size beyond that seen in non-human primates, approaching the range of modern humans. In addition, changes in the face and teeth of *H. erectus*, coupled with its more sophisticated tool technology, suggest that these hominids were consuming a higher quality and more stable diet that would have helped to fuel the increases in brain size.

Humans achieve their large brain sizes through a growth pattern that is distinct from that of other primates. In humans, very rapid brain growth that is typical of the fetal period is extended through the first year of postnatal life. The rapid rates of growth in both brain and body size contribute to very high daily energy requirements during infancy and early childhood. To accommodate these high metabolic demands, human infants are born with high levels of body fat and continue to gain fat during the first year of life.

Under conditions of nutritional stress, human infants and toddlers preserve body fat reserves for brain metabolism by reducing rates of linear growth. This process of "linear growth stunting" is also associated with reduced rates of fat oxidation and increased rates of fat storage. Thus, humans appear to show important adaptations in fat metabolism to accommodate the high energy demands of the brain early in life.

Ongoing research is providing new insights into variation in patterns of brain growth and its behavioral correlates, as well as differences in the metabolic costs of brain growth. For example, recent population-based studies of normal brain growth using magnetic resonance imaging (MRI) are now allowing researchers to explore more directly the biological and behavioral correlates of differences in brain development.[45] Similarly, through the use of positron emission tomography (PET scans), researchers can quantify patterns of variation in the metabolic costs of the brain and other organs by measuring rates of blood flow and glucose uptake.[46] Thus, the broader use of newer medical imaging techniques is expanding our understanding of brain growth and function, and providing key insights into the evolution of human brain size and behavioral complexity.

GLOSSARY

Allometry (scaling): The change in size of one biological measure with respect to another (often body size).

Altricial: Being relatively "underdeveloped" for one's chronological age.

Encephalization: Brain size in relation to body size. In general, primates are more encephalized than other mammals.

Hominids: Living humans and our fossil ancestors that lived after the last common ancestor between humans and apes.

Macronutrients: Dietary compounds required in large amounts, which can be used as sources of energy (calories). These include proteins, fats and carbohydrates.

REFERENCES

1. Holliday MA. Body composition and energy needs during growth. In: Falkner F, Tanner JM, editors. *Human growth: a comprehensive treatise.* 2nd ed, Vol. 2. New York: Plenum Press; 1986. p. 101–17.
2. Kety SS. The general metabolism of the brain in vivo. In: Richter D, editor. *Metabolism of the central nervous system.* New York: Pergamon; 1957. p. 221–37.
3. Leonard WR, Robertson ML. Evolutionary perspectives on human nutrition: the influence of brain and body size on diet and metabolism. *Am J Hum Biol* 1994;**6**:77–88.
4. Leonard WR, Robertson ML, Snodgrass JJ, Kuzawa CW. Metabolic correlates of hominid brain evolution. *Comp Biochem Physiol A* 2003;**135**:5–15.
5. Popovich DG, Jenkins DJA, Kendall CWC, Dierenfeld ES, Carroll RW, Tariq N, et al. The western lowland gorilla diet has implications for the health of humans and other hominoids. *J Nutr* 1997;**127**:2000–5.
6. Cordain L, Watkins BA, Mann NJ. Fatty acid composition and energy density of foods available to African hominids. *World Rev Nutr Diet* 2001;**90**:144–61.
7. Crawford MA, Bloom M, Broadhurst CL, Schmidt WF, Cunnane SC, Galli C, et al. Evidence for unique function of docosahexaenoic acid during the evolution of the human brain. *Lipids* 1999;**34**:S39–47.
8. Martin RD. *Human brain evolution in ecological context. 52nd James Arthur lecture on the evolution of the human brain.* New York: American Museum of Natural History; 1983.
9. Martin RD. *Primate origins and evolution: a phylogenetic reconstruction.* Princeton, NJ: Princeton University Press; 1989.
10. Rosenberg KR. The evolution of modern human childbirth. *Yrbk Phys Anthropol* 1992;**35**:89–124.
11. Sailer LD, Gaulin SJC, Boster JS, Kurland JA. Measuring the relationship between dietary quality and body size in primates. *Primates* 1985;**26**:14–27.
12. Cordain L, Brand-Miller J, Eaton SB, Mann N, Holt SHA, Speth JD. Plant to animal subsistence ratios and macronutrient energy estimations in world-wide hunter–gatherer diets. *Am J Clin Nutr* 2000;**71**:682–92.
13. Richard AF. *Primates in nature.* New York: WH Freeman; 1985.
14. Stanford CB. The hunting ecology of wild chimpanzees: implications for the evolutionary ecology of Pliocene hominids. *Am Anthropol* 1996;**9**:96–113.
15. Briefel RR, Johnson CL. Secular trends in dietary intake in the United States. *Annu Rev Nutr* 2004;**24**:401–31.
16. Tutin CEG, Fernandez M. Insect eating by sympatric lowland gorillas (*Gorilla gorilla*) and chimpanzees (*Pan troglodytes*) in the Lope Reserve, Gabon. *Am J Primatol* 1992;**28**:29–40.
17. Tutin CEG, Fernandez M. Composition of the diet of chimpanzees and comparisons with that of sympatric lowland gorillas in the Lope Reserve, Gabon. *Am J Primatol* 1993;**30**:195–211.
18. Crawford MA. The role of dietary fatty acids in biology: their place in the evolution of the human brain. *Nutr Rev* 1992;**50**:3–11.
19. Antón SC, Leonard WR, Robertson ML. An ecomorphological model of the initial hominid dispersal from Africa. *J Hum Evol* 2002;**43**:773–85.
20. Wolpoff MH. *Paleoanthropology.* 2nd ed. Boston, MA: McGraw-Hill; 1999.
21. McHenry HM, Coffing K. *Australopithecus* to *Homo*: transformations in body and mind. *Annu Rev Anthropol* 2000;**29**:125–46.
22. Ruff CB, Trinkaus E, Holliday TW. Body mass and encephalization in Pleistocene *Homo. Nature* 1997;**387**:173–6.

23. Bunn HT. Meat made us human. In: Unger PS, editor. *Evolution of the human diet: the known, the unknown, and the unknowable.* New York: Oxford University Press; 2006. p. 191−211.

24. Plummer T. Flaked stones and old bones: biological and cultural evolution at the dawn of technology. *Yrbk Phys Anthropol* 2004;**47**:118−64.

25. Asfaw B, Beyene Y, Suwa G, Walter RC, White TD, WoldeGabriel G, et al. The earliest Acheulean from Konso-Gardula. *Nature* 1992;**360**:732−5.

26. Barton RA, Capellini I. Maternal investment, life histories, and the costs of brain growth in mammals. *Proc Natl Acad Sci USA* 2011;**108**:6169−74.

27. Leigh SR. Brain growth, life history, and cognition in primate and human evolution. *Am J Primatol* 2004;**62**:139−64.

28. Baumgartner RN, Roche AF, Himes JH. Incremental growth tables: supplementary to previously published charts. *Am J Clin Nutr* 1986;**43**:711−22.

29. Food and Agriculture Organization/World Health Organization/United Nations University (FAO/WHO/UNU). *Human energy requirements. Report of a joint FAO/WHO/UNU expert consultation.* Geneva: WHO; 2004.

30. Casey CE, Hambridge KM. Nutritional aspects of human lactation. In: Neville MC, Neifert MR, editors. *Lactation: physiology, nutrition, and breast-feeding.* New York: Plenum; 1983. p. 94.

31. Prentice A. Regional variation in the composition of human milk. In: Jensen RG, editor. *Handbook of milk composition.* San Diego, CA: Academic Press; 1995. p. 155−221.

32. Prentice AM, Whitehead RG, Roberts SB, Paul AA. Long-term energy balance in child-bearing Gambian women. *Am J Clin Nutr* 1981;**34**:2790−9.

33. Koletzko B, Agostoni C, Carlson SE, Clandinin T, Hornstra G, Neuringer M, et al. Long chain poly-unsaturated fatty acids (LC-PUFA) and perinatal development. *Acta Paediatr* 2001;**90**:460−4.

34. Yuhas R, Pramuk K, Lien EL. Milk fatty acid composition from nine countries varies most in DHA. *Lipids* 2006;**41**:851−8.

35. Kuzawa CW. Adipose tissue in infancy and childhood: an evolutionary perspective. *Yrbk Phys Anthropol* 1998;**41**:177−209.

36. Dewey KG, Heinig MJ, Nommsen LA, Peerson JM, Lonnerdal B. Breast-fed infants are leaner than formula-fed infants at 1 y of age: the Darling Study. *Am J Clin Nutr* 1993;**52**:140−5.

37. Grantham-McGregor S. A review of studies of the effect of severe malnutrition on mental development. *J Nutr* 1995;**125**:S2233−8.

38. Levitsky DA, Strupp BJ. Malnutrition and the brain: changing concepts, changing concerns. *J Nutr* 1995;**125**:S2212−20.

39. Martorell R, Habicht J-P. Growth in early childhood in developing countries. In: Falkner F, Tanner JM, editors. *Human growth: a comprehensive treatise.* 2nd ed, Vol. 3. New York: Plenum Press; 1986. p. 241−62.

40. Foster Z, Byron E, Reyes-García V, Huanca T, Vadez V, Apaza L, et al. Physical growth and nutritional status of Tsimane' Amerindian children of lowland Bolivia. *Am J Phys Anthropol* 2005;**126**:343−51.

41. Hoffman DJ, Sawaya AL, Verreschi I, Tucker KL, Roberts SB. Why are nutritionally stunted children at risk of obesity? Studies of metabolic rate and fat oxidation in shantytown children of Sao Paulo, Brazil. *Am J Clin Nutr* 2000;**72**:702−7.

42. Sawaya AL, Martins PA, Grillo LP, Florêncio TT. Long term effects of early malnutrition on body weight regulation. *Nutr Rev* 2004;**62**:S127−33.

43. Hussain MA, Schmintz O, Mengel A, Glatz Y, Christiansen JS, Zapf J, et al. Comparison of the effects of growth hormone and insulin-like growth factor I on substrate utilization and on insulin sensitivity in growth hormone-deficient humans. *J Clin Invest* 1994;**94**:1126−33.

44. Frisancho AR. Reduced fat oxidation: a metabolic pathway to obesity in developing nations. *Am J Hum Biol* 2003;**15**:35−52.

45. Brain Development Cooperative Group. Total and regional brain volumes in a population-based normative sample from 4 to 18 years: the NIH MRI study of normal brain development. *Cereb Cortex* 2012;**22**:1−12.

46. Chugani HT. A critical period of brain development: studies of cerebral glucose utilization with PET. *Prev Med* 1998;**27**:184−8.

SUGGESTED READING

Barton RA, Capellini I. Maternal investment, life histories, and the costs of brain growth in mammals. *Proc Natl Acad Sci USA* 2011;**108**:6169−74.

Kuzawa CW. Adipose tissue in human infancy and childhood: an evolutionary perspective. *Yrbk Phys Anthropol* 1998;**41**:177−209.

Leigh SR. Brain growth, life history and cognition in primate and human evolution. *Am J Primatol* 2004;**62**:139−64.

Leonard WR. Food for thought: dietary change was a driving force in human evolution. *Sci Am* 2002;**287**:106−15.

Martin RD. Scaling of the mammalian brain: the maternal energy hypothesis. *News Physiol Sci* 1996;**11**:149−56.

INTERNET RESOURCES

Becoming human: http://www.becominghuman.org/

Better brains for babies: http://www.fcs.uga.edu/ext/bbb/

Brain net: http://www.brainnet.org/

CDC growth charts: http://www.cdc.gov/growthcharts/clinical_charts.htm

MRI study of normal brain development: http://www.brain-child.org/home.htm

My Plate (USDA dietary guidelines): http://www.choosemyplate.gov/

Resources for the study of the brain: http://brainmuseum.org/

World Health Organization breastfeeding: http://www.who.int/topics/breastfeeding/en/

WHO child growth standards: http://www.who.int/childgrowth/en/index.html

Zero to three (National Center for Infants, Toddlers & Families): http://www.zerotothree.org/

CHAPTER 16

Saltation and Stasis

Michelle Lampl★,★★

★Predictive Health Institute, Atlanta, GA 30308, USA
★★Department of Anthropology, Emory University, Atlanta, GA 30322, USA

Contents

16.1 INTRODUCTION

How, exactly, does the single cell conceptus become a small but perfectly formed human being of some 35 cm by 25 weeks of gestation? And by what process does the average 50 cm newborn grow to become, on average, a five to six foot adult individual? How is this flexibility in outcome possible from the same system? The nature of the process by which individual human growth proceeds has yet to be clearly elucidated. Our understanding of the precise mechanisms, or the cascade of events by which this increase in size unfolds, remains a fuzzy outline at the present time. This is one of the most exciting questions in human biology. It is also one of the most troublesome in terms of our lack of knowledge. There is a pressing need to clarify the process of normal human growth: How should we proceed to treat short stature with growth hormone? How can we best use body size as a marker for intervention efforts in international health? Could we be better informed in our attempts to help premature infants to develop normally?

N. Cameron & B. Bogin (eds): Human Growth and Development, Second edition.
ISBN 978-0-12-383882-7, Doi: 10.1016/B978-0-12-383882-7.00016-7

This chapter posits that a better understanding of individual growth biology is fundamental to answering questions such as these. The first part of the chapter focuses on a review of the discovery that individual growth proceeds by a process of saltation and stasis. This is followed by an overview of the challenges involved in documenting the details of individual saltatory growth patterns, including methodological aspects of sampling protocol, attention to measurement error (growth increments must be greater than the errors of measurement) and data-analytical approaches. The importance of using methods that neither impose assumptions about the nature of the time between saltatory events nor create artifactual patterns is described, as are the benefits of using mathematical approaches that permit comparisons between competing hypotheses representing alternative models of growth based on the empirical data. Potential mechanistic bases for saltation and stasis as the fundamental growth process are considered in the final portion of the chapter. Animal models and cell-level studies raise intriguing questions regarding the nature of the underlying biology that is expressed as integrated growth at the whole-body level.

16.2 SALTATION AND STASIS: HOW CHILDREN GROW

During the past 30 years, data have been collected that identify the process of human growth as *saltatory*. Specifically, increase in the size of the body (e.g. length/height), as well as circumferential growth of the head, is achieved through unique time-constrained growth episodes that occur only intermittently. Following the vocabulary for similar biological processes previously identified in neural tissue, in this text this growth pattern is called saltatory and the unique growth accretions, *saltations*.[1] These saltations of growth occur as increments of variable amounts within and between individuals, and the amount of growth per saltation varies by anatomical site (legs grow more than the head at each growth episode).

These saltations were originally identified in humans when the total body length of infants was measured daily. At observations within this time-frame, careful measurement techniques identify unique growth increment events that stand out by contrast with surrounding time intervals when no growth, or increase in size, occurs. These intervening time durations of no measurable accretion, or *stasis* in terms of incremental growth, separate the individual growth saltation events. Thus, growth in size can be visualized as a stepwise function, with pulsatile increases in size resulting in unique steps of variable heights (Figure 16.1).

For example, in infancy, total body length/height increments range from 4 to 20 mm during 24 hours, while head circumference saltations are about 2–3 mm in 24 hours. This is not an everyday event: saltations are separated in time by intervals ranging from 1 day to more than 60 days when no measurable growth occurs. This time interval of no

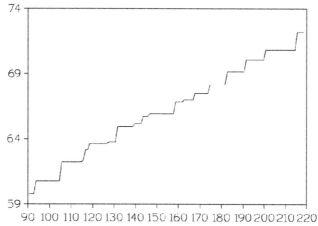

Figure 16.1 Saltation and stasis growth pattern from daily data (day of age, *x*-axis) of total body length measurements (cm, *y*-axis). The subject was a male infant followed from 90 to 218 days of age. Twelve statistically significant growth saltations contribute to the total growth during the interval. These occur at days 93 (0.98 cm), 105 (1.47 cm), 116 (0.81 cm), 118 (0.46 cm), 131 (1.2 cm), 143 (0.55 cm), 158 (0.92 cm), 167 (0.5 cm), 174 (0.68 cm), 191 (0.91 cm), 200 (0.72 cm) and 215 (1.39 cm) Thus, total growth was accrued on 12 days when variable amplitude growth saltation occurred after intervening stasis intervals of 2–16 days. Data were obtained after parental informed consent under a University of Pennsylvania approved human subject protocol.

measurable increment varies both within and between individuals. Developmental age is important: individual children have more frequent saltations when they are infants than during mid-childhood (author's unpublished data).

The present data outline that at the level of the whole body, growth is a saltatory process, occurring by episodic saltations according to a temporal clock whose parameters have yet to be clarified. To date, it appears that growth saltations occur at aperiodic (non-cyclical or unequal) but non-random time intervals, observations that suggest the growth process is an expression of a non-linear dynamic program.[2,3] This provides the ultimate flexibility for a biological system: it is likely that the variability in saltation amplitude and frequency is the mechanism that underlies the observable differences in size between individuals. The specific paths by which children achieve the same height are characterized by different series of unique increments in terms of the amount of growth per saltation and the total number of growth saltation events. Furthermore, it is likely that height differences in adulthood are the result of the accumulation of different numbers of saltation events and the amount of growth at these saltations. It is hypothesized that saltatory growth is the mechanism by which variability throughout development is achieved and is the pathway by which genetics and environment orchestrate the unique growth patterns of individual children.

16.3 GENERATION OF THE SALTATION AND STASIS HYPOTHESIS

It is said that often in the history of science, everyday observations lead to common knowledge about events that precede scientific discovery of these same occurrences by many years. Patterns of children's growth are an excellent example of this dictum. The parents of the children in these studies often state that their grandmothers knew perfectly well that children grow in spurts, and they ask how it is that scientists need to study something so obvious. The fact is that scientific documentation of day-to-day growth patterns is uncommon, and much remains to be discovered about how such a saltatory process is attained at the level of the whole body.

The study of human growth and development is the purview of many different disciplines. Cell biologists study the basic mechanisms by which cells divide and differentiate, thus contributing in their summation to growth increments. Clinicians focus on identifying and treating the abnormally growing child. In the science of auxology and human biology, the greater part of the history of the science is characterized by the collection of data from worldwide populations. These studies measured children at relatively infrequent intervals, most often at annual and semi-annual time-frames. These data have provided the basis for the commonly used growth reference charts, representing percentile distributions of height by age. These charts were intended to provide clinicians and public health professionals with a reference by which to assess the normality of an individual's progression through growth. The charts were constructed by applying a best-fitting curvilinear function to a set of sequential annual or biannual measurements, and these graphic representations became accepted as good approximations of the growth process during the time between data collection. According to the constructed charts, growth appears as a smooth and continuous daily accretion. This was assumed to be an accurate representation of the biology of growth. The error in this assumption is now clear. More frequently collected data elucidate the nature of the missing data, and thus, the growth process between annual data points. More frequently collected data do not illustrate a linear and continuous daily accretion in size. Instead, they document that growth occurs by a non-linear and discontinuous, saltatory process.

16.4 METHODS: HOW THE GROWTH PROCESS IS IDENTIFIED

When the scientific question is the process and pattern by which individual children grow, the primary issues that must be considered include measurement protocol, assessment of measurement error and the methods employed for data analysis.

16.4.1 Measurement Protocol

In order to begin to understand the underlying biological mechanisms that drive the growth process, it is necessary to conduct a time-intensive longitudinal study, following the growing organism at high-frequency intervals. The phrase "short-term growth" has been coined to refer to such studies. While emphasizing the historically unique focus on growth over less than biannual intervals, the concept of "short-term growth" asserts that the goal of the study is to document growth during short intervals, and presupposes that growth is always occurring, waiting to be documented. We prefer to think of investigations aimed at careful descriptions of individual patterns of the growth process as those involving time-intensive protocols, or time-intensive growth studies.

A serious consideration in designing a study aimed at elucidating the growth process is the sampling protocol. A measurement frequency must be chosen that will provide adequate data in relation to the timing of the underlying growth events.[4] This is not always known in advance. Thus, pilot studies are useful and often essential: the investigator chooses an initial window for sampling and changes the time-frame as appropriate.

As an illustration, Figure 16.2 presents infant growth data as they would look if collected at monthly, two-week, weekly and daily intervals. These different amounts of

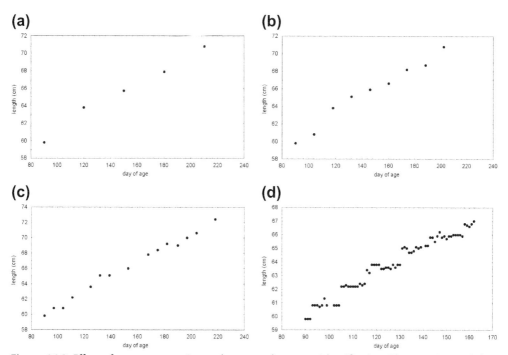

Figure 16.2 Effect of measurement interval on growth pattern identification. The experimental data from the infant described in Figure 16.1 are represented in four time-frames: (a) data at 30 day intervals beginning on study day 1; (b) data at 14 day intervals beginning on study day 1; (c) data at 7 day intervals beginning on study day 1; (d) a subset of daily data from study day 1 to 73, for clarification of the growth pattern lost by less frequent measurements.

information lead to quite different descriptions of the underlying growth process in this individual. The problem presented to the researcher who has collected such data and wishes to describe the growth pattern is this: how to identify what is happening between data points. If one has only data collected monthly, it is impossible to know the growth pattern over the course of the month. One can guess, and that is what researchers frequently do. In this example, for Figure 16.2(a) it appears that a relatively continuous line might be a reasonable approximation of the path taken by the biological process between data points. Thus, we might be tempted to connect the points and be satisfied that we understand, approximately, how this individual grew during this time interval. However, drawing a connecting line symbolically states that growth occurs each and every day between the points, at a relatively constant rate each day. This is an approach often employed in growth studies. Once such lines are drawn, they are used for deriving daily growth rates from the slope of the equation for the line.

As more data are collected, as per the example in Figure 16.2, it becomes clear that this linear proposition might not be the real or best description of how this individual is actually growing in length throughout the study interval. With data collected at two-week intervals (Figure 16.2b), the data suggest a non-linear, perhaps continuous growth pattern of variable growth rates. Weekly data (Figure 16.2c) suggest that a stepwise pattern might be the growth trajectory, and this impression is clarified by data collected daily (Figure 16.2d).

What is clear from this exercise is that with less frequent data, one cannot accurately identify the process between two data points. This example illustrates that growth data analysis problems are similar to the challenge of a connect-the-dots diagram in which the precise path becomes clarified only with increasing dots. The issue for the researcher interested in identifying the growth process is to collect sufficient information with which to reconstruct the biological process by which increase in size actually occurs.

16.4.2 Measurement Error Considerations

There is a second problem confronting the researcher when making a decision regarding measurement frequency: no measurement is free of error and certainly no human growth measurements are exempt from this consideration.[5] The sources of error at each measurement point reflect the precision of the instrument to measure an object accurately, the manner in which the object is measured, and any endogenous physiology that might contribute to variability in actual size. Thus, consideration must be given to the technology employed, the researcher's ability to take the measurement and the state of the subject.

In order to clarify the time window of observation for a study, it is necessary to conduct a pilot study to identify the actual measurement error of the particular observer and the subjects to be studied, and to compare this measurement error with the incremental process under study. If biological increment and measurement error were equal, it would not be possible to distinguish changes in the data series due to biology from those due to error. This is an often unappreciated reality of time-intensive investigations. In general, saltatory increments must be at least twice the technical error of measurement associated with data collection to be identifiable from error.[6]

Moreover, the longitudinal nature of a time-intensive growth study magnifies these issues. Because the goal of investigation involves the pattern between data points, there is a critical need to pay attention to the effects that an error at one time has on the immediately adjacent points. Formally, this is known as an issue inherent to dependent, negatively correlated data.[7] An example of the magnitude of this problem would be the following. Imagine we have taken a measurement such that we have erred, resulting in a "too long" measurement at one time-point. Through different errors, the first measurement is followed by a "too short" measurement at the subsequent data point. In the analysis, it might appear that no growth occurred between the two measurements, as an artifact of the combined errors. Serial time-intensive data must always be analyzed by a statistical method that takes this possibility into its error assessment consideration.

In this author's saltatory growth studies, pilot studies are always undertaken to ascertain the measurement error levels with the instrument and sample to be studied. It is best if an independent study of replicate measurement reliability for all parameters to be measured is conducted, and independent intrarater and interrater measurement error ranges are established. A pilot study of time-intensive serial measurements is also conducted. In the initial saltatory studies, the observation window of daily assessment was chosen after it was ascertained that technical errors of measurement were exceeded by measurement increments at the 95% confidence interval. Subsequently, careful documentation of intrarater and interrater reliability was established in the actual longitudinal studies.

16.4.3 Data Analysis

A time-intensive longitudinal study produces a time-series data set. Traditional time-series methods presented in many popular statistical packages for the computer may not, however, be the best approach to growth data analysis. Several issues intervene, the most significant of which is that many of these analytical methods are based on assumptions regarding patterns in the timing of events. These assumptions are likely not to be valid for biological data and may impose artifactual patterns, such as those resulting from Fourier time-series analysis, to be discussed below.

A simple and direct method for time-intensive data analysis is to begin with an approach that is designed to ask: where in a data series are significant differences between sequential measurements? With these identified, the occurrence of increments can be investigated for their own characteristics (duration and amplitude) and the intervals between these can be investigated for time duration, trends and random error components. This approach makes no assumptions about the presence or characteristics of increments, or the time between them. The critical aspect of time-intensive data analysis is the identification of actual growth increments from error components in the serial data. This is essential if the goal of the research is to describe the biological nature of the growth process, or the time-course and pattern of changes in size. If this step is omitted from an analysis, the results confound error and growth and may erroneously describe error components as biological growth pattern.

Research has shown that each individual's growth trajectory is unique in terms of the timing and amount of growth at saltation events (Figure 16.3). Therefore, in the author's studies, each individual's data are analyzed separately because a group analysis would obscure saltatory growth as times of saltation and stasis overlap between children (they do not occur with the same amplitude or timing). For an *incremental analysis* of individual data, the *t*-statistic for serially correlated data is applied to the sequential data.[8] This is an approach that has been used to identify significant differences in time-series endocrinological data. This statistic identifies significant differences between sequential measurements only when those differences exceed an a priori level. A 95% confidence limit is used and the *t*-statistic cut-off point is calculated, employing the individuals' pooled measurement variance, reflecting the significant individual variability in measurement error and the sample size of measurements.[7] This approach accounts for the negatively dependent nature of serial data, makes no assumptions about the underlying temporal process of growth and is relatively robust to non-normal data. Thus, increments that are greater than this calculated value have a probability of about 1 in 20 that they represent random chance rather than significant change. The *t*-statistic for stricter levels of significance can also be employed, by altering the *t*-statistic value used.

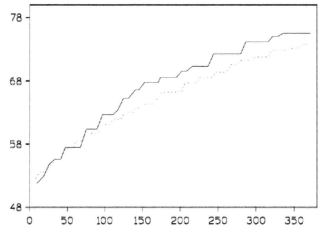

Figure 16.3 Saltation and stasis patterns for growth in length of two infants during their first year of life. Data were collected according to a University of Pennsylvania human subjects approved protocol.

This analysis permits the identification of statistically significant sequential positive and negative differences. Significant decreases that accompany significant increases are pairwise investigated for their correspondence to random error components and the remaining differences are further considered. This approach aims to focus the analysis only on sequences in the data where the measurement method identifies significant change and aims to eliminate the description of error components as part of the biology of the growth process.

In the original infant length data analysis, growth increments were identified to punctuate serial measurements during which no significant changes were documented.[8] These observations led to the hypothesis that these intervals represented times during which either growth was unresolvable from error by measurement techniques, or no incremental changes actually occurred. These alternatives were tested in this sample by a comparison of the total growth accrued by each individual during the study (size at the end of the study minus size at the beginning of the study) and the sum of the unique statistically significant growth increments found for each individual. If the stasis intervals were free of incremental growth, the two sums should be equal within measurement error, as the total growth of the child would be accounted for in the sum of the unique saltations. In this study, the total growth of the infants during their study was accounted for in the sum of discrete growth saltations. This analysis was the basis for the saltation and stasis hypothesis.

Thus, one initial strategy in analyzing time-intensive measurements is to employ a method that asks: can significant growth changes in the data be identified from measurement error? Methods for the analysis of saltatory growth must be based on discriminatory analytical methods that aim to identify significant change from error. It should be noted that data collection methods involving high error and children growing by low-amplitude increments have a lower likelihood of saltatory growth resolvability. This does not mean that the subjects do not grow by saltation and stasis; it simply means that their growth pattern is not resolvable by the methods employed.

The description of saltation and stasis outlined above was a hypothesis, or proposition about the underlying biology that is responsible for growth, based on an incremental analysis of serial infant length data and the resulting pattern of growth increments. This observation suggested that growth is a highly controlled event, not a continuous hourly and daily biological signal. The biological hypothesis generated by this observation is that growth is a two-phase process consisting of both a growth-suppressive phase (stasis), putatively controlled by growth inhibition, and a discrete growth phase (saltations) that occurs episodically owing to disinhibitory, permissive and/or activation controls. This hypothesis was in line with what has been observed to characterize cell division and differentiation[9,10] and, thus, had strong support as a developmental process whose precise proximate controls remain to be elucidated.[11]

16.4.4 Mathematical Modeling

The incremental analysis described above led to the hypothesis that growth is a process of discrete incremental events separated in time by variable intervals of no growth. This observation is a statement that growth occurs by a process unfolding in time that can be visualized as a staircase, with different heights of stairs and different plateau lengths between the steps.

The proposal of a saltation and stasis pattern in the time-series data can be statistically tested to see how well this hypothesis actually describes the experimental data. This analysis requires a mathematical statement of the proposition: a stepwise mathematical function that is flexible in the amplitude and duration of the steps would be a good approximation of such a biological process and could provide an estimate of how well the hypothesis actually describes the time-series data. The saltation and stasis mathematical model was developed by Michael Johnson, a biomathematician who had been working on similar problems in other biological systems.[12,13] The saltation and stasis mathematical model employs a pulse identification approach that is free of assumptions about both how much growth occurs at a saltation, as well as the timing between saltations. It explicitly tests the hypothesis of no growth between the events and identifies significant growth from error at a significance level set by the observer, using the entire raw data set and an error of measurement that is input by the observer.

Why use a mathematical model? Why not stop the data analysis after the incremental analysis and conclude that saltation and stasis was established? Mathematical models provide a statistically based description of how well a temporal pattern fits an entire data set, and permits statistically based comparisons between competing patterns. Thus, the researcher can ask: does a saltation and stasis pattern really describe the data better than a model of continuous daily growth? Or is the saltatory growth notion some sort of artifact from errors in an incremental analysis? Alternatively, are there really stasis intervals between discrete growth changes, or do the growth events take longer than 24 hours with some small growth continuing between each event? These questions are statements about entirely different views of the biological process of growth. These viewpoints can be directly compared to which one best describes the serial growth data of individuals. The comparative approach involves applying a mathematical function that represents each of these conceptual patterns to the raw data. The "best fit" of the patterns to the observational data is identified by a comparison of the statistical properties of the residuals, or differences between the fitted function and the experimental data points.

For example, in the infant length data, the first option can be tested by fitting a simple curvilinear function through all of the data points, testing the hypothesis that growth is a continuous daily process. The application of such a function to the raw data in Figure 16.1 results in a number of data points that are not on the mathematical line

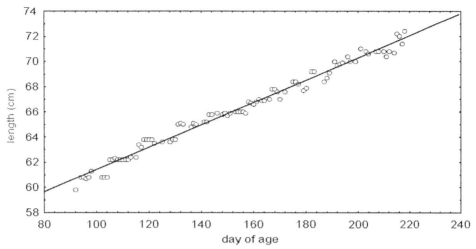

Figure 16.4 Data set for the infant in Figure 16.1 with a best fit linear approximation. Note the pattern of data on either side of the line: a non-random residual pattern.

(Figure 16.4). These are residuals, and here they occur in a non-random, wavelike pattern about the line. This illustrates that the line representing the concept of continuous daily growth is not capturing a pattern that exists in the experimental data. The pattern of the residuals suggests that a stepwise or wavelike function is being overlooked by this application.

The second proposition, that growth is continuous but characterized by growth spurts that take more than one day to complete, is tested by fitting polynomial functions to the serial experimental data. The residuals of this application are likewise investigated for their pattern and magnitude. If a mathematical descriptor fits a data set well, the residuals should be randomly distributed about the resulting model, as expected for random error, and the best fitting pattern will result in the smallest non-random error in the residuals. Statistical comparisons of the stepwise saltation and stasis mathematical model (a discontinuous function) were compared to polynomial models (functions that characterized continuous growth) (Figure 16.5) and were found to be better fits of the experimental data than the polynomial models ($p < 0.001$).[3,13]

As shown in this example, the saltation and stasis model is compared with two alternative patterns in terms of the magnitude and pattern of residuals. The saltation and stasis model is characterized by smaller, random residuals by comparison with either of these alternatives. This type of analysis does not exclude the possibility that other untested models might be equally good descriptions of the pattern of growth in these data. The analysis does identify that saltation and stasis better describes the data than models of continuous daily growth or a pattern of small and continuous mini-growth spurts.

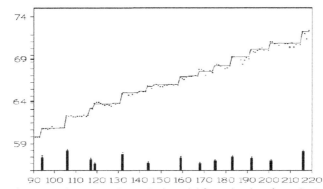

Figure 16.5 The saltation and stasis mathematical model fit to the data from the infant in Figure 16.1 with the statistically significant saltations shown below.

Thus, mathematical models are useful for investigating propositions regarding the patterns in human growth data. They are particularly useful for comparing alternative hypotheses about the nature of the process, and are helpful for hypothesis generation regarding underlying mechanisms.

To the extent that a mathematical model is a good description of the data, the raw data fit the pattern closely. Raw data that do not fit onto the pattern captured by an equation are the residuals. The better the model as a description of the data, the smaller are the residuals. A well-fitting model will have few residuals and those will represent random error. By fitting a variety of mathematical models to a serial data set, the researcher can objectively test alternative biological models of the underlying mechanism.

Some non-hypothesis-driven methods have also been applied to the same data to investigate alternative approaches for the identification of saltatory growth. Details of the methodological considerations established in these studies can be found elsewhere.[13—16]

16.4.5 Saltatory Growth Timing

An important current research question involves identifying the nature of the temporal characteristics of saltatory growth.[2] Investigations completed thus far have addressed the fundamental issue of whether growth pulses are predictable or entirely random. The question of whether growth saltations occur at consistent and regular stasis intervals was addressed using Fourier time-series analysis. This approach assumes that there are underlying mechanisms that oscillate at a constant frequency, and identified that saltations are non-periodic in occurrence.[17]

To investigate whether, by contrast, saltations occur randomly, two methods were employed. First, the observed stasis interval durations were compared to the binomial approximation for random intervals and the experimental intervals were found to be non-random. Second, Monte Carlo simulations of 1000 randomly spaced saltation events were compared with the observed stasis interval distributions in the data. The experimental stasis interval durations were identified as non-random.[2]

These analyses led to the question: if a biological system is neither predictable nor random, what sort of system is it? We posit that growth proceeds according to non-linear dynamic principles and is episodically irregular.[2,3] Such systems are typically found in complex, multinodally controlled networks,[18] which is the most likely description of the process of growth.[2]

Thus, methods for the identification of specific features of saltatory growth timing must aim to identify any saltations present without imposing assumptions about the nature of the time between events and without creating artifactual temporal patterns in the data. Fourier time-series analyses are an example of the first type of problem: these methods are a poor approach to data analysis because they assume that the data series under study exhibits periodic signaling. This method imposes this conformation on the data whether or not it is extant in the original data series.[13,17] Many complex irregular temporal patterns will be resolved by such a program to be periodic as the method attempts to characterize the unknown serial configurations as cyclical waves. The output of such an analysis does not necessarily accurately characterize the original data set. The researcher is left to compare the Fourier analysis results with those of other analytical methods to decide whether the periodicities identified are meaningful.

In the second category are smoothing methods and moving average techniques. These approaches alter the temporal characteristics of the original data series and are inappropriate approaches for the identification of saltation and stasis because of the patterns that they induce.[19] These methods are particularly inappropriate for data series characterized by short time interval changes because they attenuate, or filter, high-frequency information. This is because the approach involves altering the original data series before analysis. The raw data are subjected to a moving average replacement regime: a select number of sequential measurements are averaged and the derived mean becomes the data point in the middle of this interval, replacing the actual measured values. This is repeated for the entire series until a new time-series data set is created that consists of a series of interval averages, moving through the data set (hence the nomenclature moving average approach). In this way, a new data series is created that becomes the focus of analysis. Sometimes these approaches are used to decrease the influence of measurement error. This is not the best approach to error consideration in a saltatory growth analysis. Specifically, if a stepwise function is the actual pattern in the data, a moving average approach will create a slowly changing function and obscure the pattern of both saltation and stasis. This is particularly problematic when data collection occurs with errors of measurement that are equal to or greater than the growth saltations. Analytical smoothing results in the loss of distinctive saltation and stasis characteristics, leaving the impression of a continuous pattern. This does not mean that a continuous function is present, only that one is unable to identify the temporal structure with sufficient precision to rule it out, and/or to clarify a saltatory structure.[20] In sum, serial data analyzed by this method are subject to artifactual time characteristics of short-term changes.

16.4.6 Best Practices

From this overview, it can be summarized that in the analysis of time-intensive data, it is insufficient merely to connect all raw data points and assume that the resulting pattern is meaningful: this would result in the identification of a pattern that confounds error with biological signal. Attention to measurement error is a critical step in data acquisition and analysis, and analytical approaches must take this into consideration. While often used to correct for errors of measurement, moving averages are a faulty approach in saltatory growth analyses, since they further exacerbate the resolution between error and biological signal distinctions, particularly when measurement error is near or greater than the saltatory increment. It is optimal to employ an analytical method that considers all of the original raw measurement data without imparting any alterations or temporal characteristics. Finally, the identification of growth patterns should be amenable to statistical investigation and permit comparisons between models.

16.5 THE BIOLOGY OF SALTATORY GROWTH: MECHANISMS AND HYPOTHESIS TESTING

As parents of young children observe, children grow by leaps and bounds intermittently. While the amount of growth per saltation varies and the exact time to expect a growth saltation cannot yet be identified, each growth episode is an experienced event for the individual child. Growth saltations are accompanied by changes in behavior: agitation, sleep and appetite increase, and illness episodes co-occur with growth saltations more than can be explained by chance alone.[21,22] These observations suggest that the process of growth, which for many years has been considered to be a process restricted to increase in size, may in fact reflect much more in terms of the maturation of the organism. Thus, growth *is* maturation and saltatory growth is the manifestation of this developmental program.

In support of this proposition, variability in patterns of growth saltation amplitude and timing reflect developmental age such that infants and adolescents have more frequent growth saltations than occur in childhood, a finding that may explain the variable growth rates with age documented in the velocity curves of human growth. The data that have been considered up to this point are measurements of total body length, height and head circumference. Saltatory increments in weight have been documented to precede length,[23] and temporal coupling has been identified for head circumference and length.[20,24] The mechanisms underlying these associations of whole-organism growth remain to be elucidated by future investigation.

Once generated, scientific hypotheses must be tested on novel data collected for that purpose. The experience garnered from the above studies emphasizes the importance of measurement technique and analytical strategies that are necessary to identify saltatory growth if it is present in data.

The saltation and stasis hypothesis was generated on longitudinal data of infant recumbent length. The proposition must be tested on original data collected to investigate further the growth process and underlying biological mechanisms that are currently unknown. The protocol of daily measurements has been applied to height during childhood and adolescence with similar results.[3] Fetal ultrasound measurements have been taken at thrice-weekly intervals that clearly document intervals of stasis in fetal body parameters.[25] Daily measurements on the lower leg of human infants and children, collected by knemometric methods, have come to conflicting conclusions.[26] While these data were analyzed by approaches that were not designed to identify saltatory growth, inspection of published graphs and the authors' conclusions that the data series contain both stasis intervals and times of growth that exceed measurement error suggest that a saltatory process underlies the growth patterns in these data.

Animal studies have offered a strong evidential base that saltation and stasis is a basic pattern of growth biology at the level of the organism, and demonstrate the importance of best practices in data collection and analysis. Daily measurements on rabbit tibia illustrate a pattern of growth that is linear, much like the results of monthly human measurements.[27] Rabbits measured at 3 hour intervals,[28] by contrast, show patterns of growth that are compatible with saltatory growth: intervals of growth and intervals of no growth.[11] What is clear from the animal studies is that growth is an expression of a species-specific developmental program. Time-intensive studies of rats and rabbits illustrate that the stasis interval duration and saltation frequency reflect the overall maturation of the organism.[11,28,29] What is identifiable in a daily time-frame for humans occurs within hours in these smaller animals, whose developmental rates are some 100-fold faster than humans. Thus, the concept of "daily growth" is not a useful unit of investigation in saltatory growth studies. What is remarkable in these investigations of animal growth patterns is how informative they are about the nature of growth as a part of the maturational program. It is likely that saltatory growth is a reflection of the aging process. While sensible from a lifetime perspective, it is rare that growth and aging are united in a research program.

The mechanisms that control the whole-body growth pattern are as yet unknown. Biochemical and hormonal studies using non-invasive investigative techniques in humans, such as urinary assays, are being initiated and suggest that this may an important avenue for future investigations. The first hormonal studies with the aim of further identifying within-day and between-day patterns of growth hormone and insulin growth factor secretion have been undertaken.[30] These questions were never previously investigated. Biochemical approaches for following bone growth in urinary excretions are being developed,[31] and methods for non-invasive collection of biological specimens among infants are under way.[32]

Driving towards an integrated understanding of growth biology from the cell to the organism are animal studies documenting saltatory growth.[33–35] Observations of incremental change directly at the endochondral growth plate, the site of incremental growth of long bones,[35] aim to synthesize knowledge of cellular function and morphology with the mechanisms that may be responsible for incremental saltatory growth observed in human infants and children. These observations suggest that endochondral saltations occur on time-scales in the order of minutes to hours.[35]

In order to understand the nature of the genetic, cellular, biochemical and general hormonal mechanisms controlling the process of saltatory growth, innovative studies are needed. What seems clear is that there is a genetic basis to the timing of growth saltations: identical twins are concordant in their timing (unpublished data) and population differences in saltatory growth pattern (the timing and amplitude of saltations) may be significant.[3]

Saltatory growth is hypothesized to reflect morphogenetic processes, the pattern of growth events being a manifestation of gene expression responsible for the development of species-specific morphology and aging. A reasonable hypothesis based on a synthesis of present scientific information is that the temporal aspects of the growth/maturation saltation episodes reflect an interaction between cell intrinsic information (genes uniquely expressed in individual cell lines),[36] environmental influences with epigenetic effects, and central neural signals mediated by endocrine, paracrine and cytokine cascades.[37]

For example, it is known that growth in the nervous system and bone occurs by cells that express genes according to a pattern determined by intrinsic programs and external cues.[38–40] Coordinated organismic growth, from a single cell to a three-dimensional form, reflects the expression of proteins transcribed according to a developmental timing intrinsic to cells of similar lineage, determined during embryogenesis, that is modifiable by external input during development.[37,41,42] From this viewpoint, external input includes the metabolic signals transmitted by substances such as growth hormone to locally active hormones and cytokines, more directly involved in cellular proliferation and differentiation. Saltatory growth emphasizes the importance of further investigations into the relationship between cellular division and gene expression in determining the increase in size and differentiation of the organism.

Knowledge of the specific mechanisms by which individual growth is controlled is an imperative for scientific research. Our ability to assist in the wide range of abnormal growth experienced by children is limited by the gaps in our understanding. The very basis of normal growth of the organism not only lies at the interface of growth in the proportions and size of children, but also is the cornerstone in our understanding and, thus, ability to control abnormal growth in cancer. How growth is normally controlled is a basic question in biology and deserves intense investigation. If, in normal growth, as the saltatory proposition suggests, growth is a highly controlled and permitted event, occurring as a result of disinhibition, then the normal inhibitory control mechanisms should provide new insights into how to inhibit uncontrolled growth, employing normal cellular mechanisms.[41,42]

For many years, investigations of how to assist poorly growing children at the population level have sought to understand how and when interventions may be most beneficial. It is likely that both growth faltering and catch-up growth may reflect the resetting of saltation pulse intervals by the complex network of controls that determine the process of growth, and there may be an optimal intervention strategy that can be identified.[2]

Certainly, a clear understanding of how children grow is important in understanding the process of development in general. Parents, confronted by the behavioral changes that accompany growth, may be better able to assist their children through these episodic events if they understand the nature of the events. Changes in appetite are central to difficulties experienced by breast-feeding women, who may benefit from understanding that growth may be a biological basis for some of the episodic crying spells of their young infants. Whether or not children experience growing pains during saltation events is not clear, but there is a high incidence of painful limbs reported by children when they have been documented to be growing. Much work remains to be undertaken in the further elucidation of the normal process of human growth. Whether saltation and stasis will turn out to be the most accurate description of growth at the cellular, mechanistic level remains to be determined. It is a useful model with which to initiate study into the biology of individual growth and provides a strong theoretical framework with which to conceptualize growth from the level of the cell to the whole organism. The variability in saltation amplitude and frequency provides a mechanism by which the tremendous variability in growth rate and size documented worldwide can be explained. As the process that takes one cell to a reproductive member of the species, growth must be a flexible system that responds to multiple inputs with robust adjustments. Saltatory growth permits multiple paths to final size, moderating maturation and size in a dynamic and interactive system.

16.6 SUMMARY

This chapter reviewed the nature of the empirical evidence for saltation and stasis patterns as the fundamental growth process by which individuals grow. The challenges involved in documenting saltatory growth involve methodological rigor in both data collection (sampling frequency and attention to the measurement errors associated with equipment, observer error and subject) and data analysis. A review of a number of approaches identifies the importance of using data-analytical methods that do not impute timing characteristics to the pattern of saltations, but rather reveal the patterns inherent to the empirical data. Best practices caution against the use of smoothing procedures and advocate the use of methods that deal with the challenge of identifying biology from error by other approaches.

Variability in saltation amplitude and frequency are found among individuals, suggesting this as the basis by which individual phenotypic variability is achieved, and within individuals across age, suggesting saltatory frequency dynamics as the basis for developmental age-related growth rate changes.

Animal studies identify saltatory growth at the level of the endochondral growth plate and cell-level studies generate hypotheses about a number of potentially generative areas for future research, including the development of non-invasive methods to follow biomarkers of human development.

REFERENCES

1. Lampl M, Veldhuis JD, Johnson ML. Saltation and stasis: a model of human growth. *Science* 1992;**158**:801−3.
2. Lampl M, Johnson ML. Normal human growth as saltatory: adaptation through irregularity. In: Newell K, Molenaar P, editors. *Dynamical systems in development*. New York: Lawrence Erlbaum; 1998. p. 15−38.
3. Lampl M, Ashizawa K, Kawabata M, Johnson ML. An example of variation and pattern in saltation and stasis growth dynamics. *Ann Hum Biol* 1998;**25**:203−19.
4. Lampl M, Johnson ML. Identifying saltatory growth patterns in infancy: a comparison of results based on measurement protocol. *Am J Hum Biol* 1996;**9**:343−55.
5. Cameron N. *The measurement of human growth*. Beckenham: Croom Helm; 1984.
6. Lampl M, Birch L, Picciano MF, Johnson ML, Frongillo Jr EA. Child factor in measurement dependability. *Am J Hum Biol* 2001;**13**:548−57.
7. Winer BJ. *Statistical principles in experimental design*. New York: McGraw-Hill; 1971.
8. Lampl M. Evidence of saltatory growth in infancy. *Am J Hum Biol* 1993;**5**:641−52.
9. Edgar B. Diversification of cell cycle controls in developing embryos. *Curr Opin Cell Biol* 1995;**7**:815−24.
10. Elledge SJ. Cell cycle checkpoints: preventing an identity crisis. *Science* 1996;**274**:1664−72.
11. Lampl M. *Saltation and stasis in human growth and development: evidence, methods and theory*. London: Smith-Gordon; 1999.
12. Johnson ML, Lampl M. Methods for the evaluation of saltatory growth in infants. *Methods Neurosci* 1995;**28**:364−87.
13. Johnson ML. Methods for the analysis of saltation and stasis in human growth data. In: Lampl M, editor. *Saltation and stasis in human growth and development; evidence, methods and theory*. London: Smith-Gordon; 1999. p. 27−32.
14. Johnson ML, Veldhuis JD, Lampl M. Is growth saltatory? The usefulness and limitations of frequency distributions in analyzing pulsatile data. *Endocrinology* 1996;**137**:5197−204.
15. Schmid CH, Brown EN. A probability model for saltatory growth. In: Lampl M, editor. *Saltation and stasis in human growth and development; evidence, methods and theory*. London: Smith-Gordon; 1999. p. 121−31.
16. Johnson M, Straume M, Lampl M. The use of regularity as estimated by approximate entropy to distinguish saltatory growth. *Ann Hum Biol* 2001;**28**:491−504.
17. Johnson ML, Lampl M. Artifacts of Fourier series analysis. *Methods Enzymol* 1994;**240**:51−68.
18. Pincus SM. Quantifying complexity and regularity of neurobiological systems. *Methods Neurosci* 1995;**28**:336−63.

19. Lampl M, Johnson ML. Wrinkles induced by the use of smoothing procedures applied to serial growth data. *Ann Hum Biol* 1998;**25**:187–202.
20. Caino S, Kelmansky D, Adama P, Lejarraga H. Short-term growth in head circumference and its relationship with supine length in healthy infants. *Ann Hum Biol* 2010;**37**:108–16.
21. Lampl M. Leaps and bounds: how children grow. *Pediatr Basics* 1996;**72**:10–6.
22. Lampl M. Saltatory growth and illness patterns. *Am J Phys Anthropol* 1996;(Suppl. 22):145.
23. Lampl M, Thompson AL, Frongillo EA. Sex differences in the relationships among weight gain, subcutaneous skinfolds, and saltatory length growth spurts in infancy. *Pediatr Res* 2005; **58**:1238–42.
24. Lampl M, Johnson ML. Infant head circumference growth is saltatory and coupled to length growth. *Early Hum Dev* 2011;**87**:361–8.
25. Bernstein IM, Badger GJ. The pattern of normal fetal growth. In: Lampl M, editor. *Saltation and stasis in human growth and development; evidence, methods and theory.* London: Smith-Gordon; 1999. p. 27–32.
26. Hermanussen M, The analysis of short-term growth. *Horm Res* 1998;**49**:53–63.
27. Oerter Klein K, Munson PJ, Bacher JD, Culter Jr GB, Baron J. Linear growth in the rabbit is continuous, not saltatory. *Endocrinology* 1994;**134**:1317–20.
28. Hermanussen M, Bugiel S, Aronson S, Moell C. A non-invasive technique for the accurate measurement of leg length in animals. *Growth Dev Aging* 1992;**56**:129–40.
29. Hermanussen M, de los Angeles Rol de Lama M, Burmeister J, Fernandez-Tresguerres A. Mikro-knemometry: an accurate technique of growth measurement in rats. *Physiol Behav* 1995; **2**:347–52.
30. Gill MS, Thalange NKS, Diggle PJ, Clayton PE. Rhythms in urinary growth hormone, insulin like growth factor-1 (IGF-1) and IGF binding protein-3 excretion in children of normal stature. In: Lampl M, editor. *Saltation and stasis in human growth and development; evidence, methods and theory.* London: Smith-Gordon; 1999. p. 59–70.
31. Branca F, Robins S. Saltatory growth: evidence from biochemical measurements. In: Lampl M, editor. *Saltation and stasis in human growth and development; evidence, methods and theory.* London: Smith-Gordon; 1999. p. 90–9.
32. Thompson AL, Whitten PL, Johnson ML, Lampl M. Noninvasive methods for estradiol recovery from infant fecal samples. *Front Physiol* 2010;**1**:1–8.
33. Wilsman NJ, Farnum CE, Leiferman EM, Lampl M. Growth plate biology in the context of growth by saltations and stasis. In: Lampl M, editor. *Saltation and stasis in human growth and development; evidence, methods and theory.* London: Smith-Gordon; 1999. p. 71–87.
34. Goldsmith MI, Fisher I, Waterman R, Johnson SL. Saltatory control of isometric growth in the zebrafish caudal fin is disrupted in long fin and rapunzel mutants. *Dev Biol* 2003;**259**:303–7.
35. Noonan KJ, Farnum CE, Leiferman EM, Lampl M, Markel MD, Wilsman NJ. Growing pains: are they due to increased growth during recumbency as documented in a lamb model? *J Pediatr Orthoped* 2004;**24**:726–31.
36. Raff M. Intracellular developmental timers. *Cold Spring Harb Symp Quant Biol* 2007;**72**:431–5.
37. Rougvie AE. Intrinsic and extrinsic regulators of developmental timing: from miRNAs to nutritional cues. *Development* 2005;**132**:3787–98.
38. Qian X, Goderie SK, Shen Q, Stern JH, Temple S. Intrinsic programs of patterned cell lineages in isolated vertebrate CNS ventricular zone cells. *Development* 1998;**125**:3143–52.
39. Poliard A, Ronziere MC, Freyria AM, Lamblin D, Herbage D, Kellermann O. Lineage-dependent collagen expression and assembly during osteogenic or chondrogenic differentiation of a mesoblastic cell line. *Exp Cell Res* 1999;**15**(253):385–95.
40. Shen Q, Wang Y, Dimos JT, Fasano CA, Phoenix TN, Lemischka IR, et al. The timing of cortical neurogenesis is encoded within lineages of individual progenitor cells. *Nat Neurosci* 2006; **9**:743–51.
41. Dugas JC, Ibrahim A, Barres BA. A crucial role for p57(Kip2) in the intracellular timer that controls oligodendrocyte differentiation. *J Neurosci* 2007;**6**:6185–96.
42. Artavanis-Tsakonas S, Rand MD, Lake RJ. Notch signaling: cell fate control and signal integration in development. *Science* 1999;**284**:770–6.

SUGGESTED READING

Lampl M, Johnson ML. Wrinkles induced by the use of smoothing procedures applied to serial growth data. *Ann Hum Biol* 1998;**25**:187–202.

Lampl M, Ashizawa K, Kawabata M, Johnson ML. An example of variation and pattern in saltation and stasis growth dynamics. *Ann Hum Biol* 1998;**25**:203–19.

Lampl M, Thompson AL, Frongillo EA. Sex differences in the relationships among weight gain, subcutaneous skinfolds, and saltatory length growth spurts in infancy. *Pediatr Res* 2005;**58**:1238–42.

Noonan KJ, Farnum CE, Leiferman EM, Lampl M, Markel MD, Wilsman NJ. Growing pains: are they due to increased growth during recumbency as documented in a lamb model? *J Pediatr Orthoped* 2004;**24**:726–31.

Thompson AL, Whitten PL, Johnson ML, Lampl M. Noninvasive methods for estradiol recovery from infant fecal samples. *Front Physiol* 2010;**1**:1–8.

INTERNET RESOURCES

Pulse_XP data analysis software: http://mljohnson.pharm.virginia.edu/home.html

CHAPTER 17

Lectures on Human Growth

Peter C. Hindmarsh

Developmental Endocrinology Research Group, Institute of Child Health, University College London,
London WC1N 1EH, UK

Contents

17.1 INTRODUCTION

Postnatal growth can be considered to consist of at least three distinct phases, infancy, childhood and puberty. The infancy component is largely a continuation of the longitudinal growth process observed in utero. This displays a peak growth velocity around 27—28 weeks of gestation with a decline in growth rate during the last trimester of pregnancy. Birth, in a sense, is incidental to this declining growth rate, which continues during the first 3 years of life, reaching a plateau at or around the fourth year of life and

N. Cameron & B. Bogin (eds): Human Growth and Development, Second edition.
ISBN 978-0-12-383882-7, Doi: 10.1016/B978-0-12-383882-7.00017-9

435

remaining at this level until the commencement of the pubertal growth spurt. The factors influencing these distinct growth periods are different. We know little of the factors influencing fetal and early infant growth but know from animal experiments that nutrition plays a key role. With the appearance of the growth hormone (GH) receptor in the growth plate at around 6 months of postnatal life the GH-dependent growth assumes greater importance and during the childhood years growth is largely dependent on the GH secretory status of the individual. The final step in the growth process, the pubertal growth spurt, comprises a 50% contribution from sex steroids and 50% contribution from GH.

This chapter describes abnormalities of growth and how they are assessed, placing the discussion within the framework of our understanding of factors involved in regulating fetal growth, with particular emphasis on the insulin-like growth factor (IGF) axis. Failure of physical growth is an important sign of systemic disease but it is also the hallmark of endocrine disease since pituitary, thyroid, adrenal and gonadal hormones are all involved in this process. It is especially important to be clear that the categorization of an individual child depends upon growth assessment and clinical examination. It does not depend, at least not in the first instance, on laboratory investigations, which should not be employed unless auxological data indicate them to be necessary.

17.2 ENDOCRINOLOGY OF GROWTH

The endocrinology of growth will now be discussed.

17.2.1 Growth Hormone Secretion: Cellular and Molecular

The pituitary gland develops as an outpouching of the stomatodeum — Rathke's pouch. This process takes place between 30 and 35 days postconception and is tightly regulated by a series of homeobox genes. Close apposition between this structure, which is destined to form the anterior pituitary, and the base of the hypothalamus takes place and this leads to descent of neural tissue with the pouch to form the posterior pituitary. Stalk vascular cannulization completes the process. Differentiation of the anterior pituitary mass into the recognizable cell types is influenced in part by the homeobox genes involved in development and by cell-specific homeobox gene expression (Figure 17.1). Somatotroph cell differentiation also utilizes the expression of two genes, Prop-1 and Pit-1. Enlargement of the somatotroph cell number requires the induction of the growth hormone-releasing hormone (GHRH) receptor by Pit-1. This allows the hypothalamic peptide to stimulate somatotroph cells, leading to the synthesis and release of GH. GHRH stimulation also leads to somatotroph hyperplasia.

The human GH gene is located on chromosome 17, along with two genes for human somatomammotrophin. Pituitary GH is coded for by the hGH-N gene and transcription leads to the synthesis of GH with a molecular weight of 22 kDa. The excision of the

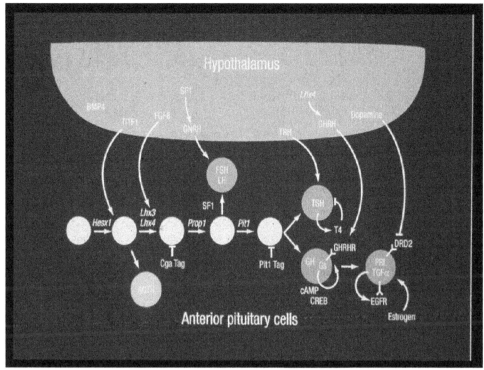

Figure 17.1 Sequence cascade of factors involved in pituitary development. Gradient factors and transcription factors are involved in the determination and situation of pituitary cell types. This figure is reproduced in the color plate section.

second intron of hGH-N leads to an alternative splicing site resulting in deletion of the message for amino acid residues 32–46 – the 20 kDa GH variant. This forms 10% of the circulating GH.

The synthesis of GH is largely regulated by the levels of GHRH impinging on the anterior pituitary somatotrophs. GHRH acts on the somatotroph by binding to its own specific receptor which activates a secondary messenger system via cyclic synthesis. This receptor is characterized by seven transmembrane loops and internal coupling to the G (guanine)-protein system. In their resting state, the G-proteins exist as heterodimeric complexes with α-, β- and γ-subunits. In practice, the β- and γ-subunits associate with such a high affinity that the functional units are $G\alpha$ and $G\beta\gamma$. After association of the G-protein complex with the occupied receptor, conformational changes in the α-subunit lead to an increased rate of dissociation of guanosine diphosphate (GDP), which is replaced by guanosine triphosphate (GTP). This guanine nucleotide exchange, in turn, causes the α-subunit to dissociate from the heterotrimeric complex. The liberated α-subunit, together with its activating GTP, then binds to a downstream catalytic unit adenylate cyclase.

Hydrolysis of the GTP bound to Gα due to its intrinsic GTPase activity liberates the Gα-subunit from the catalytic subunit and allows reassociation of GαGDP with the Gβ/γ. This newly reformed heterotrimer then returns to the G-protein pool in the membrane. In this way an individual G-protein complex is recycled, so that it can respond to further receptor occupation by ligand.

Gsα activates membrane-bound adenylate cyclase, which catalyzes the conversion of adenosine triphosphate (ATP) to the potent second messenger cyclic adenosine monophosphate (cAMP). This cyclic nucleotide, in turn, activates a cAMP-dependent protein kinase (PKA), which modulates multiple aspects of cell function. PKA phosphorylates a transcription factor called CREB (cAMP response element binding protein). This is then translocated to the nucleus, where it binds to a short palindromic sequence in the promotor region of the GH genes. It is this process that leads to transcription and synthesis of GH. The transcription of the GH gene is regulated, in turn, by a number of other hormones such as thyroxine and cortisol.

17.2.2 Physiology of Growth Hormone Secretion

Prior to the consideration of the endocrine regulation of different stages of human growth, it is worth considering the physiology of the GH—IGF axis. Figure 17.2 shows a typical GH profile in a 9-year-old boy generated by taking blood samples for GH measurement every 20 minutes. Frequent sampling is essential in order to define clearly the true heights of the peaks. If sampling were too infrequent the true peak heights might be underestimated or peaks missed altogether. The profile is characterized by episodes of GH release generating peak GH concentrations interspersed with periods when GH secretion is effectively switched off and GH concentrations are undetectable. This appears to be the predominant pattern in males, whereas in females in varying species although the peak concentrations tend to be similar the most striking difference is that there is an elevation in the trough concentrations so that at all times concentrations are detectable. One further point to note is that the pulses occur as fairly frequent intervals of one every 3 hours, suggesting that most of the GH signal is contained in the amplitude of the pulses.

In both rodents and humans there is evidence to support the concept of an inverse relationship between the secretion of the two hypothalamic peptides: somatostatin (SS) and GHRH. GHRH is involved in both the release and synthesis of GH, while SS inhibits GH release. Normal GH pulsatility requires endogenous GHRH, although GH responses to exogenous GHRH are variable and reveal varying periods of responsiveness and refractoriness. There are several possible explanations for this. First, the phenomenon may be intrinsic to the GH-secreting cells. Second, acute down-regulation of the GHRH receptors or their intracellular signaling systems may take place. This appears to be an unlikely explanation as down-regulation only takes place at very high GHRH levels, certainly well above those usually encountered physiologically. Third, depletion of the readily releasable pool of GH may occur. The final and most likely explanation is that

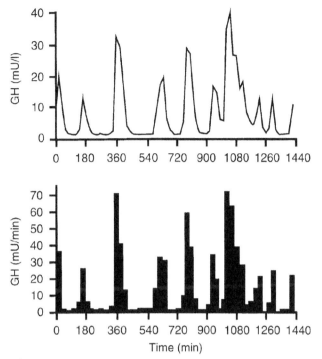

Figure 17.2 Twenty-four-hour serum growth hormone (GH) concentration profile in the upper panel with a deconvoluted estimate of the pituitary secretion of GH needed to generate the serum concentrations shown in the lower panel.

the pattern reflects variation in endogenous SS tone imposing an ultradian rhythm in GHRH responses. Evidence for this comes from the observation that continuous GHRH administration leads to pulsatile GH release, implying modulation by another factor, i.e. SS. Although SS readily suppresses pulsatile GH secretion in rats and humans, its effects are short lived and rapid release of GH takes place on SS removal. This rebound secretion can be detected in vitro but is even more pronounced in vivo. In general, GHRH administration alone leads to the gradual attenuation of the GH response with time. SS withdrawal can produce GH rebound secretion in the human subject but for regular repeatable GH release to take place it is the combination of SS withdrawal coupled with GHRH administration that is the most efficacious (Figure 17.3).

This close relationship between GHRH and SS acting as integrators of other signals, for example sleep, is complicated by the recent discovery of a family of GH secreto-gogues, which differ substantially from GHRH in both their structure and receptors. These GH secretagogues behave in a similar manner to GHRH in terms of their physiology, but the important difference is that they act synergistically with GHRH to generate GH release. Recent work has identified the endogenous "third factor" as ghrelin, which is present in the stomach. The precise physiology of this substance

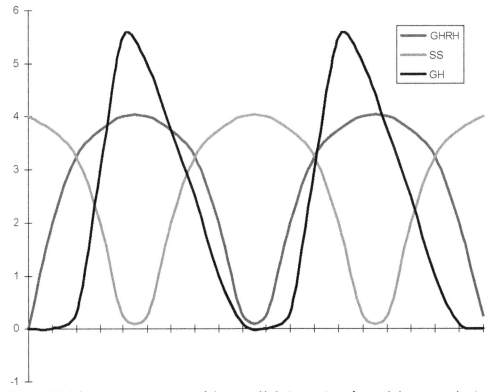

Figure 17.3 Schematic representation of the most likely interaction of growth hormone-releasing hormone (GHRH) and somatostatin (SS) in the generation of a growth hormone (GH) pulse. This figure is reproduced in the color plate section.

remains to be determined, although its role would appear from both animal and human studies to be minimal. A number of endogenous agents affect GH pulsatility, including opioids, calcitonin and glucagon, but these have not been used to manipulate GH secretory patterns to alter growth and their physiological relevance to the control of endogenous pulsatility remains unclear.

There is good evidence that GH feeds back to inhibit its own release and this may have a bearing on the temporal control of GH pulsatility. Experiments indicate that exogenous GH acts directly on the hypothalamus rather than on the pituitary. The most likely mode of action of GH is through increased secretion of SS into the portal blood, but there is evidence that GH also leads to an inhibition of GHRH production. GH can also feed back indirectly by the generation of IGF-1 which, in turn, inhibits GH synthesis and release, chiefly at the pituitary level.

The secretion of GH into the circulation is pulsatile. The entry rate of endogenous GH is governed by the kinetics of GH release from the somatotrophs and the removal from the circulation is largely determined by the amount of GH bound to its binding protein and internalized by the GH receptor on the target organ cells. The precise role of the binding proteins in humans, at least, is far from clear. There is a correlation with GH status but only at the extremes, and the effect of GH treatment is highly variable. Although GH binding protein might increase the amount of GH available for constant delivery of GH to the receptor, this is not at all clear. There is no evidence to suggest that the preferred mode of presentation of GH to the receptor is continuous. A body of evidence exists to suggest that the pulsatile mode is most optimal. An alternative role for the binding proteins might be to buffer the system from overexposure to GH. Given the high affinity of GH for its binding protein, this might be a more likely explanation.

The pulsatile signal appears to be important in determining a number of target organ effects. For example in the rodent, the pattern of hormone secretion, either pulsatile (male mode) or continuous (female mode), has an impact on the growth of the animal, expression of a number of liver enzymes, the determination of the level of GH binding protein in the circulation and GH receptor expression. The tissue response is also variable, in that the liver will generate IGF-1 in response to GH irrespective of mode of administration, whereas adequate expression of IGF-1 in the muscle is highly dependent on the pulsatile mode of administration.

There is increasing evidence in humans that the mode of GH secretion is important in determining target organ response. In humans, IGF-1 generation occurs best when GH is present in the pulsatile rather than the continuous mode, at least in the physiological situation. This may not be the case when GH treatment uses the subcutaneous route, as the pharmacokinetics of subcutaneous GH tend towards a more continuous exposure. On the other hand, the pattern of fat distribution around the abdomen is more influenced by the trough concentrations of GH in humans.

17.2.3 Growth Hormone Receptor and Target Organ Signaling

Receptors for GH, together with those for cytokines such as the interleukins and erythropoietin, have a major structural feature in common, in that they have four long α-helices arranged in an anti-parallel fashion. As a consequence, this subgroup is commonly referred to as the cytokine/hemopoietic receptors. The structure of human GH with its receptor is a ternary complex consisting of a single molecule of the hormone and two receptors. After GH has bound to one molecule of receptor, association of this complex with a second receptor molecule occurs. The dimerization of the cytoplasmic region in the ternary complex is particularly important for signal transduction.

The GH receptor uses an unusual intracellular signaling system: janus-associated kinase-2 (JAK-2). The JAK system is coupled to further intracellular proteins: the signal transducer and activator of transcription (STAT) proteins. These are transcription factor proteins. They contain a crucial tyrosine residue located in the carboxy-terminal in a homologous position in all STAT proteins (residue 694), and phosphorylation of this is essential for STAT activation. STAT proteins have a dual function: signal transduction in the cytoplasm followed by activation of transcription in the nucleus. The family members of STAT proteins were named in the order of their identification. GH induces tyrosine phosphorylation of STAT proteins 1, 3, 5a and 5b, but STAT-5 is the major component of the JAK—STAT cascade. Dimerization of the STAT proteins appears to be essential for their final translocation to the nucleus, where they activate immediate-early response genes, which regulate proliferation, or more specific genes that determine the differentiation status of the target cell.

17.3 THE GROWTH HORMONE—INSULIN-LIKE GROWTH FACTOR-1 AXIS AND ITS ASSOCIATION WITH FETAL GROWTH

It is interesting to contrast the situation in the fetus with that in the child and the adolescent. The fetus is subject, particularly in the latter half of pregnancy, to a fairly constant delivery of metabolites via the placenta. The flow of metabolites continues through the umbilical vein and the distribution thereafter utilizes a circulatory pattern in which blood is predominantly diverted in its oxygenated form to the developing brain. Given the fact that the fetus is highly dependent on this mode of delivery of substrate then perhaps it is not too surprising that the growth process differs. In addition, the growth of the individual organs also differs from that observed in postnatal life, with different patterns of growth and development exhibited by many of the differing tissues. Figure 17.4 depicts the changes in size of the adrenal gland and serves to remind us that organ size is dependent on functional need at different stages of development.

Studies in larger domestic animals show that pulsatile secretion of GH and other pituitary hormones is already demonstrable in fetal life and is sensitive to nutrition. Little is known about the evolution of GH secretion in the human fetus. Studies have demonstrated a gradual increase in circulating GH concentration during the first 12 weeks of pregnancy, reaching a peak at 20—24 weeks and declining towards birth. These early changes to serum GH concentrations appear to parallel the known development of the hypothalamic peptides GHRH and SS. The human fetal pituitary is able to respond to these two factors and it is proposed that the GHRH effect predominates, with SS increasing in effect towards term. Even at term, GH levels are 20—30 times higher than those observed in childhood; but perhaps of greater importance is that the levels are continuously elevated and lack the pulsatile pattern observed in childhood and adult life. However, these high GH concentrations are not associated with elevated levels of IGF-1 in the fetus, implying that there is "relative resistance" to the effects of GH in the fetus. The effect may diminish towards term but it can be assumed that GH is not the predominant determinant of fetal growth. This is also borne out by experiments of nature in which the GH gene is deleted or where the GH receptor is non-functional. These individuals are normal size at birth when due account is taken of maternal size.

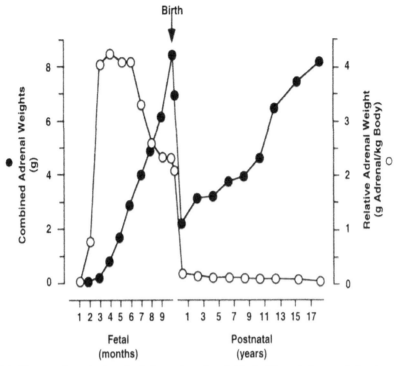

Figure 17.4 Changes in actual and relative adrenal weight at different stages of development.

Because of the problems associated with accessibility to human fetal tissue the role of endocrine factors in determining fetal growth is largely inferred from studies in animals. The most elegant series of these studies involves the use of transgenic animals in which the various components of the IGF axis (Table 17.1) have been knocked out. It must be appreciated that these studies reveal quite major effects of the whole gene and they tell us that the peptide is of particular importance in the determination of size of the fetus. That they are clearly important comes from the observation that many of the knockout offspring die in the first few hours of life. Table 17.1 shows the type of knockouts that have been constructed and from this it can be clearly seen that both IGF-1 and IGF-2 play important roles in the determination of body size in the mouse. It is likely that a similar situation also pertains in the human because there is a clear relationship between birth weight and levels of both these growth peptides (Figure 17.5); in addition, a boy with an IGF-1 gene defect was born with low birth weight. Perhaps rather surprisingly, loss of the insulin gene did not appear to alter body size. This would at first sight appear contradictory to clinical observations of macrosomia associated with maternal hyper-glycemia and the condition of congenital hyperinsulinemia, where excess fetal and neonatal insulin production leads to fetal overgrowth. It is likely that in these situations the effects of hyperinsulinemia in the fetus are mediated via the IGF receptors rather than

Table 17.1 Results of insulin-like growth factor (IGF) and insulin knockout mice studies

Inactivation	Fetal size
IGF-1/IGF-2	30%
Insulin receptor	100%
Type 1 IGF receptor	45%
+IGF-1	Fatal
+IGF-2	30%

a direct effect of insulin via its own receptor. The IGF receptor knockout studies indicate the importance of the type 1 IGF receptor in mediating the growth effects of IGF-1 and IGF-2. All these studies demonstrate a pivotal role for the IGF family in the determination of fetal growth.

In the newborn, studies have revealed markedly amplified GH secretory episodes which occur throughout the day and night. Preterm infants have even higher secretory profiles than term babies. The high GH secretion at birth is sensitive to inhibition by dopamine and by stimulation by intravenous GHRH. The GH response to GHRH is, in turn, modified by the birth size of the baby, with greater responses seen in those of lower birth weight. As IGF-1 levels are lower in these babies, this might imply that the

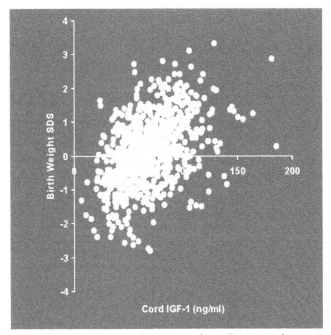

Figure 17.5 Relationship between birth weight expressed as a Z-score and serum insulin-like growth factor-1 (IGF-1) concentration in cord blood. This figure is reproduced in the color plate section.

feedback effect of IGF-1 is also operative at this age. However, as IGF-1 levels are generally lower at birth and increase thereafter through childhood into adolescence, it is possible that elevated GH values may represent, in part, "immaturity" in this part of the feedback loop. Although the GH response to SS is blunted the components for generating episodic secretion are clearly present and become more operative during the neonatal period.

17.4 ENDOCRINOLOGY OF PREPUBERTAL GROWTH

After 2–3 months of life, clinical evidence suggests that GH is necessary for sustaining normal growth. The postnatal elevated GH levels observed subside so that by 3–6 months of age values approach those observed in childhood. During the prepubertal years GH secretion gradually increases, primarily in terms of the amplitude of the GH pulses. Although many reports demonstrate differences in GH secretion between tall normal and short stature children and between normal stature and short stature, the differences pertain more to the growth rates of the individuals than to the stature observed. These differences in GH secretion are reflected in the serum levels of IGF-1 seen in these groups. IGF-1 values increase gradually from birth throughout childhood and relate well to the levels of GH secreted. Apart from pathological conditions, it remains difficult to relate any measure of GH pulsatility to the observed growth rate in individual children with short stature, probably because the variability in growth rate is so small.

The situation is complicated further by the interaction of body composition with GH secretion. Particularly in the area of short stature, which is a heterogeneous condition, we can imagine a number of diagnoses impinging upon the growth process which will also influence the relationship between growth and GH. Generally speaking, height at its extremes can be related to GH secretory status. In situations where body mass index is controlled for, the more important relationship is that between GH secretion and growth rate. This relationship has been documented by several groups and is probably described as a curvilinear relationship (Figure 17.6). These observations tend to relate to long-term growth, usually over a period of 1 year, with GH secretion on a particular day. More detailed studies using repeated estimations of GH secretion in urine have linked GH secretion not only to growth rate but also to the intraindividual variation in growth rate that occurs on a week-by-week basis.

Considerable interest has centered on the components of the GH profile that contribute to the effect on growth. Early studies demonstrated that the growth process was pulse amplitude modulated and that GH pulse frequency did not change, remaining relatively fixed at a 200 minute periodicity. Changes in pulse frequency are largely confined to pathophysiological states such as poorly controlled diabetes mellitus. Paradoxically, poor growth in chronic renal failure is associated with high GH secretion, or at

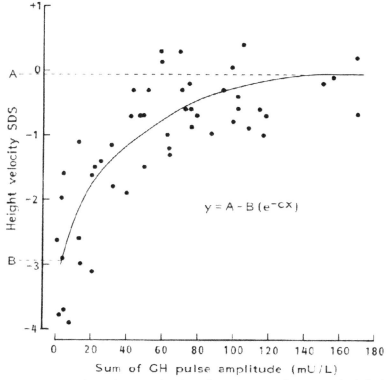

Figure 17.6 Asymptotic relation between height velocity expressed as a standard deviation score (SDS) and the sum of growth hormone (GH) pulses secreted over a 24 hour period.

least high GH levels in the circulation, which in part may result from reduced GH clearance, although a degree of GH hypersecretion probably also exists.

The pulse amplitude is determined predominantly by the rate of entry of GH to the circulation. As the duration of the pulse is relatively fixed, the rate of change of GH in the circulation becomes an important factor. Several studies have suggested that the rate of rise of the pulse is the actual growth signal, so that the main information is contained in the rate of change of hormone concentration than in the level achieved. This rather presupposes that there is an actual level above which growth is likely to take place and that any further modulation is due to the rate of rise of the hormone secreted. This has not been tested formally but remains an intriguing possibility. The suggestion is not too farfetched and it appears to be borne out to a certain extent by receptor studies. Rapid receptor turnover would be a prerequisite in pulsatile systems and this is certainly the case with the insulin receptor in fat and muscle. Fairly rapid internalization takes place with the GH receptor, and the intracellular signaling system functions optimally with a 3 hourly change in ambient GH concentrations.

One further component of the profile that appears to be important is the trough concentration of GH achieved in a secretory profile. Mention has already been made of the importance of the secretory pattern in the rodent and to an extent in humans on influencing the generation of IGF-1 and for the maintenance of body composition. The precise role of trough concentrations in altering, or rather influencing, growth rate in humans is still far from clear. However, there is evidence to suggest that although the predominant effect on growth is determined by the amplitude of the GH secretory pulses, the effect of this pulse is modulated to a certain extent by the level of trough concentration. In the situation where the GH pulse amplitude is sufficient to generate normal growth, alterations in trough concentration appear to have little effect on the overall growth rate. In the situation where GH secretion is attenuated owing to a low GH pulse amplitude with consequent reduction in growth rate, the presence or absence of trough levels of GH has a profound effect on the growth rate observed. A situation of low GH pulse amplitude combined with a high trough concentration is associated with an extremely poor growth rate, compared to that observed with a similar GH pulse amplitude and a lower or normal trough concentration. These effects on growth rate are also mirrored in the levels of serum IGF-1 concentration measured (Table 17.2).

Surprisingly little is known about the effect of alterations in GH receptor status and its effects on growth in normal individuals. Apart from the situation of GH receptor deficiency due to genetic abnormalities in the GH receptor gene, little is understood about differences in sensitivity to GH between individuals. This is a rather surprising situation given the fact that GH has been used for the treatment of a number of growth disorders over many years. The syndrome of GH resistance due to abnormalities in the GH receptor leads to an individual who produces considerable quantities of GH but very small amounts of IGF-1 or the other GH-independent protein, IGF binding protein-3.

Table 17.2 Growth rate and insulin-like growth factor-1 (IGF-1) levels in 50 children with respect to peak and trough growth hormone (GH) concentrations

	GH peak <50th	GH peak >50th	Total
(a) Height velocity SDS			
GH trough <50th	−1.38	−0.82	−1.05
GH trough >50th	−1.93	−0.72	−1.37
Total	−1.70	−0.77	
(b) Serum IGF-1 concentration μ/l)			
GH trough <50th	0.44	0.53	0.49
GH trough >50th	0.21	0.43	0.31
Total	0.31	0.49	

Data are shown as mean values.
SDS: standard deviation score.

The net result is a growth phenotype that is similar to but probably more severe than individuals with GH gene deletion. The treatment of these individuals with IGF-1 is only partially successful, probably because GH is also required in its own right in the commitment of stem cells to the proliferative and hypertrophic zones of the cartilage. What happens during IGF-1 treatment is that any endogenous GH is suppressed, hence any chance of stem cells entering the proliferative zone is reduced and the effect of IGF-1 is simply to proliferate those cells that are available and gradually reducing in number during the course of therapy.

A few studies have suggested differences in GH sensitivity in the general population, but the lack of good dose—response curves to define the terms adequately has seriously hampered the development of concepts in this area. A polymorphism in the GH receptor (GHR) gene leading to retention (full length, fl) or deletion of exon 3 (d3), which encodes a 22-amino acid residue sequence in the extracellular domain, has been associated with the degree of height increase in response to GH replacement in children born short for gestational age, those with idiopathic short stature and in a GH-deficient population. Patients with at least one d3 allele had a significantly better first year response leading to an improved adult height on GH treatment than patients with homozygosity for GHR-fl. However, reported studies are not all consistent, which may reflect differing populations and conditions. False-positive findings are more likely with small sample sizes and for quantitative trait loci phenotypic variations tend to be overestimated with small sample sizes.

As suggested above, the GH axis acts as a final common pathway in childhood for a number of pathophysiological situations affecting growth. In acquired hypothyroidism there is a general permissive effect of thyroid hormone on the whole growth axis. In hypothyroidism there is a reduction in the efficacy of GHRH-stimulated GH release, probably as a result of a reduction in the transcription of the GH gene. Any GH that is secreted has probably less of an effect on the target tissues as there is quite good evidence to suggest that thyroxine is particularly important for mediating GH action at target tissue level. This is in addition to effect of post-GH receptor of thyroxine on cartilage growth.

Although GH plays an important role in prepubertal growth there is probably an interaction with other factors. The mid-childhood growth spurt is a good example of this. If an individual's growth chart is examined an increase in growth rate can be detected around the age of 7—8 years. The precise etiology of the spurt is unclear but it is likely that adrenal androgens, which are increasing in circulatory concentration at this time, play a role. Supportive evidence comes from patients with early-onset Addison's disease, where adrenal function is lost. These patients do not manifest a mid-childhood growth spurt, or at least it is attenuated. These patients, and indeed anyone who has suppressed adrenal androgen secretion, have delay in the timing of the onset of puberty. This suggests that adrenal androgen production is involved not only in the mid-childhood growth spurt but also in priming the hypothalamopituitary axis for puberty.

17.5 ENDOCRINOLOGY OF PUBERTY

The pubertal growth spurt in human subjects represents the contribution of sex steroids and GH contributing 50% of the height gained. Augmentation of GH secretion occurs during puberty with an approximate two- to three-fold increase in amplitude of the secretory bursts, whereas the frequency of GH pulses does not change. Many cross-sectional studies have demonstrated that the increase in GH pulse amplitude coincides with the pubertal growth spurt and confirmation of this observation has come from detailed longitudinal studies where puberty has been induced with the hypothalamic peptide gonadotrophin-releasing hormone.

Sex steroids play an important role in regulating the physiology of GH secretion in childhood. Estrogen has long been known to alter GH responses to stimuli and small doses of estrogen when given to girls with gonadal dysgenesis enhance GH secretion, as does testosterone administration to boys with delayed puberty. Conversely, suppressing puberty, as in the situation of precocious puberty, with a gonadotrophin-releasing hormone analogue leads to a decrease in GH secretion. Both testosterone and estradiol stimulate GH, so the question arises as to whether they act as independent agents. Non-aromatizable androgens such as oxandrolone do not greatly affect GH secretion. The effect of testosterone is certainly time and is probably dose dependent. Androgen receptor blockade has little effect on GH secretion, whereas the anti-estrogen tamoxifen blocks testosterone stimulation of GH secretion. This suggests that aromatization of testosterone to estradiol, which then acts via the estrogen receptor, may be more important for testosterone's effects on GH.

The precise mechanisms by which the sex steroids alter the interaction of GHRH and SS and the generation of GH pulses in childhood are unclear. There is, however, a large amount of experimental data, primarily from the rat, suggesting that neonatal exposure to steroids has an imprinting effect on the hypothalamic systems regulating GHRH and SS, but their full expression requires continued gonadal steroid exposure in adult life. Prepubertal gonadectomy can markedly alter the expression of the sexually dimorphic GH secretory pattern in the adult rat. In addition to the hypothalamic effects, the raised basal GH release by estradiol in normal rats may be due to effects directly on pituitary GH synthesis or GH cell number.

Estrogen has additional effects on the growth process. The dose—response effect between estrogen and growth is biphasic in nature, with a peak acceleration in growth observed with ethinylestradial doses of approximately 10 µg/day, which is equivalent to the mid-point of pubertal development. Increasing the estrogen dose leads to a reduction in growth rate and an advance in skeletal maturation. It would appear that estrogen has two effects in puberty. The first is to augment GH secretion in lower doses, leading to generation of the pubertal growth spurt, and at higher doses, certainly in the rat, to suppression of or at least a reduction in GH release. The second is to accelerate ossification and closure of the growth plate. The latter process appears to be highly dependent on the density of estrogen receptors in the growth plate. In experiments of nature where estrogen is either deficient owing to aromatization defects or unable to act owing to estrogen receptor mutation then growth continues, albeit slowly, and the epiphyses do not close.

17.6 SHORT STATURE

We shall now discuss short stature.

17.6.1 Definition

The definition of shortness is arbitrary. The general rule has been that any child whose height falls below the 3rd centile for his or her community should be considered short. This immediately raises problems over the definition of the term and the most appropriate standards for the assessment of height. It is important to realize that the height standards are only a statistical description of the general population. Three per cent of children will have heights below the 3rd centile regardless of whether or not there is anything wrong with them. Different ethnic groups have different growth standards. Secular trends in heights are well documented and are more marked in certain groups, and up-to-date growth charts need to be used to account for such changes. In the UK there is a secular trend in height so that each generation tends to be 1—1.5 cm taller than the previous one.

17.6.2 Growth Assessment

There is a tendency with height charts to simply look at stature as a static process rather than a dynamic one. This has led to the use of height velocity charts to assist in decision making. However, because of the cyclical nature of growth, constructing simple decision models on the basis of these charts is difficult.

Figure 17.7 illustrates the British 1990 Growth Reference Standards for boys. The equispacing of the new centiles, which more adequately define standard deviations about the mean, coupled with the introduction of two new centiles, the 0.4 and 99.6, allow decision making to be more easily made using a single chart.

Parental heights give an estimate of the genetic height potential of the individual, which is a useful measure against which to judge the current height of the individual. Adjustments need to be made to parental heights depending upon whether a boy or girl is the subject in question. Males are taller than females by about 14 cm, so if a boy's height is

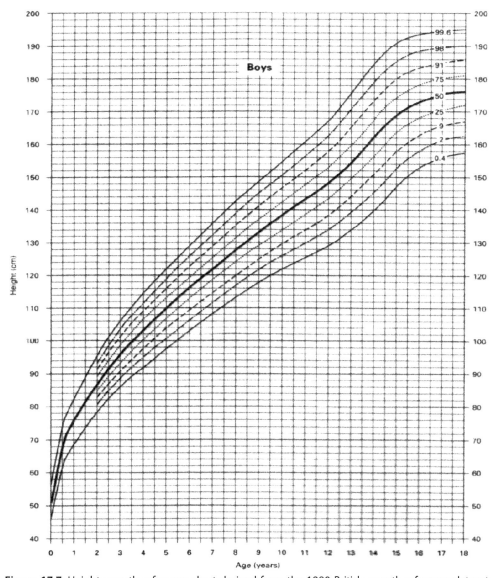

Figure 17.7 Height growth reference chart derived from the 1990 British growth reference dataset.

being considered the father's height can be plotted on the "boy's chart" directly; but the mother's height needs to be adjusted by adding 14 cm and this derived value entered on the chart. The mid-parental height can be derived [in this case: (paternal height + corrected maternal height)/2] as well as the target height (mid-parental height ± 1 standard deviation ∼3 cm) and plotted. When the father's height is plotted on a "girl's chart", 14 cm needs to be deducted from his current height before plotting.

Unless the current height falls above or below a centile trigger point then it is likely that further measurements are going to be required. The time interval between these measurements will be determined by the precision of height measurement, but the minimum time interval will be about 3 months and at least 6–12 months will need to elapse before a growth pattern can be established.

Making decisions using new charts has been simplified. Using the old cut-off of the 3rd height centile means that in each year in the UK 24,000 individuals (the vast majority of whom will be normal) might require further monitoring or investigation. Introduction of the 0.4th centile is important because the likelihood that in this situation the individual is normal is markedly diminished. The diagnostic return on investigating individuals whose heights are below the 0.4th centile is quite high and is much higher than the 2 per 1000 return when the cut-off value was around the 3rd height centile. Identification of the short child is, therefore, based on diagnostic return, not on an arbitrary height centile.

The charts also allow for decisions to be made about individuals whose height may be normal but whose growth pattern is crossing centile bands. The correlation between heights measured at different ages is extremely high. For example, a height measured at the age of 3 years has a correlation coefficient associated with another measurement made at 6 years of 0.91. Knowing this type of correlation, 2% of children might be expected to fall half a centile band or more between 5 and 6 years; a similar percentage should rise by half a centile band. More conservative changes of 1 centile or more over a similar period are extremely rare, with a rate well below 1 per 1000 individuals. Less stringent criteria can be applied if there are more than two measurements, as the chance of two successive growth rates being low is very small.

One final point worth considering relates to when growth problems may evolve. Congenital abnormalities, e.g. GH gene deletion, will present early in life and as a general rule any period of fast growth, such as infancy and puberty, may unmask a growth disorder. There is, however, in the totality of growth disorders a law of diminishing returns with the age of the child. For example, by school entry 60–70% of growth disorders should have presented/been identified, with pubertal growth disorders contributing a further 15–20%. Growth surveillance programs are best concentrated, therefore, in the preschool/early school years.

17.6.3 Classifications on Growth Assessment

Using these approaches, small children can be classified into three groups. A flowchart for the differential diagnoses of short stature is shown in Figure 17.8.

Short Normal Children

This group of children contains at least two subgroups. The first are characterized by a height in the lower centiles and a normal growth velocity. The height of one or both of the parents is likely to be on a similar centile. The second group, which may include

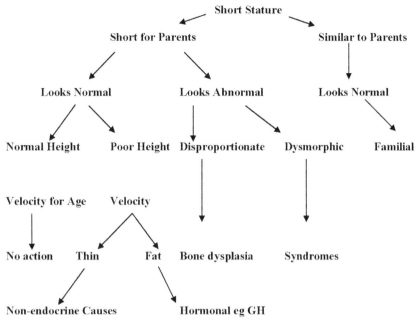

Figure 17.8 Algorithm for assessing short stature.

some children from the first, consists of children whose stature may be anywhere between the 0.4th and 25th centiles and whose growth rate is normal during the childhood years, but who lose ground because of the late onset of puberty and a delayed pubertal growth spurt. Skeletal maturation becomes delayed in children towards the end of the first decade of life and a final height appropriate for the parents is likely to ensue.

Short Stature Due to an Early Event

In most cases children with a syndrome who are short have been of low birth weight. The problem is not confined to intrauterine growth restriction. Poor nutritional support of prenatal infants will produce similar effects. It is important to remember that growth in the first 6 months of life is independent of GH and is largely dependent on nutrition, so poor nutritional intake during this period of rapid growth can have profound and long-lasting effects. Once GH-dependent growth becomes established, normal growth follows, but stature lost at this stage is not easily recovered.

Two syndromes deserve mention. The first is the association of low birth weight and its resultant short stature with dysmorphic characteristics, the Silver–Russell syndrome. The clinical features are triangular-shaped faces, body symmetry, thinness and clinodactyly of the fifth fingers. However, the most characteristic feature is the formidable difficulty encountered by the parents in feeding such children in infancy. Postnatal growth beyond 12 months of age is characteristically normal in these children. Most cases are sporadic, but familial cases have been described. Evidence from mice and humans suggests that uniparental disomy may be an important explanation for many cases of Silver—Russell syndrome and this is certainly worth looking for. Knockout experiments in mice of the IGF family also lead to low birth weight and evidence for a human parallel has recently been presented.

The second syndrome of note is that of Turner, which results from abnormalities — absence, mosaicism, rings or isochromosomes — of the X chromosome. Mosaicism is common, but the effect of Turner syndrome on growth is to add persistent low growth velocity to a prenatal growth deficit. Growth over the first 3 years of life tends to be relatively normal in terms of growth velocity, but thereafter a decline in growth rate can be discerned with no obvious pubertal growth spurt.[9] The condition is common (1 in 3000 female births) and there may not necessarily be the well-known dysmorphic features. Short stature is common and is due to loss of a statural gene on the X chromosome (SHOX), but the major implication of diagnosis relates to pubertal development and reproductive capacity.

Short Stature with a Remedial Condition

These children may or may not be short, but more importantly their growth velocity is slow. This group includes a wide spectrum of disease. An explanation for the poor growth rate needs to be found and appropriate treatment recommended. A carefully-taken history and examination may point to a remediable cause. Full investigation encompassing the renal, gastrointestinal, cardiac, respiratory, hematological and neurological systems is often required. Finally, careful attention needs to be paid to the social and family history. Psychological deprivation can take forms, ranging from emotional deprivation to anorexia by proxy, as well as extremes of physical and sexual abuse.

17.6.4 Classification of Growth Hormone Secretory Disorders

The characteristic clinical picture of severe GH insufficiency is of a very short, rather plump child with a round, immature face. Birth weight is usually normal, and poor growth is apparent from about 6 months of age. The insufficiency may be isolated or associated with other pituitary hormone deficiencies. Small genitalia are especially characteristic of associated gonadotrophin deficiency. Hypoglycemia in the newborn period is often a feature of adrenocorticotrophic hormone (ACTH) deficiency. Prolonged neonatal jaundice, particularly a conjugated hyperbilirubinemia, raises the question of thyroxine or cortisol deficiency.

Pituitary Gland

The genetic causes of GH deficiency, in isolation or in association with multiple pituitary hormone deficiency, are summarized in Table 17.3. As there is a cascade of genes

Table 17.3 Genes implicated in isolated growth hormone (GH) deficiency and combined pituitary hormone deficiencies in mice and humans

Gene (murine/human)	Protein (murine/human)	Murine loss of function phenotype	Human phenotype	Inheritance murine/human
Hesx1/HESX1	Hesx1/HESX1	Anophthalmia or microphthalmia, agenesis of corpus callosum, absence of septum pellucidum, pituitary dysgenesis or aplasia	Variable: SOD, CPHD, IGHD with EPP/dominant or recessive	Dominant or recessive in both
Sox3/SOX3	Sox3/SOX3	Unknown	Isolated GH deficiency with mental retardation	Unknown in mouse; X-linked in human
Lhx3/LHX3	Lhx3/LHX3	Hypoplasia of Rathke's pouch	GH, TSH, gonadotrophin deficiency with pituitary hypoplasia. Corticotrophs spared. Short, rigid cervical spine with limited rotation	Recessive in both
Lhx4/LHX4	Lhx4/LHX4	Mild hypoplasia of anterior pituitary	GH, TSH, cortisol deficiency, persistent craniopharyngeal canal and abnormal cerebellar tonsils	Recessive in mouse, dominant in human
Prop1/PROP1	Prop1/PROP1	Hypoplasia of anterior pituitary with reduced somatotrophs, lactotrophs, thyrotrophs and gonadotrophs	GH, TSH, prolactin and gonadotrophin deficiency. Evolving ACTH deficiency. Enlarged pituitary with later involution	Recessive in both
Pit1/POU1F1 (PIT1)	Pou1f1/POU1F1	Anterior pituitary hypoplasia with reduced somatotrophs, lactotrophs and thyrotrophs	Variable anterior pituitary hypoplasia with GH, TSH and prolactin deficiencies	Recessive in mouse, dominant/recessive in human
Ghrhr/GHRHR	Ghrhr/GHRHR	Reduced somatotrophs with anterior pituitary hypoplasia	GH deficiency with anterior pituitary hypoplasia	Recessive
Gh-1/GH-1	Growth hormone (GH)		GH deficiency	Recessive, dominant or X-linked in human

SOD: septo-optic dysplasia; CPHD: combined pituitary hormone deficiency; IGHD: isolated growth hormone deficiency; EPP: ectopic posterior pituitary; TSH: thyroid-stimulating hormone; ACTH: adrenocorticotrophic hormone.

regulating anterior pituitary development, this list is likely to expand considerably in the future.

Onset of GH deficiency in children with gene deletion is extremely early, and poor growth can be detected as early as the sixth month of postnatal life. Pituitary aplasia is commonly associated with multiple pituitary hormone deficiency. Growth failure occurs early, although the effects of hypoglycemia, hypothyroidism and the consequent persistent conjugated hyperbilirubinemia bring the disorder to the attention of the clinician at an earlier stage. Failure of the migration of the anterior pituitary, derived embryologically from Rathke's pouch, leads to characteristic high-resolution computed tomographic (CT) scan or magnetic resonance image (MRI) appearances in children with this disorder. The pituitary fossa is very small and the neurohypophysis does not descend, remaining at the base of the infundibulum. This may be evident on CT and MRI as a small enhancing nodule.

Hypoplastic pituitary glands are often seen on both CT and MRI in children with GH secretory abnormalities.

The clinical and auxological features of GH deficiency are also manifest in a situation where abnormal polymers of GH are secreted from the pituitary gland. Such material is bioinactive, but measurable by immunoassay. The phenotype features of absent GH with severe growth failure are also seen in individuals who have mutations of the GH receptor causing GH insensitivity syndromes. Individuals with this disorder, which is inherited in an autosomal recessive pattern, have high plasma levels of GH but low levels of IGF-1.

Acquired causes of GH deficiency are outlined in Table 17.4. Acquired destruction of the anterior pituitary gland is most often associated with the presence of

Table 17.4 Causes of growth hormone deficiency

Congenital	Acquired
Genetic	*Trauma*
See Table 17.3	Perinatal trauma
Associated with structural defects of the brain	Postnatal trauma
Agenesis of the corpus callosum	*Infection*
Septo-optic dysplasia	Meningitis/encephalitis
Holoprosencephaly	*CNS tumors*
Encephalocele	Craniopharyngioma
Hydrocephalus	Pituitary germinoma
Associated with midline facial defects	Histiocytosis
Cleft lip/palate	*Postcranial irradiation*
Single central incisor	*Postchemotherapy*
Idiopathic	*Pituitary infarction*
	Neurosecretory dysfunction
	Transient
	Peripubertal
	Psychosocial deprivation
	Hypothyroidism

a craniopharyngioma. Visual disturbances or headaches may be the first signs. Infiltration of the pituitary gland in histiocytosis X may result in GH deficiency, as may transection of the pituitary stalk as a result of severe head injury.

Hypothalamus and Disturbances of Pulsatile Growth Hormone Secretion

GH secretion is characterized by secretory bursts, which raise the serum concentration from extremely low concentrations to a peak value. Secretion then ceases and plasma values return to undetectable values. This process is regulated, as already discussed, by GHRH and SS with a complex set of neurotransmitter pathways; in addition, various peripheral metabolic and hormonal factors influence GH secretion.

The vast majority of short children formerly labeled as GH deficient appear to have GH *insufficiency* on the basis of 24 hour GH secretory profiles. Disorders in secretion of GHRH would explain most of the GH pulse amplitude problems observed.

Growth Hormone Secretion and Growth Hormone Insufficiency

GH secretion is low during many hours of the day. As a result, provocative tests were designed to test the integrity of the hypothalamopituitary axis. The standard test, the insulin-induced hypoglycemia stimulation test [insulin tolerance test (ITT)], has been the mainstay of pharmacological assessment of GH secretion. Clonidine, L-dopa, glucagon and arginine have also been used. Table 17.5 depicts some of the more common tests used in

Table 17.5 Tests of growth hormone secretion

Stimulus	Dose	Sampling protocol (min)	Notes
Levodopa	<15 kg: 125 mg; 15–30 kg: 250 mg; >30 kg: 500 mg	Every 15 min for 90 min	Nausea
Clonidine	0.15 mg/m^2	Every 30 min for 90 min	Tiredness; postural hypotension
Arginine HCl	0.5 g/kg (max. 30 g) i.v., given as 10% arginine HCl in 0.9% NaCl over 30 min	Every 15 min for 90 min	May cause insulin release
Insulin	0.05–0.1 U/kg i.v.	Every 15 min for 120 min	Hypoglycemia; requires supervision. Can also measure cortisol reliably
Glucagon	0.1 mg/kg i.m. (max. 1 mg)	Every 30 min for 180 min	Nausea
GHRH	1 μg/kg i.v.	Every 15 min for 120 min	Flushing. Only assesses pituitary reserve, not whole hypothalamopituitary axis

the assessment of GH secretion. The accepted criteria in the UK were to label a child *severely insufficient* if the peak GH response to pharmacological stimuli was below 5 ng/ml and *partially insufficient* if the value lay between 5 and 7 ng/ml.

These values were based largely on the GH responses to pharmacological stimuli in adults. These cut-off values present many problems of interpretation: for example, a peak GH response of 9 ng/ml in a 5-year-old child may be acceptable as a normal GH response, but such a value in a child at the height of the pubertal growth spurt is probably inadequate. Using strict cut-off values may seem to be inappropriate in view of the evidence that the peak serum GH concentration response to pharmacological stimuli is continuous in a large group of short children with varying growth rates. The wide range of methods now available for measuring serum GH concentrations makes comparison between laboratories difficult and the use of universal cut-off values impossible.

One of the main problems in pediatrics in interpreting the response to these tests is the lack of normal data as standards would be needed for tall, normal and short children, because their GH secretion differs. If age is an important determination of GH secretion, and puberty clearly is, then values for these would have to be included as well.

In general, there is little to choose between the tests when compared with the ITT. Such an analysis assumes that the ITT obtains a correct diagnosis in all cases. The cut-off chosen for the definition of GH deficiency/insufficiency depends on the sensitivity and specificity of the test. It will also depend, in part, on the type of assay used.

Although great emphasis has been placed on the physiological tests of GH secretion, neither sleep nor 24 hour profile studies are easy to perform. Urinary GH measurements may be a step forward in this situation but variability is high and two or three overnight collections need to be performed to overcome this problem. An alternative approach might be to measure the serum concentrations of IGF and/or their binding proteins. The advantages of a single blood test are obvious. Serum IGF-1 concentrations are low in GH deficiency but, because normal IGF-1 values are also low in early childhood, poor discrimination ensues. IGF-1 binding protein 3 (IGFBP-3) is largely regulated by the circulating GH concentration and could be used to reflect GH status. These observations, coupled with recent field studies, suggest that these measures may be overestimated in terms of specificity and sensitivity, particularly following cranial irradiation.

It is unlikely that any tests will improve on the 80–90% sensitivity and specificity reported. The reasons for the false-negative results are easier to understand than false positives. Height velocity of necessity is measured over a long period, between 6 and 12 months. The pharmacological and even the physiological tests are performed over a very short period, a day at the most. This must mean, in view of our knowledge of seasonal growth, that some stimulatory tests are performed in the period of relatively good growth but of necessity would have to be compared with the overall growth period of 6 months to 1 year. Explaining the false-positive results is more difficult. It is a well-recognized phenomenon that when a stimulation test is performed and the 0 minute specimen is high, any further elevation in response to the stimulus is unlikely to occur during the period of the test. This has widely been ascribed to the effects of stress resulting from the insertion of the intravenous cannula. A more likely explanation is that the stimulation test has been performed upon a background of endogenous GH secretory activity. The ITT was devised for performance in adult patients in whom GH secretory episodes are unlikely to occur during the morning. This is not the case in children, where GH secretory episodes between 09.00 and 12.00 hours are commonplace, particularly if the individual is in puberty. GH secretory bursts probably take place as a result of a withdrawal of SS inhibitory tone with or without concomitant stimulation of GHRH. It is quite likely, therefore, that endogenous secretory bursts in some individuals are followed by an increase in SS tone preventing further release of GH.

17.7 CONCLUSION

The human growth process can be broken down into at least three distinct phases of growth: infancy, childhood and puberty. There are probably further subdivisions that could be made but these three phases, at least, are regulated by different aspects of the endocrine system. Very little is known about antenatal growth and growth in the first year of life. The information available, however, strongly suggests that growth in utero and probably in the first 6 months of life is largely GH independent. The precise factors that determine this growth process are still unclear. The nutritional status of the individual is clearly an important determinant but the precise factors that translate nutrient input into growth remain to be defined. Transgenic technology coupled with knockout studies strongly suggests that the IGF axis plays an important role in this process. During the first year of life there is a gradual switch from this nutrition-dependent growth process to GH dependency. Full dependence on GH for the growth process appears to be attained towards the second year of life and thereafter the majority of childhood growth can be explained in terms of the amount of GH secreted by the individual. GH appears to be the final common pathway for integrating the effects of a number of growth signals, and in pathophysiological situations where growth is affected abnormality in the GH axis can be expected. The pubertal growth spurt is made up by a contribution of sex steroids coupled with GH. The most important component appears to be estrogen in females and the aromatization of testosterone to estrogen in males.

A practical approach to the diagnosis of a child with GH deficiency is grounded on clinical assessment with allocation of pretest probability of disease presence. In the prepubertal child with abnormal growth, serum concentrations of IGF-1 and IGFBP-3 provide a means for excluding a diagnosis of GH deficiency. Provocative GH testing will provide information on GH secretory capability. All data need to be interpreted together with known test performances and integrated with the pretest probability to generate a post-test probability, which would then lead to a decision as to whether intervention is required.

FURTHER READING

Karlberg J. On the modelling of human growth. *Stat Med* 1987;**6**:185–92.

Ulijaszek SJ, Johnston FE, Preece MA. *The Cambridge encyclopedia of human growth and development*. 1st ed. Cambridge: Cambridge University Press; 1998. p. 182–4.

Robinson ICAF, Hindmarsh PC. The growth hormone secretory pattern and statural growth. In: Kostyo JL, editor. *Handbook of physiology. Section 7: The endocrine system. vol. 5. Hormonal control of growth*. New York: Oxford University Press; 1999. p. 329–95.

Brook CGD, Clayton P, Brown R, editors. *Brook's clinical paediatric endocrinology*. 5th ed. Oxford: Blackwell; 2009.

Dattani MT, Hindmarsh PC. Growth hormone deficiency in children. In: De Groot LJ, Jameson JL, editors. *Endocrinology*. 6th ed. Philadelphia, PA: Saunders Elsevier; 2010.

CHAPTER 18

Body Composition During Growth and Development

Babette S. Zemel

Perelman School of Medicine, University of Pennsylvania, Division of Gastroenterology, Hepatology and Nutrition, The Children's Hospital of Philadelphia, Philadelphia, PA 19104, USA

Contents

N. Cameron & B. Bogin (eds): Human Growth and Development, Second edition.
ISBN 978-0-12-383882-7, Doi: 10.1016/B978-0-12-383882-7.00018-0

18.1 INTRODUCTION

The body is comprised of water, lipids, protein and minerals. The absolute amounts and relative proportions of these compounds change throughout the life cycle. Growth, maturation and aging, as well as other factors such as disease, nutrition and behavior, alter the chemical composition of the body. This chapter will review methods of assessing body composition, the changes in body composition associated with growth and maturation, the role of body composition in determining nutritional needs, and factors that influence body composition, such as diet, heredity, environment, behavior and disease.

18.2 BASIC CONCEPTS

We will take a look at some of the basic concepts of growth and development.

18.2.1 Chemical Maturation and the Life Cycle

The proliferation, differentiation, expansion and replacement of cells from conception onwards result in changes in the relative and absolute amounts of chemical compounds in the body. For example, at birth, the brain and other organs comprise a large proportion of both lean and total body mass. As the infant grows, the skeletal muscle compartment expands. Although the brain and other organ tissues continue to grow, they gradually come to represent proportionately less of the lean and total body mass. Similarly, at birth, many bones are present as ossification centers and are gradually filled in with bone matrix of hydroxyapatite. During pregnancy, lactation and senescence, bone mass fluctuates and declines. Thus, the calcium content of the body shifts as the body matures and ages. These examples illustrate the process of chemical maturation and how the composition of the body changes during the life cycle.

18.2.2 Body Composition Models

The body is formed of basic elements such as carbon, oxygen and hydrogen, which combine into molecules (e.g. water, lipids, protein), tissue compartments (e.g. fat, muscle, bone) and the whole organism.[1] These levels of biological complexity provide a conceptual framework for addressing body composition questions and selecting appropriate methods. For example, to study changes in bone mass during growth and development, an anthropometric measure such as height provides a general measure of the amount of bone for the *whole organism*. Measurement of bone mineral content by a technique such as dual-energy X-ray absorptiometry (DXA) would provide an excellent measure of the size of the *tissue compartment*. Since the bone is the primary reservoir for calcium, total body calcium (*elemental compartment*) can also be estimated from total body bone mineral content based on the chemical composition of hydroxyapatite (calcium $= 0.34 \times$ bone mineral content).[2]

This conceptual framework for body composition using levels of biological hierarchy is particularly useful in understanding the assumptions that underlie body composition methods and for defining the limitations of current knowledge of body composition in the life cycle. Infants and children present special challenges in measuring various compartments safely, accurately and reliably in vivo.

Many of the methods used to measure body composition are described below. The most commonly used methods partition the body compartments into two compartments (fat-free mass or fat mass). Some newer methods use three (lean body mass, fat mass or bone mass) or four (water, protein, fat, mineral) compartments. Although fat-free mass and lean body mass often are used interchangeably, fat-free mass consists of body weight minus the ether-extractable lipid fraction of the body (fat mass), whereas lean body mass also contains a small amount of essential lipid (2—3%). Bone mineral is part of the fat-free mass compartment. DXA techniques of bone densitometry (described below) allow for separate assessment of bone mineral mass from lean and fat tissue. Advanced imaging techniques such as computed tomography (CT) and magnetic resonance imaging (MRI) offer new possibilities for understanding, for example, the size and composition of organ and tissue compartments.

18.2.3 Nutrition, Adaptation and Functional Outcomes

The chemical changes in the body during growth depend on the availability of nutrient substrate. Essential nutrients in adequate quantities are required to assure that genetically programmed cell growth, proliferation and differentiation proceed unhindered. Furthermore, the relationship between body composition and nutrition is synergistic in that nutrient requirements are determined, in part, by the composition of the body. For example, lean body mass (mainly the smooth and skeletal muscle compartments) is the most metabolically active part of the body and is the primary determinant of energy requirements in normal individuals. With a deficit in energy intake, both fat and lean body mass are used as fuel. The resulting weight loss is composed of losses in both these tissue compartments, thereby altering the energy required for weight maintenance.

Body composition is also sensitive to behavioral patterns in a similarly synergistic manner. For example, physical activity promotes development and growth of muscle and bones, and prolonged inactivity results in muscle wasting and bone loss. Likewise, physical activity and endurance may be limited by inadequate muscle development. Thus, body composition both reflects and contributes to human adaptation to lifestyles, activity and work patterns, and the social and physical environment.

18.2.4 Tempo of Growth

As a child ages and sexual maturity approaches, the hormonal changes associated with puberty produce rapid changes in body composition. Body composition, especially in late childhood and adolescence, is regulated by a "biological clock" rather than chronological time. Children of similar age may be very different in terms of their physical maturation. Similarly, children of the same body size (stature or weight) may be of different ages or stages of maturation due to the biological clock or "tempo of growth". Variability in body composition, between and within populations, may be mediated by differences in the tempo of growth.

18.2.5 Fatness Versus Fat Patterning

Adipose tissue is an important body component for survival. It serves as a reservoir for energy during periods of nutritional deprivation, and it insulates the body from the environment to maintain thermal homeostasis. Excess adipose tissue, or obesity, is associated with a cascade of physiological abnormalities that can threaten health and well-being. A further consideration is the distribution of fat on the body, or fat patterning. A centralized fat distribution, one which has a greater proportion of fat on the trunk of the body compared to the extremities, is associated with metabolic abnormalities and therefore represents a risk of health complications independent of the actual amount of excess fat. Typically, fat patterns are characterized as "android", with a greater amount of fat on the trunk and less on the extremities, versus "gynoid", with greater amounts of fat on the hips and extremities than on the trunk of the body.

18.2.6 Heredity Versus Environment

Body composition has a strong heritable component, although individual genes have not been identified. Measures of body size, such as height, weight and body mass index (BMI) have a strong genetic component. For example, over 40 loci have been identified in genome-wide association studies of adult stature,[3] and genes for BMI, adiposity and fat patterning have also been identified.[4–6] Familial aggregation studies also show significant heritability of fat, fat-free mass and fat patterning.[7] There are also ethnic differences in fatness and fat patterning,[8,9] body density[10] and bone mineral density. Despite these findings there is a large margin for environmental influences. The role of the environment in the development of excess adiposity is evidenced by the increasing rate of obesity in the USA[11] and in some developing countries.[12] Studies of monozygotic and dizygotic twins have been useful in characterizing the gene—environment interactions; in particular, the similar responses of twins in studies of overfeeding and negative energy balance (weight loss) show that there is a genetic basis to the way in which the body responds to changing environments.[7]

18.3 METHODS

The techniques used to assess body composition are an important part of the field itself, because of the shortcomings and limitations of all methods. With the exception of cadaveric studies, nearly all other body composition methods are indirect and involve assumptions that may introduce a bias in the results. Therefore, each method needs to be evaluated in the context of the quality of information obtained, the level of expense and risk involved, and the biological issue of interest for any given research study or clinical investigation.

18.3.1 Anthropometry

With measurements obtained by a well-trained anthropometrist, anthropometry under many circumstances can be a highly suitable method for body composition assessment in population-based studies or a screening tool for disease risk. The tools are moderately simple, precise, portable and inexpensive, and the anthropometric examination is rapid and non-invasive. The tools used for anthropometric evaluations include scales, stadiometers, anthropometers, tape measures and skinfold calipers. Weight and length or stature, the most basic information used to assess growth and nutritional status, are also used to form indices that provide an approximate representation of body composition. The most commonly used index is the BMI, calculated as weight (kg)/stature (m)^2. In infants, weight for length is commonly used. BMI is useful as a screening tool for both excess adiposity and malnutrition, although it has several drawbacks. During adolescence, BMI is influenced by the timing of puberty and may poorly represent adiposity. In addition, a high BMI may be due to high lean body mass and normal adiposity. To screen for overweight and obesity, BMI growth charts are available from the US Centers for Disease Control and Prevention,[13] the International Obesity Task Force,[14] and from the WHO Multicenter Growth Study[15] for infants and young children.

Upper arm anthropometry is also widely used as an indicator of the composition of the whole body. Mid-upper arm circumference and triceps skinfold thickness measures are used to compute the total area, fat area, muscle area and muscle circumference of the upper arm (Table 18.1). At the population level, these measures correlate well with whole-body measures of body fatness and muscularity even though they are measured at a single site. In addition, there are reference data available for these derived measures,[16] so that it is possible to assign a percentile rank or standard deviation score to an individual's measure which indicates whether that person is relatively muscular or fat in comparison to same age-and-sex peers.

Table 18.1 Formulae for computation of upper arm indicators of body composition
Upper arm muscle circumference (mm) $= C - \pi T$
Upper arm area (A) (mm^2) $= (\pi/4)(C/\pi)^2$
Upper arm muscle area (M) (mm^2) $= ([C - \pi T]^2)/4\pi$
Upper arm fat area (F) (mm^2) $= A - M$

C: upper arm circumference; T: triceps skinfold.
Note: check the units; convert arm circumference to millimeters by multiplying by 10.
Source: From Frisancho.[16]

Waist circumference is a simple anthropometric measure that may be a good measure of both overall body fat and fat distribution, which may have implications for metabolic complications such as hypertriglyceridemia.[17] However, the measurement technique for waist circumference is poorly standardized, making it difficult to compare across studies and establish reference ranges.[18] At least one study demonstrated that the waist circumference measurement techniques vary in their association with metabolic syndrome in adolescents.[19]

Anthropometric measures can also be used to estimate whole-body fat-free mass, fat mass and percentage body fat. This technique is based on prediction equations established from comparisons of skinfold measures with a criterion method such as hydro-densitometry (see below). These models assume that the prediction equations are generalizable from the samples from which they were derived, and that body density is the same across age and sex groups. Despite these assumptions, the fat-free mass and fat mass estimates correlate well with independently derived estimates such as DXA.[20]

Table 18.2 Equations for predicting body composition from anthropometry

Two-skinfold method for prediction of percentage body fat★

Prepubescent white males:	% Body fat $= 1.21\,(T + S) - 0.008\,(T + S)^2 - 1.7$
Prepubescent black males:	% Body fat $= 1.21\,(T + S) - 0.008\,(T + S)^2 - 3.2$
Pubescent white males:	% Body fat $= 1.21\,(T + S) - 0.008\,(T + S)^2 - 3.4$
Pubescent black males:	% Body fat $= 1.21\,(T + S) - 0.008\,(T + S)^2 - 5.2$
Postpubescent white males:	% Body fat $= 1.21\,(T + S) - 0.008\,(T + S)^2 - 5.5$
Postpubescent black males:	% Body fat $= 1.21\,(T + S) - 0.008\,(T + S)^2 - 6.8$
All females:	% Body fat $= 1.33\,(T + S) - 0.013\,(T + S)^2 - 2.5$
When sum of triceps and subscapular skinfolds is >35 mm, use:	
All males:	$0.783\,(T + S) + 1.6$
All females:	$0.546\,(T + S) + 9.7$

Four-skinfold method for prediction of percentage body fat★★

Prepubertal children (1−11 years)	
Males:	Density $= 1.1690 - 0.0788$ log sum of 4 skinfolds
Females:	Density $= 1.2063 - 0.0999$ log sum of 4 skinfolds
Adolescent children (12−16 years)	
Males:	Density $= 1.1533 - 0.0643$ log sum of 4 skinfolds
Females:	Density $= 1.1369 - 0.0598$ log sum of 4 skinfolds
Percentage body fat $= [(4.95/\text{body density}) - 4.5]\,100$	

★ T: triceps; S: subscapular; from Slaughter et al.[21]
★★ Sum of four skinfolds = (triceps + biceps + subscapular + suprailiac); from Brook[22] and Durnin and Rahaman.[23]
Source: Adapted from Zemel et al.[24]

Table 18.2 provides sets of prediction equations that illustrate age, gender and ethnicity specific equations. Anthropometric measures are also used to derive indicators of fat patterning, such as the waist—hip ratio (using waist and hip circumference) or the centripetal fat ratio, defined as subscapular skinfold/(triceps + subscapular skinfold), using the triceps and subscapular skinfold measures.

Body breadth measures, such as biacromial, biiliac, elbow and wrist diameters, can be informative as part of the anthropometric description of body composition,[25] although they have not been used extensively. Because they quantify frame size, they correlate well with measures such as bone mineral density, or can be used to distinguish between large- and small-framed individuals when BMI is being used to characterize adiposity.

18.3.2 Densitometric Methods

Densitometric methods utilize the principle that body density can be determined as body mass divided by volume. Body density is then used to estimate fat-free mass, fat mass and percentage body fat using conversion formulae. The method is based on several assumptions, including the assumption that the densities of the major tissue compartments (density of fat $= 0.900 \text{ g/cm}^3$ and fat-free mass $= 1.100 \text{ g/cm}^3$) are relatively constant across individuals. However, these constants vary with growth, maturation, illness, degree of obesity and aging. The Siri and Brozek formulae (Table 18.3) are the most widely used conversion formulae in adults. Lohman[28] and more recently Wells[29] have published age- and sex-specific constants for children to be used in equations similar to that of Siri, which take into account the chemical immaturity of the growing child. In children and adolescents, the chemical composition of the body changes, particularly with respect to the decreasing water and increasing mineral content of fat-free mass. For example, the density of fat-free mass in 8-year-old boys is 1.0877 g/cm^3 and for girls it is $1.0900 \text{ g/cm}^{3,29}$ as opposed to the value for adults of 1.100 g/cm^3.

Hydrodensitometry, or underwater weighing, was at one time the most readily available criterion method for assessment of body composition (fat-free mass and fat mass). It has been used mainly in adults and adolescents, and can be used in children (≥ 8 years) who are healthy, ambulatory and have normal cognitive status. Body volume is determined from measurement of body mass in air and while immersed in water using Archimedes' principle. According to Archimedes' principle, the apparent weight of an object immersed in water, relative to its weight in air, is decreased by an amount equal to the weight of the displaced water. One milliliter of water has a mass almost exactly equal

Table 18.3 Prediction of body fat using body density measurements

Siri, 1956[26]	% Body fat $= (4.95/D_b - 4.50) \times 100$
Brozek, 1963[27]	% Body fat $= (4.570/D_b - 4.142) \times 100$
	$D_b = $ body density

to one gram. Therefore, the difference between the mass in air and the mass under water (in grams) is equivalent to the volume (in milliliters) of the object. The density is then calculated as mass divided by volume. Corrections are needed for the volume of air in the lungs and intestines, and for the density of air and water.

Air-displacement plethysmography is similar to hydrodensitometry in using mass and volume to measure body density. This method uses the displacement of air to estimate body volume. Figure 18.1 shows a Bod Pod® (Life Measurement Instruments, Concord, CA, USA) body composition analyzer, containing a two-compartment chamber of known size. Using a pulsating diaphragm between the two chambers to vary the pressure, the displacement of air when a subject is seated in the outer chamber is measured. A breathing apparatus is built into the device to estimate lung volume for a more accurate estimate of body density. Once body density is determined, the calculations are similar to those for hydrodensitometry. A similar device, the PeaPod, is used to determine body composition in infants.

One of the major sources of bias in the densitometric methods involves the assumptions about the water and mineral content of the fat-free mass. Multicompartment approaches that include other measures such as total body water (TBW) to measure

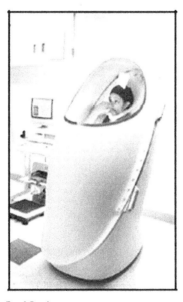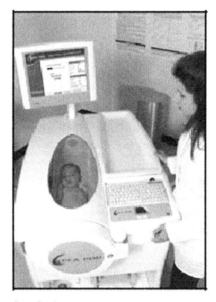

Bod Pod   Pea Pod

Figure 18.1 Air-displacement plethysmography (ADP). ADP measures body density through measurement of body mass and volume. The Bod Pod is a commercially available device for use in children and adults. It requires the subject to be minimally clothed in spandex shorts or bathing suit, with a spandex hair covering, to minimize air trapping around the body. The Pea Pod is designed for infants, who are placed in the chamber without clothing or diapers. The measurement time is approximately 3–5 minutes.

the water content of the fat-free mass, and dual-energy absorptiometry to measure bone mineral content, greatly improve the accuracy of body composition estimates, especially in growing children.[30]

18.3.3 Isotope Dilution Methods

Stable isotopes are used to estimate the size of various compartments of the body using the classic dilution principle. Provided proper sampling, dosing and storage procedures are followed, this method is very accurate[31] and measurement error is mainly related to the laboratory analysis of the isotopic concentrations. The stable isotopes, deuterium oxide (^2H_2O) or oxygen-18 (^{18}O), are naturally occurring isotopes. They are a safe, effective and non-invasive means of measuring the size of the TBW pool in infants and children. These isotopes are naturally occurring so they are already present in the body, and a baseline body fluid sample (such as blood, urine or saliva) must be obtained. Then, a small but concentrated dose of isotope is administered orally which elevates the concentration of the isotopes in the body above that observed from drinking water. After an equilibration period during which the isotope mixes with the total body water pool (usually about 4 hours), further sample collections are acquired. Samples are analyzed by mass spectrometry. The rise in isotopic concentrations from baseline to the postdose equilibrium is proportional to the total amount of water in the body. Owing to mixing of the isotopes with non-aqueous fractions of the body, ^2H_2O overestimates TBW by about 4%, and ^{18}O overestimates TBW by about 1%. Fat-free mass is derived from the TBW measurement using hydration factors[29] which estimate the fraction of the TBW in fat-free mass. Once fat-free mass is estimated, fat mass and percentage body fat can be derived (fat mass = body weight − fat-free mass; percentage body fat = fat mass/body weight × 100).

The compartmentalization of water in the body can also be determined using the isotope dilution method. Bromide and isotopic chloride dilution are used to estimate the extracellular water compartment so that the distribution of the TBW pool into the intracellular and extracellular water compartments can be determined.

18.3.4 Bioelectrical Methods

Several body composition techniques have been developed based on the electrical properties of water and electrolytes in the body. Total body electrical conductivity (TOBEC) devices for infants and adults provide accurate, rapid, non-invasive estimates of fat-free mass, fat mass and percentage body fat. The TOBEC device consists of a low-energy electromagnetic coil through which the body passes on a gantry table. Disturbances in the measured conductance as the body traverses the coil, caused by the water and electrolytes in the body, are measured and the signal is converted to body composition estimates by computerized prediction equations developed for this methodology.

The adult device can be used with children as young as 4 years of age; however, its accuracy in children under the age of 8 is uncertain. The infant TOBEC is suitable for children up to 1 year of age (see Figure 18.2). TOBEC devices are no longer commercially available.

Bioelectrical impedance analyzers (BIAs) are another class of devices that utilizes the electrical properties of water and electrolytes in the body. BIAs measure the impedance of a low-energy electrical signal as it passes through the body. The body fluid compartment, rich in electrolytes, has the least impedance to the flow of an electrical signal, whereas the lipid and bone compartments have greater impedance. Original devices used source and detector electrodes placed on the hand and foot to measure impedance of the entire body, or at other locations to determine impedance of body segments. Some newer BIA models measure foot-to-foot impedance and appear similar to bathroom scales with metal foot pads for bare feet. Others use hand-to-hand impedance or a combination of hands and feet. In all models, there are assumptions about the shape and distribution of the tissues being measured and calibration equations are needed to convert the resistance signal to estimates of body composition. Resistance, one of the key components of impedance, is disproportionately affected by the resistance in the limbs; arms and legs contribute close to

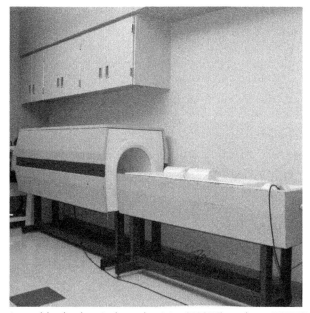

Figure 18.2 Pediatric total body electrical conductivity (TOBEC) analyzer. TOBEC uses a low-energy electromagnetic coil with a gantry table on which the subject moves in and out of the electromagnetic field. The human body contains water and electrolytes that create a disturbance in the field that is proportional to fat-free mass and total body water. Shown is the pediatric TOBEC for body composition assessment in infants.

50% of whole-body resistance in some BIA systems, whereas the trunk contributes 5–12% to whole-body resistance measures.[32] These findings underscore the potential bias that can result from shifts in fluid distribution in the body.

Prediction equations for BIA devices usually require measurement of height and weight. Care should be taken to use the prediction equations devised for children and for the appropriate ethnic group, as ethnic-specific equations are required.[33] Differences by disease group, such as cystic fibrosis, also occur.[34]

Multifrequency BIA and bioelectrical impedance spectroscopy operate on similar principles. They have the added advantage of using both lower and higher frequencies to estimate the intracellular and extracellular water compartments. At low frequencies (up to 5 kHz), the current is unable to penetrate the cell membrane, so impedance is attributed to the extracellular water compartment. At frequencies of 100–500 kHz, both the intracellular and extracellular water compartments impede current flow, so TBW is estimated. Since TBW and extracellular water are estimated in these two ranges, the intracellular water compartment can be derived. Bioelectrical impedance spectroscopy devices sample across the full range of frequencies and use mathematical models to determine water compartments.[32]

18.3.5 Potassium-40

Potassium is found mostly in the intracellular fluid, and is used to estimate body cell mass. Body cell mass is the fat-free intracellular space and the most metabolically active part of the body.[35] It consists of the intracellular fluids and a smaller proportion of intracellular solids of the organs and muscles, and excludes extracellular fluids and solids (such as bone mineral and collagen). A constant ratio of intracellular fluid to body cell mass is assumed, so measurement of total body potassium can be used to estimate body cell mass (body cell mass = total body K (mmol) × 0.0083). Potassium (^{40}K) is a naturally occurring stable isotope found in human tissue. ^{40}K emits a strong gamma ray which can be counted in a lead-shielded room (^{40}K counter) with a gamma ray detector for determination of the whole-body content of ^{40}K. ^{40}K occurs as a very small percentage of the non-radioactive ^{39}K also present in the body, and total body potassium occurs in the ratio of ^{40}K/0.0118%. Since potassium is within the intracellular space, ^{40}K also can be used in combination with TBW to estimate the intracellular and extracellular fluid compartments of the body.

18.3.6 Absorptiometry Methods

The early absorptiometry methods used dual-photon absorptiometry (DPA) with a radionuclide source and digital detector to determine body composition. Dual-energy X-ray absorptiometry (DXA), using a low energy X-ray source, is now more widely used because of its greater accuracy. DXA measures three body compartments: bone mass, lean body mass and fat mass. Each of these tissues varies in density, and therefore they attenuate the energy beams differently. DXA involves radiation exposure, although the exposure is extremely low (3.5 mR). Whole-body estimates of body composition for infants, children and adolescents can be obtained in less than 5 minutes (Figure 18.3). For subjects with metal implants, who are not able to lie supine or who are unable to complete a measurement without movement, DXA-derived body composition estimates are suspect. However, since variability in bone mineral density is one of the primary sources of error in the estimation of the density of fat-free mass, DXA measurements are more accurate in estimating fat-free mass than techniques that use a two-compartment model such as densitometry. Compared to the bioelectrical and anthropometric methods of body composition assessment described above, it has the added advantage of being independent of sample-based prediction equations.

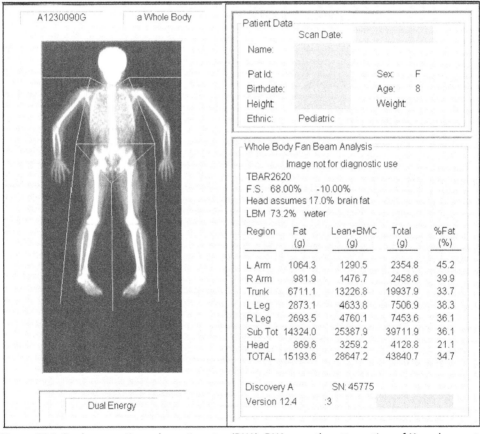

Figure 18.3 Dual-energy X-ray absorptiometry (DXA). DXA uses the attenuation of X-ray beams to assess body composition. Fat, muscle and bone differ in density, so they attenuate the X-ray beams in varying amounts. The image shows the fat, muscle and bone of an 8-year-old female, and the body composition analysis results for each subregion of the body. This figure is reproduced in the color plate section.

Body composition determined by different DXA systems are not interchangeable as there are slight differences in equipment design and results between manufacturers.[36] For children, the accuracy of DXA body composition estimates is an even greater concern because changes in hardware and software by the same DXA manufacturer have had a large effect on body composition estimates.[37–39] Few validation studies of DXA body composition have been performed in children. One study compared percentage body fat by DXA with percentage body fat from the "gold standard" four-compartment model in over 400 healthy children 6–18 years of age.[40] They found that DXA underestimated fatness in subjects with lower percentage body fat and overestimated fatness in subjects with higher percentage body fat. However, there was a strong relationship between the two measures ($R^2 = 0.85$). In a similar study of obese children, DXA overestimated fat mass and underestimated lean mass.[41] Nevertheless, because DXA is widely available, relatively safe and precise,[42] it is often the technique of choice in body composition assessment in children.

18.3.7 Neutron Activation

In vivo neutron activation analysis is a very specialized method for measuring atomic-level components of the body. The major elements are calcium, carbon, chlorine, hydrogen, nitrogen, sodium, oxygen and phosphorus, and trace elements of aluminum, cadmium, copper, iron and silicon are also measured. There is only a handful of research centers worldwide that use neutron activation analysis and it is not an acceptable technique for infants and children. However, it is extremely accurate and the only in vivo method for this kind of body composition assessment. While resting in a shielded chamber, the subject is bombarded with a dose of fast neutrons. The neutrons interact with the nuclei of the element or elements of interest (e.g. carbon or nitrogen), forming unstable isotopes which emit gamma radiation. The whole-body gamma radiation counter is then able to determine the total quantity of the element in the body.[43]

18.3.8 Computed Tomography

CT scans give three-dimensional images for regional analysis of body composition. While whole-body analysis is possible, it is quite impractical owing to the expense, time and radiation exposure involved with the technique. CT systems use an X-ray source and a detector, and the attenuation of the X-rays is used to construct the image of the tissue area. Image production can be generated from "slices" , or spiral images. CT scans have been used effectively to estimate visceral adipose tissue, organ volumes, the area and density of vertebral bodies, and skeletal muscle mass. Peripheral CT devices are also available for imaging the bone (especially cortical versus trabecular bone), muscle and fat at appendicular sites such as the distal radius or tibia.[44] Muscle density can also be determined by peripheral CT devices.[45] Figure 18.4 shows the bone, muscle and fat from a pQCT image obtained at the 66% distal tibia site.

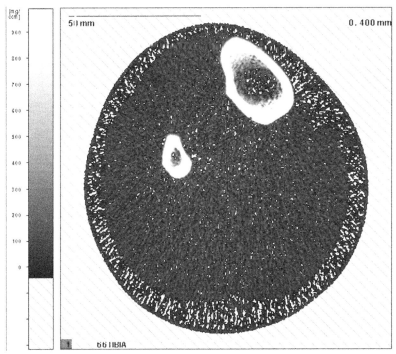

Figure 18.4 Peripheral quantitative computed tomography (pQCT) image of muscle, subcutaneous fat and bone at the 66% distal tibia. pQCT is a table-top device that is able to measure trabecular bone mineral density at the ends of long bones, and cortical density and dimensions in the mid-shaft of long bones. It is also able to quantify the amount of muscle, fat and bone in specific regions. Shown here is the pQCT image taken at 66% of tibia length in the mid-shaft. It is also the site of maximal muscle circumference. This figure is reproduced in the color plate section.

18.3.9 Magnetic Resonance Imaging

MRI generates detailed images of organs and tissues using a powerful magnetic field combined with radiofrequency pulses that excite the hydrogen atoms in tissues. It has wide application in body composition for purposes such as determining the volume of visceral adipose fat, organ volume, skeletal muscle size, quantification of intermuscular adipose tissue[46] and water content of bone.[47] More recently, nuclear magnetic resonance spectroscopy has been used to measure the intramyocellular and intrahepatic lipid fractions.[48]

18.4 BODY COMPOSITION AND GROWTH

18.4.1 Infancy

Water content of the human fetus is high and is about 75% of body weight at birth. Following birth, there are rapid changes in hydration. During the first few days of life, the term infant loses 5—10% of body weight, much of which is water (mainly extracellular). The extracellular water as a percentage of body weight declines from 42.5% on the first day of life to 26% by 10 years of age. Intracellular water increases from 27% to 35% over the same period.[49] The composition of lean tissue is significantly affected. At birth, the hydration of fat-free mass is approximately 80%, and declines to 78% by 3 months of age.[50]

Human infants are born with a large head relative to the size of the total body. At birth, the brain represents 13% of total body weight (compared to 2% of total body weight in adulthood). Other organs (heart, lung, liver, etc.) also comprise a large percentage of infant body mass. Thus, organ tissue makes a greater contribution to body weight and lean body mass during infancy.

Infancy is one of the most rapid periods of growth during the human life cycle. Weight and length increase rapidly and birth weight is usually doubled by 4—5 months of age. Fat and fat-free mass increase, and fat as a percentage of total body weight, peak at about 3—6 months. A comprehensive study of body composition changes in the first year of life showed that by 6 months of age, average percentage body fat for boys was 29.1 ± 4.7, and for girls it was 32.0 ± 4.4%.[51] Boys also had significantly greater fat-free mass, TBW, total body potassium and bone mineral content than girls throughout infancy. Thus, girls and boys differ in body composition, even during infancy. In addition, feeding patterns influenced the changes in the amount and relative proportions of the fat and fat-free mass compartments in these infants; breast-fed infants were not as heavy and had less fat at 9 and 15 months of age compared to formula-fed babies.[52] However, recent data suggest that differences in growth between breast-fed and formula-fed babies do not persist into childhood.[53]

One of the unique features of newborns is the greater amount of brown adipose tissue compared to the amount found in children and adults. It is a highly vascularized and enervated tissue, rich in mitochondria, found in the interscapular, supraclavicular, axillary, neck and suprarenal regions of the body. It is highly metabolically active and functions to maintain body temperature. It develops during gestation and its abundance peaks around the time of birth, and begins to decline during infancy. Brown adipose tissue may be an adaptation to the low amount of skeletal muscle in infants, which reduces their ability to maintain temperature homeostasis through shivering.[54]

18.4.2 Childhood

During childhood growth proceeds at a far slower pace than in infancy and adolescence. Likewise, growth in body compartments and changes in chemical composition also proceed at an unremarkable pace. During this period, sex differences in body fatness are apparent, with girls, on average, having a higher percentage body fat than boys even before the onset of sexual maturation.

The mid-childhood growth spurt, which occurs in many children around the age of 6—8 years, is a small increase in the rate of gain in weight, height and body breadth. At approximately the same age, the body mass "rebound" occurs. BMI peaks near the end of infancy and declines from early childhood (1 year of age), reaching a nadir at about the age of 5—6 years. BMI then begins to increase continuing through adolescence and into adulthood. The "rebound" refers to the turnaround in BMI,[55] the timing of which may be genetically regulated.[56] Children who experience this rebound at an earlier age are more likely to have a higher BMI and become obese, although they are not necessarily obese at the time of rebound.[57] Recent studies suggest that they are also at increased risk of complications such as type 2 diabetes.[58]

18.4.3 Adolescence

The onset of sexual maturation is associated with profound and rapid changes in the body compartments and chemical maturation. These changes are primarily due to the effect of gonadal steroids on the tissues (muscle, fat, bones and organs). Although sex differences in body composition are present during infancy and childhood, they become far more pronounced during adolescence.

In part, the body composition changes are due to the rapid increases in body mass associated with the adolescent growth spurt. Organs such as the heart and brain also increase in size during this period. Fat and lean mass change in absolute amount and relative proportion. Girls gain steadily in fat and fat-free mass through childhood, but there are more rapid gains in these compartments and in percentage body fat associated with puberty. Growth of breast tissue contributes to the gain in overall fat mass and percentage body fat, as does the gradual attainment of a mature female fat distribution with additional fat at the hips and thighs. Many boys experience a prepubertal fat spurt. The subsequent adolescent growth spurt in boys results in significant gains in lean body mass, reductions in fat at the extremities (such as at the triceps skinfold site) and increasing fat deposition on the trunk (such as at the subscapular site). Cross-sectional patterns of changes in fat and lean mass associated with age and puberty are shown in Figures 18.5 and 18.6.

Bone mineralization also changes significantly during adolescence. Approximately 40% of peak bone mass (the maximum amount of bone in the body during one's lifetime) is attained during adolescence.[59,60] During the adolescent growth spurt, the expansion of the skeleton is due to both increasing height and breadth, and the gain in bone mass continues after cessation of height growth. The density of cortical and trabecular bone compartments also increases during puberty.[61,62]

18.4.4 Adulthood and Senescence

Body mass and composition change during adulthood, although generally the changes are not as pronounced as during adolescence or infancy. In most westernized countries, adults continue to gain weight through adulthood. The age-associated increase in BMI is mainly increased fatness. Recent cross-sectional DXA-based body composition results

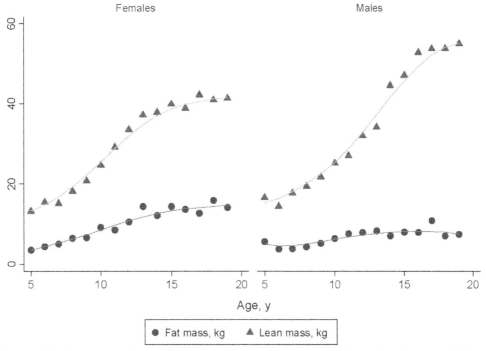

Figure 18.5 Age-related changes in fat and lean mass by dual-energy X-ray absorptiometry (DXA) in males and females. Males and females have distinct age-related patterns of growth in lean and fat mass. Data shown are based on a cross-sectional study of over 800 youth in Philadelphia (the Reference Project on Skeletal Development). Lean mass increases rapidly around the ages when the adolescent growth spurt occurs, but males gain considerably greater amounts of lean mass than females. Fat mass increases with age in females throughout childhood and adolescence. This figure is reproduced in the color plate section.

from the US National Health and Nutrition Examination Survey show that percentage body fat and fat mass index continue to increase through adulthood in males and females.[63] For example, median percentage body fat values for males of European ancestry are 23% at age 20 and 31% at age 70, and for women are 35% at age 20 and 43% at age 70. Median values for lean mass index peak around the age of 40–50 years for both males and females and decline thereafter. Traditional non–westernized societies do not experience similar gains in adult weight or BMI, although with increasing modernization, the age-related increases in weight and BMI mimic westernized countries. Towards the end of life, there is often a loss of body weight, particularly the fat-free mass compartment, which can be due to underlying illness, reduced physical activity, poor nutritional intake and poor nutrient absorption.

In women, changes in reproductive status (pregnancy, lactation and menopause) are also associated with rapid and significant shifts in body composition. Fluctuations in bone

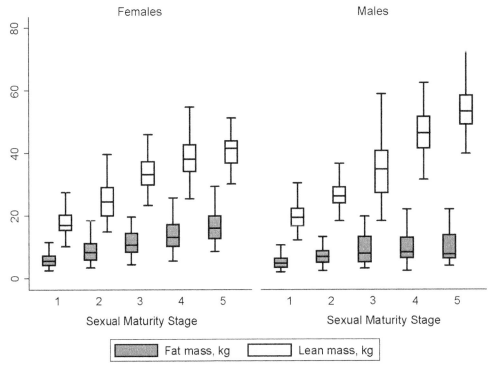

Figure 18.6 Puberty-related changes in fat mass and lean mass by dual-energy X-ray absorptiometry (DXA) in males and females. Fat and lean mass increase significantly as puberty advances. Data shown are based on a cross-sectional study of over 800 youth in Philadelphia (The Reference Project on Skeletal Development). Puberty stage was determined by a self-assessment pictograph and questionnaire that described the five stages of breast (girls) and genital (boys) development.

and fat mass can be particularly pronounced, owing to the effect of hormonal changes on these body compartments. During pregnancy, women gain in TBW (4–6 kg) and fat (2–4 kg), in addition to the gains associated with the fetus, placenta and amniotic fluid. Hydration of fat-free mass increases from 72.5% at 10 weeks' gestation to 74% at term,[64] and water shifts from the intracellular to the extracellular compartment. These changes alter some of the assumptions that underlie standard body composition techniques. During the teenage years, the body composition changes associated with pregnancy can include the combined effects of pregnancy and ongoing growth of the mother, simultaneously. For these very young women, pregnancy may have long-lasting effects on body composition in terms of increased adiposity and reduced bone mass.[65] Loss of bone mass and density are associated with pregnancy and lactation; however, recovery of bone mass appears to be fairly rapid and complete with the onset of postpartum menses and cessation of lactation.[66]

Menopause is another period of body composition change for women. There is an increase in fatness and changes in fat distribution following the mid-life hormonal changes.[67,68] Skeletal changes also occur. In the years just before and after the onset of menopause, there is significant bone loss under natural conditions (i.e. without hormone replacement therapy). One longitudinal study estimated a 10% loss of bone density at the lumbar spine region.[69] The loss of trabecular bone, especially of the spine, can be particularly profound.[70] In premenopausal women, the trabecular bone loss has been estimated at −0.45 mg/ml per year; for perimenopausal women it was −4.39 mg/ml per year and postmenopausal women was −1.99 mg/ml per year. These changes can result in osteoporosis and increased risk of hip fracture, with its associated mortality risk.

18.5 BODY COMPOSITION IN HEALTH AND DISEASE

Body composition is influenced by heredity and the environment. Body composition patterns in childhood can have lifelong consequences in terms of overall health, physical activity patterns and work productivity. Several examples of the importance of body composition are described below.

18.5.1 Chronic Undernutrition, Physical Activity and Work Capacity

Chronic undernutrition results in smaller body size and delayed maturation. Provided the undernutrition is chronic and not acute, weight-for-height relationships, such as BMI, are preserved in children. However, in the long run, chronic undernutrition can result in lower physical activity levels, work productivity, morbidity and mortality.[71] For example, among school-aged boys in Colombia, the group of poorly nourished boys had less spontaneous physical activity than adequately nourished boys.[72] Similarly, among children with sickle cell disease who have reduced fat-free mass and fat mass, total energy used for physical activity was lower than in healthy controls.[73] In studies of populations at risk for chronic undernutrition, there is a significant, positive association between BMI (a measure of current nutritional status) and height (an indicator of nutritional history) and the amount of time devoted to work. In other words, both higher levels of BMI and stature were correlated with an increased capacity to carry out work.[74]

18.5.2 Body Composition and Diet

There is still much to be learned about the relationship between body composition and diet. The essential nutrients required for normal growth and cell functioning assure normal body composition. Clearly, when severe nutrient deficiencies exist, body composition is altered. For example, severe protein malnutrition leads to muscle wasting and altered fluid balance. Milder nutrient deficiencies may have direct or indirect effects. Inadequate calcium intake during growth results in inadequate bone mineral accrual, thereby lowering peak bone mass and increasing the risk of fracture in childhood[32] and later in life.[75] Depletion of total body iron stores from inadequate iron intake resulting in iron-deficiency anemia in children is associated with lethargy and poor cognitive development, which in turn may limit a child's engagement in usual childhood physical activities that would promote muscle and bone growth.[76]

The association between diet patterns and body composition is even less well defined. On average, vegetarians have less body fat then omnivores, and vegans, who exclude all animal products from their diet, are leaner still than both vegetarians and omnivores. Vegetarians and vegans are also known to have lower blood pressure and cholesterol. It is uncertain whether these health effects are directly related to diet or mediated by differences in body composition and lifestyle. At the other extreme, excess adiposity is associated with high energy intake, consumption of energy-dense, highly processed foods, low fruit and vegetable intake, and consumption of sugared beverages.[77] Both lower body fat and improved bone mineral accrual are associated with diets rich in dark green and deep yellow vegetables and low in fried foods.[78]

18.5.3 Health Consequences of Obesity

Among children, obesity is defined as having a BMI greater than the 95th percentile for age and sex, and overweight is defined as having a BMI between the 85th and 95th percentiles for age and sex.[79,80] Excess adiposity is reaching epidemic proportions among children and adults in many industrialized and industrializing nations. Results of the 2007−2008 National Health and Nutrition Examination Survey in the USA showed that 9.5% of infants and toddlers had a weight for length greater than or equal to the 95th percentile, and 16.9% of children aged 2−19 years had a BMI greater than or equal to the 95th percentile. While prevalence differed by age and population ancestry group, the survey showed that the prevalence of obesity has reached a plateau since 1999 except for the very heaviest children (BMI percentile ≥97th percentile).[81] The prevalence of pediatric obesity is increasing in industrializing nations too. For example, a comparison of national survey data for 7−17-year-old children measured in 1995 to 2005 in Shandong Province, China, showed that overweight increased from 8% to 14% in boys, and from 5% to 8% in girls; obesity increased from 3% to 11% in boys and from 2 to 6% in girls.[82] For boys, this was a 75% increase in the prevalence of overweight and a 266% increase in the prevalence of obesity. Of note, children in 2005 were significantly taller and heavier, and the prevalence of anemia was lower, suggesting positive shifts in nutritional status.

The health consequences of overweight and obesity in childhood and adolescence are also beginning to emerge. In children, the health risks associated with obesity include bone and joint disease, increased blood pressure, serum cholesterol and insulin resistance, as well as increased risk of non-insulin-dependent diabetes.[83] For example, in a study of over 4000 Chinese children, 7−12 years of age, the prevalence of elevated blood pressure among children in the 95th percentile for fat mass index was 56% for boys and 51% for girls, compared to 4% and 7% for boys and girls with a fat mass index less than the 5th percentile.[84]

Perhaps the greatest health risk of childhood overweight and obesity is the increased risk of morbidity and mortality later in life.[85] Adult obesity is associated with increased risk of heart disease, hypertension, stroke, coronary heart disease, non-insulin-dependent diabetes, lipid disorders, gallbladder disease, orthopedic problems, sleep apnea and some cancers. A recent systematic review of studies showed that childhood overweight and obesity were associated in adulthood with a significantly elevated risk of premature mortality and increased risk of cardiometabolic morbidity (diabetes, hypertension, ischemic heart disease and stroke). Results for cancer were not consistent. Other health issues related to childhood overweight and obesity were disability, asthma and polycystic ovary syndrome.[86]

18.5.4 Fat Distribution as a Correlate of Diseases

The distribution of fat on the body represents a risk factor for certain diseases that is independent of total body fat. In particular, the tendency to accumulate fat on the upper trunk is associated with non-insulin-dependent diabetes, hypertension and gallbladder disease. Both adiposity and fat patterning were associated with cardiovascular risk factors (lipid profiles and blood pressure) in a large, multiethnic sample of boys and girls participating in the National Heart, Lung, and Blood Institute Growth and Health Study.[87,88] Another study showed that waist circumference, a simple indicator of excess trunk fat, predicts elevated blood pressure, triglycerides and high-density lipoprotein concentrations in addition to BMI.[89] Increased intra-abdominal adipose tissue has been associated with metabolic syndrome in children.[90]

18.5.5 Physical Activity and Body Composition

As noted above, body composition, fatness and fat patterning have a strong hereditary component. However, behavioral and lifestyle factors, most notably physical activity, can influence body composition. Physical activity, especially weight-bearing activity, is important for growth and maintenance of the muscle and bone compartments. The effects of physical activity during growth can have beneficial effects into adulthood.[91] The energy demands of intense physical activity also influence fatness levels. These are illustrated most easily at the extremes of physical activity. Children with severe quadriplegic cerebral palsy who are unable to walk have markedly reduced growth of lower limbs, reduced muscle and fat stores, and low bone mineral content and bone density. Children with diplegic or hemiplegic cerebral palsy have deficits commensurate with their ability to ambulate and bear weight. Even among previously healthy individuals with normal physical activity patterns, prolonged bed rest results in muscle and bone atrophy. Astronauts in space living in a weightless environment experience similar problems.

Milder limitations of physical activity may promote increased fatness. Increased hours of television viewing and reduction in hours of physical activity are associated with increased risk of overweight in children.[92] However, increased body mass in itself increases the weight-bearing stress of usual activities, and is thereby associated with an increase total body lean mass. Therefore, among overweight and obese children, both fat-free mass and fat mass are often increased. At the other extreme, intense physical activity is associated with reduced fatness. Body composition profiles vary with different sports activities. Long-distance running and ballet dancing, and other activities known for prolonged and intense training, are associated with significantly reduced fat mass. In females, these athletes often become amenorrheic and develop osteoporosis related to estrogen insufficiency. Sports that involve resistance training and high-impact, weight-bearing physical activity generally promote higher bone density.

18.6 SUMMARY

The composition of the human body is regulated by genes, but is sensitive to environmental, behavioral and nutritional factors. Interaction between the genetic and non-genetic influences contributes to the variability in body composition observed within and between populations. Body composition is also an integral part of human growth, maturation and senescence, and has a wide range of health implications. There is a broad array of body composition assessment techniques that can be used in clinical, research and field settings to understand further the life-cycle changes in body composition and their role in health and disease.

REFERENCES

1. Wang Z, Pierson RN, Heymsfield SB. The five level model: a new approach to organizing body composition research. *Am J Clin Nutr* 1992;**56**:19—28.
2. Ellis KJ. Human body composition: in vivo methods. *Physiol Rev* 2000;**80**:649—80.
3. Lettre G. Genetic regulation of adult stature. *Curr Opin Pediatr* 2009;**21**:515—22.
4. Zhao J, Bradfield JP, Zhang H, Sleiman PM, Kim CE, Glessner JT, et al. Role of BMI-associated loci identified in GWAS meta-analyses in the context of common childhood obesity in European Americans. *Obesity (Silver Spring)* 2011;**19**:2436—9.
5. Lindgren CM, Heid IM, Randall JC, Lamina C, Steinthorsdottir V, Qi L, et al. Genome-wide association scan meta-analysis identifies three loci influencing adiposity and fat distribution. *PLoS Genet* 2009;**5**:e1000508.
6. Wang K, Li WD, Zhang CK, Wang Z, Glessner JT, Grant SF, et al. A genome-wide association study on obesity and obesity-related traits. *PLoS ONE* 2011;**6**:e18939.
7. Bouchard C. Genetic influences on human body composition and physique. In: Roche A, Heymsfield SB, Lohman TG, editors. *Human body composition*. Champaign, IL: Human Kinetics; 1996. p. 305—27.
8. Malina R. Regional body composition: age, sex, and ethnic variation. In: Roche AF, Heymsfield SB, Lohman TG, editors. *Human body composition*. Champaign, IL: Human Kinetics; 1996. p. 217—55.
9. Malina R, Huang YC, Brown KH. Subcutaneous adipose tissue distribution in adolescent girls of four ethnic groups. *Int J Obes Relat Metab Disord* 1995;**19**:793—7.
10. Schutte J, Townsend EJ, Hugg J, Shoup RF, Malina RM, Blomqvist CG. Density of lean body mass is greater in blacks than in whites. *J Appl Physiol* 1984;**56**:1647—9.

11. Troiano R, Flegal KM, Kuczmarski RJ, Campbell SM, Johnson CL. Overweight prevalence and trends for children and adolescents. The National Health and Nutrition Examination Surveys 1963 to 1991. *Arch Pediatr Adolesc Med* 1995;**149**:1085−91.

12. Martorell R, Kettel Khan L, Hughes ML, Grummer-Strawn LM. Overweight and obesity in preschool children from developing countries. *Int J Obes Relat Metab Disord* 2000;**24**:959−67.

13. Kuczmarski R, Ogden CL, Grummer-Strawn LM. *June 8, CDC growth charts − United States. Advance data from vital and health statistics.* Hyattsville, MD: National Center for Health Statistics; 2000.

14. Cole T, Bellizzi MC, Flegal KM, Dietz WH. Establishing a standard definition for child overweight and obesity worldwide: international survey. *BMJ* 2000;**320**:1240−3.

15. WHO Multicentre Growth Reference Study Group. *WHO child growth standards: growth velocity based on weight, length and head circumference: methods and development.* Geneva: World Health Organization; 2009.

16. Frisancho A. New norms of upper limb fat and muscle areas for assessment of nutritional status. *Am J Clin Nutr* 1981;**34**:2540−5.

17. Esmaillzadeh A, Mirmiran P, Azizi F. Clustering of metabolic abnormalities in adolescents with the hypertriglyceridemic waist phenotype. *Am J Clin Nutr* 2006;**83**:36−46. quiz 183−4.

18. Wang J. Standardization of waist circumference reference data. *Am J Clin Nutr* 2006;**83**:3−4.

19. Johnson ST, Kuk JL, Mackenzie KA, Huang TT, Rosychuk RJ, Ball GD. Metabolic risk varies according to waist circumference measurement site in overweight boys and girls. *J Pediatr* 2010;**156**:247−52. e241.

20. Sentongo TA, Semeao EJ, Piccoli DA, Stallings VA, Zemel BS. Growth, body composition, and nutritional status in children and adolescents with Crohn's disease. *J Pediatr Gastroenterol Nutr* 2000;**31**:33−40.

21. Slaughter MH, Lohman TG, Boileau RA, Horswill CA, Stillman RJ, Van Loan MD, et al. Skinfold equations for estimation of body fatness in children and youth. *Hum Biol* 1988;**60**:709−23.

22. Brook CG. Determination of body composition of children from skinfold measurements. *Arch Dis Child* 1971;**46**:182−4.

23. Durnin J, Rahaman M. The assessment of the amount of fat in the human body from measurements of skinfold thickness. *Br J Nutr* 1967;**21**:681−9.

24. Zemel BS, Riley EM, Stallings VA. Evaluation of methodology for nutritional assessment in children: anthropometry, body composition, and energy expenditure. *Annu Rev Nutr* 1997;**17**:211−35.

25. Frisancho A. *Anthropometric standards for the assessment of growth and nutritional status.* Ann Arbor, MI: University of Michigan Press; 1990.

26. Siri WE, The gross composition of the body. In: Lawrence JH, Tobias CA, editors. *Advances in biological and medical physics*, vol. IV. New York: Academic Press; 1956. p. 239−80.

27. Brozek J, Grande F, Anderson J, Keys A. Densitometric analysis of body composition: revision of some quantitative assumptions. *Ann NY Acad Sci* 1963;**110**:113−40.

28. Lohman T. Assessment of body composition in children. *Pediatr Exerc Sci* 1989;**1**:19−30.

29. Wells JC, Williams JE, Chomtho S, Darch T, Grijalva-Eternod C, Kennedy K, et al. Pediatric reference data for lean tissue properties: density and hydration from age 5 to 20 y. *Am J Clin Nutr* 2010;**91**:610−8.

30. Wells JC, Fuller NJ, Dewit O, Fewtrell MS, Elia M, Cole TJ. Four-component model of body composition in children: density and hydration of fat-free mass and comparison with simpler models. *Am J Clin Nutr* 1999;**69**:904−12.

31. Parker L, Reilly JJ, Slater C, Wells JC, Pitsiladis Y. Validity of six field and laboratory methods for measurement of body composition in boys. *Obes Res* 2003;**11**:852−8.

32. National Institutes of Health Obesity Education Initiative Expert Panel on the Identification. *Evaluation, and Treatment of Overweight and Obesity in Adults. Clinical guidelines on the identification, evaluation, and treatment of overweight and obesity in adults: the evidence report.* Bethesda, MD: National Institutes of Health, US Department of Health and Human Services; 1998.

33. Haroun D, Taylor SJ, Viner RM, Hayward RS, Darch TS, Eaton S, et al. Validation of bioelectrical impedance analysis in adolescents across different ethnic groups. *Obesity (Silver Spring)* 2010;**18**:1252−9.

34. Puiman PJ, Francis P, Buntain H, Wainwright C, Masters B, Davies PS. Total body water in children with cystic fibrosis using bioelectrical impedance. *J Cyst Fibros* 2004;**3**:243–7.

35. Heymsfield S, Wang Z, Baumgartner RN, Ross R. Human body composition: advances in models and methods. *Annu Rev Nutr* 1997;**17**:527–58.

36. Laskey MA. Dual-energy X-ray absorptiometry and body composition. *Nutrition* 1996;**12**:45–51.

37. Ellis KJ, Shypailo RJ. Bone mineral and body composition measurements: cross-calibration of pencil-beam and fan-beam dual-energy X-ray absorptiometers. *J Bone Miner Res* 1998;**13**:1613–8.

38. Shypailo RJ, Butte NF, Ellis KJ. DXA: can it be used as a criterion reference for body fat measurements in children? *Obesity (Silver Spring)* 2008;**16**:457–62.

39. Shypailo RJ, Ellis KJ. Bone assessment in children: comparison of fan-beam DXA analysis. *J Clin Densitom* 2005;**8**:445–53.

40. Sopher AB, Thornton JC, Wang J, Pierson Jr RN, Heymsfield SB, Horlick M. Measurement of percentage of body fat in 411 children and adolescents: a comparison of dual-energy X-ray absorptiometry with a four-compartment model. *Pediatrics* 2004;**113**:1285–90.

41. Wells JC, Haroun D, Williams JE, Wilson C, Darch T, Viner RM, et al. Evaluation of DXA against the four-component model of body composition in obese children and adolescents aged 5–21 years. *Int J Obes (Lond)* 2010;**34**:649–55.

42. Shepherd JA, Wang L, Fan B, Gilsanz V, Kalkwarf HJ, Lappe J, et al. Optimal monitoring time interval between DXA measures in children. *J Bone Miner Res* 2011;**26**:2745–52.

43. Ryde S. In vivo neutron activation analysis: past, present and future. In: Davies P, Cole TJ, editors. *Body composition techniques in health and disease. Society for the Study of Human Biology, Symposium 36.* Cambridge: Cambridge University Press; 1995.

44. Zemel B, Bass S, Binkley T, Ducher G, Macdonald H, McKay H, et al. Peripheral quantitative computed tomography in children and adolescents: the 2007 ISCD pediatric official positions. *J Clin Densitom* 2008;**11**:59–74.

45. Farr JN, Funk JL, Chen Z, Lisse JR, Blew RM, Lee VR, et al. Skeletal muscle fat content is inversely associated with bone strength in young girls. *J Bone Miner Res* 2011;**26**:2217–25.

46. Lee SY, Gallagher D. Assessment methods in human body composition. *Curr Opin Clin Nutr Metab Care* 2008;**11**:566–72.

47. Rad HS, Lam SC, Magland JF, Ong H, Li C, Song HK, et al. Quantifying cortical bone water in vivo by three-dimensional ultra-short echo-time MRI. *NMR Biomed* 2011;**24**:855–64.

48. Shen W, Liu H, Punyanitya M, Chen J, Heymsfield SB. Pediatric obesity phenotyping by magnetic resonance methods. *Curr Opin Clin Nutr Metab Care* 2005;**8**:595–601.

49. Bechard L, Wroe E, Ellis K. Body composition and growth. In: Duggan C, Watkins J, Walker W, editors. *Nutrition in pediatrics.* 4th ed. Hamilton, BC: Decker; 2008. p. 27–39.

50. Olhager E, Flinke E, Hannerstad U, Forsum E. Studies on human body composition during the first 4 months of life using magnetic resonance imaging and isotope dilution. *Pediatr Res* 2003;**54**:906–12.

51. Butte NFHJ, Wong WW, Smith EO, Ellis KJ. Body composition during the first 2 years of life: an updated reference. *Pediatr Res* 2000;**47**:578–85.

52. Butte NFWW, Hopkinson JM, Smith EO, Ellis KJ. Infant feeding mode affects early growth and body composition. *Pediatrics* 2000;**106**:1355–66.

53. Hediger ML, Overpeck MD, Ruan WJ, Troendle JF. Early infant feeding and growth status of US-born infants and children aged 4–71 mo: analyses from the third National Health and Nutrition Examination Survey, 1988–1994. *Am J Clin Nutr* 2000;**72**:159–67.

54. Tews D, Wabitsch M. Renaissance of brown adipose tissue. *Horm Res Paediatr* 2011;**75**:231–9.

55. Rolland-Cachera M, Deheeger M, Bellisle F, Sempe M, Guilloud-Bataille M, Patois E. Adiposity rebound in children: a simple indicator for predicting obesity. *Am J Clin Nutr* 1984;**39**:129–35.

56. Sovio U, Mook-Kanamori DO, Warrington NM, Lawrence R, Briollais L, Palmer CN, et al. Association between common variation at the FTO locus and changes in body mass index from infancy to late childhood: the complex nature of genetic association through growth and development. *PLoS Genet* 2011;**7**:e1001307.

57. Rolland-Cachera MF, Deheeger M, Maillot M, Bellisle F. Early adiposity rebound: causes and consequences for obesity in children and adults. *Int J Obes (Lond)* 2006;**30**(Suppl. 4):S11–7.

58. Eriksson JG. Early growth and coronary heart disease and type 2 diabetes: findings from the Helsinki Birth Cohort Study (HBCS). *Am J Clin Nutr* 2011;**94**:17995–8025.

59. Bailey DA, Martin AD, McKay HA, Whiting S, Mirwald R. Calcium accretion in girls and boys during puberty: a longitudinal analysis. *J Bone Miner Res* 2000;**15**:2245–50.

60. McKay H, Bailey DA, Mirwald RL, Davison KS, Faulkner RA. Peak bone mineral accrual and age at menarche in adolescent girls: a 6-year longitudinal study. *J Pediatr* 1998;**133**:682–7.

61. Leonard MB, Elmi A, Mostoufi-Moab S, Shults J, Burnham JM, Thayu M, et al. Effects of sex, race, and puberty on cortical bone and the functional muscle bone unit in children, adolescents, and young adults. *J Clin Endocrinol Metab* 2010;**95**:1681–9.

62. Kirmani S, Christen D, van Lenthe GH, Fischer PR, Bouxsein ML, McCready LK, et al. Bone structure at the distal radius during adolescent growth. *J Bone Miner Res* 2009;**24**:1033–42.

63. Kelly TL, Wilson KE, Heymsfield SB. Dual energy X-ray absorptiometry body composition reference values from NHANES. *PLoS ONE* 2009;**4**:e7038.

64. Institute of Medicine (US). *Subcommittee on Nutritional Status and Weight Gain during Pregnancy. Institute of Medicine (US). Subcommittee on Dietary Intake and Nutrient Supplements during Pregnancy. Nutrition during pregnancy: Part I, Weight gain; Part II, Nutrient supplements*. Washington, DC: National Academy Press; 1990.

65. Hediger M, Scholl TO, Schall JI. Implications of the Camden Study of adolescent pregnancy: interactions among maternal growth, nutritional status, and body composition. *Ann NY Acad Sci* 1997;**817**:281–91.

66. Kalkwarf HJ. Lactation and maternal bone health. *Adv Exp Med Biol* 2004;**554**:101–14.

67. Toth M, Tchernof A, Sites CK, Poehlman ET. Menopause-related changes in body fat distribution. *Ann NY Acad Sci* 2000;**904**:502–6.

68. Poehlman ET, Tchernof A. Traversing the menopause: changes in energy expenditure and body composition. *Coronary Artery Dis* 1998;**9**:799–803.

69. Recker R, Lappe J, Davies K, Heaney R. Characterization of perimenopausal bone loss: a prospective study. *J Bone Miner Res* 2000;**15**:1965–73.

70. Block JE, Smith R, Gluer CC, Steiger P, Ettinger B, Genant HK. Models of spinal trabecular bone loss as determined by quantitative computed tomography. *J Bone Miner Res* 1989;**4**:249–57.

71. Pelletier DL, Frongillo EA. Changes in child survival are strongly associated with changes in malnutrition in developing countries. *J Nutr* 2003;**133**:107–19.

72. Spurr G, Reina JC. Patterns of daily energy expenditure in normal and marginally undernourished school-aged Colombian children. *Eur J Clin Nutr* 1988;**42**:819–34.

73. Barden EM, Zemel BS, Kawchak DA, Goran MI, Ohene-Frempong K, Stallings VA. Total and resting energy expenditure in children with sickle cell disease. *J Pediatr* 2000;**136**:73–9.

74. Kennedy E, Garcia M. Body mass index and economic productivity. *Eur J Clin Nutr* 1994;**48**:S45–55.

75. Kalkwarf HJ, Khoury JC, Lanphear BP. Milk intake during childhood and adolescence, adult bone density, and osteoporotic fractures in US women. *Am J Clin Nutr* 2003;**77**:257–65.

76. Olney DK, Pollitt E, Kariger PK, Khalfan SS, Ali NS, Tielsch JM, et al. Young Zanzibari children with iron deficiency, iron deficiency anemia, stunting, or malaria have lower motor activity scores and spend less time in locomotion. *J Nutr* 2007;**137**:2756–62.

77. Gillis LJ, Kennedy LC, Gillis AM, Bar-Or O. Relationship between juvenile obesity, dietary energy and fat intake and physical activity. *Int J Obes Relat Metab Disord* 2002;**26**:458–63.

78. Wosje KS, Khoury PR, Claytor RP, Copeland KA, Hornung RW, Daniels SR, et al. Dietary patterns associated with fat and bone mass in young children. *Am J Clin Nutr* 2010;**92**:294–303.

79. Ogden CL, Flegal KM. Changes in terminology for childhood overweight and obesity. *Natl Health Stat Report* 2010;**25**:1–5.

80. Cole TJ, Bellizzi MC, Flegal KM, Dietz WH. Establishing a standard definition for child overweight and obesity worldwide: international survey. *BMJ* 2000;**320**:1240–3.

81. Ogden CL, Carroll MD, Curtin LR, Lamb MM, Flegal KM. Prevalence of high body mass index in US children and adolescents, 2007–2008. *JAMA* 2010;**303**:242–9.

82. Zhang YX, Wang SR. Changes in nutritional status of children and adolescents in Shandong, China from 1995 to 2005. *Ann Hum Biol* 2011;**38**:485–91.

83. Dietz WH, Robinson TN. Assessment and treatment of childhood obesity. *Pediatr Rev* 1993;**14**:337—44.

84. Zhang YX, Wang SR. Relation of body mass index, fat mass index and fat-free mass index to blood pressure in children aged 7—12 in Shandong, China. *Ann Hum Biol* 2011;**38**:313—6.

85. Must A, Jacques PF, Dallal GE, Bajema CJ, Dietz WH. Long-term morbidity and mortality of overweight adolescents. A follow-up of the Harvard Growth Study of 1922 to 1935. *N Engl J Med* 1992;**327**:1350—5.

86. Reilly JJ, Kelly J. Long-term impact of overweight and obesity in childhood and adolescence on morbidity and premature mortality in adulthood: systematic review. *Int J Obes (Lond)* 2011;**35**:891—8.

87. Morrison JA, Sprecher DL, Barton BA, Waclawiw MA, Daniels SR. Overweight, fat patterning, and cardiovascular disease risk factors in black and white girls: the National Heart, Lung, and Blood Institute Growth and Health Study. *J Pediatr* 1999;**135**:458—64.

88. Morrison JA, Barton BA, Biro FM, Daniels SR, Sprecher DL. Overweight, fat patterning, and cardiovascular disease risk factors in black and white boys. *J Pediatr* 1999;**135**:451—7.

89. Lee S, Bacha F, Arslanian SA. Waist circumference, blood pressure, and lipid components of the metabolic syndrome. *J Pediatr* 2006;**149**:809—16.

90. Lee S, Bacha F, Gungor N, Arslanian S. Comparison of different definitions of pediatric metabolic syndrome: relation to abdominal adiposity, insulin resistance, adiponectin, and inflammatory biomarkers. *J Pediatr* 2008;**152**:177—84.

91. Baxter-Jones AD, Kontulainen SA, Faulkner RA, Bailey DA. A longitudinal study of the relationship of physical activity to bone mineral accrual from adolescence to young adulthood. *Bone* 2008;**43**:1101—7.

92. Jackson DM, Djafarian K, Stewart J, Speakman JR. Increased television viewing is associated with elevated body fatness but not with lower total energy expenditure in children. *Am J Clin Nutr* 2009;**89**:1031—6.

SUGGESTED READING

Butte NF, Hopkinson JM, Wong WW, Smith EO, Ellis KJ. Body composition during the first 2 years of life: an updated reference. *Pediatr Res* 2000;**47**:578—85.

Ellis KJ. Human body composition: in vivo methods. *Physiol Rev* 2000;**80**:649—80.

Roche AF, Heymsfield SB, Lohman TG, editors. *Human body composition*. Champaign, IL: Human Kinetics; 1996.

Wang Z, Pierson RN, Heymsfield SB. The five level model: a new approach to organizing body composition research. *Am J Clin Nutr* 1992;**56**:19—28.

Wells J, Williams JE, Chomtho S, Darch T, Grijalva-Eternod C, Kennedy K, et al. Pediatric reference data for lean tissue properties: density and hydration from age 5 to 20 y. *Am J Clin Nutr* 2010;**91**:610—8.

INTERNET RESOURCES

Excellent interactive sites describing body composition methods, http://nutrition.uvm.edu/bodycomp/ http://www.bcm.edu/bodycomplab/mainbodycomp.htm

For a table listing the elemental composition of the human body, http://web2.iadfw.net/uthman/ elements_of_body.html

CHAPTER 19

The Measurement of Human Growth

Noël Cameron

Centre for Global Health and Human Development, School of Sport, Exercise and Health Sciences, Loughborough University, Leicestershire LE11 3TU, UK

Contents

19.1 INTRODUCTION

This chapter deals with measuring growth using anthropometric methods to assess the external dimensions of the body such as height, weight, and measures of subcutaneous fat. Chapter 18 deals with the assessment of body composition in relation to growth, and some

N. Cameron & B. Bogin (eds): Human Growth and Development, Second edition.
ISBN 978-0-12-383882-7, Doi: 10.1016/B978-0-12-383882-7.00019-2

of the methods required to do this, such as measures of subcutaneous fat using skinfolds, will also be covered in this chapter. The anthropometric methods used to assess human growth have been described by this author in a number of publications including in a book, *The Measurement of Human Growth*,[1] and various book chapters[2–4] which may be available in your university library. This chapter will necessarily borrow heavily from those publications to maintain the need for standardization of techniques and consistency of description.

Clearly it is vitally important that all those who assess human growth do so using similar methods so that their data are comparable between individuals in the same study and between samples in different studies. The universal standardization of anthropometric measurement has been a desire of those who measure the human body since the nineteenth century. However, the anthropometry peculiar to human growth that we use today had its recent development in the American longitudinal studies of the first half of the twentieth century. From 1904 to 1948, 17 such studies were started and 11 completed. Their complexity varied from the relatively simple study of height and weight to data yielding correlations between behavior, personality, social background and physical development.[1] Researchers were aware of the need for comparability of measurements and published precise accounts of their methods and techniques, with suitable adaptations for the measurement of growth. The three most important and informative accounts from this period are those of Frank Shuttleworth for the Harvard Growth Study of 1922,[5] Harold Stuart for the Centre for Research in Child Health and Development Study of 1930,[6] and Katherine Simmons in her reports of the Brush Foundation's Studies of 1931.[7] In Britain, few longitudinal growth studies were undertaken before 1949. However, the Harpenden Longitudinal Growth Study, managed by J.M. Tanner and R.H. Whitehouse from 1949 to 1972, became the strongest influence on British studies of human growth and did much to advance the anthropometry applied to human growth (often referred to as "auxological anthropometry"). The team of Tanner and Whitehouse radically altered the approach to auxological anthropometry. Whitehouse, for example, was dissatisfied with the instrumentation available and developed the "Harpenden" range of instruments that eliminated graduated rules for measuring linear distances (e.g. height) and instead used counter mechanisms.[8] These counters were turned by a simple ratchet system and displayed the measurement in millimeters, which reduced reading errors.

The series of longitudinal growth studies coordinated through the International Children's Centre in Paris between 1960 and 1980 had a major effect on standardizing anthropometric measurements and growth study design. Research teams from growth studies in Belgium, Britain, France, Senegal, Sweden, Switzerland, Uganda and the USA met every two years to initially discuss their methods and eventually their results, which resulted directly or indirectly in a bibliography of 948 references.[9] The International Biological Project from 1962 to 1972 brought together scientists from all over the world under the umbrella of research in "human biology" and gave rise to one of the standard texts for research into the human biological sciences. The original *IBP Handbook*,[10] revised and renamed *Practical Human Biology* in its second edition,[11] forms the source for many scientists who wish to use standard techniques to measure growth. Its value lies in its acceptance, by many, as the source, not only for techniques of anthropometric measurement but also for many other techniques that are applied in research in human biology such as dermatoglyphics, physiological measures, salt and water balance, thermal comfort, nutritional status, etc. The development of digital technology has brought with it advantages in anthropometric instrumentation which will make the practice of reading instruments, calling out numbers and recording these numbers by hand virtually obsolete in the future. However, the measurement of human growth is still characterized by an observer using non-digital instruments to measure children, although software developments have made data recording, editing and analysis an automated process and put sophisticated analytical procedures at the fingertips of most research workers.

A final note in this context must concern the problems inherent in measuring growth. The fact that accurate instrumentation has overcome many of the errors inherent in anthropometry does not mean that the greatest source of error is not still the measurer him- or herself. The qualities outlined 60 years ago by the anthropologist Aleš Hrdlička[12] are still relevant today: good eyesight for distance and color, freedom from halitosis and other unpleasant odors, sympathy, perseverance, orderliness, thorough honesty and carefulness. The anthropometrist should be "careful of the sensibilities of his subjects; careful in technique, careful in reading the scale of his subject, careful in recording and capable of concentration on his work".[12] To Hrdlička's essential qualities must be added basic qualifications in the sciences of anatomy, physiology and physics. Also, because of the extremely numerical nature of anthropometry, the anthropometrist must be numerate and have studied some basic statistical methods. So, accurate measurement is not just about the instrument, it is also about the person using the instrument.

19.2 THE CONTEXT OF MEASUREMENT: SCREENING, SURVEILLANCE AND MONITORING

Our knowledge of the process of human growth and development is directly dependent on the methods employed to measure that process and the scientific and/or clinical context within which those methods are employed. The context will usually be one of three types: screening, surveillance or monitoring.

Screening is concerned with the identification of a particular subset of the population with certain prescribed characteristics. Most usually they will be outside (above or below) a certain cut-off point for height, weight, or a combination of the two such as weight for height or body mass index (BMI). Such cut-off points are replete within the literature pertaining to growth. Stunting, for example, is defined as a height-for-age Z-score of less than -2.[13] Overweight, using the UK reference charts,[14] is defined as a BMI for age greater than the 91st centile and obesity as a BMI for age greater than the 98th centile. These values correspond to BMIs of 25 and 30 at 18 years of age but to lesser values at younger ages. In the USA a "danger of overweight" is defined by a BMI greater than the 85th centile and "obesity" by a BMI greater than the 95th centile of CDC 2002 charts.[15] To identify children outside all of these cut-offs it is necessary to measure the dimension of interest with known accuracy and reliability. Without such knowledge the possibility of falsely classifying a child as normal when they are not (false negative), or unwell when they are not (false positive), is unknown. Screening is a cross-sectional, one-off activity and usually involves large samples of children. Once identification is complete it will lead on to further assessment or intervention. In this context of large sample sizes and the need for high levels of measurement efficiency, growth status may be characterized by only a few dimensions such as height and weight. Instrumentation tends to be basic and portable and assessment is undertaken by more than one observer with each one working independently.

Surveillance is concerned with assessing the process of growth on more than one occasion on a sample of children and may well follow a screening exercise. For instance, the screening process may have identified a subsample of stunted children below −2 Z-scores for height for age. It is decided that such children are to partake in a nutritional supplementation program and that they will be reassessed at some future date to see if their height-for-age Z-scores have improved. This situation is now termed surveillance and may involve reassessment over a number of months and/or years. Clearly this context requires rather more attention to detail than the first. The samples tend to be smaller, the need for a more comprehensive description of growth requires more dimensions to be assessed, and the number of observers will be fewer with perhaps greater expertise. In addition, and because assessment on two or more occasions is required, accuracy and reliability of measurement will become more of an issue.

It may be that this process of surveillance leads to the identification of a child with an unusual pattern of growth and the decision is made to regularly reassess the child in a clinical situation with access to treatment regimes if a treatable abnormality is diagnosed. This final context is termed monitoring and involves the long-term high-frequency assessment of a single child or small sample of children. This context requires a specialist setting with dedicated space, high-quality instrumentation and a broad range of dimensions to properly assess all aspects of the growth pattern.

Screening provides only cross-sectional data but in the latter two scenarios the outcome data includes both cross-sectional and longitudinal components. Knowledge is gained both of the current growth status and of the rate of change of growth with time or "growth velocity". The usefulness of the velocity data will be dependent on the period of time between assessments: too short and the outcome is marred by diurnal variation and short-term fluctuations characteristic of saltatory growth,[16] too long and any meaningful shorter-term fluctuations will not be noticed.

19.3 ACCURACY, PRECISION, RELIABILITY AND VALIDITY

Within these contexts it is important to understand the terms accuracy, precision, reliability and validity because an understanding of these concepts is fundamental to obtaining anthropometric estimations of growth that can be used to provide a true picture of the growth process. Although a dictionary definition of accuracy may be "careful, precise, in exact conformity with truth" (Oxford English Dictionary), within an anthropometric context we do not know what "truth" is. For example, when we measure a child's height we only have the estimation of that height, we do not know what the actual, true height is. We can improve the accuracy and decrease error by ensuring that we use an appropriate, specific, purpose-made, valid, calibrated instrument to measure the child and by using a properly trained observer. Accuracy may be determined from a test—retest experiment in which the standard error of measurement (Smeas) or technical error of measurement (TEM) is calculated.[1,17] Precision is the proximity with which an instrument can measure a particular dimension. Note that this is different from accuracy in that it relates to the smallest unit of measurement possible with a particular instrument. So a stadiometer that measures height to the nearest centimeter is not as precise as one that measures to the nearest millimeter. Clearly, accuracy and precision are related in that the proximity of the measurement to the true value can only be within the limits imposed by the precision of the instrument. Thus, two observers may appear to have equal accuracy when they use a stadiometer to measure height to the last completed centimeter but a more precise instrument, which measures in millimeters, might demonstrate that one of the observers is more accurate than the other. Reliability is the extent to which an observer or instrument consistently and accurately measures a particular dimension. Reliability involves three sources of error; the observer, the instrument and the subject. Also, reliability involves two characteristics: consistency and accuracy. Reliability can be assessed in a test—retest experimental design by calculating both the Smeas (or TEM) and the standard deviation of differences (Sd).[1,17] Finally, validity is the extent to which a measurement procedure measures or assesses the variable of interest. For instance, the length of the body may be assessed by measuring height or by deriving an estimate of height from the known relationship of height to another physical dimension such as arm span (the distance from fingertip to fingertip with the arms held horizontally parallel to the floor). Arm span is generally a reasonable indicator of height but it is not as valid an indicator of the length of the body as an actual measurement of height.

19.4 FREQUENCY OF MEASUREMENT

Within the measurement of human growth it is important to know one's accuracy and reliability both for the absolute (cross-sectional) determination of a child's height and for the determination of growth velocity (rate of change with time). Obviously, when taking two measurements of height to calculate growth velocity the estimation is affected by two sources of error, one for each height estimation. The interval between measurement occasions will be determined, to some extent, by the reliability of the observer. If, for instance, the observer has a reliability, or Sd (or TEM) of 0.3 cm and the child is growing at a rate of 4 cm/year then growth will not be certain to have occurred (with 95% confidence limits) until the difference in heights between two measurement occasions is greater than 1.96×0.3 cm, or 0.59 cm. It will take this child 54 days to grow 0.6 cm, thus measurements taken on a monthly or even 2-monthly basis will be subject to observer error. The minimal time between measurement occasions for this child should be at least 3 months to ensure that false growth due to observer error is minimal. Thus, the minimum frequency of measurement is four times per year.

19.5 MEASUREMENTS

This chapter will provide descriptions of how to measure the common dimensions that describe human growth. Lengthy discussions of instrumentation and variation in measuring techniques may be found in Cameron[1-3] and Lohman et al.[17] The organization of this chapter is such that a measurement technique is described with the instrumentation recommended for the most accurate and reliable results. The measurements chosen for description are those that are recommended either as baseline measurements (e.g. height, weight, skinfolds) or as examples of measurement techniques that may be applied to other similar dimensions. Thus, where appropriate, such as for skinfolds and girths or circumferences, a generic technique is described that can be applied to any assessment site as well as specific techniques for the recommended sites. For example, a generic technique for the assessment of subcutaneous fat is described, as well as specific techniques for the triceps, biceps, subscapular and suprailiac sites. These four sites are the most commonly measured but you may wish to increase the number of sites because you are undertaking a more extensive investigation of fat patterning or have a particular interest in fat distribution. In that case, the generic technique can be applied to any skinfold site as long as the site is accurately identified. The following measurements will be described followed by the recommended instruments to obtain accurate values: linear dimensions: stature, sitting height, biacromial diameter, biiliac diameter; circumferences/girths: head, arm, waist and hip; skinfolds: triceps, biceps, subscapular and suprailiac; and weight. Finally, a glossary of anatomical surface landmarks is provided.

The accuracy with which the following measurements may be obtained can be maintained at a high level by following a few simple rules of procedure:

1. Ensure that the subject is in the minimum of clothing or at least in clothing that in no way interferes with the identification of surface landmarks.

2. Familiarize the subject with the instrumentation, which may appear frightening to the very young subject, and ensure that he or she is relaxed and happy. If necessary, involve the parents to help in this procedure by conversing with the child.

3. Organize the laboratory so that the minimum of movement is necessary and so that the ambient temperature is comfortable and the room well lit.

4. When possible, use another person to act as a recorder who will complete a measurement form. Place the recorder in such a position that he or she can clearly hear the measurements and is seated comfortably at a desk with enough room to hold recording forms, charts, and so on.

5. Measure the left-hand side of the body unless the particular research project dictates that the right-hand side should be used or unless the comparative projects have used the right-hand side.

6. Mark the surface landmarks with a water-soluble felt-tip pen prior to starting measurements.

7. Apply the instruments gently but firmly. The subject will tend to pull away from the tentative approach but will respond well to a confident approach.

8. Call out the results in whole numbers; for example, a height of 112.1 cm should be called out as "one, one, two, one" not as "one hundred and twelve point one" nor as "eleven, twenty-one". Inclusion of the decimal point may lead to recording errors and combinations of numbers may sound similar, for example, "eleven" may sound similar to "seven".

9. Establish the reliability of measurement (accuracy and repeatability) prior to the study using subjects with similar characteristics (age, sex, etc.) to the participants in the study proper.

10. During the course of the study create a quality-control procedure by repeating measurements on randomly selected subjects. At the end of each session these can be compared to the a priori reliability figures obtained in the pilot study.

11. When measurements are repeated, e.g. skinfolds, the recorder should check that the repeat value is within the known reliability of the observer. If it is not, then a third measurement is indicated. For a final value average the two that fall within the limits.

12. Do not try to measure too many subjects in any one session. Fatigue will detract from reliable measurement, for which concentration is vital.

19.5.1 Stature (Harpenden Stadiometer, Martin or GPS Anthropometer, Wall-Mounted or Rigid Stadiometer)

The subject presents for the measurement of stature dressed in the minimum of clothing, preferably just underclothes, but if social custom or environmental conditions do not permit this then at the very least without shoes and socks. The wearing of socks will not, of course, greatly affect height, but socks may conceal a slight raising of the heels that the observer from his or her upright position may not notice.

The subject is instructed to stand upright against the stadiometer such that the heels, buttocks and scapulae are in contact with the backboard, and the heels are together (Figure 19.1). If the subject suffers from "knock-knees" then the heels are slightly spread so that the knees touch, but do not overlap. As positioning is of the greatest importance the observer should always check that the subject is in the correct position by starting with the feet and checking each point of contact with the backboard as he or she moves up the body. Having reached the shoulders, he or she then checks that they are relaxed, by running his or her hands over them and feeling the relaxed trapezius muscle. The observer then checks that the arms are relaxed and hanging loosely at the sides. The head should be positioned in the "Frankfurt plane" (the lower orbits of the eyes and the external auditory meatus are in a horizontal line), and the headboard of the instrument is then moved down to make contact with the vertex of the skull. With the subject in the correct position he or she is instructed: "Take a deep breath and stand tall". This is done to straighten out any kyphosis or lordosis and produce the greatest unaided height. It is at this point that the observer applies pressure to the mastoid processes — not to physically raise the head but to hold it in the position that the subject has lifted it to by breathing deeply. The subject is then instructed to "Relax" or to "Let the air out" and "Drop the shoulders". The shoulders are naturally raised when the subject takes a deep breath and

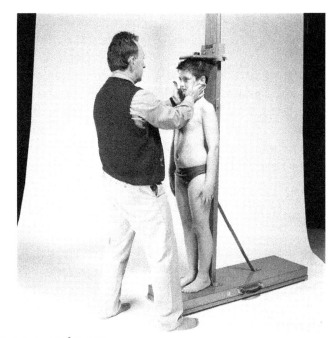

Figure 19.1 Measurement of stature.

thus tension is increased in the spinal muscles and prevents total elongation of the spine. Relaxing or breathing out releases this tension and commonly produces an increase of about 0.5 cm in absolute height. The effect of this pressure or traction technique is to counteract the effect of diurnal variation that works to reduce stature during the normal course of a day.[18,19] Stature is read to the last completed unit whether from a counter or a graduated scale. Height is not rounded up to the nearest unit as this will produce statistical bias and almost certainly invalidate estimates of height velocity.

19.5.2 Sitting Height (Sitting-Height Table or Martin or GPS Anthropometer)

Accurate measurement of sitting height requires impeccable positioning from the subject and, therefore, great attention to detail from the observer. Whether on a specially constructed sitting-height table or a suitable flat surface, the subject is positioned so that the head is in the Frankfurt plane, with the shoulders relaxed, the back straight, the upper surface of the thighs horizontal and the feet supported so that a right angle is formed with the thighs and the tendons of the biceps femoris, at the back of the knee, are just clear of the table. The arms are loose at the sides and the hands rest in the subject's lap (Figure 19.2).

Each part of this position may be checked, as for stature, by starting at the feet and moving up to the head. It is essential to have the knees raised away from the table, but

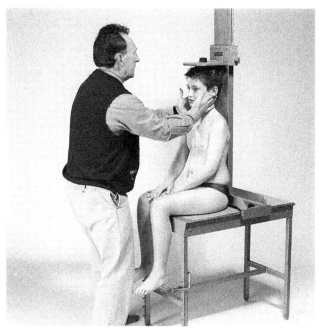

Figure 19.2 Measurement of sitting height.

only to the point where a right angle is formed with the thighs. Too acute an angle tends to cause the subject to roll backwards, and too obtuse an angle or the thighs touching the table tends to roll the subject forwards. Both situations reduce sitting height, and prevent the subject from maintaining a straight back. If the subject's hands are not resting in the lap or along the thighs then he or she may use them to push themself from the surface of the table and falsely increase sitting height. Similarly, contraction of the thighs or buttocks will raise the subject away from the table. The straightness of the back is the most difficult part to accomplish but may be achieved by giving the subject clear instructions to, for instance, "Sit up straight" or "Sit tall" and at the same time running the fingers up the spine from the sacrum to the thoracic vertebrae, causing the subject to involuntarily move away from the observer's hand and thus straighten any slouching or rounding of the spine common to most subjects when asked to sit. Having lowered the headboard the subject is instructed to "Take a deep breath and relax". Pressure is applied to the mastoids to prevent the head from falling on relaxation and the measurement read to the last completed unit.

19.5.3 Biacromial Diameter (Harpenden, Martin or GPS Anthropometer)

Biacromial diameter is the distance between the tips of the acromial processes (see Figure 19.3). It is measured from the rear of the subject with the anthropometer.

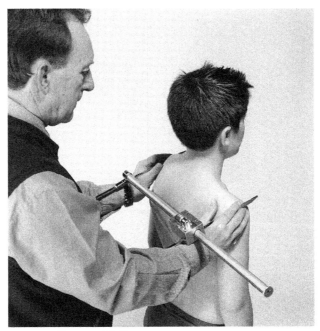

Figure 19.3 Measurement of biacromial diameter.

The position of the lateral tips of the acromials is slightly different in each subject and it is therefore necessary for the observer to carefully palpate their exact position in each subject before applying the instrument. This is most easily done with the subject standing with his or her back to the observer such that the observer can run his hands over the shoulders of the subject. This tactile awareness of the positioning of the acromials is an important part of the measurement procedure because it allows the observer to be confident of the measurement points when he applies the instrument. Having felt the position of the acromia and that the subject's shoulders are relaxed, the observer applies the anthropometer blades to the lateral tips of the processes.

The anthropometer is held so that the blades rest medially to the index fingers and over the angle formed by the thumb and index finger. The index fingers rest on top of the blades, to counteract the weight of the bar and countermechanism, and the middle fingers of each hand are free to palpate the measurement points immediately prior to measurement. In this position the observer can quite easily move the blades of the anthropometer so long as it is of the counter type. Other anthropometers have too great a frictional force opposing such easy movement and must be held by the main bar so that the blades are remotely applied to the marked acromial processes. The blades must be pressed firmly against these protuberances so that the layer of tissues that covers them is minimized. To ensure that the correct measurement is being made it is a simple matter to roll the blades up and over the acromia and then outwards and downwards so that the observer feels the blades drop over the ends of the acromia.

19.5.4 Biiliac Diameter (Harpenden, Martin or GPS Anthropometer)

The subject stands with his or her back to the observer, feet together and hands away from his or her sides to ensure a clear view of the iliac crests (see Figure 19.4). For measurements depending on the identification of any bony landmark (e.g. biacromial diameter) it is a good procedure for the observer to feel the position and shape of the landmark prior to measurement. Thus, in this case the iliac crests should be palpated prior to applying the instrument, especially when the subject has considerable fat deposits in that region. The anthropometer is held so that the blades rest medially to the index fingers and over the angle formed by the thumb and index finger. The index fingers rest on top of the blades, to counteract the weight of the bar and countermechanism, and the middle fingers of each hand are free to palpate the measurement points immediately prior to measurement. In this position the observer can quite easily move the blades of the anthropometer so long as it is of the counter type. Other anthropometers have too great a frictional force opposing such movement and must be held by the main bar so that the blades are remotely applied to the most lateral points of the iliac crests. This will be more easily accomplished if the anthropometer is slightly angled downwards and the blades are applied to the crests at a point about 2–3 cm from the tips.

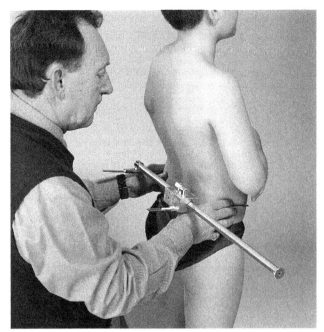

Figure 19.4 Measurement of biiliac diameter.

To ensure that the most lateral points have been obtained it is a useful point of technique to "roll" the blades over the crests. It will be seen on the counter of the instrument that at a particular point the distance between the crests is greatest; this is the point of measurement.

19.5.5 Circumferences/Girths

Methods for measuring circumferences/girths of the body are now discussed.

Head Circumference (Tape Measure)

Head circumference used to be described as a fronto-occipital circumference or a Frankfurt plane circumference, but both techniques have largely been replaced by simply measuring the maximum head circumference (see Figure 19.5).

The subject stands sideways to the observer such that the left-hand side is closer. The head is held straight or in the Frankfurt plane but the plane is of little consequence as long as the head is straight and the eyes looking forward. The subject's arms are relaxed. It is easier for the observer if he or she is positioned so that his or her eyes are level with the subject's. The tape is opened and passed around the head from left to right. The free and fixed ends are then transferred to the opposite hands so that the tape now passes completely around the head and crosses in front of the observer. Using the middle finger

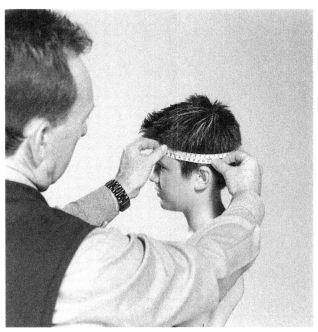

Figure 19.5 Measurement of head circumference.

of the left hand the observer presses the loose tape to the forehead of the child and moving the finger up and down determines the most anterior part of the head. Having done this he pulls the tape tighter and repeats this procedure with the middle finger of the right hand to determine the most posterior part of the occiput. Once determined, the tape is pulled tight to compress the hair and the measurement read to the last completed unit. Head circumference is the only circumference in which the tape is pulled tight and even then should cause the child no undue discomfort.

Arm Circumference or Mid-Upper Arm Circumference (Tape Measure)

The subject stands in the same position as for head circumference measurement; sideways with the left arm hanging loose at his or her side (see Figure 19.6). The mid-arm level is determined as the mid-point between the acromion and olecranon with the arm flexed at a right angle. The tape is passed around the arm from left to right and the free and fixed ends are transferred from one hand to the other. Ensuring that the tape is at the same level as the mid-upper-arm mark, it is tightened so that it touches the skin all round the circumference but does not compress the tissue to alter the contour of the arm. The circumference is then read to the last completed unit.

Because the arm in cross-section is not an exact circle but rather oval, some difficulty is usually met in ensuring that the tape actually touches the skin on the medial side of the

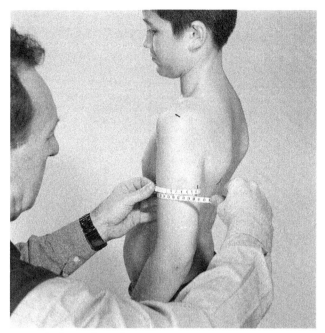

Figure 19.6 Measurement of arm circumference.

arm. To ensure that this is so the middle finger of the left hand can be used to gently press the tape to the skin.

Waist Circumference or Abdominal Circumference (Tape Measure)

The measurement is taken at the minimum circumference between the iliac crests and lower ribs. The general technique is for the subject to stand erect facing the observer with the arms away from the body. The tape is passed around the body and tightened at the required level, ensuring that it is horizontal and not compressing the soft tissue.

Hip Circumference (Tape Measure)

Hip circumference should be measured at the level of the greatest protrusion of the buttocks when the subject is standing erect with the feet together. The subject stands sideways to the observer with the feet together and arms folded. The observer passes the tape around the body at the level of the most prominent protrusion of the buttocks so that it lightly touches but does not compress the skin.

19.5.6 Skinfolds (Harpenden[20], Holtain or Lange Calipers[22,23])

The technique of picking up the fold of subcutaneous tissue measured by the skinfold caliper is often referred to as a "pinch", but the action to obtain the fold is to sweep the index or middle finger and thumb together over the surface of the skin from about

6 to 8 cm apart.[21] This action may be simulated by taking a piece of paper and drawing a, say, 10 cm line on its surface. If the middle finger and thumb are placed at either end of this line and moved together such that they do not slide over the surface of the paper but form a fold of paper between them then that is the action required to pick up a skinfold. To "pinch" suggests a small and painful pincer movement of the fingers and this is not the movement made. The measurement of skinfolds should not cause undue pain to the subject, who may be apprehensive from the appearance of the calipers and will tend to pull away from the observer; in addition, a pinching action will not collect the quantity of subcutaneous tissue required for the measurement.

Triceps Skinfold

The level for the triceps skinfold[24] is the same as that for the arm circumference: midway between the acromion and the olecranon when the arm is flexed at a right angle (see Figure 19.7). It is important that the skinfold is picked up both at a mid-point on the vertical axis of the upper arm and at a mid-point between the lateral and medial surfaces of the arm. If the subject stands with his or her back to the observer and bends the left arm the observer can palpate the medial and lateral epicondyles of the humerus. This is most easily done with the middle finger and thumb of the left hand, which will eventually grip the skinfold. The thumb and middle finger are then moved upwards, in contact with the skin, along the vertical axis of the upper arm until they are at a level

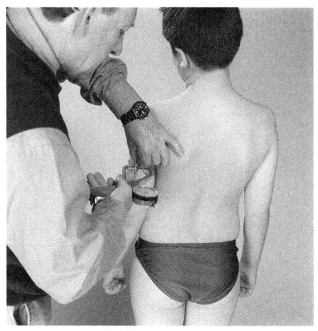

Figure 19.7 Measurement of triceps skinfold.

about 1.0 cm above the marked mid-point. The skinfold is then lifted away from the underlying muscle fascia with a sweeping motion of the fingers to the point at which the observer is gripping the "neck" of the fold between middle finger and thumb.

The skinfold caliper, which is held in the right hand with the dial upwards, is then applied to the neck of the skinfold just below the middle finger and thumb at the same level as the marked mid-point of the upper arm. The observer maintains his or her grip with the left hand and releases the trigger of the skinfold caliper with the right to allow the caliper to exert its full pressure on the skinfold. In almost every case the dial of the caliper will continue to move but should come to a halt within a few seconds, at which time the reading is taken to the last completed 0.1 mm. In larger skinfolds the caliper may take longer to reach a steady state but it is unusual for this to be longer than 7 seconds. Indeed, if the caliper is still moving rapidly it is doubtful that a true skinfold has been obtained and the observer must either try again or admit defeat. This situation is only likely to occur in the more obese subject with skinfolds greater than 20–25 mm; that is, above the 97th centile of British charts. Within the 97th and 3rd centiles skinfolds are relatively easy to obtain but they do require a great deal of practice.

Biceps Skinfold

The biceps skinfold (Figure 19.8) is the exact opposite of the triceps skinfold, being on the anterior aspect of the arm and at the same mid-point level as previously described for

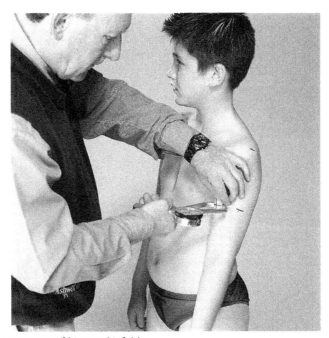

Figure 19.8 Measurement of biceps skinfold.

triceps skinfold. It is picked up with the subject facing the observer and the left arm hanging relaxed but with the palm facing forwards.

The middle finger and thumb sweep together at a point 1 cm above the marked mid-point level, coming together at the vertical axis joining the center of the anticubital fossa and the head of the humerus. It is unusual for the movement of the dial to present any problem with this skinfold measurement, as it is not a site for major fat deposits.

Subscapular Skinfold

The point of measurement is located immediately below the inferior angle of the scapula (see Figure 19.9). The subject stands with his or her back to the observer and the shoulders relaxed and arms hanging loosely at the sides of the body. This posture is most important to prevent movement of the scapulae; if the subject folded his arms, for instance, the inferior angle of the scapula would move laterally and upwards and therefore no longer be in the same position relative to the layer of fat. The skinfold is picked up, as for triceps skinfold, by a sweeping motion of the middle finger and thumb, and the caliper applied to the neck of the fold immediately below the fingers.

The fold will naturally be at an angle laterally and downwards and will not be vertical. Once again, the dial of the caliper will show some movement that should soon cease.

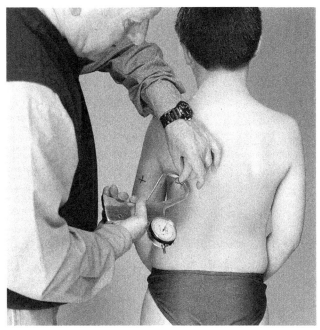

Figure 19.9 Measurement of subscapular skinfold.

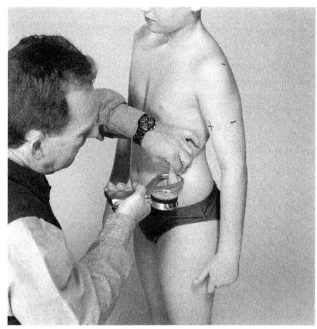

Figure 19.10 Measurement of suprailiac skinfold.

Suprailiac Skinfold

The point of measurement for the suprailiac skinfold is 1 cm superior and 2 cm medial to the anterior superior iliac spine (see Figure 19.10). This is best palpated with the subject standing facing the observer. The skinfold is picked up with a sweep of the middle finger and thumb and is a vertical skinfold.

Once again the caliper is applied below the fingers and, after the dial has stopped moving, the measurement is read to the last completed 0.1 mm. It should be noted that Durnin and Rahaman (1967) and Durnin and Womersley (1974),[25,26] who developed linear regression equations that are frequently used to predict total body fat from skinfolds, do not use this conventional suprailiac site for their measurement to derive total body fat. Instead, they use a mid-axillary suprailiac skinfold, just above the iliac crest but still vertical (Figure 19.11).

19.5.7 Weight (Digital Weighing Scales or Beam Balance)

The measurement of weight should be the simplest and most accurate of the anthropometric measurements. Assuming that the scales are regularly calibrated, the observer ensures that the subject is dressed either in the minimum of clothing or in a garment of known weight that is supplied by the observer. The subject stands straight, but not rigid or in a "military position", and is instructed to "stand still". If the instrument is a beam

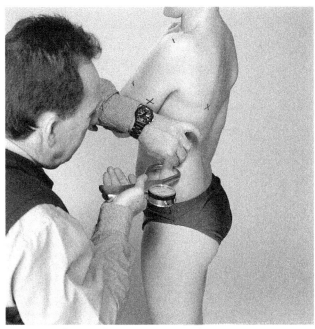

Figure 19.11 Measurement of mid-axillary suprailiac skinfold.

balance then the observer moves the greater of the two counterweights until the nearest 10 kg point below the child's weight is determined. The smaller counterweight is then moved down the scale until the nearest 100 g mark below the point of overbalance is reached and this is recorded as the true weight. This procedure is necessary to determine weight to the last completed unit. If the weight is taken as the nearest 100 g above true weight then that 100 g is greater than actual weight and the last unit has not been completed.

Determining the weight of neonates can be a noisy and tearful procedure but need not be if the help of the mother is solicited. The observer simply weighs mother and child together and then transfers the baby to the assistant's arms and weighs the mother by herself. The baby's weight can thus be determined by difference [(weight of mother + baby) − (weight of mother)] and the child is left relatively undisturbed.

19.5.8 Surface Landmarks

Acromion process (lateral border of the acromion) (Figure 19.12): The acromion projects forwards from the lateral end of the spine of the scapula, with which it is continuous. The lower border of the crest of the spine and the lateral border of the acromion meet at the acromial angle, which may be the most lateral point of the acromion. Great diversity in the shape of the acromion between individuals means

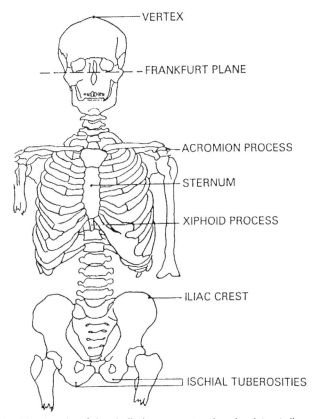

Figure 19.12 Skeletal landmarks of the skull, thorax, pectoral and pelvic girdles.

that sometimes the acromial angle is not the most lateral point. Palpation of the most lateral part may best be performed by running the anthropometer blades laterally along the shoulders until they drop below the acromia. If the blades are then pushed medially the most lateral part of the acromia must be closest to the blades and may be felt below the surface marks left by the blades. There is the possibility of the inexperienced anthropometrist confusing the acromioclavicular joint with the lateral end of the acromion. Great care must be taken to distinguish between these two landmarks prior to measurement.

Anterior superior iliac spine (Figure 19.13): This is the anterior extremity of the ilium which projects beyond the main portion of the bone and may be palpated at the lateral end of the fold of the groin. It is important to distinguish the iliac crest from the anterior spine when measuring biiliac diameter (see Biiliac diameter, section 5.4).

Biceps brachii: The biceps brachii is the muscle of the anterior aspect of the upper arm. Its two heads, the short and the long, arise from the coracoid process and the supraglenoid tubercle of the scapula, respectively, and are succeeded by the muscle

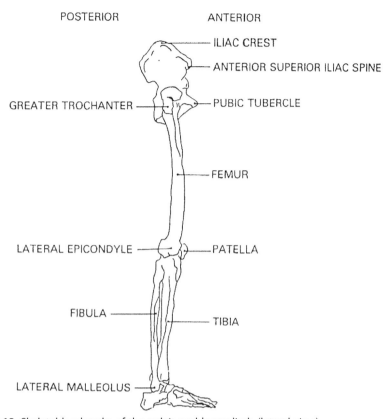

Figure 19.13 Skeletal landmarks of the pelvis and lower limb (lateral view).

bellies before they end in a flattened tendon that is attached to the posterior part of the radial tuberosity. When relaxed the muscle belly has its greatest bulge towards the radius but when contracted with the arm flexed the belly rises to a point nearer the shoulder. Thus, relaxed and contracted arm circumferences, taken at the maximum bulge of the muscle, are not at exactly the same level.

Distal end of the radius (Figure 19.14): This is the border of the radius proximal to the distal—superior borders of the lunate and scaphoid and medial to the radial styloid. It may be palpated by moving the fingers medially and proximally from the radial styloid (see Radial styloid, below).

External auditory meatus: This landmark, used to obtain the Frankfurt plane, is also called the external acoustic meatus and leads to the middle ear from the external auricle. In terms of a surface landmark it is therefore simply present as a hole in the external ear and may therefore be easily seen. The tragus, the small curved flap that extends posteriorly from the front of the external ear, overlaps the orifice of the meatus and may be used to gauge the level of the orifice.

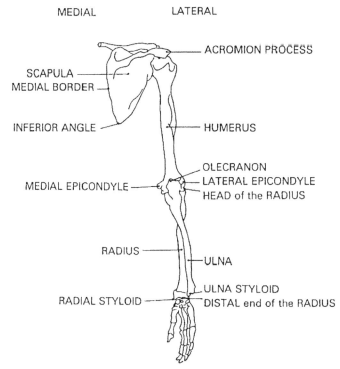

MEDIAL   LATERAL

ACROMION PROCESS

SCAPULA
MEDIAL BORDER

INFERIOR ANGLE

HUMERUS

OLECRANON
LATERAL EPICONDYLE
HEAD of the RADIUS

MEDIAL EPICONDYLE

RADIUS

ULNA

ULNA STYLOID
DISTAL end of the RADIUS

RADIAL STYLOID

Figure 19.14 Skeletal landmarks of the scapular and upper limb (posterior view).

Femur epicondyles (Figure 19.13): The lower end of the femur consists of two prominent masses of bone called the condyles, which are covered by large articular surfaces for articulation with the tibia. The most prominent lateral and medial aspects of the condyles are the lateral and medial epicondyles. These may be easily felt through the overlying tissues when the knee is bent at a right angle, as in the sitting position. If the observer's fingers are then placed on the medial and lateral aspects of the joint the epicondyles are the bony protuberances immediately above the joint space.

Frankfurt plane (Figure 19.12): This plane, used extensively in anthropometric measurement, is obtained when the lower margins of the orbital openings and the upper margins of the external acoustic (auditory) meatus lie in the same horizontal plane. The supinated Frankfurt plane, used in the measurement of recumbent and crown—rump length, is vertical rather than horizontal.

Gastrocnemius: This is the most superficial of the group of muscles at the rear of the lower leg and forms the belly of the calf.

Glabella (Figure 19.15): This landmark is in the midline of the forehead between the brow ridges and may be used as the most anterior point of the head.

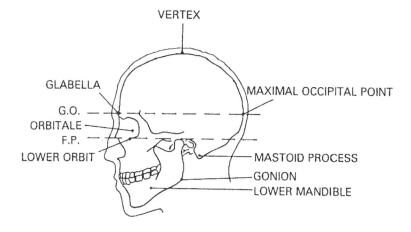

VERTEX

GLABELLA

MAXIMAL OCCIPITAL POINT

G.O. —
ORBITALE —
F.P. —
LOWER ORBIT

MASTOID PROCESS
GONION
LOWER MANDIBLE

F.P. = FRANKFURT PLANE

G.O. = GLABELLA-OCCIPITAL PLANE

Figure 19.15 Skeletal landmarks of the skull.

Gluteal fold: This fold or furrow is formed by the crossing of the gluteus maximus and the long head of the biceps femoris and semitendinosus. It may therefore be viewed from the lateral aspect or the posterior aspect as the crease beneath the buttock. In some subjects, perhaps because of a lack of gluteal development, a crease may not be present. In this case the level of the gluteal fold is judged from the lateral profile of the buttocks and posterior thigh.

Head of the radius (Figure 19.14): This may be palpated as the inverted, U-shaped bony protuberance immediately distal to the lateral epicondyle of the humerus when the arm is relaxed with the palm of the hand facing forwards.

Humeral epicondyles (Figure 19.14): These are the non-articular aspects of the condyles on the lower surface of the humerus. The medial epicondyle forms a conspicuous blunt projection on the medial aspect of the elbow when the arm is held at the side of the body with the palm facing forward. The lateral epicondyle may be palpated opposite and a little above the medial epicondyle.

Iliac crest (Figure 19.13): This may be palpated as the most superior edge of the ilium and may be easily felt through the overlying soft tissue. Greater difficulty will be experienced with the more obese subject but it is quite possible with the anthropometer blades to compress the tissue and feel the crest.

Malleoli (Figure 19.13): The medial malleolus is the bony protuberance on the medial side of the ankle. It is the inferior border of this malleolus that is palpated and used as a landmark for the measurement of tibial length.

Mastoid process (Figure 19.15): This is the conical projection below the mastoid portion of the temporal bone. It may be palpated immediately behind the lobule of the ear and is larger in the male than in the female.

Mid-axillary line: The axilla is the pyramidal region situated between the upper parts of the chest wall and the medial side of the upper arm. The mid-axillary line is normally taken as the line running vertically from the middle of this region to the iliac crest.

Mid-inguinal point (inguinal crease): The inguinal ligament runs from the anterior superior iliac spine to the pubic tubercle at an angle of 35 to 40 degrees and is easily observed in all individuals. The mid-point between the anterior spine and the pubic tubercle on the line of the inguinal ligament is taken as the mid-inguinal point.

Mid-point of the arm: The mid-point of the arm, used for arm circumference, is taken as the point on the lateral side of the arm midway between the lateral border of the acromion and the olecranon when the arm is flexed at 90 degrees. This may be most easily determined by marking the lateral border of the acromion and applying a tape measure to this point. If the tape is allowed to lie over the surface of the arm, the mid-point may easily be calculated and marked. Alternatively, tape measures exist with a zero mid-point that are specifically designed to determine this landmark. It has been common to refer to this point, and the circumference or girth at this level, as the "mid-upper arm" landmark/circumference.

Occiput (Figure 19.15): The occipital bone is situated at the back part and base of the cranium. The occiput is the most posterior part of this bone and may be clearly seen from the side view of the subject.

Olecranon (Figure 19.14): The olecranon is the most proximal process of the ulna and may be easily observed when the arm is bent as the point of the elbow.

Patella (Figure 19.13): The patella is the sesamoid bone in front of the knee joint embedded in the tendon of the quadriceps muscle. It is flat, triangular below and curved above. When the subject is standing erect its lower limit lies above the line of the knee joint and its upper border may be palpated at the distal end of the quadriceps muscle.

Pinna of the ear: The pinna of the ear is more correctly called the lobule and is the soft part of the auricle that forms the earlobe.

Radial styloid (Figure 19.14): The radial styloid is the distal projection of the lateral surface of the radius. It extends towards the first metacarpal and may be palpated as a bony projection on the lateral surface of the wrist when the hand is relaxed.

Scapula (Figure 19.14): The scapula is the large, triangular flattened bone on the posterolateral aspect of the chest, and is commonly known as the shoulder blade. Its medial border slopes downwards and laterally to the inferior angle that may be easily palpated, and lies over the seventh rib or seventh intercostal space when the arm is relaxed.

Sternum (Figure 19.12): The sternum or breastbone is the plate of bone inclined downwards and a little forwards at the front of the chest. It is composed of three parts: the manubrium at the top, the body or mesosternum at the center and the xiphoid process at the lower end. The mesosternum and xiphoid process are important landmarks in anthropometry. The mesosternum is marked by three transverse ridges or sternabrae and the junction between the third and fourth sternabrae form a landmark in chest measurement. The fourth sternabra may not be easily palpated but the junction lies below the more easily palpated third sternabra. The xiphoid process may be palpated by following the line of the sternum to its end. The sternum is considerably larger in males than in females.

Trapezius: The trapezius is a flat, triangular muscle extending over the back of the neck and the upper thorax.

Triceps: The triceps muscle is the large muscle on the posterior side of the upper arm. When the arm is actively extended two of the three triceps heads may be seen as medial and lateral bulges.

Trochanters (Figure 19.13): The greater and lesser trochanters are projections at the proximal end of the femur. The lesser trochanter cannot be palpated on the living subject because it lies on the posterior surface of the femur and is covered by the large gluteal muscles. The greater trochanter, however, is palpable as the bony projection on the lateral surface of the upper thigh approximately a hand's breadth below the iliac crest.

Ulna styloid (Figure 19.14): The styloid process of the ulna is present as a short, rounded projection at the distal end of the bone. It may be easily palpated on the posterior—medial aspect of the wrist opposite and about 1 cm above the styloid process of the radius.

Umbilicus: The umbilicus, or naval, is clearly observable in the center of the abdomen. It is variable in position, lying lower in the young child owing to the lack of abdominal development.

Vertex of the skull (Figure 19.15): This is the top-most point of the skull and theoretically comes into contact with the stadiometer headboard when height is being properly measured. With the head in the Frankfurt plane the vertex is slightly posterior to the vertical plane through the external auditory meatus and may be easily palpated.

19.6 SUMMARY

Human growth is measured in the contexts of screening, surveillance and monitoring. Whether growth is assessed in a cross-sectional or longitudinal research design it is important to understand the importance of accuracy, precision, reliability and validity. The precision of the instrument and the reliability of the observer will dictate the frequency of measurement and thus the resulting pattern of growth. It is vital to understand anatomy in order to properly locate the measurement landmarks on the surface of the body.

REFERENCES

1. Cameron N. *The measurement of human growth*. London: Croom-Helm; 1984.
2. Cameron N. The methods of auxological anthropometry. In: Faulkner F, Tanner JM, editors. *Human growth*. New York: Plenum; 1978. p. 35−90.
3. Cameron N. The methods of auxological anthropometry. In: Faulkner F, Tanner JM, editors. *Human growth*. 2nd ed. New York: Plenum; 1986. p. 3−46.
4. Cameron N. *Methods in human growth research*. Cambridge: Cambridge University Press; 2004.
5. Shuttleworth FK. Sexual maturation and the physical growth of girls aged six to sixteen. *Child Dev Monogr* 1937;**2**:1.
6. Stuart HC. Studies from the Center for Research in Child Health and Development, School of Public Health Harvard University, 1. *The Center, the group under observation, sources of information and studies in progress. Child Dev Monogr* 1939;**4**:1.
7. Simmons K. The Brush Foundation study of child growth and development. *Child Dev Monogr* 1944;**9**:1.
8. Tanner JM, Whitehouse RH. The Harpenden skinfold caliper. *Am J Phys Anthropol* 1955;**13**:743−6.
9. International Children's Centre. *Growth and development of the child*. Paris: International Children's Centre; 1980.
10. Weiner JS, Lourie JA. *Human biology: a guide to field methods*. Oxford: Blackwell Scientific Publications; 1969. IBP Handbook No. 9.
11. Weiner JS, Lourie JA. *Practical human biology.*. London: Academic Press; 1981.
12. Hrdlicka A. *Hrdlicka's practical anthropometry* Stewart TD. 3rd ed. Philadelphia, PA: Wistar Institute; 1947.
13. Ong KKL, Ahmed ML, Emmett PM, Preece MA, Dunger DB. ALSPAC Study Team. Association between postnatal catch-up growth and obesity in childhood: prospective cohort study. *BMJ* 2000;**320**:967−71.
14. Cole TJ, Bellizzi MC, Flegal KM, Dietz WH. Establishing a standard definition for child overweight and obesity worldwide: international survey. *BMJ* 2000;**320**:1240−3.
15. Barlow SE, Dietz WH. Obesity evaluation and treatment: expert committee recommendations. The Maternal and Child Health Bureau, Health Resources and Services Administration, and the Department of Health and Human Services. *Paediatrics* 1998;**102**:E29.
16. Lampl M, Veldhuis JD, Johnson ML. Saltation and stasis: a model of human growth. *Science* 1992;**258**:801−3.
17. Lohman TG, Roche AF, Martorell R. *Anthropometric standardization reference manual*. Champaign, IL: Human Kinetics Books; 1988.
18. Strickland AL, Shearin RB. Diurnal height variation in children. *Pediatrics* 1972;**80**:1023.
19. Whitehouse RH, Tanner JM, Healy MJR. Diurnal variation in stature and sitting-height in 12−14 year old boys. *Ann Hum Biol* 1974;**1**:103.
20. Edwards DAW, Hammond WH, Healy MJR, Tanner JM, Whitehouse RH. Design and accuracy of calipers for measuring subcutaneous tissue thickness. *Br J Nutr* 1955;**9**:133−43.
21. Harrison GG, Buskirk ER, Carter JEL, Johnston FE, Lohman TG, Pollock ML, et al. Skinfold thicknesses and measurement technique. In: Lohman TG, Roche AF, Martorell R, editors. *Anthropometric standardization reference manual*. Champaign, IL: Human Kinetics; 1991. p. 55−70.
22. Lange KO, Brozek J. A new model of skinfold caliper. *Am J Phys Anthropol* 1961;**19**:98−9.
23. Schmidt PK, Carter JE. Static and dynamic differences among five types of skinfold calipers. *Hum Biol* 1990;**623**:369−88.
24. Ruiz L, Colley JRT, Hamilton PJS. Measurement of triceps skinfold thickness. *Br J Prev Soc Med* 1971;**25**:165.

25. Durnin JVGA, Rahaman MM. The assessment of the amount of fat in the human body from measurements of skinfold thickness. *Br J Nutr* 1967;**21**:681.
26. Dumin JVGA, Womersley J. Body fat assessed from total body density and its estimation from skinfold thickness: measurements on 481 males and females aged from 16 to 72 years. *Br J Nutr* 1974;**32**:77.

SUGGESTED READING

Alder K. *The measure of all things*. London: Little, Brown; 2002.
Cameron N. *The measurement of human growth*. London: Croom-Helm; 1984.
Lohman TG, Roche AF, Martorell R. *Anthropometric standardization reference manual*. Champaign, IL: Human Kinetics; 1988.

INTERNET RESOURCES

Harpenden instruments: http: //www.fullbore.co.uk/holtain/medical/welcome.html
Lange caliper: http: //www.langecaliper.com/
The reliability of measurement discussed in relation to a large longitudinal growth study conducted by the World Health Organization: http://www.who.int/childgrowth/standards/Reliability_anthro.pdf
Descriptions of anthropometric measurements in a large cross-sectional surveys of American youth: http:// www.cdc.gov/nchs/data/nhanes/nhanes3/cdrom/nchs/manuals/anthro.pdf
International Society for the Advancement of Kinanthropometry (ISAK) – an association specializing in anthropometric measurement methods: http://www.isakonline.com/

CHAPTER 20

Assessment of Maturation

Noël Cameron

Centre for Global Health and Human Development, School of Sport, Exercise and Health Sciences, Loughborough University, Leicestershire LE11 3TU, UK

Contents

20.1 INTRODUCTION

The process of maturation is continuous throughout life — it begins at conception and ends at death. This chapter will concentrate on the assessment of the process of maturation from birth to childhood, i.e. that part of the total process that is intimately linked to physical growth. It is important, therefore, to differentiate between "growth" and "maturation". Bogin[1] defines the former as "a quantitative increase in size or mass" such as increases in height or weight. Development or maturation, on the other hand, is

N. Cameron & B. Bogin (eds): Human Growth and Development, Second edition.
ISBN 978-0-12-383882-7, Doi: 10.1016/B978-0-12-383882-7.00020-9

defined as "a progression of changes, either quantitative or qualitative, that lead from an undifferentiated or immature state to a highly organized, specialized, and mature state". The endpoint of maturation, within the context of the growth, is the attainment of adulthood, which is defined here as being a "functionally mature individual". Functional maturation, in a biological context, implies the ability to successfully procreate and raise offspring who themselves will successfully procreate. In addition to the obvious functional necessities of sperm and ova production, reproductive success within any mammalian society is dependent on a variety of morphological characteristics such as size and shape. The too short or too tall, the too fat or too thin, are unlikely to achieve the same reproductive success as those within an "acceptable" range of height and weight values that are themselves dependent on the norms in a particular society. Thus, in its broadest context, maturation and growth are intimately related and both must reach functional and structural endpoints that provide the opportunity for successful procreation.

20.2 INITIAL CONSIDERATIONS

In order to understand how maturation can be assessed it is important first to appreciate that maturation is not linked to time in a chronological sense. In other words, one year of chronological time is not equivalent to one year of maturational "time". This is perhaps best illustrated in Figure 20.1, in which three boys and three girls of precisely the same chronological ages demonstrate dramatically different degrees of maturity as evidenced by the appearance of secondary sexual characteristics. In addition, they exhibit changes in the proportion and distribution of subcutaneous fat, and the development of the skeleton and musculature that result in sexually dimorphic body shapes in adulthood. Although each individual has passed through the same chronological time span they have done so at very different rates of maturation.

Second, maturation is most often assessed by the identification of "maturity indicators". Such indicators are discrete events or stages recognizable within the continuous changes that occur during the process of maturation. Thus, the maturity indicators that identify breast or pubic hair development divide the continuous changes that occur into discrete stages.

Third, there is variability of maturation within the individual. For instance, while skeletal and secondary sexual maturation are associated they are not correlated so significantly that one can categorically associate a particular stage of sexual maturation with a particular skeletal "age".[3,4] In the closest association, of skeletal age to menarcheal age, it is possible to state that a girl with a skeletal age less than 12 years is unlikely to have experienced menarche and that one with a skeletal age of 15 years is likely to be post-menarcheal. We cannot state with any real degree of confidence that the association of these two maturational processes is closer than that.

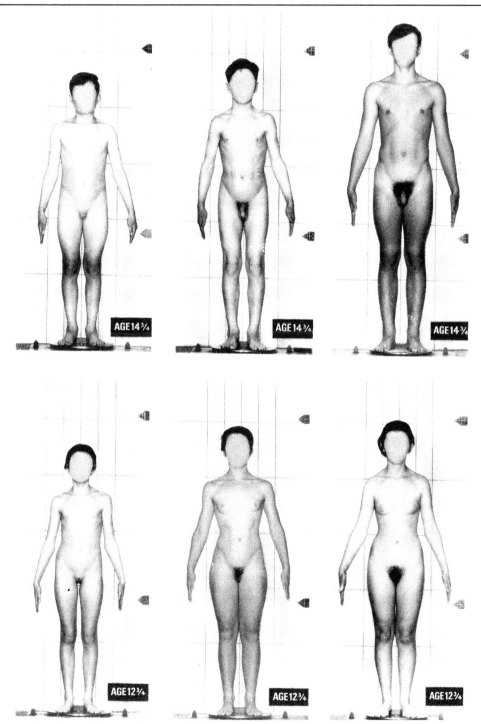

Figure 20.1 Three boys and three girls photographed at the same chronological ages within sex: 12.75 years for girls and 14.75 years for boys. (Source: *Tanner.[2]*)

Fourth, within a particular maturational process, such as sexual maturation, it is apparent that different structures, e.g. genitalia and pubic hair, will not necessarily be at precisely the same level of maturity. Thus, we have a process of "uneven maturation".

Fifth, there is clear sexual dimorphism within human growth and maturation such that females tend to be advanced relative to males at any particular chronological age. In Figure 20.1, for instance, the females are aged exactly 12.75 years and the males are aged 14.75 years, yet their levels of secondary sexual development are similar.

Sixth, maturation is not related to size except in very general terms; a small individual is likely to be a child and thus less mature than a large individual, who is more likely to be an adult. As the ages of the two individuals approach each other so the distinction between size and maturity narrows and disappears such that, within a group of similar maturity, there will be a range of sizes and within a group of similar size there will be a range of maturity levels. Thus, when maturation is assessed size must be controlled for or excluded from the assessment method.

These six considerations, the relationship of maturity to time, the quantification of the continuous process of maturation by using discrete events, the relative independence of different processes of maturation within the individual, the appreciation of uneven maturation, sexual dimorphism, and the lack of a relationship between maturity and size, have governed the development of techniques for the assessment of maturation.

20.2.1 The Concept of Time

Roy M. Acheson[5] (1921–2003) elegantly described the problem of "time" within the development of skeletal maturity assessment methods:

> *Because maturation is distinct from growth it merits a distinct scale of measurement, indeed the whole basis of the medical and scientific interest it attracts is that it does not proceed at the same rate in the various members of a random group of healthy children. The corollary of this is that the unit of measurement, "the skeletal year", does not have the same meaning for any two healthy children, nor even … does a skeletal year necessarily have the same meaning for two bones in a single healthy child.[5, p. 471]*

The core problem is the use of an age scale to represent maturity. This fails at the extreme because no particular age can be associated with full maturity, and prior to full maturity because of the lack of a constant relationship between maturity and time both between and within the sexes. Thus, when using the Greulich–Pyle atlas technique for skeletal maturity,[6] one is faced with the final "standards" for males and females which correspond to an "age" of 18 years but which in fact represent full maturity or the maturity to be found in any individual who has achieved total epiphyseal fusion regardless of his or her actual chronological age.

To overcome this problem in the assessment of skeletal maturity by moving away from an "age"-based method, Acheson[7,8] and Tanner and his colleagues[9–11] developed the "bone-specific scoring" techniques in which numerical scores were assigned to each bone rather than a bone "age". Acheson's earlier attempt, which became known as the "Oxford method", simply gave scores of 1, 2, 3, etc., to each stage. However, this scoring method did not account for the fact that the differences between stages were not equivalent; the "difference" between stage 1 and stage 2 was not necessarily equivalent, in terms of the advancement of maturity, to the difference between stages 2 and 3. Tanner's basic principle was that the development of each single bone, within a selected area, reflected the single process of maturation. Ideally, the scores from each bone in a particular area should be the same and the common score would be the individual's maturity. However, such scores would not be the same because of the large gaps between successive events in a single bone. Thus, the scoring process would need to minimize the overall disagreement between different bones. The disagreement is measured by the sum of squares of deviations of bone scores about the mean score, and it is this sum which is minimized. In order to avoid what Tanner described as the "trivial solution" of perfect agreement by giving the same scores to each stage, the scores were constrained on a scale of 0 to 100, i.e. each bone starts at zero and matures at 100. In essence, each maturity indicator is rated on a maturity scale from 0% maturity to 100% maturity. Without dwelling on the mathematics, which are given in detail by Tanner in his 1983 publication,[11] the principle is an important one and should be applied to any new system of assessing maturity. In addition, the bone-specific scoring approach can be applied to an appropriate sample of radiographs from any population to derive maturity norms.

The principle of scoring maturity indicators was later applied to the assessment of dental maturity by Demirijanand and colleagues[12] but, to date, has not been applied to other attempts at maturational assessment such as secondary sexual development. The reason for this apparent neglect may be that we still use the staging system originally developed by Nicholson and Hanley[13] and modified by Tanner and his colleagues in 1962.[10] Only five stages are used within any particular area and these are often difficult to assess accurately. Also, secondary sexual development takes place over a relatively short period, say between 10 and 17 years in girls, compared to the birth to adulthood temporal basis of skeletal maturity. Thus, one is faced with fewer maturity indicators within a short period and the application of a scoring technique has seemed inappropriate. However, other aspects of skeletal maturity may lend themselves to a scoring system. Cranial suture closure, for instance, has rarely been investigated as an indicator of maturity in children. Yet this latter technique is important in biological anthropology in which the maturity of skeletal remains is of forensic interest to determine chronological age, and of course, in paleoanthropology, in which the maturity of the "subadult" fossil has a bearing on the interpretation of the morphology of the individual. Meindl and Lovejoy[14] have described a "revised method" for determining skeletal age using the lateral—anterior sutures. They use a scoring system which is the equivalent of Acheson's scoring system for the Oxford method and in so doing repeat the erroneous concept that differences between scores are equivalent, i.e. that the difference between stage 1 and 2 is the same as that between stage 2 and 3. Suture closure is a suitable area for the application of the methods developed by auxologists and would have broad relevance within biological anthropology.

20.2.2 Maturity Indicators

The development of the concept of maturity indicators by Wingate Todd,[15] based on the pioneering work of Milo Hellman in 1928,[16] was fundamental in developing methods to accurately assess skeletal maturity. Prior to the identification of maturity indicators, skeletal maturity was assessed by the "number of ossification centers" method, in which a count was made, either from the hand and wrist[17,18] or from a skeletal survey of each child,[19] of the number of centers that were present or absent in the total skeleton. Alternatively, planimetry was used to assess the total amount of bony tissue apparent in radiographs.[20−22] The former method failed because of a lack of appreciation of the fact that the order of appearance of ossification centers is largely under genetic control[23] and the latter method because only the carpus was used which, as we now appreciate, is not representative of overall maturity.

This author defines a maturity indicator as a definable and sequential change in any part or parts of the body that is characteristic of the progression of the body from immaturity to maturity. Skeletal development provides the clearest example of such maturity indicators.

Figure 20.2 illustrates the maturity indicators for the developing radius used in the atlas and bone-specific scoring methods of Greulich and Pyle (GP)[6] and Tanner, Whitehouse and Healy (TW).[10] Both groups examined the development of the radius apparent in radiographs of the left hand and wrist of children from birth to adult maturity. The former group identified 11 indicators while the latter described eight. It is apparent, however, that Tanner et al.[10] and Greulich and Pyle[6] concurred in their description of these maturity indicators. Indeed, it is extremely important that they did concur. If the two groups of researchers had disagreed in their description of maturity indicators within the same skeletal area, then each group would have been identifying different aspects of maturation and would cast doubt on our ability to recognize unequivocal indicators of the process of maturation. Regardless of the particular maturational process under investigation, the identification of maturity indicators is fundamental to quantifying that process and arriving at measures of individual and population variation.

Maturity indicators must, however, conform to certain prerequisites if they are to be useful. They must possess the quality of "universality" in that they must be present in all normal children of both sexes and they must appear sequentially, and in the same sequence, in all children. They must have the ability to discriminate between different stages of maturation and reflect a continuous process of maturation through to adult maturity and thus be complete in adulthood. As they are to be used as measures of

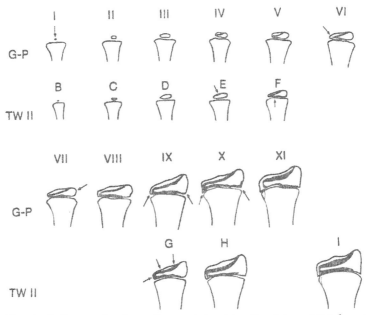

Figure 20.2 Maturity indicators for the radius as defined by Greulich and Pyle[6] and Tanner and Whitehouse.[11]

maturation they must be reliable both within and between different observers and validly reflect the process of maturation.

While such criteria may appear obvious, it is possible to find examples of maturity indicators that simply do not conform to these desiderata. For instance, "age at semenarche" was reported in terms of a mean and standard deviation, for a sample of boys from the Transkei region of South Africa who had completed a self-administered questionnaire.[24] The question each boy responded to was, "How old were you when you had your first wet dream (ejaculation)?" One can appreciate that the accuracy of the response to such a question is at best dubious, and indeed the authors maintained that their estimate of mean age at semenarche, "… was crude, and relied on recall of a fairly nebulous isolated event that in most cases is difficult to recall precisely" (Buga, 1996, personal communication). Such estimates ignore the fundamental rationale of maturity indicators and are not constructive to the accurate determination of sexual maturity.

Without maturity indicators we cannot develop methods to assess the process and thus when we search for new methods the "holy grail" of that search is the identification of appropriate indicators of maturity.

20.2.3 Maturational Variation

Maturational variation covers two aspects: (1) the variation of maturation within a process and (2) the variation of maturation between processes. The former aspect may be observed within sexual maturation from the data published by Marshall and Tanner[3,4] on British children. They illustrated variation by investigating the percentage of girls or boys within any particular stage of development of one indicator of maturation when they entered a particular stage of another indicator. For instance, 84% of girls were in at least stage 2 of breast development when they entered stage 2 of pubic hair development. In other words, they did not enter pubertal maturation in both breast and pubic hair development simultaneously. Breast development for the vast majority was the first stage of puberty, followed by pubic hair development. Similarly, 39% of girls were already adult for breast development when they became adult for pubic hair development.

A similar pattern of variation was observed in males, with 99% of boys starting genitalia development prior to pubic hair development. This variation is critical in that it requires any modification of the method to allow for intraindividual variation. Within clinical situations, for instance, the difficulties in accurately rating the various stages of breast, genitalia or pubic hair development within the Tanner five-point classification have led to the combination of the stages into a three- or four-point "pubertal" staging technique. In the three-point technique, stage P1 represents the prepubertal state (B1/ G1; PH1) and stage P3 the postpubertal state (B5/G5; PH5). All indicators of maturational change between these two extremes have been combined into the P2 stage. Thus variations within individuals between the different aspects of secondary sexual development are impossible to quantify and, in terms of research to investigate variability in maturation, the pubertal staging technique loses significant sensitivity. The variation of maturation between different aspects of maturity presents difficulties in implying a general maturational level to the individual. For instance, entry into the early stages of puberty is apparently not associated with any particular level of skeletal maturity except in the broadest sense. The only real exception to this rule, with regard to skeletal and sexual maturation, is menarcheal age, in which skeletal age and chronological age are associated at a level of 0.35 and in which menarche tends to occur at a skeletal age of 12.5—14.0 "years" regardless of chronological age.

It may appear strange that these two maturational processes do not appear to be more closely related. As one would expect, there is evidence that skeletal maturity is related to changing levels of growth hormones, such as insulin-like growth factor-1 (IGF-1), during puberty.[25] Sex steroids also increase dramatically during puberty and the action of these hormones (growth hormone, IGF-1, testosterone, estrogen) clearly affects the mineralization of the skeleton.[26] It may be that our ability to assess secondary sexual development is simply not sophisticated enough to accurately reflect the hormonal changes that cause the morphological changes associated with puberty. Menarche, on the other hand, is usually an obvious and accurately timed maturational event that therefore can be statistically associated with skeletal maturity.

Maturity indicators derived from mathematical functions that describe the growth curve might be far more useful than morphological indicators because much closer associations are evident between markers of somatic growth and skeletal maturity. For instance, skeletal and chronological ages are known to be uncorrelated at 95% of mature height and therefore skeletal age is more or less fixed at 95% mature height regardless of chronological age. Thus, function parameters that have a real biological meaning have the potential to be appropriate maturity indicators. The problem is that most existing functions have resulted from attempts to smooth the growth curve or to reduce data rather than to understand the biology of growth. For example, the family of models proposed by Preece and Baines[27] (see Chapter 3) resulted from attempts to model the total growth curve with the fewest parameters. Previous attempts by Bock et al.[28] had resulted in double or triple logistic curves involving nine or more parameters for which clear biological meanings were not apparent. Preece and Baines[27] were able to model using just five parameters to which they were also able to assign biological meaning, e.g. age and height at peak height velocity (PHV). That "biological" meaning resulted from the fact that high correlations were apparent between the function parameter and the maturational event. For example, in model 1, θ correlated with age at PHV at a 0.99 level for boys and a 0.97 level for girls, and Hθ at a 0.99 level with height at PHV in both males and females. But one needs to be cautious about implying direct associations between the function parameters and the maturational events; the rate constants S0 and S1 for instance, correlated most strongly with velocity at take-off and velocity at PHV but only at the 0.50 and 0.55 levels. However, the possibility of using function parameters as maturity indicators is an attractive prospect, particularly as mathematical modeling moves us closer to a more accurate depiction of the pattern of human growth.

20.2.4 Sexual Dimorphism

Ideally, any method that assesses maturity should be able to assess the same process of maturation in both males and females. That criterion is true of skeletal and dental maturity assessment methods and also of methods that might be developed from mathematical models of the pattern of human growth. It is not, of course, true of all aspects of secondary sexual development, although the gender-specific assessment methods have a great deal in common. In the former methods sexual dimorphism is accounted for by having gender-specific scores for each bone or tooth and in the latter by identifying equivalent functional processes in the different sexes. However, the interpretation of maturation, or the meaning of the attainment of a particular level of maturity, may be different within the sexes. For instance, it could be argued that spermarche and menarche are equivalent stages of maturation in males and females yet their position within the pattern of growth is quite different and thus their association with other aspects of maturation also differs. Extensive data on menarche demonstrate that it occurs following PHV and towards the latter part of secondary sexual development, i.e. in breast stage 3, 4 or 5. Relatively sparse data on spermarche identifies its occurrence at approximately 14 years in boys, which would be in the early or middle part of the adolescent growth spurt and thus indicative of an earlier stage of pubertal maturation.

20.2.5 Maturity and Size

The fact that a large individual is likely to be older and thus more mature than a small individual was emphasized earlier in this chapter. This might indicate that size should in some way be included in a consideration of maturation. Indeed, the early methods of skeletal maturity assessment by planimetry used precisely that reasoning. It is now clearly recognized that, except in very general terms, size does not play a part in the assessment of maturation. Size does, however, enter assessment as a maturity indicator as a ratio measure. For example, the maturity indicator for stage D in the radius of the TWII system is the fact that the epiphysis is "half or more" the width of the metaphysis, i.e. the size is relative to another structure within the same area. However, except for such a ratio situation, the only maturity assessment method that uses a quantitative indicator of maturity is testicular volume, 4 ml represents the initiation of pubertal development and 12 ml mid-puberty. This is not to say that there is no variation in testicular volume. Like all aspects of growth and development, variability is an inherent aspect of testicular growth. Clinicians, however, use the above measures as indicators of normal testicular growth and of the initial and middle stages of pubertal development.

20.3 METHODS OF ASSESSMENT

Maturation is assessed using a combination of processes and events. Maturational "processes" include secondary sexual development, dental development and skeletal development. Maturational "events" include those aspects of maturation that occur once and provide an unambiguous signal that the individual has reached a particular level of maturity. Examples are the exact age at which menarche (the first menstrual period) is experienced in girls or the exact age of PHV during the adolescent growth spurt.

20.3.1 Secondary Sexual Development

Secondary sexual development is assessed using maturity indicators that provide discrete stages of development within the continuous process of maturation. The most widely accepted assessment scale is described as the Tanner scale or the Tanner staging technique. It was developed by Tanner[29] and was based on the work of Reynolds and Wines[30] and Nicholson and Hanley.[13] Tanner[29] divided the processes of breast development in girls, genitalia development in boys and pubic hair development in both sexes into five stages, and axillary hair development in both sexes into three stages. The usual terminology is to describe breast development in stages B1—B5, genitalia development in stages G1—G5, pubic hair development in stages PH1—PH5 and axillary hair development in stages Al—A3.

Breast Development (Figure 20.3)

Stage 1: Preadolescent: elevation of papilla only.

Stage 2: Breast bud stage: elevation of breast and papilla as small mound. Enlargement of areolar diameter.

Stage 3: Further enlargement and elevation of breast and areola, with no separation of their contours.

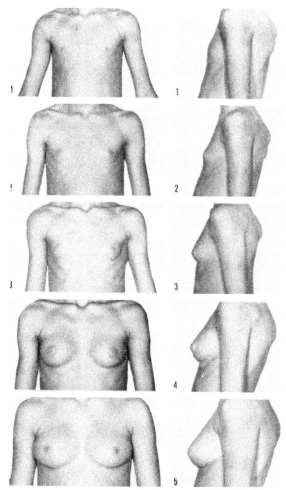

Figure 20.3 Breast standards from the Tanner method. *(Source: Tanner.[29])*

Stage 4: Projection of areola and papilla to form a secondary mound above the level of the breast.

Stage 5: Mature stage: projection of papilla only, due to recession of the areola to the general contour of the breast.

Genitalia Development (Figure 20.4)

Stage 1: Preadolescent: testes, scrotum and penis are of about the same size and proportion as in early childhood.

Stage 2: Enlargement of scrotum and testes: the skin of the scrotum reddens and changes in texture. There is little or no enlargement of penis at this stage.

Stage 3: Enlargement of penis: this occurs first mainly in length. Further growth of testes and scrotum.

Stage 4: Increased size of the penis with growth in breadth and development of glans. Further enlargement of testes and scrotum; increased darkening of scrotal skin.

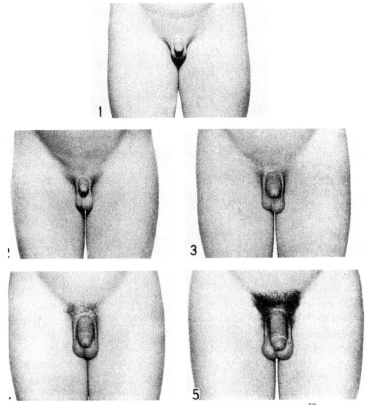

Figure 20.4 Genitalia standards from the Tanner method. *(Source: Tanner.[29])*

Stage 5: Genitalia adult in size and shape.

Pubic Hair Development (Figure 20.5)

Stage 1: Preadolescent: the vellus over the pubes is not further developed than that over the abdominal wall, i.e. no pubic hair.

Stage 2: Sparse growth of long, slightly pigmented downy hair, straight or only slightly curled, appearing chiefly at the base of the penis or along the labia.

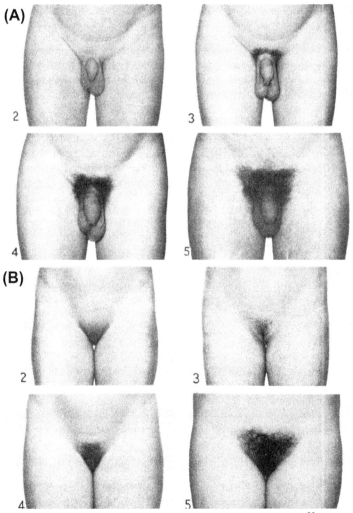

Figure 20.5 Pubic hair standards from the Tanner method. *(Source: Tanner.[29])*

Stage 3: Considerably darker, coarser and more curled. The hair spreads sparsely over the junction of the pubes.

Stage 4: Hair now resembles adult in type, but the area covered by it is still considerably smaller than in the adult. No spread to the medial surface of the thighs.

Stage 5: Adult in quantity and type with distribution of the horizontal or classically feminine pattern. Spread to the medial surface of the thighs but not up the linea alba or elsewhere above the base of the inverse triangle.

20.3.2 Clinical Evaluations

The assessment of secondary sexual development is a standard clinical procedure and at such times the full Tanner scale is used. There are some practical problems with the Tanner stages, however, in that the unequivocal observation of each stage is often dependent on having longitudinal observations. In most situations, outside the clinical setting, the observations are cross-sectional. This practical difficulty has led to the amalgamation of some of the stages to create pubertal stages. These pubertal stages are on either a three- or four-point scale and combine breast/genitalia development with pubic hair development.[31–33] Assessing breast/genitalia development with pubic hair development is obviously much easier than assessing these maturity indicators separately, but inevitably leads to a lack of sensitivity in the interpretation of the timing and duration of the different stages of pubertal development. Indeed, the intrasubject variation in the synchronous appearance of pubic hair and breast/genitalia stages, illustrated in British children by Marshall and Tanner,[3,4] suggests that it may be misleading to expect stage synchronization in as many as 50% of normal children.

20.3.3 Self-Assessment of Pubertal Status

The assessment of secondary sexual characteristics is, to some extent, an invasive procedure in that it invades the privacy of the child or adolescent involved. Thus, such assessments on normal children who participate in growth studies, as opposed to those being clinically assessed, are problematical from both ethical and subject compliance viewpoints. In order to overcome this problem the procedure of "self-assessment" has been developed and validated in a number of studies.

The self-assessment procedure requires the child to enter a well-lit cubicle or other area of privacy in which are provided pictorial representations of the Tanner scales and suitably positioned mirrors on the wall(s). The pictures may be either in photographic or line drawing styles as long as the contents are clear. To each picture of each stage is appended an explanation, in the language of the participant, of what the stage represents. The participant is instructed to remove whatever clothing is necessary in order for them to be able to properly observe their pubic hair/genitalia or pubic hair/breast development in the mirrors. The participant then marks on a separate sheet their stage of development and seals that sheet within an envelope on which is marked the study identity number of the participant. The envelope is either left in the cubicle or handed to the observer on leaving the cubicle.

The results of validation studies vary greatly depending upon the age of the participants (e.g. early or late adolescence),[33] gender,[33,34] the setting in which assessments are performed (e.g. school or clinic),[35,36] ethnicity,[37] and whether they are a distinct diagnostic group such as cystic fibrosis[38] or anorexia nervosa[39] or socially disadvantaged.[40] Younger, less developed children tend to overestimate their development and older more developed children tend to underestimate. Boys have been found to overestimate their development while girls have been more consistent with expert assessment.[34] The amount of attention given to explaining the required procedure appears to be of major importance. Thus, excellent rating agreement between physicians and adolescents has been found in clinical settings, with kappa coefficients between 0.66 and 0.91,[36,41,42] but rather less agreement in school settings (kappa = 0.35–0.42; correlations = 0.25–0.52).[35,36] Improved agreement in clinical settings probably reflects the more controlled environment of a physician's surgery as opposed to a school. The main reason for low correlations and thus poor validity in any setting with any group of participants is likely to be centered on the amount of explanation that is provided to the child. When the participant has been the subject of a clinical trial, and the scientist/clinician has spent considerable time and effort ensuring that the child is completely apprised of what he or she has to do, then validity is high. Less effort in explaining procedures leads to lower validity.

The procedure that should be adopted is that the observer should explain the procedure thoroughly to the participant using appropriate (non-scientific) language and invite questions to ensure that the participant fully understands the procedure. Only when the observer is sure that understanding is total should the child be allowed to follow the procedure. Randomized reliability assessments by the observer would, of course, be ideal but would also be ethically difficult to substantiate.

20.3.4 Dental Development

Dental development is best assessed by taking panoramic radiographs of the mandible and maxilla and scoring the stages of formation and calcification of each tooth using the method developed by Demirjian, Goldstein and Tanner[12] and Demirjian and Goldstein.[43] Scores are assigned to the stages of development of the seven mandibular teeth on the left side (there are no significant between-side differences) and these lead to a dental maturity score comparable to the skeletal maturity scores resulting from the Tanner–Whitehouse skeletal maturity technique described below. This score can be translated into the dental age. A similar system is available for sets of four teeth, seen on apical radiographs, notably M1, M2, PM1, PM2 or alternatively I1, M2, PM1, PM2.

Concern over the exposure of normal children to radiation has resulted in tooth emergence as the most commonly used method to obtain estimates of dental maturity. The emergence of the teeth above the level of the gum is recorded either by oral inspection or in a dental impression. Most observers have considered that a tooth has emerged if any part has pierced the gum, but some have used the criterion of the tooth being halfway between gum and final position.[44] Three types of standard have been developed that give the number of teeth emerged at specific ages, or the average age when 1, 2, 3, etc., teeth have emerged, or the median age in a population for the emergence of a specific tooth or pair of teeth. The latter technique is considered the best for permanent teeth because of the individual variation in the order of emergence of each tooth pair.[44]

20.3.5 Skeletal Development

Although a number of techniques exist to assess skeletal maturity, assessment procedures have been dominated by two different approaches to the problem; the "atlas" technique of Greulich and Pyle[6] and the Tanner—Whitehouse "bone-specific scoring" technique.[11] Both use the left hand and wrist to estimate a skeletal age or bone age, yet they are different both in concept and in method. Greulich—Pyle bone ages are most commonly assessed by comparing a radiograph to a series of standard radiographs photographically reproduced in the atlas. The chronological age assigned to the standard most closely approximating the radiograph is the bone age of the subject. In practice, a more precise estimate of bone age may be obtained by assessing each bone in the hand and wrist separately, but this is rarely done. Thus, there are errors in most Greulich—Pyle estimations because the dysmaturity present in the hand and wrist is not acknowledged. The system is based on subjects from Cleveland, Ohio, who were assessed during the 1920s and 1930s. The Tanner—Whitehouse system requires 20 bones of the hand and wrist to be assessed individually and a score to be assigned to each. The summation of the scores results in a bone maturity score which is equivalent to a particular bone age. This technique originally used subjects from a variety of studies conducted in the south of England during the 1950s and 1960s. Although the latter is more recent, the effect of positive secular trends and population differences in average maturity status means that estimates of skeletal age based on either technique must be viewed with some degree of caution, although later editions of the Tanner—Whitehouse method (TW3) have corrected for secular trends by using an updated source sample of European children from the 1980s and 1990s.[45] However, the statistical rationale of the bone-specific scoring technique can be applied to any series of radiographs from a representative sample of a population. In contrast to the atlas technique, it is thus possible to develop specific national references for the assessment of skeletal maturity using a bone-specific approach, which would result in a more sensitive clinical appraisal.

20.3.6 Age at Menarche

Age at menarche is usually obtained in one of three ways: status quo, retrospectively or prospectively. Status quo techniques require the girls to respond to the question, "Do you have menstrual cycles ('periods')?" The resulting data on a sample of girls will produce a classical dose—response sigmoid curve that may be used to graphically define an average age at menarche. More commonly, the data are analyzed using logit or probit analysis to determine the mean or median age at menarche and the parameters of the distribution such as the standard error of the mean or the standard deviation. Retrospective techniques require the participants to respond to the question, "When did you have your first period?" Most adolescents can remember to within a month, and some to the day, when this event occurred. Others may be prompted to remember by reference to whether the event occurred during summer or winter, whether the girl was at school or on holiday, and so on. One interesting result of such retrospective analyses is that there appears to be a negative association between the age of the women being asked and the age at which they report menarche — the older the women the younger they believe they were. Such results have been found in both developed and developing countries and cast a seed of doubt about the reliability of retrospective methods beyond the teenage and early adult years. Prospective methods are normally only used in longitudinal monitoring situations such as repeated clinic visits or longitudinal research studies. This method requires the teenager to be seen at regular intervals (usually every 3 months) and to be asked on each occasion whether or not she has started her periods. As soon as the response is positive an actual date on which menarche occurred can be easily obtained.

There is little doubt that the prospective method is the most accurate in estimating menarcheal age but it has the disadvantage of requiring repeated contact with the subjects. That is seldom possible except in clinical situations and it is thus more likely that status quo and retrospective methods are the techniques of choice. Status quo techniques that rely on logit or probit analysis require large sample sizes because the analysis requires the data to be grouped according to age classes. With few subjects broader age ranges are required such as whole or half years, with a consequent loss of precision in the mean or median value. Retrospective methods result in parametric descriptive statistics but have the problem of the accuracy of recalled ages at this particular event.

20.3.7 Secondary Sexual Events in Boys

While status quo, prospective and retrospective methods may easily obtain age at menarche, assessments of secondary sexual development in boys are complicated by the lack of a similar clearly discernible maturational event. Attempts to obtain information on the age at which the voice breaks, or on spermarche, are complicated by the time taken for the voice to be consistently in a lower register, and the logistical complications involved in the assessment of spermarche. Testicular volume, using the Prader orchidometer,[46] is commonly the only measure of male secondary sexual development outside the rating scales previously mentioned, although other measurement techniques have been described to estimate testicular volume.[47]

The detection of spermatozoa in the urine has been proposed as a quick, non-invasive method to assess the functional state of the maturing gonad and may be useful as a screening technique in population studies.[48–53] Its use, however, may be limited because longitudinal[52,54] and cross-sectional[51] studies have shown that spermaturia is a discontinuous phenomenon.

20.3.8 Landmarks on the Growth Curve

The identification of landmarks on the human growth curve that can be used for comparison between individuals or groups started with age at PHV. This is the most distinctive feature of the velocity curve during adolescent growth and may be determined in individual longitudinal data as a change in acceleration from positive to negative values. Use of other landmarks on the curve, such as age at take-off and at the cessation of growth, and the magnitude of height or weight velocity at these ages, did not become prevalent until the implementation of mathematical curve-fitting techniques became possible using personal-computer-based software. Initial curve-fitting techniques used only part of the growth curve (e.g. from birth to the start of adolescence) or involved the addition of different functions. The major problem with these early techniques, apart from their mathematical complexity and biological interpretation, was their relative inability to cover the transition between developmental periods such as preadolescence to adolescence. This was solved to a certain extent by the development of single curves that described growth from birth through to adulthood.[27,28] However, long-term parametric models have the advantage that the researchers preselect the shape of the resulting growth curve. The choice of the model necessitates the acceptance of its form as being representative of the pattern of growth. Individuals or samples departing from the standard pattern of growth in height or weight would not be fitted well by any of these parametric functions. Estimates of landmarks on the growth curve are difficult to determine. Tanner and Davies,[55] for instance, when developing the clinical longitudinal standards for American children, relied on empirically derived values for the magnitude of peak velocity because "parametric curves … are insufficiently flexible to accommodate the full rise of the observed curves during adolescence".[55, p. 328] The widely used Preece—Baines curve,[27] for example, is known to underestimate the peak velocity.

Non-parametric models, such as the smoothing spline function[56] and kernel estimation,[57,58] have been proposed to overcome the problems inherent in preselection of a pattern of growth. Non-parametric techniques are usually short-term functions that smooth adjacent data points rather than fit a function to data from birth to adulthood. They have been useful in demonstrating the sensitivity of growth analysis using acceleration[57,58] but cannot result in mathematically derived values for adolescent landmarks such as the age and magnitude of peak velocity. Such landmarks, if taken from curves that have been smoothed using non-parametric techniques, may be more accurately determined than if derived from a parametric function. However, that accuracy is dependent on the frequency of data points during the period of adolescence. The inability of preselected parametric functions to fit abnormal growth and the retrospective nature of growth assessment of non-parametric methods make these techniques useful as research tools but not for diagnosis and monitoring the value of treatment.

20.4 SUMMARY

The assessment of maturation depends on the identification of a series of maturity indicators that characterize the transition from immaturity to maturity in all children. These maturity indicators must be universal, sequential, discriminant, reliable, valid and complete in their characterization of maturation. The study and assessment of maturity has been confined to processes and events: the processes of skeletal, dental and sexual maturity; and the events of menarche, spermarche and inflection on the growth curve, such as peak height velocity.

REFERENCES

1. Bogin B. *Patterns of human growth*. Cambridge: Cambridge University Press; 1988.
2. Tanner JM. Growth and endocrinology of the adolescent. In: Gardner L, editor. *Endocrine and genetic diseases of childhood*. 2nd ed. Philadelphia, PA: WB Saunders; 1975.
3. Marshall WA, Tanner JM. Variations in the pattern of pubertal changes in girls. *Arch Dis Child* 1969;**44**:291−303.
4. Marshall WA, Tanner LM. Variations in the pattern of pubertal changes in boys. *Arch Dis Child* 1970;**45**:13−23.
5. Acheson RM. Maturation of the skeleton. In: Falkner F, editor. *Human development*. Philadelphia, PA: Saunders; 1966. p. 465−502.
6. Greulich WW, Pyle SI. *Radiographic atlas of the skeletal development of the hand and wrist*. Palo Alto, CA: Stanford University Press; 1959.
7. Acheson RM. A method of assessing skeletal maturity from radiographs. *J Anat Lond* 1954;**88**:498−508.
8. Acheson RM. The Oxford method of assessing skeletal maturity. *Clin Orthop* 1957;**10**:19−39.
9. Tanner LM, Whitehouse RH. *Standards for skeletal maturity. Part I*. Paris: International Children's Centre; 1959.
10. Tanner JM, Whitehouse RH, Healy MJR. *A new system for estimating the maturity of the hand and wrist, with standards derived from 2600 Health British Children. Part II. The scoring system*. Paris: International Children's Centre; 1962.
11. Tanner JM, Whitehouse RH, Cameron N, Marshall WA, Healy MJR, Goldstein H. *Assessment of skeletal maturity and prediction of adult height*. 2nd ed. London: Academic Press; 1983.
12. Demirjian A, Goldstein H, Tanner JM. A new system of dental age assessment. *Hum Biol* 1973;**45**:211−27.
13. Nicholson AB, Hanley C. Indices of physiological maturity. *Child Dev* 1952;**24**:3−38.
14. Meindl RS, Lovejoy CO. Ectocranial suture closure: a revised method for the determination of skeletal age at death and blind tests of its accuracy. *Am J Phys Anthropol* 1985;**68**:57−66.

15. Todd TW. *Atlas of skeletal maturation. Part L. The hand.* St Louis, MO: CV Mosby; 1937.
16. Hellman M. Ossification of epiphyseal cartilages in the hand. *Am J Phys Anthropol* 1928;**11**:223—57.
17. Rotch TM. A study of the development of the bones in childhood by the Roentgen method, with the view of establishing a developmental index for the grading of and the protection of early life. *Trans Am Assoc Phys* 1909;**24**:603—30.
18. Bardeen CR. The relation of ossification to physiological development. *J Radiol* 1921;**2**:1—8.
19. Sontag LW, Lipford J. The effect of illness and other factors on appearance pattern of skeletal epiphyses. *J Pediatr* 1943;**23**:391—409.
20. Lowell F, Woodrow H. Some data on anatomical age and its relation to intelligence. *Pedagog Semin* 1922;**29**:1—15.
21. Carter TM. Technique and devices in radiographic study of the wrist bones of children. *J Educ Psychol* 1926;**17**:237—47.
22. Flory CD. Osseous development in the hand as an index of skeletal development. *Monogr Soc Res Child Dev* 1936;**1**:3.
23. Pryor JW. The hereditary nature of variation in the ossification of bones. *Anat Rec* 1907;**1**:84—8.
24. Buga GAR, Amoko DHA, Ncayiyana DJ. Sexual behaviour, contraceptive practice and reproductive health among school adolescents in rural Transkei. *S Afr Med J* 1996;**86**:523—7.
25. Masoud M, Masoud I, Kent RL, Gowharji N, Cohen LE. Assessing skeletal maturity by using blood spot insulin-like growth factor 1 (IGF-1) testing. *Am J Orthodont Dentofac Orthop* 2008;**134**:209—16.
26. Mauras N, Rogol AD, Haymond MW, Veldhuis JD. Sex steroids, growth hormone, insulin-like growth factor-1: neuroendocrine and metabolic regulation in puberty. *Horm Res* 1996;**45**:74—80.
27. Preece MA, Baines MI. A new family of mathematical models describing the human growth curve. *Ann Hum Biol* 1978;**5**:1—24.
28. Bock RD, Wainer H, Peterson A, Thissen LM, Roche A. A parameterization for individual human growth curves. *Hum Biol* 1973;**45**:63—80.
29. Tanner JM. *Growth at adolescence.* Oxford: Blackwell Scientific Publications; 1962.
30. Reynolds EL, Wines JV. Physical changes associated with adolescence in boys. *Am J Dis Child* 1948;**75**:329—50.
31. Kulin BE, Bwibo N, Mutie D, Santner SJ. The effect of chronic childhood malnutrition on pubertal growth and development. *Am J Clin Nutr* 1982;**36**:527—36.
32. Chaning-Pearce SM, Solomon L. A longitudinal study of height and weight in black and white Johannesburg children. *S Afr Med J* 1986;**70**:743—6.
33. Varona-Lopez W, Guillemot M, Spyckerelle Y, Deschamps JP. Self assessment of the stages of sex maturation in male adolescents. *Pediatrie* 1988;**43**:245—9.
34. Sarni P, de Toni T, Gastaldi R. Validity of self-assessment of pubertal maturation in early adolescents. *Minerva Pediatr* 1993;**45**:397—400.
35. Wu WH, Lee CH, Wu CL. Self-assessment and physician's assessment of sexual maturation in adolescents in Taipei. *Chung Hua Min Kuo Hsiao ErhKo I Hsueh Hui Tsa Chih* 1993;**4**:125—31.
36. Schlossberger NM, Turner RA, Irwin Jr CE. Validity of self-report of pubertal maturation in early adolescents. *J Adolesc Health* 1992;**13**:109—13.
37. Hergenroeder AC, Hill RB, Wong WW, Sangi-Haghpeykar H, Taylor W. Validity of self-assessment of pubertal maturation in African American and European American adolescents. *J Adolesc Health* 1999;**24**:201—5.
38. Boas SR, Falsetti D, Murphy TD, Orenstein DM. Validity of self-assessment of sexual maturation in adolescent male patients with cystic fibrosis. *J Adolesc Health* 1995;**17**:42—5.
39. Hick KM, Kutzman DK. Self-assessment of sexual maturation in adolescent females with anorexia nervosa. *J Adolesc Health* 1999;**24**:206—11.
40. Hardoff D, Tamir A. Self-assessment of pubertal maturation in socially disadvantaged learning-disabled adolescents. *J Adolesc Health* 1993;**14**:398—400.
41. Duke PM, Litt IF, Gross RT. Adolescents' self-assessment of sexual maturation. *Pediatrics* 1980;**66**:918—20.
42. Brooks-Gunn J, Warren MP, Russo J, Gargiulo J. Validity of self-report measures of girls' pubertal status. *Child Dev* 1987;**58**:829—41.

43. Demirjian A, Goldstein H. New systems of dental maturity based on seven and four teeth. *Ann Hum Biol* 1976;**3**:411–21.

44. Eveleth PB, Tanner JM. *Worldwide variations in human growth.* 2nd ed. Cambridge: Cambridge University Press; 1990.

45. Tanner JM, Healy MJR, Goldstein H, Cameron N. *Assessment of skeletal maturity and prediction of adult height (TW3 method).* 3rd ed. London: WB Saunders; 2001.

46. Prader A. Testicular size: assessment and clinical importance. *Triangle* 1966;**7**:240.

47. Daniel WA, Feinstein RA, Howard-Pebbles P, Baxley WD. Testicular volume of adolescents. *J Pediatr* 1982;**101**:1010–2.

48. Schaefer F, Marr J, Seidel C, Tilgen W, Scharer K. Assessment of gonadal maturation by evaluation of spermaturia. *Arch Dis Child* 1990;**65**:1205–7.

49. Baldwin B. The determination of sex maturation in boys by a laboratory method. *J Comp Psychol* 1928;**8**:39–43.

50. Richardson D, Short R. Time of onset of sperm production in boys. *J Biosoc Sci* 1978;**5**:15–25.

51. Hirsch M, Shemesh J, Modan M. Emission of spermatozoa: age of onset. *Int J Androl* 1979;**2**:289–98.

52. Nielson CT, Skakkebaek NS, Richardson DW. Onset of the release of spermatozoa (spermarche) in boys in relation to age, testicular growth, pubic hair, height. *J Clin Endocrinol Metab* 1986;**62**:532–5.

53. Kulin HE, Frontera ME, Demers LD, Bartholomew MJ, Lloyd TA. The onset of sperm production in pubertal boys. *Am J Dis Child* 1989;**143**:190–3.

54. Hirsch M, Lunenfeld B, Modan M, Oradia J, Shemesh J. Spermarche — the age of onset of sperm emission. *J Adolesc Health Care* 1985;**6**:35–9.

55. Tanner JM, Davies PSW. Clinical longitudinal standards for height and height velocity for North American children. *J Pediatr* 1985;**107**:317–29.

56. Largo RH, Gasser TH, Prader A, Stuetzle W, Huber PJ. Analysis of the human growth spurt using smoothing spline functions. *Ann Hum Biol* 1978;**5**:421–34.

57. Gasser T, Kohler W, Muller HG, Kneip A, Largo R, Molinari L, et al. Velocity and acceleration of height growth using kernel estimation. *Ann Hum Biol* 1984;**11**:397–411.

58. Gasser T, Kohler W, Muller HG, Largo R, Molinari L, Prader A. Human height growth: correlational and multivariate structure of velocity and acceleration. *Ann Hum Biol* 1985;**12**:501–15.

SUGGESTED READING

Acheson RM. Maturation of the skeleton. In: Falkner F, editor. *Human development.* Philadelphia, PA: Saunders; 1966. p. 465–502.

Cameron N, Jones LL. Growth, maturation and age. In: Black S, Aggrawal A, Payne-James J, editors. *Age estimation in the living: the practitioner's guide.* Oxford: Wiley-Blackwell; 2010.

Greulich WW, Pyle SI. *Radiographic atlas of the skeletal development of the hand and wrist.* Palo Alto, CA: Stanford University Press; 1959.

Marshall WA, Tanner JM. Variations in the pattern of pubertal changes in girls. *Arch Dis Child* 1969;**44**:291–303.

Marshall WA, Tanner JM. Variations in the pattern of pubertal changes in boys. *Arch Dis Child* 1970;**45**:13–23.

Mauras N, Rogol AD, Haymond MW, Veldhuis JD. Sex steroids, growth hormone, insulin-like growth factor-1: neuroendocrine and metabolic regulation in puberty. *Horm Res* 1996;**45**:74–80.

Susman EJ, Houts RM, Steinberg L, Belsky J, Cauffman E, DeHart G, et al. Longitudinal development of secondary sexual characteristics in girls and boys between ages 9 ½ and 15 ½ years. *Arch Pediatr Adolesc Med* 2010;**164**:166–73.

INTERNET RESOURCES

A variety of internet sources of information on maturity assessment is apparent when using a major search engine such as Google. However, the reader is warned that searches that include the term sexual, as in secondary sexual development, will inevitably link to websites featuring adult content.

CHAPTER 21

Growth References and Standards

T.J. Cole
UCL Institute of Child Health, University College London, London WC1N 1EH, UK

Contents

N. Cameron & B. Bogin (eds): Human Growth and Development, Second edition.
ISBN 978-0-12-383882-7, Doi: 10.1016/B978-0-12-383882-7.00021-0

537

21.1 INTRODUCTION

21.1.1 Purpose

Growth assessment is comparison. To measure the height of an individual an accurately calibrated instrument called a ruler is used. The height status of an individual is assessed in just the same way, with a form of calibrated instrument called a growth reference. Without some form of reference, growth assessment is arbitrary and unsatisfactory.

But there is an important difference between measuring a child's height and assessing their growth status. On the whole, rulers agree about how long a meter is, but an individual's growth rate depends on a wide variety of factors including their sex, age, pubertal stage, parental size, ethnicity, health, socioeconomic status and so on. The ruler to assess it needs to be multidimensional to take all the relevant factors into account. This is the role of the growth reference, to provide a way of displaying expected growth as a function of (some of) these other factors in a compact, accessible and visually appealing form.

21.1.2 Growth and Size

It is important to be clear about the distinction between growth on the one hand and size on the other. Strictly speaking, growth is a form of *velocity*, the rate of change in size over time, and it requires measurements on at least two occasions to assess it. The term "growth chart" is unfortunate as most such charts do not assess growth at all, they measure size. Tanner has proposed that, by analogy with growth and velocity, charts measuring size should be called "distance" charts — they measure the *distance* the child has traveled on the journey from conception to adulthood. As shown later, most distance charts not only fail to assess growth, but their underlying reference data also lack any information about growth. Nevertheless, this chapter follows common practice by referring to size or distance charts as growth charts where necessary.

21.1.3 Chart Form

A growth reference is essentially a database defining the statistical distribution of one or more measures of size and/or growth, indexed by sex, age and/or other factors. The information may be summarized in a table, but for clinical purposes is usually presented as a chart plotted against age. This form of presentation has developed over the past hundred years or so.[1] An important principle of growth charts is that the curves making up the chart should appear smooth.

The frequency distribution of each measurement can be summarized in several ways. At its simplest, the mean and standard deviation (SD) are tabulated by age and sex. Then, assuming a normal or Gaussian distribution, the mean and SD define the entire distribution and its centiles (see below).

Curves are drawn on the chart to represent the distribution at each age. A common pattern is to draw seven curves, corresponding to the mean, 1 SD above and below the mean, then 2 SDs and 3 SDs similarly. So there are seven curves spaced 1 SD apart, the pattern used for the World Health Organization (WHO) 2006 growth standard from 0 to 5 years of age.[2] The British 1990 reference[3] and its successor, the UK−WHO growth charts for age 0−4 years based on the WHO standard,[4] use a format based on nine curves that are spaced two-thirds of an SD apart. This spacing gives a set of curves very similar to the centile-based curves that are described below, but with a greater distributional range.[5] With a normal distribution SD-based curves appear as equally spaced on the chart at each age.

When the distribution is not normal, so that the mean and SD are insufficient by themselves to define it, the frequency distribution can be specified in terms of empirical centiles. A centile is a point on the distribution that splits the population into two specified fractions. The 50th centile, also known as the median, is the mid-point of the distribution, with 50% to the left of it and 50% to the right. The 3rd centile has 3% to the left of it and 97% to the right. For non-normal data centiles are estimated directly from the data (empirical centiles), whereas with normal data the centiles are calculated from the mean and SD. The latter approach is more efficient, as the mean and SD are estimated more precisely than individual centiles.

A set of several centiles is used in the growth chart to represent the range of the distribution, the values chosen to be symmetric about the median. A common set is the seven centiles 3rd, 10th, 25th, 50th, 75th, 90th and 97th, approximately two-thirds of an SD apart for a normal distribution. For comparison, the nine-centile set for the UK−WHO growth chart, based on an exact two-thirds SD spacing, is the 0.4th, 2nd, 9th, 25th, 50th, 75th, 91st, 98th and 99.6th, with the middle seven centiles very similar.[3] Other sets may include three or five centiles. Cole[5] discusses reasons why particular centile sets are used.

A centile *curve* is a curve joining up the values of a specified centile at different ages. So the percentage chance of an individual child's value lying below a given centile curve is given by the value of the centile, e.g. 3% chance below the 3rd centile curve. In addition, this chance is the same at all ages, assuming that the child comes from the reference population on which the chart is based.

21.1.4 Assessment

The curves on the chart represent either centiles or fractions of an SD above or below the mean. The assessment of individual subjects follows the same principle. The subject's measurement is plotted on the chart and the corresponding centile or SD relative to the mean is read off it. Take a girl aged 3 years who is 90 cm tall: her height is on the 9th centile of the UK—WHO reference, just above the 9th centile on the chart, and it corresponds to 1.33 SDs below the mean. By convention, the child's SD position on the chart is known as the SD score, or SDS or Z-score for short.

There has been much debate about the pros and cons of centiles and SD scores for assessing growth. Centiles are on a scale from 1 to 99, centered on 50 (0 and 100 are off-scale), and they correspond to the percentage chance of a reference child having a smaller value than the subject. The SD score scale is centered on zero, with an SD of 1, and is normally distributed.

Centiles are easier for subjects and their parents to understand, whereas SD scores are preferred by researchers as they are better behaved statistically. In addition, they provide greater resolution than centiles in the tails of the distribution. The WHO growth standard, for example, uses SDs rather than centiles to quantify the size of malnourished individuals who lie well below the 3rd centile. The 3rd, 1st and 0.1th centiles for example correspond to SD scores of -1.9, -2.3 and -3.1, so that the region between the -2 and -3 SD curves on the chart corresponds to a very narrow centile range.

A third form of assessment is percentage of the median, where the measurement is expressed as a percentage of the median value for the child's age and sex. This is used mainly in the developing world to assess nutritional status, and is a simpler version of the SD score. An SD score of 0 always corresponds to 100% of the median, while an SD score of -2 corresponds to 92% of the median for height and only 80% for weight. The percentage of the median does not take into account the variability of the measurement, while the SD score does.

So a child's position on the chart can be expressed as a centile, an SD score or a percentage of the median. The secondary purpose of the chart is to follow the child over time and see how their position changes as they grow. Often they stay close to their previous position but they can change position quite dramatically, i.e. *cross centiles* up or down. It is useful to know how much centile crossing to expect, but the distance chart does not contain this information. A velocity chart is needed to assess centile crossing (see later).

21.1.5 Unconditional and Conditional References

Some growth references are called "conditional", meaning that the reference data are *conditional* on, or adjusted for, some specified factor. Examples are references conditional on growth tempo or mid-parental height. But it is a misleading term as all growth references are conditional to some extent — on age and sex if nothing else. "Conditional" is taken here to mean conditional on factors over and above age and sex. References for velocity are a particular and important case of conditional references.

21.1.6 Structure of the Chapter

The process of developing references consists of four main stages, involving first the choice of the reference population, then the drawing of the sample, then data collection, cleaning and analysis, and finally the design and production of the chart. The chapter follows the same structure. Conditional references involve different statistical principles from unconditional references, and are discussed separately.

21.2 DEFINING THE REFERENCE POPULATION

The choice of reference population is one of the most important decisions to make when developing a growth reference. It relates to how the reference will be used, by whom and on which subjects. There are two key questions: Is the reference to be used primarily as a clinical or as a public health tool? And is it intended to reflect "optimal" or "typical" growth?

21.2.1 Clinical or Public Health Tool

For doctors involved in the care of individual patients the growth chart is an essential part of their clinical toolkit. The child's measurement centiles are a direct measure of health, and the medical assessment of the child involves interpreting the centiles. If the child is not representative of the chart's reference population, the centiles will be biased and the growth assessment may be invalid.

For public health purposes the applicability of the chart to the individual child is less important. The aim is to summarize the nutritional status of a *group* of children, with a view to comparing the group with other groups (e.g. by socioeconomic status), so that the position of the individual child, or indeed the group, on the chart is not the primary concern.

These alternative aims are contradictory. The first requires the chart to be appropriate for the child, while the second applies to different groups of children, and it cannot be appropriate for them all. This fundamental contradiction lies behind many arguments about the use of growth charts. In practice, there is a compromise where the chart can be useful both clinically and in public health terms (see below).

For clinical use the chart's reference population needs to be clearly defined in geographical, cultural and/or social terms. An example is the British 1990 reference,[3] representative of ethnic white children living in England, Scotland and Wales in 1990. "White" was originally specified because ethnic groups differ in their growth potential. An obvious disadvantage of this definition is that British ethnic minority children are excluded from the reference, which implies that they need ethnic-specific references of their own. But in practice there are many different ethnic minorities, and the alternatives are compounded by ethnic mixing of the races, so that separate charts for all are quite impractical. The compromise is to use the British chart for everybody, irrespective of their ethnic make-up, but to introduce ethnic-specific adjustments where necessary to extend the coverage to ethnic minorities.[6] These adjustments can be estimated on relatively small samples of children, and far fewer than needed to derive a full growth chart.

Other examples of charts for clinical use are syndrome-specific charts, e.g. for children with Down's[7] or Turner's syndrome.[8] Their growth is known to differ from that of children without the disorder so the syndrome-specific chart is appropriate. Another example is a chart for breast-fed infants, who grow differently from formula-fed infants. But this is a less clear-cut example, as the mother's decision whether or not to breast feed is not only a social but also a health issue. And this relates to the question of references versus standards (see below).

For public health use the chart does not need to be based on any particular group, so long as it is politically acceptable. Charts representative of the national population are obviously popular, like the British 1990 chart, but in principle a single chart for the whole world could be just as useful. Unfortunately, politics tends to intrude at this point, and developed countries have in the past preferred to use their own charts rather than work towards international comparability. However, the advent of the WHO 2006 growth standard has changed attitudes in this respect, and many countries (including several developed countries) have now either endorsed or actively adopted the WHO standard for national growth assessment in the early years of life.

21.2.2 Reference or Standard

In addition to selection by geography or ethnic background, the reference population can be identified on health grounds, excluding children for example with a growth disorder. The assumption here is that the growth portrayed by the chart is better than for the unselected population, and so is to some degree optimal. In this case, the growth reference is known as a growth *standard* rather than a reference. (Growth references based on unselected populations are also often called standards, but strictly this is incorrect.)

References based on healthy subgroups, i.e. standards, can be contentious. Three examples are birth weight in premature babies, height in children with Down's syndrome and growth in elite groups in the developing world.

Among babies who are born preterm, those who are induced tend to be sicker and hence lighter than babies delivering spontaneously at the same gestation. For this reason some argue that birth-weight standards should exclude preterm babies who have been induced. But this ignores the fact that all preterm babies are to some extent unhealthy — they are born earlier than expected. Cole et al.[9] discuss this in more detail.

Similarly, up to one-third of children with Down's syndrome are born with cardiac defects that can materially affect their growth. Most defects are now corrected within the first year, but some children remain appreciably growth retarded later in childhood. Again the case is made to exclude from the reference those with serious cardiac defects.

Children in the developing world of high socioeconomic or "elite" status are known to grow better than their poorer contemporaries. Indeed, they can grow as well as children in the developed world.[10] This has been the motivation for some developing countries, e.g. India, to base their national growth standards on elite children.

Although there are advantages in restricting the reference population on health grounds, there are also disadvantages. First, if the chart is restricted to a healthy subset of children, how, logically, does one assess the growth of those children that have been excluded? By definition, the chart is not appropriate for them, as it portrays growth in children who have been selected to grow better on average.

Against this, elite standards, it is argued, show how the child *might* grow if their socioeconomic status were to be raised. The chart documents the *potential* for growth. This is the basis of the WHO 2006 growth standard, which aims to define normal growth in terms of breast-fed children with unconstrained growth. But this same argument does not apply to induced preterm babies or Down's syndrome children with cardiac defects; in each case their status is immutable and the chart can never be appropriate for them.

The second disadvantage of growth standards is the need to define *health*. Criteria are required to identify which reference subjects to include and which to exclude on health grounds, and this is usually arbitrary. The examples above happen to have fairly clear-cut criteria (though the severity of cardiac defect in Down's syndrome children needs to be specified), but often this does not hold. For example, should one exclude children with asthma, or renal disease, from the reference population? These conditions affect growth in some children but not others. The case can be made either way, to include or exclude them, but ultimately it is arbitrary.

For growth assessment in the developed world it is in some ways simpler to use a reference than a standard, so that all children are eligible for the reference data set irrespective of their health status. Against that, the advent of the WHO growth standard for age 0−5 has strengthened the case for using a standard, and several countries including the UK have included it as part of their national charts. Its strengths are that (1) it represents the growth of breast-fed infants, whereas a reference necessarily documents the growth of a mix of breast- and formula-fed infants, which leads to bias when breast-fed growth is the norm; and (2) it defines a relatively low "plane of nutrition" in the first 2 years, as the WHO reference children were relatively light and thin. Thus, it tends to highlight overweight rather than underweight when applied to other populations, and at a time when child obesity is an increasing concern, this represents an appropriate shift in emphasis.

21.3 DRAWING THE SAMPLE

Having settled on the reference population, the next stage is to decide the study design. This involves answering such questions as: is the focus of the study growth distance or growth velocity? How big should the sample be? How should the sample be chosen?

21.3.1 Design: Cross-Sectional, Longitudinal, Mixed-Longitudinal?

The most common form of growth study is the cross-sectional survey. This collects data on children over a range of ages, each child contributing a measurement at a single moment in time. Such a design is conventionally called a growth survey, but it contains no information about growth velocity as each child is seen only once.

To assess growth velocity the survey needs to measure subjects more than once. Where all subjects are measured repeatedly this is a longitudinal design, whereas if only some of the subjects are measured again this is a mixed longitudinal design. Longitudinal designs are more costly than cross-sectional designs for several reasons: they last longer, they have to maintain subject contact, and often it is cheaper to keep highly trained staff employed than to recruit new staff at each measurement occasion.

Longitudinal designs provide information not only on mean growth velocity, but also and more importantly on its variability. Cross-sectional designs can estimate, say, the annual growth rate of the mean through the difference in size of successive year groups, but they provide no information about the variability of growth velocity.

The main difference between a longitudinal and a mixed longitudinal design is the period over which velocity is assessed. Longitudinal designs cover longer periods. If annual velocities are the only concern, then two successive cross-sectional surveys 1 year apart, with say 50% of subjects measured on both occasions, is a mixed longitudinal design that provides all the required growth velocity information. Tanner[11] discussed this design in some detail and highlighted its statistical efficiency.

Longitudinal designs, with infants recruited at or before birth and followed up for extended periods, have been popular in the past but are less so now. See, for example, the coordinated studies carried out in London, France, Switzerland and elsewhere.[12] Their main advantage is that they provide complete growth curves for individual children, which cannot be obtained in any other way. But longitudinal designs are expensive, for the reasons described above, and mixed designs have tended to take their place.

The age range of the data is another important aspect of the design. Should it start at birth? If so, should it include some premature births? Should it extend to adulthood? When *is* adulthood: 18, 20, 25 years? The answers to these questions may relate to the ease or difficulty of obtaining the sample at particular ages, and there are statistical arguments for and against different age ranges. These issues need to be considered and agreed at the outset.

21.3.2 Sample Size

The estimation of sample size through a power calculation is standard practice in medical research, yet it is surprisingly difficult to apply to growth studies. Traditional sizes of study have developed over the years, but they are difficult to justify statistically. The problem is often that the data can be used to estimate both distance and velocity, and it is not clear which should determine the sample size.

A common rule of thumb is a sample containing 200–300 subjects per age group. This is broadly speaking the size of the European longitudinal growth studies of the 1950s and 1960s, with larger numbers recruited and rather smaller final sample sizes. Yet the number is hard to justify statistically. In addition, it is not helpful in a cross-sectional study when the width of the age group can be anything from 1 month to 1 year.

The WHO Multicentre Growth Reference Study[13] used the criterion of 200 subjects per 3-month age–sex group for a longitudinal design from birth to 2 years (i.e. 200 subjects per sex altogether), and a similar cross-sectional design over the age range 18 months to 6 years, giving a sample size of 800 per year. Extrapolated to 0–20 years this implies a sample of 16,000 subjects per sex, which is very substantial. In practice, many surveys are appreciably smaller, even down to one-tenth the size. So what is lost when the smaller sample sizes are used?

Put simply, the loss is in the resolution. The information needed to construct a growth reference distance curve is the mean and the standard deviation at each age, possibly plus information about the shape of the distribution (see later). Empirical evidence from fitting growth centiles to surveys of different sizes suggests that a survey of say 2000, or 50 per year per sex from 0–20 years, estimates the mean with high precision, and the standard deviation with moderate precision, but provides little information about the distribution. Conversely, a survey of 16,000 provides ample information on all three. This provides a "ball-park" figure for the sample size.

21.3.3 Weighting by Age, or Extending Age Range

The precision with which the curve is estimated varies with age: it is highest in the middle of the age range and lowest at the extremes owing to the presence of "edge effects". To compensate for edge effects two strategies are available: one is to oversample at the extremes, with say two or three times as many subjects in the youngest and oldest age groups compared to other ages, and the other is to extend the age range. The latter approach is not available at birth, but at the upper end the age can be increased by, say, 1–4 years. This ensures that at the original upper age the curves are estimated with much greater precision, while at the new upper age the relatively low precision does not matter.

Another age-related issue is whether measurements should be taken at precise pre-specified ages (e.g. at 3 months or 12 months), or whether they should be distributed uniformly within a given age range. In the past the statistical techniques available to fit centiles to growth data required measurements to be grouped at specific ages, but this is no longer an issue. Even so, longitudinal studies are usually designed around a set of fixed ages when measurements are made, whereas cross-sectional surveys are likely to recruit subjects within specified age ranges. The advantages of fixed ages in longitudinal studies are partly administrative, i.e. subjects are all measured at the same ages, and also that velocities can be calculated over constant periods, e.g. monthly or yearly.

21.3.4 Sampling

The subjects in the reference sample should be selected from the target population in a way that ensures generalizability, ideally by random sampling. The sample may be a simple random sample, or it may be a complex multistage design involving clusters or strata. For example, a national sample might be based on randomly selected clusters of geographical areas, then random households within areas, then random children within households. The advantages and disadvantages of the different designs involve a trade-off between complexity (i.e. cost) and efficiency (i.e. precision of the estimates). See Armitage and Berry[14] for a fuller description of sampling designs.

Children of school age can be ascertained through their schools, which is more efficient than working through households. Conversely, children of preschool age or those who have left school are difficult to sample randomly, and this can pose problems for obtaining representative samples over age ranges extending beyond school.

21.4 COLLECTING THE DATA

Having identified the reference population and sample, the next stage is to measure the subjects. This requires decisions about which measurements to make, how to make them and how to ensure their quality.

21.4.1 Selecting Measurements

The choice of measurements depends on the aim of the study. Weight and height (or length in infancy) are obvious choices as they are the two "whole-body" measurements, they require relatively simple equipment to measure them, and taken together they provide a measure of weight for height such as body mass index (BMI: $weight/height^2$).

Skinfold thicknesses are a useful proxy for regional body fat, and the contrast of, for example, triceps and subscapular skinfold gives a measure of fat distribution in limbs relative to the trunk. The main disadvantage of skinfolds is the considerable interobserver variation, which reduces their generalizability. In addition, they can be particularly difficult to measure in obese subjects.

Body circumferences, e.g. arm or waist, are simpler to measure than skinfolds, and provide information over and above the other measurements described so far. Arm circumference (also known as mid-upper arm circumference or MUAC) is a useful alternative to weight for height for assessing wasting in malnutrition.[15] In addition, arm circumference and triceps skinfold together provide an estimate of arm muscle area.[16] Waist circumference is increasingly important as a risk factor for obesity and its sequelae in children and adults.[17]

Head circumference acts as a proxy for brain size, and is usually of interest primarily in infancy when head growth is maximal.

Body proportions can be studied by measuring sitting height or cristal height and expressing it as a fraction of height. Leg length is increasingly seen as a proxy for growth in early life, and is important in life-course studies.[18] Body widths, e.g. biiliac width or biacromial width, are for more specialized anthropometry studies.

21.4.2 Choosing Location and Personnel

The next questions are: who will take the measurements, and where will these observers be based? Depending on the size of the survey, the observers may be existing anthropometrists, school nurses, practice nurses or research nurses, or alternatively they may be recruited specifically for the survey. There may be one measuring team based centrally, which travels to each region in turn, or alternatively there may be teams based in each region that make use of local staff. The measurements can be made in subjects' houses, or alternatively the subjects can be invited to a central meeting place like a clinic or school. The choices will depend on the ages of the subjects, their geographical spread and the cost and availability of staff. The choice between central and local measuring teams will depend on the number of measurement regions, the number of subjects, the time available for the survey and the likely compliance of subjects and/or their parents.

21.4.3 Equipment and Technique

Anthropometry is the primary focus of growth studies, so it is essential for the measurements to be of the highest possible quality. This requires attention to the instruments used, their calibration and maintenance, to the training of observers in terms of technique, precision and accuracy, and in particular to quality control throughout the study. For long-term studies this involves regular training sessions where observers meet together to assess intraobserver and interobserver variation by measuring and remeasuring small groups of children.

For details of measurement technique and quality control, Weiner and Lourie,[19] Cameron[20] and Lohman et al.[21] give comprehensive accounts.

21.5 CLEANING THE DATA

There is a grave temptation, once the data have been collected, to start analyzing them immediately. This is a mistake. An important preliminary stage is to *look* at the data, to search out errors of measurement or coding and fix them. Left untended they can seriously affect the validity of the analysis.

21.5.1 Diagnostic Plots: Marginal and Conditional

The key to data cleaning is the inspection of diagnostic plots, which fall into two categories: marginal and conditional. A marginal plot is a plot of one variable on its own, typically a histogram, showing the distribution of the measurement. This plot highlights the presence of any outliers, and also indicates the broad distributional shape of the variable, i.e. whether it is normally distributed (bell shaped), or if there is some skewness (one tail, usually the right, longer than the other) or perhaps kurtosis (heavy tails) or bimodality (with two peaks). With suitable software it is possible to draw the histogram, highlight each outlier, check the corresponding data, and correct or eject as appropriate, all very quickly. An example of such a package is Data Desk (Data Description Inc.), which though now rather old is designed on exploratory data analysis (EDA) principles and has a strongly visual philosophy.

Marginal diagnostic plots are useful for identifying the most obvious outliers, but they fail to pick up many others. For example, consider a height of 160 cm incorrectly coded as 60 cm. In a data set covering 0−20 years this corresponds to an adult height appearing as an infant length, so a marginal plot will fail to spot it. However, a scatterplot of height versus age will cause it to stand out as an obvious outlier. This is a *conditional* diagnostic plot of height on age.

The conditional plot works well when two variables are reasonably highly correlated (i.e. height and age here). It can be very sensitive, and for measurements like height which have a small coefficient of variation (see later) and where the correlation is high, it will spot outliers that are far less extreme than the example above.

Weight versus age works nearly as well, except that weight usually has a positively skewed distribution, with the right tail longer than the left. Spotting outliers here is more difficult, as the individual points in the extreme right tail (i.e. at the top of the scatterplot of weight versus age) are spread farther apart, and so appear to be more extreme, than those to the left. It is tempting to treat all those in the right tail as outliers, but this is generally not wise.

Ideally, the scatterplot needs to be redrawn so that the variable is approximately normally distributed, which ensures that the two tails are spread out roughly equally. This can be done using a power transformation (see later). The search for outliers is then more balanced in the two tails.

To help in the identification of outliers it is useful to have an objective criterion, particularly for outliers that are not obvious. Working with SD scores rather than the original measurements is useful as the age and sex differences are adjusted out, and weight or height for the entire data set can be plotted as a marginal diagnostic plot without involving age.

What is a reasonable range of values for the SD scores? A useful cut-off is ±5: the chance of a genuine point outside this range is vanishingly small (3 in 10 million), so even for large sample sizes of 10,000−20,000 such points are highly likely to be wrong. For appreciably smaller sample sizes a tighter cut-off of say ±4 can be used, corresponding to a chance of 3 in 100,000.

The way to deal with outliers is to go back to the original coding form and look for something obviously wrong. In longitudinal studies it should also be possible to check the consistency of the subject's measurements on other occasions. Often an error will be found which can be corrected. But there will be occasions when the measurement is apparently correct, and the question then is whether or not to retain it in the data set.

There ought to be very few such points, so that they can be described and justified on an individual basis when the analysis is written up. Statistically there is a case for omitting them even though they are correct, as they may challenge the statistical assumptions made by the analysis. If the analysis is not robust to outliers it may be seriously affected; an example is regression analysis, where a single outlying point can alter a multiple regression equation dramatically.

21.6 ESTIMATING DISTANCE CENTILES

Once the data are clean, work can start on fitting the centiles. The process of fitting distance centiles to data involves a series of choices. Is age to be treated as continuous or grouped, e.g. by whole years? Can the measurement be assumed to be normally distributed at all ages, or is some form of adjustment needed? And how should the centile curves be modeled — what form of equation is to be used? These issues are discussed in turn.

21.6.1 Age Grouped or Continuous

Splitting the data into distinct age groups is the traditional approach to centile fitting. Within each group age is less critical and the distribution can be characterized across all the data, e.g. as the mean and SD. These summary statistics can be adjusted for minor age effects,[22] plotted against age to represent the whole age range, and summarized by smooth curves drawn through them, one curve for the mean and one for the SD. Together the two curves allow any centiles to be drawn assuming that the distribution is normal, using the formula:

$$Centile_{100\alpha} = Mean + SD \times z_\alpha \qquad [21.1]$$

where z_α is the normal equivalent deviate for the required distance centile, and *Mean* and *SD* are values read off the curves at a particular age. For the median or 50th centile (*Centile$_{50}$*) $z_{0.5} = 0$, while for the 3rd centile (*Centile$_3$*) $z_{0.03} = -1.88$. Values of *Centile$_{100\alpha}$* for a series of ages can be plotted against age to give the required centile curve.

But splitting the data into age groups is inefficient and arbitrary, and age is better treated as a continuous variable. This then becomes a form of regression analysis, where a curve representing the mean is fitted to the data plotted against age. A separate curve needs to be fitted representing the SD, and several iterative methods have been proposed for this.[23–26] Again the outcome is two curves, *Mean* and *SD*, which can be plugged into eqn [21.1] to provide any required centile curves.

21.6.2 Distributional Assumptions

The previous section assumes that the data are normally distributed, so that the mean and SD are summary statistics for the distribution at each age. Usually this applies to measurements with relatively low variability, such as height or head circumference, where the coefficient of variation is less than 5%. But other more variable measurements, e.g. weight or skinfold thickness, have distributions with some degree of right skewness. Here the assumption of normality does not hold and a different approach is needed.

In the past the centiles for such measurements were obtained empirically, i.e. grouping the data by age, sorting the data into order and reading off the required centiles. The resulting centile values for each age group could then be plotted against age and smooth curves fitted through them. This led to a set of centile curves which avoided assuming an underlying distribution, normal or otherwise. The American National Center for Health Statistics (NCHS) reference[27] was constructed in this way. But it is an inefficient process as it requires age to be grouped. When age is treated as continuous the methodology goes under the general name "quantile regression",[28] and in recent years it has been extended.[29,30] One disadvantage of quantile regression is that the centile curves may touch or even in extreme cases cross.

Another way to handle skewness is to transform the data in some way to bring the distribution closer to normal, two common transformations being the Box–Cox power transform[31] and the shifted log transform.[32] Effectively this introduces a third parameter, in addition to the mean and SD, to compensate for skewness in the distribution at each age. The value of the parameter may be constant (e.g. transforming all the data to logarithms), or it may change with age in the same way that the mean and SD change.

The concept of an age-varying adjustment for skewness was first proposed by Van't Hof et al.[33] and extended and formalized by Cole.[34] Based on the Box–Cox transformation Cole called it the LMS method, the three letters L–M–S representing, respectively, λ, the Box–Cox power, μ, the median, and σ, the coefficient of variation. The median is estimated from the mean on the transformed scale, where the distribution is symmetrical as the skewness has been removed. The coefficient of variation (CV) is preferred to the SD because the SD, like the mean, tends to increase with age, whereas the CV is more constant through childhood, and indeed is often similar in infancy and adulthood. The quantities λ, μ and σ have corresponding smooth curves plotted against age, estimated by maximum likelihood, and these L, M and S curves together define any required centile curve using the equation:

$$Centile_{100\alpha} = M(1 + LSz_\alpha)^{1/L} \qquad [21.2]$$

where, as before, z_α is the normal equivalent deviate corresponding to the required centile. Substituting $L = 1$ for a normal distribution gives the simpler formula:

$$Centile_{100\alpha} = M(1 + Sz_\alpha) = M + MSz_\alpha$$

which, bearing in mind that M is the mean, S the coefficient of variation and MS the standard deviation, is the same as eqn [21.1].

An immediate spin-off of this approach is that skew data, in addition to those that are normally distributed, can be expressed as SD scores simply by rearranging eqn [21.2]:

$$z = \frac{(Measurement/M)^L - 1}{LS} \qquad [21.3]$$

Setting $L = 1$ in eqn [21.3] gives the formula:

$$z = \frac{(Measurement/M) - 1}{S} = \frac{Measurement - M}{MS}$$

which is the usual formula for calculating the SD score. This ability to express measurements as SD scores, irrespective of whether or not they come from a skew distribution, leads to substantial simplifications in the analysis of anthropometry data (see later).

Cole[34] originally described the LMS method for distinct age groups, but later extended it (with Peter Green) to continuous age.[35] Variations on the same general principle, with the skewness adjusted for in different ways, have been proposed by several authors.[30,36–41]

Healy[42] suggested a quite different way of dealing with non-normality that links the two methods described above. His empirical centiles are first smoothed by scatterplot smoothing,[43] then they are fitted by low-order polynomials in time t, where the coefficients of the polynomials are themselves constrained to follow low-order polynomials in z, the normal equivalent deviates corresponding to the centiles. In this way the centiles are estimated as separate curves, but their shapes and the spacings between them are constrained to provide a consistent and regularly spaced set of centiles. Healy subsequently found that his method was less effective than a Box–Cox or shifted log transform for handling non-normally distributed data.[44]

The WHO growth standard was constructed using the GAMLSS method,[41] which was chosen after an exhaustive comparison of all the available methods.[45] GAMLSS (generalized additive models of location, scale and shape) is an extension of the LMS method which includes adjustment for kurtosis (in addition to skewness) and a choice of error distributions. In practice, the GAMLSS models fitted to the WHO curves omitted any kurtosis adjustment, and so were equivalent to models fitted by the LMS method.[35]

21.6.3 Forms of Smoothing

The simplest form of curve to use for smoothing data is a low-order polynomial, e.g. a linear, quadratic or cubic curve. The polynomial is easy to fit using regression analysis, and the regression coefficients provide a parsimonious summary of the fitted curve.

But the substantial disadvantage of polynomials, particularly higher order polynomials applied to data with complex age trends, is that they behave poorly at the extremes of the data. Edge effects mean that the polynomial is often a poor fit at the youngest and oldest ages, and the curve may be unacceptably "wiggly" in between.

Fractional polynomials[46] largely avoid these problems. The conventional polynomial, containing terms with successive integer powers of age (t, t^2, t^3, etc.) is replaced by an equation with selected powers of age, the set of permissible powers being integer powers in the range -2 to $+3$ and certain non-integral or zero powers, for example $\sqrt{t}$, $\log(t)$ or $1/t$. As an example the growth curve described by Earl Count in the 1940s[47] is a fractional polynomial:

$$Y(t) = \beta_0 + \beta_1 t + \beta_2 \log(t)$$

Fractional polynomials are extremely effective at modeling the shapes of curves where both the curve and its slope either increase or decrease monotonically, as happens with anthropometry during gestation and in the preschool period. But they are less useful over longer periods, where either the measurement or its velocity changes non-monotonically with age. Height in childhood and adolescence, or BMI during childhood, are two examples where fractional polynomials are insufficiently flexible to model the underlying trends — height because of the pubertal growth spurt and BMI because of the rise then fall, then second rise.[48]

Tailor-made parametric growth curves have been developed for certain measurements, e.g. the Jenss–Bayley curve for weight or length in infancy,[49] the Preece–Baines curve for height in puberty[50] or the JPA-2 curve for height from birth to adult.[51] In general they are parsimonious (with between four and eight parameters to be estimated) and provide a good fit to the data. As such they are useful functions to estimate the mean or median curves described in the previous section.

But these special parametric growth curves are not available for all measurements, and in any case they are not well suited to modeling age-related trends in say the SD or the skewness, except in very simple cases. For this a more flexible approach is needed.

Spline smoothing and kernel smoothing are two related techniques that have proved effective for fitting smooth curves to data. They are both forms of local moving averages of the data, where the range of data averaged at each age (bandwidth) and the weightings applied to the data (weighting function) are varied in different ways.[52,53]

Kernel regression has generally been applied to quantile regression,[30,54–56] while natural smoothing splines have been preferred for semiparametric regression applications like the LMS method.[35]

21.6.4 Available Software

Most of the techniques described here can, with more or less effort, be fitted with standard software, e.g. Stata, SAS or R. The GAMLSS package in R is a powerful tool for fitting a wide range of semiparametric GAMLSS models,[41] including the LMS method as a special case. Royston has provided Stata do-files to fit the shifted log, Box–Cox and exponential transforms with fractional polynomials. Cole's LMSchart-maker is a dedicated program to fit the LMS method using cubic smoothing splines.

21.7 VARIANTS OF THE DISTANCE CHART

The discussion so far has focused on the simplest form of distance chart. However, other forms of chart based on the distance chart also deserve a mention.

21.7.1 Puberty

Puberty is a time when children of the same age can differ dramatically in size, owing to differences in their stage of maturation.[57] In principle, this could be represented on a chart with an extra scale for stage of maturation, but it would need to be plotted in three dimensions. As a compromise, a modified version of the distance chart has been developed that partially addresses this issue.

Variability in the timing of puberty causes the median curve on the distance chart to be flattened relative to the growth curve of individual children.[58,59] The slope of the median curve at its steepest, which represents the measurement's peak velocity, is biased downwards, i.e. it is less than the peak velocity in individual children. There are broadly two ways to respond to the bias: ignore it or adjust for it.

Tanner and Whitehouse minimized it by modifying the shapes of the height centile curves during puberty.[60] They drew them as steeper than they actually were, so that instead of representing the median height at each age, they followed the growth curve of a hypothetical child of average height, average height velocity and average growth tempo (tempo indicates the rate of maturation). They called it a tempo-conditional chart, and its advantage was said to be that it minimized centile crossing as the median curve is similar in shape to a growth curve.

In practice, it does not eliminate centile crossing. Any child whose height distance, velocity or tempo is not average will cross centiles at some point during puberty. The one advantage of the tempo-conditional chart over the conventional chart is that its median curve looks more like a growth curve. Its disadvantage is that the "median" curve no longer represents the population median height in puberty, and similarly the other "centile" curves do not correspond to the population centiles.

The alternative approach to the bias is to ignore it, which is what a conventional distance chart does. The British 1990 reference is an example. Here, the centile curves provide unbiased estimates of the distribution centiles at all ages including puberty.

Neither approach satisfactorily solves the problem of assessing both distance and velocity in puberty. Opinion is divided as to which approach is better, and ultimately the choice will depend on the circumstances in which the assessment is being made.

21.7.2 Repeated Measures

The distance chart is usually based on data where each subject provides one measurement. But quite often the study design is longitudinal or mixed longitudinal and participants have more than one measurement; the question is: what to do with them? There are three schools of thought: (1) restrict the data to one measurement per subject (the first or a random choice); (2) use the mean value for each subject; or (3) use all the data for each subject, treating them as unrelated.

The first alternative is safe but conservative: it ensures that there are no repeated measures but it also wastes data. The second approach is not correct, as it introduces differential weighting. The variability of the mean of several points is smaller than for a single point, so the measurement error for subjects with averaged repeated measures is artificially reduced.

The third alternative, to retain all the data, is probably the best, although this is controversial. The issue comes back to distance versus velocity, in that a distance chart contains no information about velocity. The distance centile curves can be thought of as a series of snapshots of the measurement distribution at different ages, smoothly joined across ages. Each such snapshot is unrelated to the others, so it does not matter if a subject is represented in more than one snapshot — the shapes of the centiles are not affected.

What *is* affected is the precision with which the centile curves are calculated. Consider an extreme example, two surveys each with 1000 points, one a longitudinal study of 50 subjects measured annually from 0 to 19 years, the other a cross-sectional survey of 1000 children with ages uniformly distributed between 0 and 19 years. If the two sets of subjects are drawn randomly from the same reference population, then on average the two sets of fitted centiles will be the same: both will be unbiased estimates of the population centiles. The longitudinal centiles will be far less *precise*, in that the between-subject variability will be based on 50 subjects rather than 1000, and the confidence intervals for each centile will be wider, but the centiles themselves will (on average) be the same. Wade and Ades[61] have argued the case more formally, showing that adjusting explicitly for the correlation between repeated measures does not materially alter the shapes of the fitted centiles.

This may appear counter-intuitive, but Healy[62] has shown it to be *exactly* true for the analogous situation of paired organs in large samples. Here, subjects provide two measurements (e.g. two arms or two ears), and Healy shows that the correct approach is to ignore the pairings and treat the data as independent.

In general, if the repeated measures data are balanced, i.e. every subject has measurements at the same age (as would be the case in a longitudinal study with no missing data), then the repeated measures structure can be ignored for the purposes of constructing a distance chart. If some of the data are missing, then so long as they are missing at random the repeated measures structure can be ignored.

21.8 INTERPRETING THE CURVES

The previous section has shown how growth references can be summarized in terms of three curves summarizing the mean or median, variability and skewness of the measurement through childhood. These curves provide useful information about the growth processes underlying the measurement.

Take the LMS method for example, where the three curves are the median (M), the CV (S) and the Box—Cox power to minimize skewness (L). Figures 21.1 and 21.2 show M and S curves by sex for the British 1990 height and weight references.[9] The median curves are familiar in shape, but the CVs have an unexpected dip soon after birth. In proportional terms the SDs of height and weight fall until about 9 months after birth, then rise until puberty, 2 years earlier for girls than boys, and then fall again to values near those seen at birth. There is considerable centile crossing in both weight and length during the first year, and this process obviously reduces variability in the first few months. But then a different process starts to increase the variability, which continues until puberty and then drops away again. This process must represent heterogeneity in the rate of maturation.

The distribution of height is close to normal but for weight it is appreciably skewed, and is adjusted for using the Box—Cox power transformation. Figure 21.3 shows the L curves by sex for the weight reference, the Box—Cox power as it changes with age, and for comparison, the first derivatives of the weight M curves. The L curves in Figure 21.3(a) have a different shape from the M or S curves in Figure 21.2, falling from birth until puberty, when they rise briefly then fall again. The weight "velocity" curves in Figure 21.3(b) are calculated from weight differences over 1 year and so are not true velocity curves, but they give an idea of how velocity changes during childhood. In shape they are similar to the S curves in Figure 21.2(b), while the first derivatives of the velocity curves (i.e. weight "acceleration") broadly match the L curve shapes in Figure 21.3(a). It has been shown that in general the first and second derivatives of the M curve (i.e. "velocity" and "acceleration") correspond in shape to the second and third moments (i.e. the S and L curves, respectively) of the frequency distribution.[59]

These insights into the processes underlying height and weight growth are in addition to the practical value of the charts that they define (see also Figure 21.4).

21.9 CHARTING VELOCITY

21.9.1 Unconditional Velocity

The distance chart treats repeated measures data as unrelated, so how is velocity to be assessed?

Traditionally, growth velocity has been measured in the original units of measurement, e.g. cm/year for height velocity, which leads to a velocity chart with an interesting

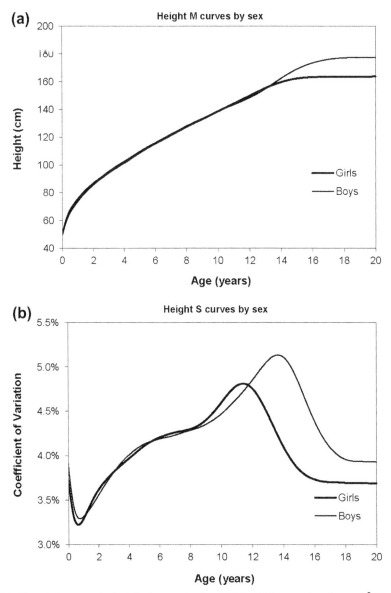

Figure 21.1 *M* and *S* curves for height by sex in the British 1990 growth reference.[3] The *M* curves (a) are median height by age, and the *S* curves (b) the coefficient of variation.

shape where the centile curves highlight the pubertal growth spurt.[63] The simplest way to construct a velocity chart is in the same way as a distance chart, except that paired data from individual subjects are needed to calculate each subject's velocity. In addition, the timing of the pairs of measurements needs to be exact, e.g. 1 year apart for height, as this affects the variability of the velocity.[64]

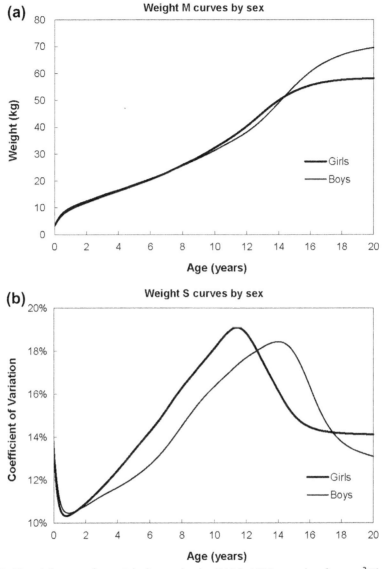

Figure 21.2 *M* and *S* curves for weight by sex in the British 1990 growth reference.[3] The *M* curves (a) are median weight by age, and the *S* curves (b) the coefficient of variation.

The effect of puberty on the distance chart applies equally to the velocity chart, and Tanner and Whitehouse[60] produced tempo-conditional velocity charts to match their distance charts.

The disadvantage of this whole approach to velocity is that it requires a separate chart from the distance chart and the data need to be plotted twice, once for distance and once for velocity. Also, the individual velocities need to be calculated before they are plotted.

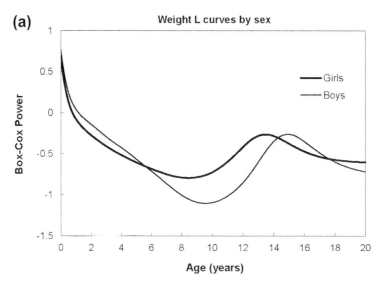

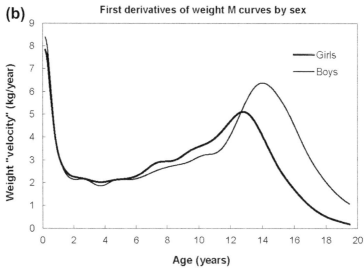

Figure 21.3 *L* curves for weight by sex in the British 1990 growth reference,[3] and for comparison slopes of the weight *M* curves. The *L* curves (a) are the Box–Cox power transformation needed to remove skewness at each age, and the slopes of the *M* curves (b) a form of weight "velocity".

An alternative is to focus on centile crossing. Velocity, compared to average velocity for age, is effectively centile crossing, so it is logical to measure velocity in centile units. Statistically, SD scores are more appropriate than centiles (as the SD score scale is not bounded at 0 and 100), so the velocity can be expressed as the rate of change in the SD score. A child whose centile (and hence SD score) remains constant over time (i.e. no centile crossing) has zero rate of change, and this corresponds to the median velocity.

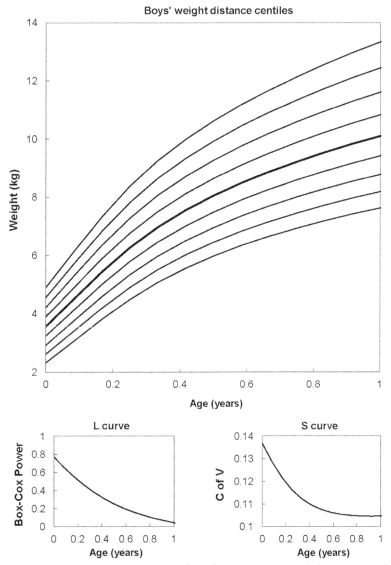

Figure 21.4 Centiles for weight in British male infants from 0 to 1 year, with the corresponding *L* and *S* curves. The *L* curve is near 1 at birth, so the centiles are normally distributed and equally spaced, while by 1 year the *L* curve has fallen to 0 indicating right skewness requiring a log transformation, and the centiles are more widely spaced above the median than below.

Measuring velocity on the SD score scale has two benefits: the change in SD score can be assessed directly, and the corresponding velocity can be represented visually on the distance chart.[65] To express centile crossing as a velocity centile, the variability of centile crossing needs to be known. This comes directly from the standard deviation of the change in SD score, which surprisingly depends only on the correlation r between SD scores at the two ages the child is measured.[66] The actual formula is:

$$\text{SD of SD score change} = \sqrt{2(1 - r)} \qquad [21.4]$$

Take, for example, weight from birth to 1 year. The correlation between weight SD scores at birth and 1 year is 0.59,[67] so the SD of the SD score change from birth to 1 year is:

$$\sqrt{2(1 - 0.59)} = \sqrt{2 \times 0.41} = \sqrt{0.82} = 0.91$$

The mean change in SD score over the year is 0, and 95% of infants are within ± 2 SDs of this, i.e. changes in the range -1.82 to $+1.82$ SD score units.

The same information can be used to express an individual infant's SD score change as a velocity centile. Take a child whose weight is on the 16th centile at birth (SD score $= -1$), but who has caught up to the median by 1 year: this centile crossing upwards is an increase of 1 unit of SD score. A change in SD score of $+1$ corresponds to $+1/0.91 = +1.1$ SDs, corresponding to the 86th centile for SD score change, or equivalently the 86th velocity centile (assuming velocity is normally distributed). So this child has grown more rapidly over the first year than 86% of similar infants, i.e. who were on the 16th centile at birth.

21.9.2 Conditional Velocity

This calculation is a simple and fairly intuitive way of assessing centile change. However, there is a complication due to the statistical phenomenon of regression to the mean. This states that on average, the centiles of individuals (or groups of individuals) followed over time will tend to become less extreme, more ordinary, closer to the median. As a result, there is a built-in negative correlation between the starting centile and the change in centile over time: a child on a low starting centile will on average cross centiles upwards (i.e. exhibit catch-up growth), whereas the opposite will occur for a child starting on a high centile. Cole[67] discusses this in the context of infant weight gain.

This form of velocity is known as *conditional* velocity, i.e. conditional on the previous measurement. Velocity as defined in the previous section is by analogy called *unconditional* velocity.

To adjust for regression to the mean, the SD score on the second occasion is compared with what would be predicted from the first occasion. The prediction comes from linear regression analysis:

$$z_2 = r \times z_1 \qquad [21.5]$$

So instead of the change in SD score over time, e.g. $z_2 - z_1$, an adjusted change is used, $z_2 - r \times z_1$, where again r is the correlation between the two SD scores z_1 and z_2. The SD of this adjusted SD score change is $\sqrt{1 - r^2}$, which is slightly smaller than [21.4].

In the example above, where the SD score increases from -1 to 0 from birth to 1 year, the adjusted increase in SD score is:

$$z_2 - r \times z_1 = 0 - 0.59 \times (-1) = 0.59$$

which is rather less than the unadjusted increase of 1 unit. The corresponding SD for velocity is $\sqrt{1 - 0.59^2} = 0.81$, so the velocity SD score is $0.59/0.81 = 0.73$, corresponding to the 76th velocity centile. So an infant with a low birth weight would be expected to show catch-up to some extent, and adjusting for this puts her velocity at the 76th rather than the 86th centile.

This process can be reversed to define a given velocity centile in terms of the SD scores at the start and end of the interval. This is useful for representing velocity on the distance chart (see below). For a notional child with SD score z_1 measured at age t_1 and z_2 at age t_2, where the correlation between z_1 and z_2 is known to be r and the required velocity centile corresponds to SD score z_v, then the four quantities are related as follows:

$$z_2 = r \times z_1 + \sqrt{1 - r^2} \times z_v$$

This relationship allows a velocity centile to be represented as a line on the distance chart, where the *slope* of the line represents the velocity. It starts at the measurement corresponding to z_1 at age t_1 and ends at the measurement corresponding to z_2 at age t_2. Several such lines can be drawn between t_1 and t_2 by choosing different values for z_1, so whatever the size of the child there will be a velocity centile line near their data plotted on the chart.

Contiguous lines can be drawn for later intervals, e.g. from age t_2 to t_3 and later ages, by choosing the calculated value of the second SD score (z_2 in the equation) for each interval as the starting SD score (z_1 in the equation) for the next interval. This gives a set of curves which cut across the distance centiles, and the slopes of the curves (*not* their positions) indicate the velocity centiles at each age. Velocity centiles below the 50th cross distance centiles downwards, while those above the 50th cross upwards. In general, the curves are not parallel at each age, showing that expected velocity depends on the size (i.e. SD score z_1) of the child. Figure 21.5 shows the distance chart for weight in infancy, with the 2nd velocity centile curves superimposed, for measurements taken 4 weeks apart.

A child's growth is assessed on the chart by comparing the slope of their line joining successive pairs of points with the slope of the nearest velocity centile curve over the same age range. If the two lines are parallel then the child's velocity is equal to the nominal velocity centile. If her line is steeper she is above the centile, while if it is shallower she is below it.

The addition of these velocity curves to the distance chart inevitably makes the chart harder to read. In principle, more than one set of curves could be provided, e.g. both the 5th and 95th velocity centiles, but the chart rapidly becomes too cluttered to be useful. An alternative is to provide each set of velocity centile curves as a transparent plastic

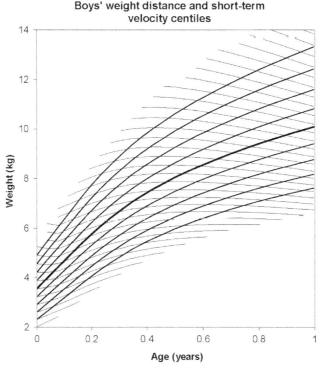

Figure 21.5 The boys' weight distance centiles of Figure 21.4 with lines superimposed representing the 2nd conditional velocity centile. The slope of the line joining an infant's successive weights measured 4 weeks apart is compared with the slope of the nearest velocity centile line. See text for details.

overlay that can be placed on the distance chart, and this keeps the chart simple while providing an assessment of velocity. It also requires the data to be plotted only once and avoids calculating velocities. Overlays for the 5th and 95th centiles of conditional weight velocity in infancy, known as "Thrive lines", are available for the UK–WHO charts (Harlow Printing, http://www.healthforallchildren.co.uk/).

21.10 CONDITIONAL REFERENCES

Conditional velocity is just one example of the family of conditional references. Equation [21.5] shows how one SD score is predicted from another, where the two SD scores correspond to consecutive measurements in one child. But the meanings of the two SD scores are quite general; they could, for example, be birth weights for two siblings, giving a reference of birth weight conditional on sibling birth weight, or heights of parent and child, i.e. a reference of height conditional on parental height.[68] By specifying the reference on the SD score scale the only extra information required for the reference is the correlation between the pairs of measurements.

21.11 DESIGNING AND PRINTING THE CHART

The final stage in the production of a growth reference is to design, print and distribute the charts. As with any other marketable commodity it pays to design the chart to suit its users, and extensive consultation is needed to ensure that the format of the chart is optimal. This involves issues to do with the choice of centiles on the chart (discussed earlier), the age ranges and combinations of different measurements to include in particular charts (e.g. weight, length and head circumference on a single chart in the first year), the choice of age scale (e.g. decimal or in months for the first year), and the chart's general appearance in terms of orientation (portrait or landscape), color scheme, line types, labeling, scales and gridlines. They make a big difference to the usability of the chart in clinical practice.

Remember that different users, e.g. endocrinologists versus community pediatricians versus health visitors, will use the charts in different ways, so it is important to canvass opinion as widely as possible before settling on the final design. Providing some forum for feedback can ensure that later printings of the charts incorporate modifications to improve them further. As an example, the UK–WHO charts were the outcome of an extensive design process involving an expert group working with graphic designers, and incorporating feedback from several rounds of focus groups.[4]

21.12 WHEN TO REPLACE

Once charts have been available for a period of time, the question arises: should they be updated? Owing to secular trends in growth, children change in size and/or shape after the chart is produced. Updating the chart ensures that the centiles continue to reflect accurately the proportions of children outside the extreme centiles. An interval of 10–15 years between updates is typical, e.g. the Dutch national growth surveys took place in 1955, 1965, 1980, 1997 and most recently 2009. But there are two disadvantages to updating charts that also need to be borne in mind.

First, it takes a very long time for a new chart to displace the old. This is due partly to ignorance (the new chart is not widely known about), partly to inertia ("I prefer the chart I'm used to") and partly to finance (purchasing officers expect existing stocks of charts to be exhausted first).

The second disadvantage is more subtle: the secular trend may be towards less optimal growth, e.g. the recent increase in obesity that has shifted weight and BMI centiles upwards. If the centiles are used to define overweight, e.g. the 91st centile on the BMI chart,[46] then this provides a nominal prevalence of 9% overweight. Such a figure was appropriate at the time the chart data were collected (1990), but with the trend towards increasing fatness the prevalence has since increased. If the chart were to be updated and the 91st centile shifted upwards the prevalence of overweight would revert to 9%, but it would not be comparable with the 9% prevalence on the previous chart, i.e. the prevalence rates before and after would not be comparable.

For this reason the British BMI chart has been "frozen" in time and will not be updated.[69] In effect the chart has become a standard rather than a reference: it reflects BMI at an earlier time when the population was less fat than it is now.

21.13 CONCLUSION

The process of producing new growth charts involves several important stages. The choice of reference population and sample, collecting and recording the anthropometry, analyzing the data and designing and printing the chart, together require the skills of specialists in many different areas. The outcome should be a set of charts that are effective in recording and assessing the growth of the children they serve.

REFERENCES

1. Tanner JM. *A history of the study of human growth*. Cambridge: Cambridge University Press; 1981.
2. World Health Organization. *WHO child growth standards: methods and development: length/height-for-age, weight-for-age, weight-for-length, weight-for-height and body mass index-for-age*. Geneva: WHO; 2006.
3. Freeman JV, Cole TJ, Chinn S, Jones PRM, White EM, Preece MA. Cross-sectional stature and weight reference curves for the UK 1990. *Arch Dis Child* 1995;**73**:17−24.
4. Wright CM, Williams AF, Elliman D, Bedford H, Birks E, Butler G, et al. Using the new UK−WHO growth charts. *BMJ* 2010;**340**. c1140.
5. Cole TJ. Do growth chart centiles need a face lift? *BMJ* 1994;**308**:641−2.
6. Chinn S, Cole TJ, Preece MA, Rona RJ. Growth charts for ethnic populations in UK (letter). *Lancet* 1996;**347**:839−40.
7. Styles ME, Cole TJ, Dennis J, Preece MA. New cross sectional stature, weight and head circumference references for Down's syndrome in the UK and Republic of Ireland. *Arch Dis Child* 2002;**87**:104−8.
8. Ranke MB, Plüger H, Rosendahl W. Turner's syndrome: spontaneous growth in 150 cases and review of the literature. *Eur J Pediatr* 1983;**141**:81−8.
9. Cole TJ, Freeman JV, Preece MA. British 1990 growth reference centiles for weight, height, body mass index and head circumference fitted by maximum penalized likelihood. *Stat Med* 1998;**17**:407−29.
10. Graitcer PL, Gentry M. Measuring children: one standard for all. *Lancet*; 1981:297−9. ii.
11. Tanner JM. Some notes on the recording of growth data. *Hum Biol* 1951;**23**:93−159.
12. Falkner F. Twenty-five years of internationally coordinated research: longitudinal studies in growth and development. *Courier* 1980;**30**:3−7.
13. Garza C, De Onis M. A new international growth reference for young children. *Am J Clin Nutr* 1999;**70**:169−172S.
14. Armitage P, Berry B. *Statistical methods in medical research*. Oxford: Blackwell; 1987.
15. Briend A, Zimicki S. Validation of arm circumference as an indicator of risk of death in one to four year old children. *Nutr Res* 1986;**6**:249−61.
16. Rolland-Cachera MF, Brambilla P, Manzoni P, Akrout M, Del Maschio A, Chiumello G. A new anthropometric index validated by magnetic resonance imaging (MRI) to assess body composition. *Am J Clin Nutr* 1997;**65**:1709−13.
17. Han TS, Van Leer EM, Seidell JC, Lean MEJ. Waist circumference action levels in the identification of cardiovascular risk factors: prevalence study in a random sample. *BMJ* 1995;**311**:1401−5.
18. Gunnell DJ, Davey Smith G, Holly JMP, Frankel S. Leg length and risk of cancer in the Boyd Orr cohort. *BMJ* 1998;**317**:1350−1.

19. Weiner JS, Lourie SA. *Practical human biology.* London: Academic Press; 1981.
20. Cameron N. *The measurement of human growth.* London: Croom-Helm; 1984.
21. Lohman T, Martorell R, Roche AF. *Anthropometric standardization reference manual.* Champaign, IL: Human Kinetics; 1998.
22. Healy MJR. The effect of age grouping on the distribution of a measurement affected by growth. *Am J Phys Anthropol* 1962;**20**:49−50.
23. Aitkin M. Modelling variance heterogeneity in normal regression using GLIM. *Appl Stat* 1987;**36**:332−9.
24. Altman DG. Construction of age-related reference centiles using absolute residuals. *Stat Med* 1993;**12**:917−24.
25. Rigby RA, Stasinopoulos DM. A semiparametric additive-model for variance heterogeneity. *Stat Comput* 1996;**6**:57−65.
26. Tango T. Estimation of age-specific reference ranges via smoother AVAS. *Stat Med* 1998;**17**:1231−43.
27. Hamill PVV, Drizd TA, Johnson CL, Reed RB, Roche AF. *NCHS growth curves for children birth−18 years. Vital and health statistics series 11: No. 165.* Washington DC: National Center for Health Statistics; 1977.
28. Koenker RW, Bassett GW. Regression quantiles. *Econometrica* 1978;**46**:33−50.
29. Koenker R, Ng P, Portnoy S. Quantile smoothing splines. *Biometrika* 1994;**81**:673−80.
30. Heagerty PJ, Pepe MS. Semiparametric estimation of regression quantiles with application to standardizing weight for height and age in US children. *J Roy Stat Soc Ser C Appl Stat* 1999;**48**:533−51.
31. Box GEP, Cox DR. An analysis of transformations. *J Roy Stat Soc Ser B* 1964;**26**:211−52.
32. Royston P. Estimation, reference ranges and goodness of fit for the 3-parameter lognormal distribution. *Stat Med* 1992;**11**:897−912.
33. Van't Hof MA, Wit JM, Roede MJ. A method to construct age references for skewed skinfold data, using Box−Cox transformations to normality. *Hum Biol* 1985;**57**:131−9.
34. Cole TJ. Fitting smoothed centile curves to reference data (with discussion). *J R Stat Soc A* 1988;**151**:385−418.
35. Cole TJ, Green PJ. Smoothing reference centile curves: the LMS method and penalized likelihood. *Stat Med* 1992;**11**:1305−19.
36. Thompson ML, Theron GB. Maximum likelihood estimation of reference centiles. *Stat Med* 1990;**9**:539−48.
37. Royston P. Constructing time-specific reference ranges. *Stat Med* 1991;**10**:675−90.
38. Wade AM, Ades AE. Age-related reference ranges − significance tests for models and confidence intervals for centiles. *Stat Med* 1994;**13**:2359−67.
39. Royston P, Wright EM. A method for estimating age-specific reference intervals ("normal ranges") based on fractional polynomials and exponential transformation. *J Roy Stat Soc Ser A Stat Soc* 1998;**161**:79−101.
40. Sorribas A, March J, Voit EO. Estimating age-related trends in cross-sectional studies using S-distributions. *Stat Med* 2000;**19**:697−713.
41. Rigby RA, Stasinopoulos DM. Generalized additive models for location, scale and shape (with discussion). *Appl Stat* 2005;**54**:507−44.
42. Healy MJR, Rasbash J, Yang M. Distribution-free estimation of age-related centiles. *Ann Hum Biol* 1988;**15**:17−22.
43. Cleveland WS. Robust locally weighted regression and smoothing scatterplots. *J Am Stat Assoc* 1979;**79**:829−36.
44. Healy MJR. Normalizing transformations for growth standards. *Ann Hum Biol* 1992;**19**:521−6.
45. Borghi E, de Onis M, Garza C, Van den Broeck J, Frongillo EA, Grummer-Strawn L, et al. Construction of the World Health Organization child growth standards: selection of methods for attained growth curves. *Stat Med* 2006;**25**:247−65.
46. Royston P, Altman DG. Regression using fractional polynomials of continuous covariates: parsimonious parametric modelling (with discussion). *Appl Stat* 1994;**43**:429−67.
47. Count E. Growth pattern of the human physique. *Hum Biol* 1943;**15**:1−32.

48. Cole TJ, Freeman JV, Preece MA. Body mass index reference curves for the UK 1990. *Arch Dis Child* 1995;**73**:25−9.
49. Jenss RM, Bayley N. A mathematical method for studying growth in children. *Hum Biol* 1937;**9**:556 63.
50. Preece MA, Baines M. JA new family of mathematical models describing the human growth curve. *Ann Hum Biol* 1978;**5**:1−24.
51. Jolicoeur P, Pontier J, Abidi H. Asymptotic models for the longitudinal growth of human stature. *Am J Hum Biol* 1992;**4**:461−8.
52. Wand M, Jones MC. *Introduction to kernel smoothing*. London: Chapman and Hall; 1994.
53. Green PJ, Silverman BW. *Nonparametric regression and generalized linear models*. London: Chapman and Hall; 1994.
54. Guo S, Roche AF, Baumgartner RN, Chumlea WC, Ryan AS. Kernel regression for smoothing percentile curves: reference data for calf and subscapular skinfold thicknesses in Mexican Americans. *Am J Clin Nutr* 1990;**51**:908−916S.
55. Rossiter JE. Calculating centile curves using kernel density estimation methods with application to infant kidney lengths. *Stat Med* 1991;**10**:1693−701.
56. Ducharme GR, Gannoun A, Guertin MC, Jequier JC. Reference values obtained by kernel-based estimation of quantile regressions. *Biometrics* 1995;**51**:1105−16.
57. Tanner JM. *Growth at adolescence*. Oxford: Blackwell; 1962.
58. Merrell M. The relationship of individual growth to average growth. *Hum Biol* 1931;**3**:37−70.
59. Cole TJ, Cortina Borja M, Sandhu J, Kelly FP, Pan H. Nonlinear growth generates age changes in the moments of the frequency distribution: the example of height in puberty. *Biostatistics* 2008;**9**:159−71.
60. Tanner JM, Whitehouse RH. Clinical longitudinal standards for height, weight, height velocity, weight velocity, and the stages of puberty. *Arch Dis Child* 1976;**51**:170−9.
61. Wade AM, Ades AE. Incorporating correlations between measurements into the estimation of age-related reference ranges. *Stat Med* 1998;**17**:1989−2002.
62. Healy MJR. Reference values and standards for paired organs. *Ann Hum Biol* 1993;**20**:75−6.
63. Tanner JM, Whitehouse RH, Takaishi M. Standards from birth to maturity for height, weight, height velocity, and weight velocity: British children 1965 Parts I and II. *Arch Dis Child* 1966;**41**:454−71, 613−5.
64. Cole TJ. The use and construction of anthropometric growth reference standards. *Nutr Res Rev* 1993;**6**:19−50.
65. Cole TJ. Growth charts for both cross-sectional and longitudinal data. *Stat Med* 1994;**13**:2477−92.
66. Cole TJ. Growth monitoring with the British 1990 growth reference. *Arch Dis Child* 1997;**76**:1−3.
67. Cole TJ. Conditional reference charts to assess weight gain in British infants. *Arch Dis Child* 1995;**73**:8−16.
68. Cole TJ. A simple chart to assess non-familial short stature. *Arch Dis Child* 2000;**82**:173−6.
69. Cole TJ, Power C, Preece MA. Child obesity and body-mass index (letter). *Lancet* 1999;**353**:1188.

INTERNET RESOURCES

There are several websites with free resources for growth assessment, some general and some dedicated to particular growth references. A few are given here.

LMSgrowth, Excel add-in software for analyzing growth data, and LMSchartmaker, software to fit centiles to reference data using the LMS method, http://www.healthforallchildren.com/index.php/shop/category-list/Software.

UK−WHO charts, information, instructions and training materials, http://www.growthcharts.rcpch.ac.uk/.

US CDC 2000. charts, data tables, educational materials, reports, http://www.cdc.gov/growthcharts/.

WHO charts, data tables, training, publications. WHO Anthro software, http://www.who.int/childgrowth/en/.

INDEX

Page references followed by *t* and *f* denote tables and figures, respectively.

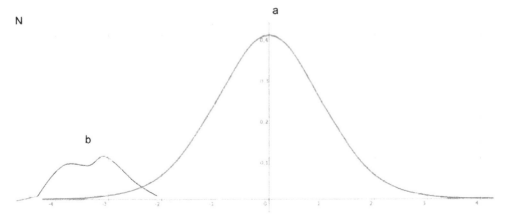

Figure 2.11 Frequency distribution (Z-scores) of heights of normal children and those presenting with a growth disorder. a: Normal children; b: growth disorders.

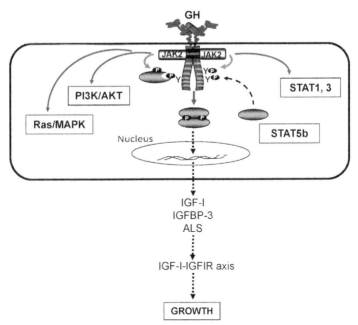

Figure 5.1 Growth hormone (GH) regulation of the insulin-like growth factor (IGF) axis. GH binds to its homodimeric receptor, resulting in association of janus kinase 2 (JAK2) and phosphorylation of key tyrosines on the GH receptor intracellular domain. These tyrosines do not appear necessary for activation of PI3/AKT, MAP kinase (MAPK) or signal transducer and activator of transcription (STAT) 1 and 3, but are obligatory for docking and phosphorylation of STAT5a and STAT5b. After phosphorylation, STAT5b dissociates from the GH receptor, dimerizes, enters into the nucleus, and transcriptionally activates the three critical GH-dependent genes, IGF1, IGFBP3 and ALS.

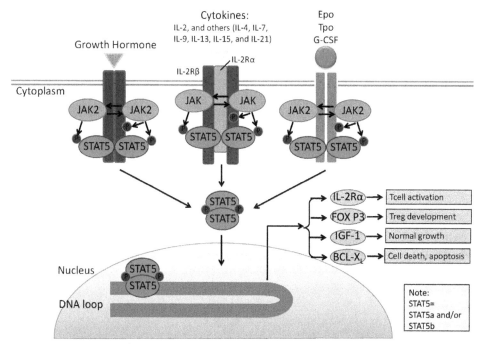

Figure 5.2 A schematic of JAK/STAT5 pathway signaling is shown for growth hormone, cytokines, such as interleukin (IL)-2, IL-4, IL-7, IL-9, IL-13, IL-15 and IL-21, and erythropoietin (Epo), thromobo-poieten (Tpo), granulocyte colony-stimulating factor (G-CSF) and their cognate receptors on the cell surface. STAT5 represents both STAT5a and STAT5b. *(Source: Nadeau et al.[25])*

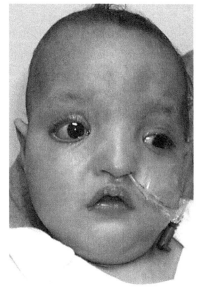

Figure 7.1 Characteristic facial features in an infant with the Wolf—Hirschhorn syndrome. The forehead is high, the nasal root is broad and the eyes are wide set.

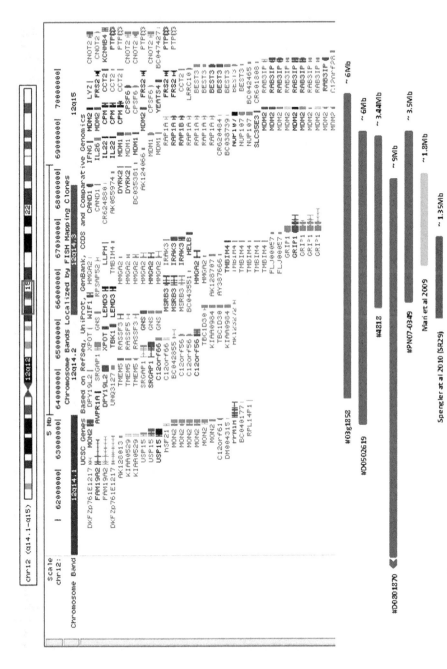

Figure 7.3 Overview of seven patients with the 12q14 microdeletion syndrome. The microdeletion in each patient is shown as a horizontal bar at the bottom of the figure. Red bars represent patients investigated by the authors. The microdeletions are plotted against the corresponding genomic region on chromosome 12q14–q15 (UCSC Genome Browser) showing the RefSeq genes within this region. With the exception of the patient reported by Mari et al., all deletions encompass the LEMD3 gene. The HMGA2 gene is deleted in all patients.

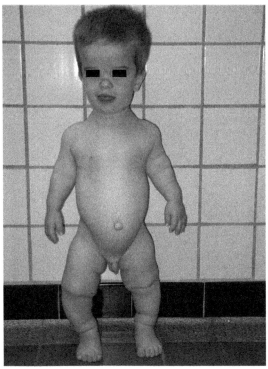

Figure 7.4 Three-year-old boy with achondroplasia. Note the short stature with short limbs and bowing of the legs.

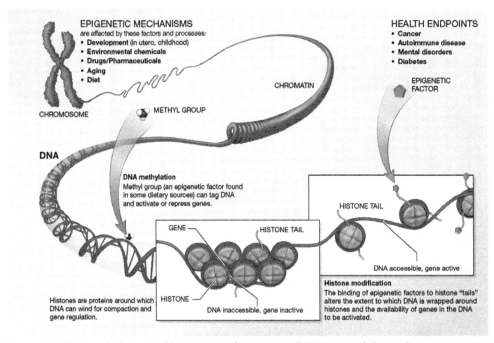

Figure 7.5 DNA methylation and histone acetylation control DNA accessibility and gene activation/inactivation. (Source: *Image from: http://nihroadmap.nih.gov/EPIGENOMICS/epigeneticmechanisms.asp*)

Insulator model – Igf2 cluster

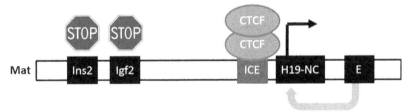

Binding of the insulator protein CTCF to imprint control element (ICE) blocks Ins2 and Igf2 mRNA transcription, enhancers activate noncoding RNA H19

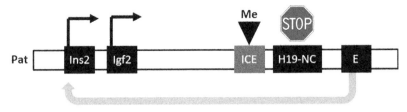

Paternal DNA methylation imprinting silences ICE and transcription of ncRNA H19, enhancers activate Ins2 and Igf2 mRNA transcription

Figure 7.6 Schematic overview of imprinting at the Igf2/H19 loci on the human chromosome 11p15.

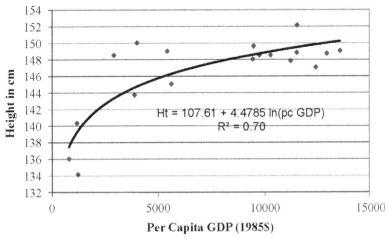

$$Ht = 107.61 + 4.4785 \ln(pc\ GDP)$$
$$R^2 = 0.70$$

Figure 9.2 Per capita GDP and height at age 12 (boys).

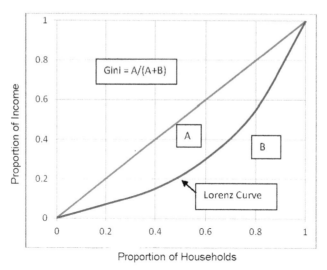

Figure 9.3 Lorenz curve and Gini coefficient.

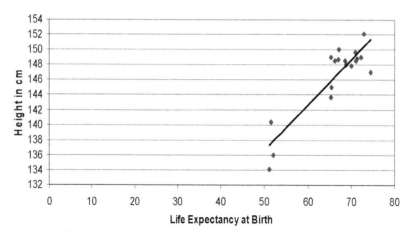

Figure 9.4 Height of boys age 12 and life expectancy at birth.

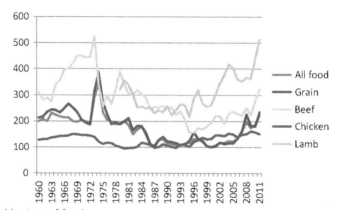

Figure 9.6 World prices of food.

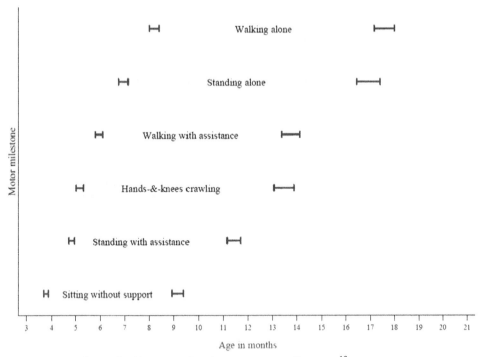

Figure 11.5 Windows of achievement for six gross motor milestones.[18]

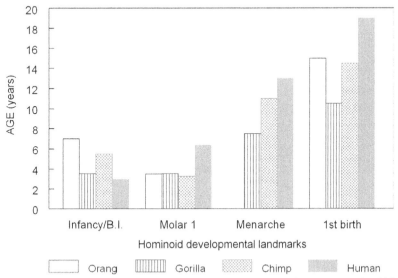

Figure 11.11 Hominoid developmental landmarks. Data based on observations of wild-living individuals or, for humans, healthy individuals from various cultures. Infancy/B.I.: period of dependency on mother for survival, usually coincident with mean age at weaning and/or a new birth (where B.I. is birth interval); Molar 1: mean age at eruption of first permanent molar; Menarche: mean age at first estrus or menstrual bleeding; 1st birth: mean age of females at first offspring delivery. Orang, *Pongo pygmaeus*; gorilla, *Gorilla gorilla*; chimp, *Pan troglodytes*; human, *Homo sapiens*.[44]

Figure 11.12 Cooperative care of children by women and juvenile girls. The example is from the Kaqchikel-speaking Maya region of Guatemala. The women perform household and food preparation duties while the juvenile girls play with and care for the children. *(Photograph by Barry Bogin.)*

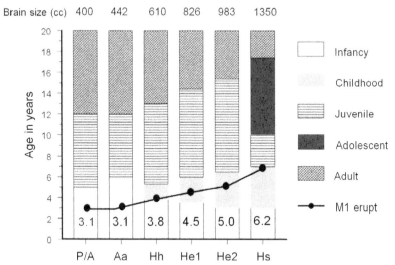

Figure 11.14 The evolution of hominin life history during the first 20 years of life. Abbreviations of the pongid and hominin taxa are: P/A: *Pan, Australopithecus afarensis*; Aa: *Australopithecus africanus*; Hh: *Homo habilis*; He1: early *Homo erectus*; He2: late *Homo erectus*; Hs: *Homo sapiens*; M1 erupt: eruption of first molar.[33]

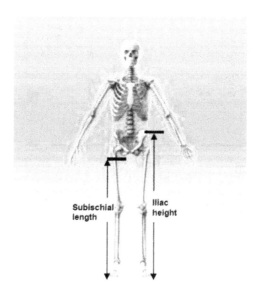

Figure 13.1 Iliac height and subischial length. *(Source: Roger Harris/Science Photo Library, royalty-free image, labeling added by the authors.)*

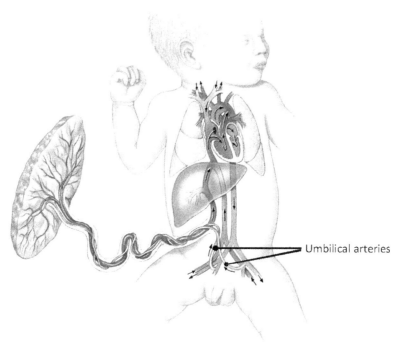

Figure 13.8 Human fetal circulation. The relative amount of oxygen in the fetal blood is greatest in the upper thorax, neck and head, indicated by the red color of the vessels ascending from the heart. Blood flowing to the abdomen and legs is less well oxygenated; indicated by the violet color of the vessels descending from the heart. *(Source: Adapted from Martini and Bartholomew.[40])*

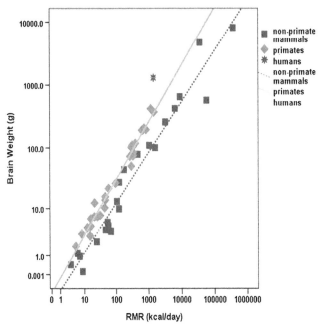

Figure 15.1 Log–log plot of brain weight (BW; g) versus resting metabolic rate (RMR; kcal/day) for humans, 36 other primate species and 22 non-primate mammalian species. The primate regression line is systematically and significantly elevated above the non-primate mammal regression. For a given RMR, primates have brain sizes that are three times those of other mammals, and humans have brains that are three times those of other primates.

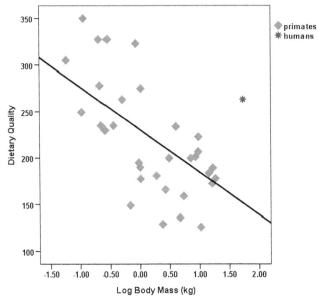

Figure 15.2 Plot of diet quality (DQ) versus log-body mass for 33 primate species. DQ is inversely related to body mass [$r = -0.59$ (total sample), -0.68 (non-human primates only); $p < 0.001$], indicating that smaller primates consume relatively high-quality diets. Humans have systematically higher quality diets than predicted for their size. *(Source: Adapted from Leonard et al.[4])*

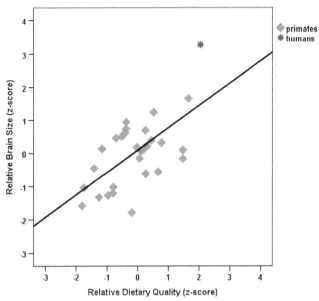

Figure 15.3 Plot of relative brain size versus relative diet quality for 31 primate species (including humans). Primates with higher quality diets for their size have relatively large brain size ($r = 0.63$, $p < 0.001$). Humans represent the positive extremes for both measures, having large brain:body size and a substantially higher quality diet than expected for their size. *(Source: Adapted from Leonard et al.[4])*

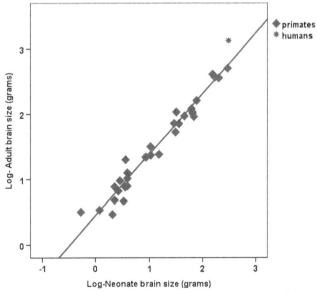

Figure 15.4 Log—log plot of adult brain size (g) versus neonate brain size (g) for humans and 31 other primate species. Brain growth of human infants is much faster than in other primate species. Consequently, adult human brain sizes average about 3.5 times those of newborns, compared to 2.3 times newborn size in other primates. *(Source: Data derived from Barton and Capellini.[26])*

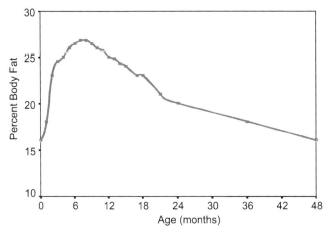

Figure 15.7 Changes in percentage body fatness of human infants from birth to 48 months. Body fatness increases from 16 to 26% during the first 9 months, and then declines to ~16% by 4 years of age. (Source: *Data from Dewey et al.*[36])

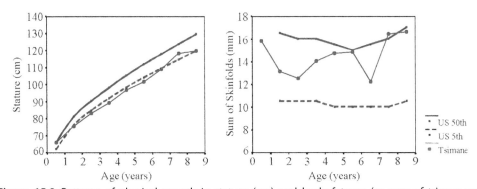

Figure 15.8 Patterns of physical growth in stature (cm) and body fatness (as sum of triceps and subscapular skinfolds, mm) in girls of the Tsimane' of lowland Bolivia. Growth of Tsimane' girls is characterized by marked linear growth stunting, whereas body fatness compares more favorably to US norms. (Source: *Data from Foster et al.*[40])

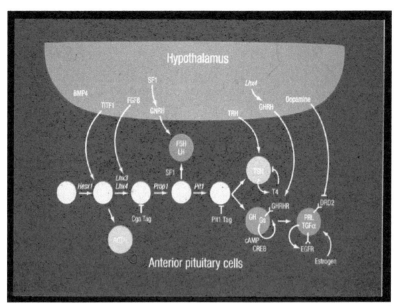

Figure 17.1 Sequence cascade of factors involved in pituitary development. Gradient factors and transcription factors are involved in the determination and situation of pituitary cell types.

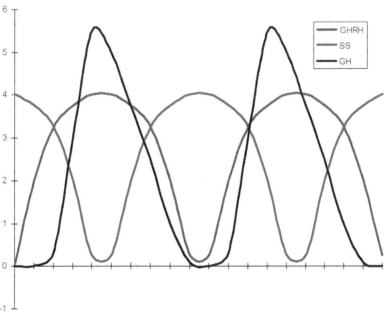

Figure 17.3 Schematic representation of the most likely interaction of growth hormone-releasing hormone (GHRH) and somatostatin (SS) in the generation of a growth hormone (GH) pulse.

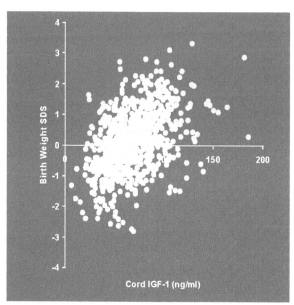

Figure 17.5 Relationship between birth weight expressed as a Z-score and serum insulin-like growth factor-1 (IGF-1) concentration in cord blood.

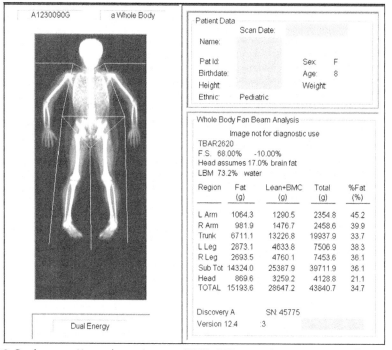

Figure 18.3 Dual-energy X-ray absorptiometry (DXA). DXA uses the attenuation of X-ray beams to assess body composition. Fat, muscle and bone differ in density, so they attenuate the X-ray beams in varying amounts. The image shows that fat, muscle and bone of an 8-year-old female, and the body composition analysis results for each subregion of the body.

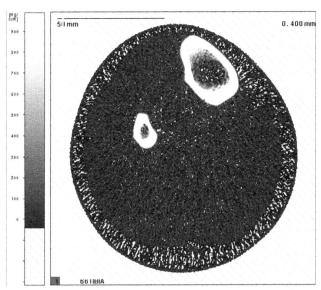

Figure 18.4 Peripheral quantitative computed tomography (pQCT) image of muscle, subcutaneous fat and bone at the 66% distal tibia. pQCT is a table-top device that is able to measure trabecular bone mineral density at the ends of long bones, and cortical density and dimensions in the mid-shaft of long bones. It is also able to quantify the amount of muscle, fat and bone in specific regions. Shown here is the pQCT image taken at 66% of tibia length in the mid-shaft. It is also the site of maximal muscle circumference.

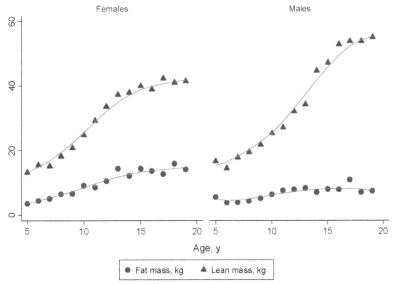

Figure 18.5 Age-related changes in fat and lean mass by dual-energy X-ray absorptiometry (DXA) in males and females. Males and females have distinct age-related patterns of growth in lean and fat mass. Data shown are based on a cross-sectional study of over 800 youth in Philadelphia (the Reference Project on Skeletal Development). Lean mass increases rapidly around the ages when the adolescent growth spurt occurs, but males gain considerably greater amounts of lean mass than females. Fat mass increases with age in females throughout childhood and adolescence.

Printed by Printforce, the Netherlands